Dates

Production, Processing, Food, and Medicinal Values

Medicinal and Aromatic Plants — Industrial Profiles

Individual volumes in this series provide both industry and academia with in-depth coverage of one major genus of industrial importance.

Series Edited by Dr. Roland Hardman

Volume 1
Valerian, edited by Peter J. Houghton
Volume 2
Perilla, edited by He-ci Yu, Kenichi Kosuna and Megumi Haga
Volume 3
Poppy, edited by Jenö Bernáth
Volume 4
Cannabis, edited by David T. Brown
Volume 5
Neem, edited by H.S. Puri
Volume 6
Ergot, edited by Vladimír Křen and Ladislav Cvak
Volume 7
Caraway, edited by Éva Németh
Volume 8
Saffron, edited by Moshe Negbi
Volume 9
Tea Tree, edited by Ian Southwell and Robert Lowe
Volume 10
Basil, edited by Raimo Hiltunen and Yvonne Holm
Volume 11
Fenugreek, edited by Georgios Petropoulos
Volume 12
Ginkgo biloba, edited by Teris A. Van Beek
Volume 13
Black Pepper, edited by P.N. Ravindran
Volume 14
Sage, edited by Spiridon E. Kintzios
Volume 15
Ginseng, edited by W.E. Court
Volume 16
Mistletoe, edited by Arndt Büssing
Volume 17
Tea, edited by Yong-su Zhen
Volume 18
Artemisia, edited by Colin W. Wright
Volume 19
Stevia, edited by A. Douglas Kinghorn
Volume 20
Vetiveria, edited by Massimo Maffei
Volume 21
Narcissus and Daffodil, edited by Gordon R. Hanks
Volume 22
Eucalyptus, edited by John J.W. Coppen
Volume 23
Pueraria, edited by Wing Ming Keung
Volume 24
Thyme, edited by E. Stahl-Biskup and F. Sáez
Volume 25
Oregano, edited by Spiridon E. Kintzios
Volume 26
Citrus, edited by Giovanni Dugo and Angelo Di Giacomo
Volume 27
Geranium and Pelargonium, edited by Maria Lis-Balchin
Volume 28
Magnolia, edited by Satyajit D. Sarker and Yuji Maruyama
Volume 29
Lavender, edited by Maria Lis-Balchin
Volume 30
Cardamom, edited by P.N. Ravindran and K.J. Madhusoodanan

Volume 31
Hypericum, edited by Edzard Ernst
Volume 32
Taxus, edited by H. Itokawa and K.H. Lee
Volume 33
Capsicum, edited by Amit Krish De
Volume 34
Flax, edited by Alister Muir and Niel Westcott
Volume 35
Urtica, edited by Gulsel Kavalali
Volume 36
Cinnamon and Cassia, edited by P.N. Ravindran, K. Nirmal Babu and M. Shylaja
Volume 37
Kava, edited by Yadhu N. Singh
Volume 38
Aloes, edited by Tom Reynolds
Volume 39
Echinacea, edited by Sandra Carol Miller Assistant Editor: He-ci Yu
Volume 40
Illicium, *Pimpinella and Foeniculum*, edited by Manuel Miró Jodral
Volume 41
Ginger, edited by P.N. Ravindran and K. Nirmal Babu
Volume 42
Chamomile, edited by Rolf Franke and Heinz Schilcher
Volume 43
Pomegranates, edited by Navindra P. Seeram, Risa N. Schulman and David Heber
Volume 44
Mint, edited by Brian M. Lawrence
Volume 45
Turmeric, edited by P. N. Ravindran, K. Nirmal Babu, and K. Sivaraman
Volume 46
Essential Oil-Bearing Grasses, edited by Anand Akhila
Volume 47
Vanilla, edited by Eric Odoux and Michel Grisoni
Volume 48
Sesame, edited by Dorothea Bedigian
Volume 49
Citrus Oils, edited by Giovanni Dugo and Luigi Mondello
Volume 50
Dates, edited by A. Manickavasagan, M. Mohamed Essa, and E. Sukumar

Dates

Production, Processing, Food, and Medicinal Values

Edited by A. Manickavasagan
M. Mohamed Essa
E. Sukumar

Medicinal and Aromatic Plants — Industrial Profiles

CRC Press
Taylor & Francis Group
Boca Raton London New York

CRC Press is an imprint of the
Taylor & Francis Group, an **informa** business

CRC Press
Taylor & Francis Group
6000 Broken Sound Parkway NW, Suite 300
Boca Raton, FL 33487-2742

Version Date: 20120302

International Standard Book Number: 978-1-4398-4945-3 (Hardback)

Visit the Taylor & Francis Web site at
http://www.taylorandfrancis.com

and the CRC Press Web site at
http://www.crcpress.com

Contents

Preface

Date palm is considered to be a wonder plant and a traditional food of the Middle Eastern countries. There are many varieties of dates, each one with unique properties. Of them, the species *Phoenix dactylifera* Linn. is cultivated in large numbers, traded, and consumed widely. The tree has a life span of about 80 years, and the female plant starts reproduction at the age of 15 years. Date fruits are nutritious and delicious and are being consumed regularly by many people all over the world. Several research papers on various aspects of date palm have been published over the years, but there is no comprehensive collection of information available for dates. Hence, we made an earnest attempt to compile a book on dates, starting with cultivation and extending to medicinal values.

The details are presented in four parts: Production, Processing, Food, and Medicinal Values. In the first part, cultural practices and their implications on date quality are discussed. Tissue culture studies, on-farm water management, fertilization, salinity problems in production, insect problems and their management, mechanization approaches in pollination and harvesting operations, and marketing aspects are covered in detail as exclusive chapters in this section. In the second part, postharvest operations such as drying, alternatives for methyl bromide fumigation, value-added products, and biofuel production from by-products and waste generated from date industry are covered. In the third part, physical, chemical, and structural characteristics of dates, fermentative products using dates as a substrate, the fruit as a substitute for added sugar in food, and date palm feeding to livestock are discussed. The final part deals with the nutritional and medicinal values including the use of dates in indigenous medicines.

We are optimistic that this book will be beneficial to students, scientists of various disciplines, and many others who are interested in this wonder fruit. We are grateful to all the contributors for promptly submitting their chapters. We also thank the staff of the editorial and production departments of Taylor & Francis Group for their unstinted support and efforts to bring out this book in present form. Last but not the least we thank the people of the Sultanate of Oman for their sentimental attachment to dates that inspired us to edit this book.

A. Manickavasagan
M. Mohamed Essa
Ethirajan Sukumar

Editors

Dr. A. Manickavasagan obtained his PhD in biosystems engineering from the University of Manitoba, Canada. He is a licensed professional engineer (PEng) in the province of New Brunswick, Canada. After earning his PhD, he worked with McCain Foods Limited (Canada) as a scientist. At present, he is working as an assistant professor at Sultan Qaboos University, Oman. He has published more than 50 scientific papers in peer-reviewed journals and international conferences. He has diversified research and management experience with academic institutions and industries.

Dr. M. Mohamed Essa obtained his PhD in biochemistry from Annamalai University, India, and at present is working as an assistant professor in nutrition at the College of Agricultural and Marine Sciences, Sultan Qaboos University, Oman. He conducts research in pharmacological approaches of natural products against neurological diseases by using different available models. He has published nearly 55 research articles in various peer-reviewed international journals and conferences and is an editor-in-chief for the *International Journal of Nutrition, Pharmacology, Neurological Diseases*. He is a fellow/life member of various international scientific bodies, including The Linnean Society, UK. Recently, he received the Young Researcher Award from the College of Agricultural and Marine Sciences, SQU and Professionals from Developing Countries Award given by International Society for Autism Research (INSAR), USA. His research work was on the pharmacological effect of walnuts on Alzheimer's disease and was press released at the International Conference on Alzheimer's Disease (IICAD), 2010 in Hawaii, USA. In addition, he was appointed as a visiting scientist in the Department of Pharmacology, University of New South Wales (until 2014), Sydney, Australia, and the Department of Neurochemistry, NYS Institute for Basic Research, USA. He has a strong academic and research background based on his work experiences in the United States, Malaysia, India, and Oman.

Dr. Ethirajan Sukumar obtained his PhD in phytochemistry and phytobiology from the University of Madras, India and conducted postdoctoral work at the University of Virginia, USA. He presently works as an assistant professor in the Department of Applied Sciences, Higher College of Technology, Muscat, Sultanate of Oman. Prior to full-time teaching, he was a research scientist in the Central Research Institute of Siddha Medicine in Chennai, India for more than 26 years. During his research career, he has worked on medicinal plants used in indigenous medical practices and published 52 research papers in peer-reviewed journals. In addition, he has presented many papers in various international and national forums and received a Best Paper Award in the Indian Pharmaceutical Congress. He has supervised 25 students for doctoral, MPhil, and master's programs in the multidisciplinary areas involving chemistry, biochemistry, pharmacognosy, pharmacology, and environmental science. He has delivered 55 invited lectures in universities, colleges, and conferences and has also served as a resource in many workshops and training programs. He has chaired scientific sessions and delivered plenary lectures at international conferences and is associated with leading international/national journals as associate editor, editorial board member, and peer reviewer. His name is included in the *International Directory of Experts in Spices, Herbs and Medicinal Plants Research* published by the University of Massachusetts and in the *Who's Who in Science and Engineering* by Marquis Inc., USA.

Contributors

Emmanuel S. Abbo
Department of Food Technology
Kaduna Polytechnic
Kaduna, Nigeria

Osman El-Sharief Abdalla
Department of Agricultural Affairs
Ministry of Environment
Doha, Qatar

Samir Al-Adawi
Department of Medicine
College of Medicine and
Health Sciences
Sultan Qaboos University
Al Khoud, Sultanate of Oman

Abdullah Al-Asmi
Department of Medicine
College of Medicine and
Health Sciences
Sultan Qaboos University
Al Koudh, Sultanate of Oman

Ahmed Al-Busaidi
Department of Soils, Water and Agricultural
Engineering
College of Agricultural and Marine Sciences
Sultan Qaboos University
Al Khoud, Sultanate of Oman

Mohammad Ali Al-Deeb
Department of Biology
United Arab Emirates University
Al-Ain, United Arab Emirates

Mohamed Ali Al-Farsi
Ministry of Agriculture
Muscat, Sultanate of Oman

Ghanum Abdul Rehman Al-Ghanum
Department of Agricultural Affairs
Ministry of Environment
Doha, Qatar

Amanat Ali
Department of Food Science and Nutrition
College of Agriculture and Marine Sciences
Sultan Qaboos University
Al Khoud, Sultanate of Oman

W. Al-Marzooqi
Department of Animal and Veterinary
Sciences
College of Agricultural and Marine Sciences
Sultan Qaboos University
Al Khoud, Sultanate of Oman

Ibrahim S. Al-Mssallem
Date Palm Genome Project
King Abdul-Aziz City for Science and
Technology
Riyadh, Kingdom of Saudi Arabia

Salim Al-Rasbi
Agricultural Research Center
Rumais, Sultanate of Oman

Salim H. Al-Rawahy
Department of Biology
College of Science
Sultan Qaboos University
Al Khoud, Sultanate of Oman

Nasser Salim Al-Wahaibi
Agricultural Research Center
Rumais, Sultanate of Oman

R. Al-Yahyai
Department of Crop Sciences
College of Agricultural and Marine Sciences
Sultan Qaboos University
Al Khoud, Sultanate of Oman

M. A. Basunia
Department of Soils, Water and Agricultural
Engineering
College of Agricultural and Marine Sciences
Sultan Qaboos University
Al Khoud, Sultanate of Oman

G. Chinnanchetty
Department of Farm Machinery
Tamil Nadu Agricultural University
Coimbatore, Tamil Nadu, India

Sankar Devarajan
Department of Cardiovascular Diseases
Fukuoka University Chikushi Hospital

and

WHO Regional Office
Fukuoka, Japan

M. Mohamed Essa
Department of Food Science and Nutrition
College of Agriculture and Marine Sciences
Sultan Qaboos University
Al Khoud, Sultanate of Oman

Kolawole O. Falade
Department of Food Technology
University of Ibadan
Ibadan, Nigeria

Talat T. Farook
Department of Biology
College of Science
Sultan Qaboos University
Al Khoud, Sultanate of Oman

Sardar A. Farooq
Department of Biology
College of Science
Sultan Qaboos University
Al Khoud, Sultanate of Oman

S. Feroz
Department of Mechanical and Industrial Engineering
Caledonian College of Engineering
Muscat, Sultanate of Oman

Gilles J. Guillemin
Department of Pharmacology
College of Medicine
University of New South Wales
Sydney, Australia

Nejib Guizani
Department of Food Science and Nutrition
College of Agriculture and Marine Sciences
Sultan Qaboos University
Al Khoud, Sultanate of Oman

H. H. Handali
Department of Soils, Water and Agricultural Engineering
College of Agricultural and Marine Sciences
Sultan Qaboos University
Al Khoud, Sultanate of Oman

T. Hariharan
Department of Mechanical and Industrial Engineering
Caledonian College of Engineering
Muscat, Sultanate of Oman

Nazir Hussain
Department of Agricultural Affairs
Ministry of Environment
Doha, Qatar

Hemantha P. W. Jayasuriya
Department of Soils, Water and Agricultural Engineering
College of Agricultural and Marine Sciences
Sultan Qaboos University
Al Khoud, Sultanate of Oman

I. T. Kadim
Department of Animal and Veterinary Sciences
College of Agricultural and Marine Sciences
Sultan Qaboos University
Al Khoud, Sultanate of Oman

S. Karthikeyan
Department of Agricultural Microbiology
Tamil Nadu Agricultural University
Coimbatore, Tamil Nadu, India

M. Mumtaz Khan
Water Research Center
Sultan Qaboos University
Al Khoud, Sultanate of Oman

Roohi S. Khan
Department of Biology
College of Science
Sultan Qaboos University
Al Khoud, Sultanate of Oman

S. G. Kulkarni
Fruit and Vegetable Technology
Central Food Technological
Research Institute, CSIR
Mysore, Karnataka, India

Chang Yong Lee
Department of Food Science and
Technology
Cornell University
Geneva, New York

O. Mahgoub
Department of Animal and Veterinary
Sciences
College of Agricultural and Marine Sciences
Sultan Qaboos University
Al Khoud, Sultanate of Oman

A. Manickavasagan
Department of Soils, Water and Agricultural
Engineering
College of Agricultural and Marine Sciences
Sultan Qaboos University
Al Khoud, Sultanate of Oman

Msafiri Daudi Mbaga
Department of Natural Resource
Economics
College of Agricultural and Marine Sciences
Sultan Qaboos University
Al Khoud, Sultanate of Oman

Mustaq A. Memon
Department of Veterinary Clinical Sciences
Washington State University
Pullman, Washington

Mohsen Ahmed El Mohandes
Faculty of Agriculture
Al-Azhar University
Cairo, Egypt

Ahmad Mostaan
Mechanics of Agricultural Machinery
Modares University
Tehran, Iran

and

Date Palm and Tropical Fruits Research
Institute
Ahwaz, Iran

Summar A. Naqvi
Department of Horticulture
University of Agriculture
Faisalabad, Pakistan

Kishor Patwardhan
Department of Kriya Sharir
Institute of Medical Sciences
Banaras Hindu University
Varanasi, Uttar Pradesh, India

S. A. Prathapar
Department of Soils, Water and Agricultural
Engineering
College of Agricultural and Marine Sciences
Sultan Qaboos University
Al Khoud, Sultanate of Oman

Muhammad Qasim
Department of Horticulture
University of Agriculture
Faisalabad, Pakistan

Mohammad Shafiur Rahman
Department of Food Science and Nutrition
College of Agriculture and Marine Sciences
Sultan Qaboos University
Al Khoud, Sultanate of Oman

Nandhagopal Ramachandiran
Department of Medicine
College of Medicine and Health Sciences
Sultan Qaboos University
Al Khoud, Sultanate of Oman

D. Ramesh
Department of Farm Machinery
Tamil Nadu Agricultural University
Coimbatore, Tamil Nadu, India

D. G. Rao
Department of Mechanical and Industrial Engineering
Caledonian College of Engineering
Muscat, Sultanate of Oman

M. Shanmugapriya
Department of Kriya Sharir
Institute of Medical Sciences
Banaras Hindu University
Varanasi, Uttar Pradesh, India

Vandita Singh
Department of Food Science and Nutrition
College of Agriculture and Marine Sciences
Sultan Qaboos University
Al Khoud, Sultanate of Oman

Nallusamy Sivakumar
Department of Biology
College of Science
Sultan Qaboos University
Al Khoud, Sultanate of Oman

Ethirajan Sukumar
Department of Applied Sciences
Higher College of Technology
Muscat, Sultanate of Oman

Hameed Sulaiman
Department of Biology
College of Science
Sultan Qaboos University
Al Khoud, Sultanate of Oman

Jun Tan
Date Palm Genome Project
King Abdul-Aziz City for Science and Technology
Riyadh, Kingdom of Saudi Arabia

Ragini Vaishnav
Department of Pharmacology
College of Medicine and Health Sciences
Sultan Qaboos University
Al Khoud, Sultanate of Oman

P. Vijayanand
Fruit and Vegetable Technology
Central Food Technological Research Institute, CSIR
Mysore, Karnataka, India

Mostafa Waly
Department of Food Science and Nutrition
College of Agriculture and Marine Sciences
Sultan Qaboos University
Al Khoud, Sultanate of Oman

Meng Yang
Date Palm Genome Project
King Abdul-Aziz City for Science and Technology
Riyadh, Kingdom of Saudi Arabia

Yuxin Yin
Date Palm Genome Project
King Abdul-Aziz City for Science and Technology
Riyadh, Kingdom of Saudi Arabia

Jun Yu
Date Palm Genome Project
King Abdul-Aziz City for Science and Technology
Riyadh, Kingdom of Saudi Arabia

Xiaowei Zhang
Date Palm Genome Project
King Abdul-Aziz City for Science and Technology
Riyadh, Kingdom of Saudi Arabia

Part I

Production

1 An Overview of Date Palm Production

R. Al-Yahyai and A. Manickavasagan

CONTENTS

INTRODUCTION

Date palm (*Phoenix dactylifera* L.) is the main fruit crop in arid and semiarid regions, particularly in the arid regions of western Asia and North Africa. The palm tree is well adapted to desert environments that are characterized by extreme temperatures and water shortage, both in quality and quantity, due to scarcity of rainfall. Beyond the arid climates, date palm can also be grown in many other countries for food or as an ornamental plant including the continents of Americas, southern Europe, Asia, Africa, and Australia. The majority of date palm-growing areas are located in developing or underdeveloped countries where dates are considered the primary food crop, thus playing a major role in the nutritional status of these communities. By-products from date palm are used in building structures, animal feed, and also in several items such as baskets and ropes. The date palm tree that has been in cultivation since 2400 BC was praised and cherished as is evident from the drawings and sculptures of ancient civilizations of the Sumerians, Assyrians, Babylonians, and Egyptians, and later by the Greeks and the Romans that inhabited the Mediterranean basin where date palm and other *Phoenix* species are also commonly grown (Pruessner 1920). Date palm still carries great religious significance in all three major religions of the world. In Islam, date palm is cited 21 times in the Holy Quran and 300 times in the Hadith of the Prophet Mohammed, making it by far the most frequently cited plant. Similarly, date palm is praised in Christian and Judaism faiths and has been linked to numerous religious ceremonies such as Passover and Palm Sunday (Musselman 2007).

BIOLOGY OF THE DATE PALM

The date palm (*Phoenix dactylifera* L.) is an angiosperm monocotyledonous plant that belongs to the family Arecaceae (syn. Palmaceae) that includes 200 genera (Dowson 1982; Uhl and Dransfield 1987). The genus *Phoenix* contains 12 of the 1500 species that belong to the date palm family. *Phoenix* palms are dioecious and are characterized by pinnate leaves and in duplicate leaflets with acute tips (Uhl and Dransfield 1987). Besides date palm, the other two most highly valued *Phoenix* palms are Canary Island Palm (*P. canariensis* Chabeaud), an ornamental palm (Gilman and Watson 2006), and the Sugar Palm (*P. sylvestris* Roxb) that is common in the Indian subcontinent for its sugar syrup (Zaid and de Wet 2002). Date palm can be distinguished from the other two species by the production of offshoots and the dull glaucous leaves instead of glossy leaves.

Date palm is a dioecious plant; that is, staminate (male) and pistillate (female) flowers are born on two different palms. There are thousands of male and female cultivars distributed throughout date palm-growing countries (Zaid and de Wet 2002). For example, in the Sultanate of Oman, there are over 200 female and 21 male cultivars (Al-Yahyai 2010). However, 75% of dates are produced only from 10 cultivars (Table 1.1). Date palm passport descriptors have been published by IPGRI to aid the characterization of date palm based on morphological and anatomical features (IPGRI et al. 2005). Using these characteristics, cultivars from Oman (Macki and Othman 1997), Saudi Arabia (Ministry of Agriculture 2006), and Algeria (Belguedj 2002), among others, have all been fully described. Zaid and de Wet (2002) provided a detailed description of Medjool and Barhee cultivars. In addition, fruit physical and chemical characteristics have also been used to describe and distinguish date palm cultivars (Jaradat and Zaid 2004; Al-Yahyai and Al-Khanjari 2008). Recently, molecular tools such as DNA fingerprinting techniques have also been utilized to describe and differentiate date palms in various parts of the world (Billotte et al. 2004; Cao and Chao 2002; Jubrael et al. 2005; Zehdi et al. 2004).

BOTANICAL DESCRIPTION OF DATE PALM

Roots

Date palm has a fasciculated fibrous root system that originates from a bulb at the trunk base. The primary roots have an average length of 4 m and may extend to 10 m in light soils. Primary roots

TABLE 1.1
Major Date Palm Varieties and Its Total Production in the Sultanate of Oman

Cultivars	2004 Yield (tons)	% of Total	Cumulative %
Um el Sela	32,696.48	14.15	14.15
Mabsli	30,583.24	13.24	27.39
Khasab	26,678.61	11.55	38.94
Naghal	24,423.38	10.57	49.51
Fardh	18,051.93	7.81	57.33
Shahl	11,435.75	4.95	62.28
Khunaizi	11,340.99	4.91	67.18
Khalas	11,139.04	4.82	72.01
Madloki	5,423.58	2.35	74.35
Barni	4,966.30	2.15	78.65
Total	231,034.91		

Source: Al-Yahyai, R. 2010. *Proceedings of the International Conference on Date Palm Production and Processing Technology*. Sultan Qaboos University, Muscat, Sultanate of Oman. 09–11 May 2006, pp. 1–6.

give rise to secondary roots that further branch to form tertiary roots that are shorter in length and diameter (Zaid and de Wet 2002). Primary roots originate from seeds but may also continue to grow if date palm is grown from an offshoot or a tissue-cultured seedling.

Trunk

The trunk or stripe of date palm is a single, vertical cylinder of equal diameter (average 1 m) throughout its length that can reach 30 m (Figure 1.1) (Zaid and de Wet 2002). The stem is covered with leaf bases that are enclosed in fiber, an evolutionary mechanism to protect the trunk from herbivorous insects and animals, as well as an insulation to reduce water loss. Water and nutrients are translocated via vascular tissue composed of tightly-stacked vascular bundles. The stem grows vertically at the terminal bud (phyllophor or phyllogen) and laterally via the fascicular cambium.

Leaves

Date palm leaves, called fronds, are pinnate, compound leaves spirally arranged around the trunk (Figure 1.1) (Uhl and Dransfield 1987). The fully mature leaf is 4 m long, but ranges from 3 to 6 m, and is 0.5 m wide at the middle midrib that narrows toward both leaf ends (Zaid and de Wet 2002). The date palm leaf is divided into three regions: the petiole, the spinal region that transitions into the blade region that is held by a geometrically shaped midrib. Angular leaflets are distributed in the blade region. The number of leaves produced annually varies from 10 to 26 and a mature palm may have from 100 to 125 leaves; 50% of them are photosynthetically active (Zaid and de Wet 2002). Leaves remain attached to the tree following their senescence and have to be manually pruned.

Flowers

Date palm is a dioecious plant where pistillate and staminate flowers are born on separate plants (Uhl and Dransfield 1987). Male and female flowers are arranged in strands that attach to a rachis forming an inflorescence called spadix. A bract, called spathe, enclosing the immature inflorescence, splits longitudinally at anthesis, which allows for the pollination of mature male and female flowers.

FIGURE 1.1 The pinnate leaves (fronds) and single stem (stipe) of the date palm.

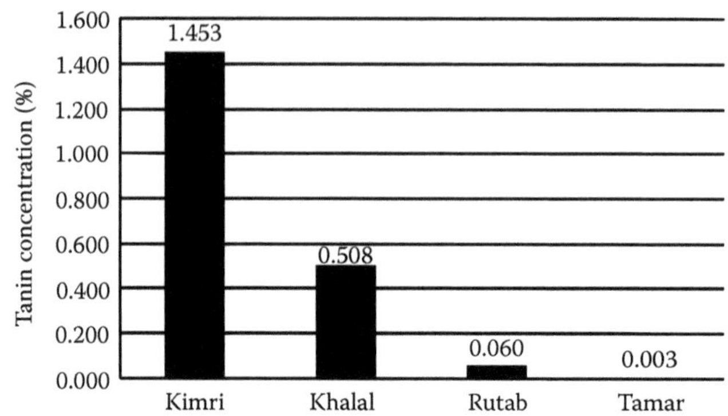

FIGURE 1.2 Tannin concentration (%) in "khalas" dates at various maturity stages (Al-Yahyai, unpublished data).

Fruits

Fruits of date palm, called dates, develop from one fertilized ovule forming one carpel, while the other two ovules are aborted but remain visible at the fruit calyx. If no fertilization occurs, three or more carpels develop simultaneously. The date fruit develops on the flowering strands and is a berry characterized by a membranous endocarp surrounding a seed. Large variations exist in the shape, size, color, and chemical composition of date fruit (Zaid and de Wet 2002; Al-Yahyai and Kharusi 2011), depending largely on varietal differences but also on climate, soil, and growing conditions. Similarly, date seeds vary in size and shape, but they are generally ventrally grooved, oblong, and range from 5 to 15 mm with an embryo born in the middle of the seed that is surrounded by the endosperm.

Following pollination and fertilization, date palm fruit grows through five distinct developmental stages characterized by physical and chemical changes at each stage. These stages are Hababouk, Kimri, Khalal, Rutab, and Tamar (Zaid and de Wet 2002). The Hababouk stage lasts 4–5 weeks after fruit set and is characterized by slow growth rate. The fruit is round and the color is cream to light green. The Kimri stage lasts 9–14 weeks and during this stage the fruit elongates and gains rapid development and increased weight, volume, and reducing sugars. The fruit at the Kimri stage is green and is marked by high tannin concentration that is reduced as the fruit develops (Figure 1.2). The Khalal stage, also known as Bisir, is when the fruit is physiologically mature and the color changes to the cultivar-specific color, usually various shades of red and yellow. This stage lasts 3–5 weeks and is marked by slow growth (3%–4% weight gain) to fruit full size and weight, and further accumulation of sugars. Several varieties with high sugars can be consumed either fresh or boiled at this stage. The Rutab stage is when the fruit apex starts to soften and the color changes to a darker color. Most dates are consumed fresh at this stage that lasts 2–4 weeks. Due to softening, dates at this stage lose water content to an average of 30%–45%, but sugars continue to accumulate. The Tamar stage is when the date palm fruit is fully ripe and has completely softened. The fruit color darkens at this stage, which is also marked by low moisture content (10%–25%) and high concentration of total sugars. This stage lasts 2–4 weeks and the dates are appropriate for long-term dry storage or processing.

ORIGIN OF DATE PALM

Date palm has been cultivated in the Middle East and North Africa (MENA region) for millennia; however, the exact origin of date palm has not been verified. Zaid and de Wet (2002) cited reports of its use in Mesopotamian times since 4000 BC (Popenoe 1973) and by the Egyptians perhaps during 3000–2000 BC (Danthine 1937). The African date palm (*P. reclinata*) or the Indian date palm (*P. sylvestris*) or both may have been the progenitor of date palm (Zaid and de Wet 2002).

GEOGRAPHICAL DISTRIBUTION OF DATE PALM

Date palm can be cultivated in all five continents of the world, largely between 39° northernmost and 20° southernmost latitudes. However, the main region of production is the Middle East and North Africa, where 89% of dates are produced (Figures 1.3 and 1.4). In recent years, there has been

FIGURE 1.3 **(See color insert.)** World map of date producing countries according to area distribution (shaded gray) and production quantity (columns).

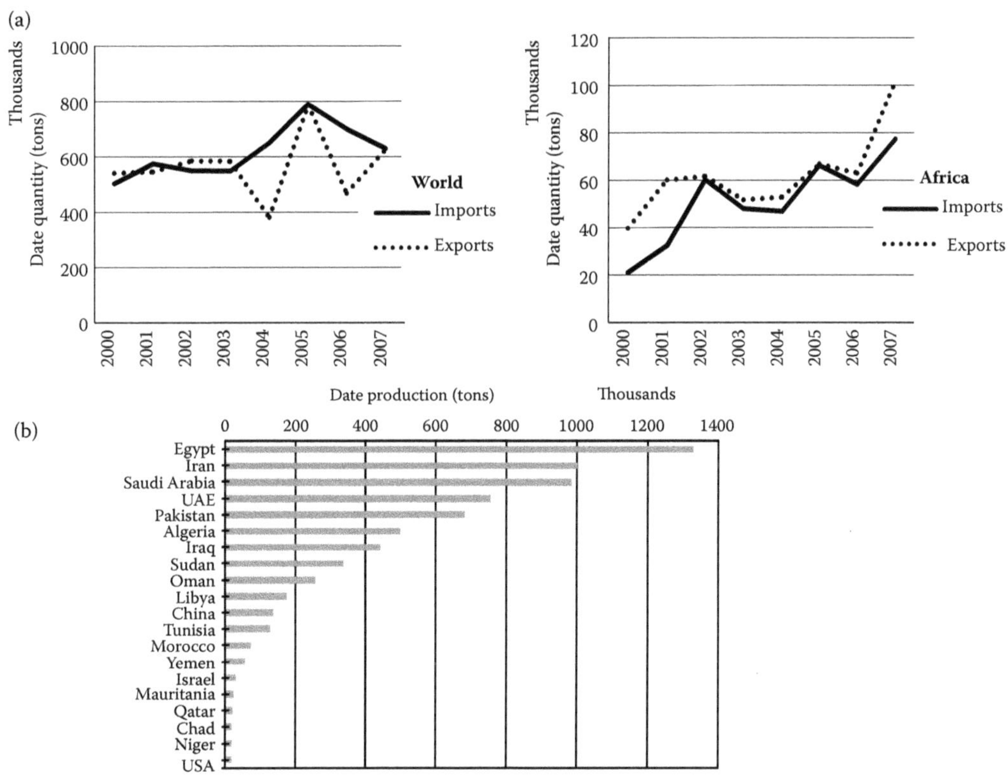

FIGURE 1.4 Imports and exports of dates (a) and leading date-producing countries (b). (Adapted from FAOSTAT. 2010. *Food and Agriculture Organization of the United Nations*. Available at: http://faostat.fao.org/. Accessed on September 1, 2010.)

a surge in the number of date palms planted in this region. The total number of date palms in the UAE, for example, has grown from 10 million in 1999 to 43 million palms in 2010. Similarly, the Sultanate of Oman is planning to add 1 million date palms by 2015. Large projects aimed at introducing date palm in dry areas that have poor resources, such as in Namibia and India, have influenced date palm distribution and expansion to new territories.

WORLD DATE PRODUCTION AND TRADE

Current world date production is 7.1 million tons as of 2008 (FAOSTAT 2010). Date exports and imports have been steadily increasing in the years since 2000 (Tables 1.2 and 1.3). Globally, date palm imports have surpassed exports in Europe, Americas, and Oceana, whereas in Africa and Asia, date trade has continued to rise simultaneously (Figure 1.5). There has been a considerable rise in the demand for dates in major date-importing countries including Morocco, the United Kingdom, and Russia (Table 1.3). This remarkable increase in date trade reflects the importance

TABLE 1.2
Leading Date-Exporting Countries

Country	2007 Exports (tons)	% of Sum	Change (tons) from 2000	% Change (2000–2007)
Iran	242,092	38.7	134,245	55
Pakistan	104,090	16.6	25,429	24
Tunisia	68,856	11.0	46,445	67
Saudi Arabia	48,762	7.8	20,514	42
UAE	38,691	6.2	−183,339	−474
Iraq	37,063	5.9	7063	19
Algeria	25,039	4.0	14,256	57
France	10,470	1.7	894	9
Israel	9,513	1.5	8184	86
Oman	9,368	1.5	−513	−5
2007 Top 10	593,944	95.0	256,517	43
2007 Total	625,439			

TABLE 1.3
Leading Date-Importing Countries

Country	2007 Imports (tons)	% of Sum	Change (tons) from 2000	% Change (2000–2007)
India	253,341	40.2	60,722	24
Morocco	50,473	8.0	45,039	89
France	27,439	4.4	3,909	14
Russia	22,375	3.6	13,541	61
Pakistan	19,777	3.1	−9,770	−49
Bangladesh	18,546	2.9	1,046	6
Indonesia	15,549	2.5	2,233	14
Malaysia	15,346	2.4	4,188	27
UK	14,357	2.3	3,927	27
Yemen	13,136	2.1	549	4
2007 Top 10	450,339	71.5	125,384	28
2007 Total	629,773			

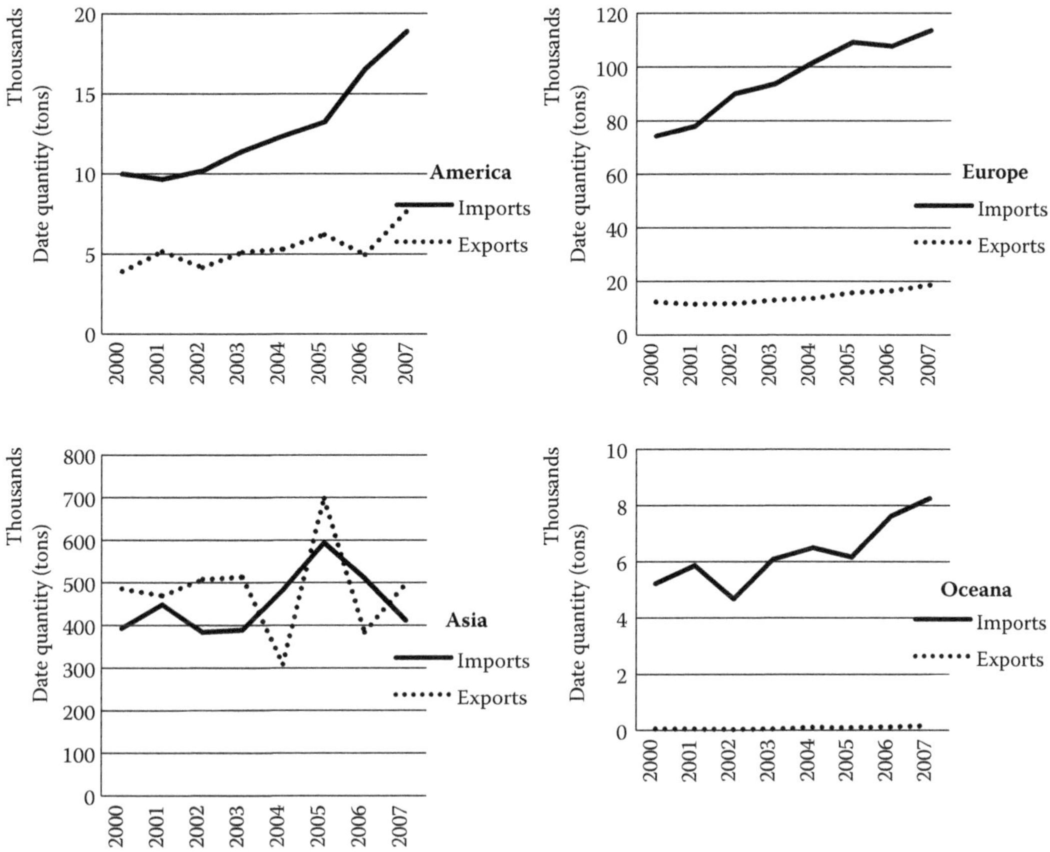

FIGURE 1.5 Global and continental date palm imports and exports from 2000 to 2007.

of dates as a trading commodity in international markets. This is perhaps due to the increased awareness among consumers of the healthy constituents of dates, such as antioxidant contents (Mansouri et al. 2005; Al-Farsi et al. 2007), as well as the use of dates as a source of sugar and medicinal alcohol in many parts of the world.

DATE PALM CULTIVATION

Date palm thrives in areas characterized by hot, low humidity, particularly during fruit development. Moisture adversely affects the quality of fruit, as high humidity leads to fruit cracking and checking. Date palm can be planted in a wide range of soils with varying amounts of organic and mineral nutrients. Date palm is known to tolerate salinity more than any other cultivated fruit crop.

Many parts of the world where date palm is grown still follow the traditional mixed planting of dates of various ages at irregular spacing. Moreover, inadequate fertilizer application and lack of proper tree and bunch management, such as pruning and fruit thinning, lead to the production of low fruit quality and thus lower market values. In Oman, for example, approximately 23% of the total dates produced are considered "surplus," meaning that they are not accounted for in human consumption (51%), livestock feed (22%), or industrial processing and export (4%). This is despite the Sultanate's production of over 230,000 tons of dates annually. Oman is ranked ninth in world date production (Figure 1.4), but exports only 2% of its total production (Al-Yahyai 2007), which makes up 1.5% of the total world date exports (Table 1.2).

Development of tissue-cultured date palms has led to further expansion of date orchards in many parts of the world, particularly in the Middle East and Africa. However, the majority of date plantations still lack modern and appropriate solutions to issues related to cultural practices, such as hand pollination, water and fertilizer requirements, by-product development and utilization, proper harvesting time and methods, and postharvest transport and storage management. Several recent studies have addressed these issues and reported improved fruit yield and quality by following proper pre- and postharvest management of date palm tree and fruits (El Mardi et al. 2007; Al-Yahyai and Al-Kharusi 2010).

Pest and disease management of date palm also needs considerable research attention. The most common pests of date palm in the Arabian Peninsula are Dubas bug (*Ommatissus lybicus* Bergevin, Homoptera: Tropiduchidae) and Red Palm Weevil (*Rhynchophorus ferrugineus* Olivier, Coleoptera: Curculionidae). Chemical control of these pests has had limited success; however, a broader ban on pesticides and public demand for chemical-free dates will require alternative methods of pest management. Biochemical control of date palm pests has been shown to reduce pest populations (El-Sufty et al. 2007), but field application has many constraints, including traditional and poor tree and plantation management as noted earlier. Additionally, diseases caused mainly by phytoplasma and fungi, such as lethal yellowing and Bayoud *Fusarium* wilt, remain a threat to the date palm-producing regions of the world.

SUMMARY AND CONCLUSIONS

Date palm is one of the most important fruit crops in many parts of the world that are usually characterized by arid, high-temperature climate. Date commodity trade has been steadily rising in recent years, reflecting greater demand from all five continents of the world. Date palm constitutes a major source of food and its by-products can be utilized in a range of items from construction to clothing. This makes date palm an ideal crop for low-income countries, where poverty and hunger prevail.

REFERENCES

Al-Yahyai, R. 2007. Improvement of date palm production in the Sultanate of Oman. *Acta Hort.* 736:337–343.

Al-Yahyai, R. 2010. Current status of date palm in the Sultanate of Oman. *Proceedings of the International Conference on Date Palm Production and Processing Technology*. Sultan Qaboos University, Muscat, Sultanate of Oman. 09–11 May 2006, pp. 1–6.

Al-Yahyai, R. and L. Al-Kharusi. 2011. Physical and chemical quality characteristics of freeze-stored dates. *Int. J. Agr. Biol.* (Accepted).

Al-Yahyai, R. and S. Al-Khanjari. 2008. Biodiversity of date palm in the Sultanate of Oman. *Afr. J. Agric. Res.* 3(6):389–395.

Al-Farsi, M., A. Morris, and M. Baron. 2007. Functional properties of Omani dates (L.). *Acta Hort.* 736:479–487.

Belguedj, M. 2002. Charactéristiques des cultivars de dattiers dans les palmeraies du Sud-Est Algérien. Dossier No. 1. Les Ressources Génétiques du Palmier Dattier. Institut national de la Recherche Agronomique, d'Algérie.

Billotte, N., N. Marseillac, P. Brottier, J.-L. Noyer, J.-P. Jacquemoud-Collet, C. Moreau, T. Couvreur, M.-H. Chevallier, J.-C. Pintaud, and A.-M. Risterucci. 2004. Nuclear microsatellite markers for the date palm (*Phoenix dactylifera* L.): Characterization and utility across the genus Phoenix and in other palm genera. *Mol. Ecol. Notes* 4:256–258.

Cao, B.R. and C.T. Chao. 2002. Identification of date cultivars in California using AFLP markers. *Hort. Sci.* 37(6):966–968.

Gilman, E. and D. Watson. 2006. *Phoenix canariensis*: *Canary Island Date Palm*. Bulletin ENH-598. Institute of Food and Agricultural Sciences, University of Florida, FL.

Danthine, H. 1937. *Le palmier dattier et les arbres sacrés dans l'iconographie de l'asie occidentale ancienne.* Librairie Orientaliste Paul Geuthner, Paris, pp. 227.

Dowson, V. H. W. 1982. Date production and protection with special reference to North Africa and the Near East. *FAO Technical Bulletin* No. 35, pp. 294.

El Mardi, M., F. Al Said, C. Sakit, L. Al Kharusi, I. Al Rahbi, and K. Al Mahrazi. 2007. Effects of pollination method, fertilizer and mulch treatments on the physical and chemical characteristics of date palm (*Phoenix dactylifera*) fruit. I: Physical characteristics. *Acta Hort.* 736:317–328.

El-Sufty, R. S., A. Al-Awash, A. M. Amiri, A. S. Shahbad, A. H. Al Bathra, and S. A. Musa. 2007. Biological control of Red Palm Weevil, *Rhynchophorus ferrugineus* (Col.: Curculionidae) by the entomopathogenic fungus *Beauveria bassiana* in United Arab Emirates. *Acta Hort.* 736:399–404.

FAOSTAT. 2010. *Food and Agriculture Organization of the United Nations.* Available at: http://faostat.fao.org/. Accessed on September 1, 2010.

IPGRI, INRAA, INRAM, INRAT, FEM, PNUD. 2005. Descripteurs du Palmier dattier (*Phoenix dactylifera* L.). Institut international des resources phytogénétiques, Rome, Italie; Fonds pour I'Environment Mondial, Washington, Etas-Unis; Programme des Nations Unies pour le Développement, New York, Etas-Unis; Institut national de la Recherche Agronomique, d'Algérie, du Maroc et de Tunisie.

Jaradat, A. and A. Zaid. 2004. Quality traits of date palm fruits in center of origin and center of diversity. *Food Agr. Env.* 2(1): 208–2017.

Jubrael, J. M., S. M. Udupa, and M. Baum. 2005. Assessment of AFLP-based genetic relationships among date palm (*Phoenix dactylifera* L.) varieties of Iraq. *J. Am. Soc. Hort. Sci.* 130(3):442–447.

Macki, M. and M. Othman. 1997. *'al Nakheel fi Sultanat Oman' (Date palms in the Sultanate of Oman).* Diwan of Royal Court, Sultanate of Oman.

Mansouri, A., G. Embarek, E. Kokkalou, and P. Kefalas. 2005. Phenolic profile and antioxidant activity of the Algerian ripe date palm fruit (*Phoenix dactylifera*). *Food Chem.* 89:411–420.

Ministry of Agriculture. 2006. *The Famous Date Varieties in the Kingdom of Saudi Arabia.* Deposit No. 4944/1427. Ministry of Agriculture, Saudi Arabia.

Musselman, L. J. 2007. *Figs, Dates, Laurel and Myrrh: Plants of the Bible and the Quran.* Timber Press, Inc., Portland, OR, pp. 114–119.

Popenoe, P. B. 1973. The date palm. In *Field Research Projects*, H. Field (ed.), Coconut, Miami, FL, 247 pp.

Pruessner, A. H. 1920. Date culture in ancient Babylonia. *Am. J. Semitic Lang. Literatures* 36(3):213–232.

Uhl, N. W. and J. Dransfield. 1987. *Genera Palmarum: A Classification of Palms Based on the Work of Harold E. Moore, Jr*, L.H. Baily Hortorium and the International Palm Society, Lawrence, KS, pp. 214–217.

Zaid, A. and P. F. de Wet. 2002. Botanical and systematic description of the date palm. In A. Zaid and E. J. Arias-Jiménez (eds.), *Date Palm Cultivation. FAO Plant Production and Protection Paper 156 Rev. 1.* Rome, Italy.

Zehdi, S., M. Trifi, N. Billotte, M. Marrakchi, and J. C. Pintaud. 2004. Genetic diversity of Tunesian date palms (*Phoenix dactylifera* L.) revealed by nuclear microsatellite polymorphism. *Hereditas* 141:278–287.

2 Tissue Culture Studies in Date Palm

Sardar A. Farooq, Roohi S. Khan, and Talat T. Farook

CONTENTS

INTRODUCTION

Date palm (*Phoenix dactylifera* L.) is one of the oldest plantation crops, mainly cultivated for its sweet and nutritious fruits. It is tolerant to high temperature and drought. It is one of the most economically important perennial plants in arid regions of the Middle East, North Africa, and the Sahara. Date palm is thought to have originated either in the Persian Gulf or in western India and domesticated more than 5000 years ago. Present-day date palms are a result of centuries of successful selection. Date palm production declined in recent years, due to irrigation using scarce water resources, soil and water salinity, pests and diseases, increased production costs as well as limited market outlets (Al-Yahyai, 2007). The improvement of the existing cultivars for superior characters is not an easy task due to its long life cycle, highly heterozygous nature, and difficulty to determine the sex before flowering (5–7 years). Date palm produces a limited number of offshoots that are traditionally used for propagation, while the seed-derived plants are undesirable, because they show high variability; because of this, *in vitro* propagation is the best option and an effective means to ensure the renewal and extension of date palm plantations (Smith and Aynsley, 1995).

BOTANY

Date palm (*P. dactylifera* L.) ($2n = 36$) belongs to the class Monocotyledoneae, the family Arecaceae (formerly Palmaceae). Its arborescent, erect, unbranched stem reaches 10–15 m height and is covered with persistent grayish leaf bases; leaves are pinnately compound, 3–5 m long with spines on the petiole; leaflets are 30 cm long and 2 cm broad. Phyllotaxy can be useful in the identification of some cultivars. Numerous offshoots (suckers) arise from the base in 5–10-year-old plants. Date palm is dioecious with small fragrant flowers that are borne on a branched spadix. The female flowers are white, whereas the male are cream colored and waxy. The fruit is oblong, grooved down at one side, 1–3 in. long, dark brown, reddish, or yellowish brown when ripe with thin or thick skin; the pulp is thick, fleshy, and sweet with a single, cylindrical, very hard stony seed. It is very nutritious and is a rich source of sugars, minerals, and vitamins and is considered as the most important economical product of date palm (Anonymous, 2008). The date palm tree tolerates extreme adverse climatic conditions of the desert and salinity and can even grow in marshes (Al-Utbi and Al-Husaibi, 2008), but succumbs at low temperatures of 8°C or less. It can tolerate low humidity and rainfall. The plants can grow in a wide variety of soil types but prefer alkaline-aerated soils. The flowering occurs in long, hot, and completely dry periods.

POLLINATION

The date palm is dioecious, having separate male and female plants. Dates are wind pollinated, but in cultivation they are pollinated manually. For natural pollination, 1:1 male and female plants are required, while in manual pollination one male can pollinate up to 100 females. Since the yield depends on fruit-bearing females, the growers do not want to waste resources in growing unproductive male plants (El-Mardi et al., 1995). Pollination is an expensive practice due to long flowering season, and requires the climbing of trees several times to pollinate to ensure good yield (Al-Baker, 1972; Hussein et al., 1979; Hussein, 1982). Reuveni (1970) and Shaheen (1986) have shown that pistils do not remain receptive for a long period and the receptivity differs among the cultivars. Delaying pollination to 4, 6, or 8 days after cracking of spathe significantly reduces the percentage of fruit set. The increase in the percentage of fruit set occurs by early pollination because the pollen tube can easily germinate and elongate to penetrate the stigma and style of the female flower, resulting in better fertilization and fruit set (Brown et al., 1969). After pollination, the fruit undergoes four distinct developmental phases: the Hababouk (fertilized rudimentary fruit), Kimri, Bisir (khalal), Rutab, and Tamr stages. These stages are classified on the basis of morphological differences. For example, the Hababouk stage, the fruits are small, conical-shaped, and light green in color with green stripes. The Kimri stage is larger than Hababouk, ovate and dark green in color. With the Bisir stage, they have reached a maximum size, and are red and yellow. The Rutab stage shows a browning of the top of fruit (distal end). The other two-third portion is red and yellow. Finally, at the Tamr stage, all fruits turn dark brown in color. Complete development takes about 200 days to reach the maturity (Kwaasi, 2003).

DATE PALM TISSUE CULTURE

Plant tissue culture is the aseptic culture of plant cells, tissues, or organs *in vitro*. Plants cells have the unique property of totipotency, which is the ability of a plant cell to perform all the functions of development to produce a plant, just like a zygote. Many plant species have demonstrated this ability and regenerated into a whole plant. Of late, tissue culture is gaining importance because a large number of true-to-type plantlets can be produced from elite genotypes in a relatively shorter time. The technique can be used in multiplying virus- and disease-free plants. Even though it is a tedious process requiring technical skills, it is definitely advantageous in tree species that have a long life cycle and suffer from heterozygosity of sexual reproduction. In date palm, clonal propagation is preferred because it is dioecious and propagation by seeds does not produce true-to-type plants. The use of offshoots is limited by their fewer numbers, takes a long time and is cost effective. Therefore, tissue culture is the most dependable and preferred choice for the mass propagation of elite types with greater genetic uniformity. Several review articles have been published on micropropagation of date palm (Benbadis, 1992; Omar et al., 1992; Tisserat, 1984; Al-Khayri, 2007; Khierallah and Bader, 2007).

Tissue culture studies in date palm began in the early 1970s and progressed slowly due to slow growth response in culture and limited research resources (Reuveni, 1979; Reynolds and Murashige,

1979; Tisserat, 1979). Several attempts have been made to employ tissue culture techniques for date palm propagation. Extensive efforts have been made for the large-scale micropropagation of date palm using techniques of somatic embryogenesis and organogenesis (Drira, 1983; Drira and Benbadis, 1985; Tisserat, 1979; El Hadrami et al., 1998). Somatic embryogenesis is known to be a quick and efficient method for large-scale propagation of date palm and is highly useful for breeding programs (El Hadrami et al., 1998). The use of cell suspension cultures coupled with somatic embryogenesis could also be used for genetic transformation (Levee et al., 1999) and biolistic gene transfer. The cell suspension would allow easier single cell or protoplast isolation which would facilitate genetic engineering and mutagenesis. Cell suspension cultures have been successfully used for the regeneration of date palm by several researchers: Sharma et al. (1986), Daguin and Letouze (1988), Bhaskaran and Smith (1992), Al-Khayri (2002), Al-Khayri and Al-Bahrany (2004), Fki et al. (2003), and Zouine et al. (2005). Rashid and Quraishi (1994) reported the initiation of whole plant from embryo callus of date palm. Organogenesis or somatic embryogenesis was obtained from vegetative or inflorescence explants cultured *in vitro* (Drira and Benbadis, 1985; Bhaskaran and Smith, 1992; Fki et al., 2003). Propagation through somatic embryogenesis has also been reported by Eenwens (1978), Drira and Benbadis (1985), Gabr and Tisserat (1985), Raj Bhansali et al. (1988), and Branton and Blake (1989). Production of shoots from bud cultures was also reported by proliferation of axillary buds or by adventitious bud development (Masmoudi-Allouche et al., 2010). Regeneration of plantlets from juvenile leaf explants of Tunisian date palm was obtained (Othmani et al., 2009). *In vitro* organogenesis by the reversion of floral buds was described (Drira and Benbadis, 1985). Somatic embryogenesis has been successfully obtained and appears to be a very promising approach for the large-scale propagation of date palm (Daguin and Letouze, 1988; El Hadrami et al., 1995). Application of somatic embryos in the production of synthetic seeds is reported by Bekheet et al. (2002). The origin and nature of embryogenic cultures are of critical importance for the *in vitro* production of somatic embryos (Veramendi and Navarro, 1997). Indeed, these authors reported that only the nodular callus originating from leaf primordia bases could produce viable somatic embryos. However, abnormal development, low maturation, and germination frequencies of somatic embryos are serious limitations (Daguin and Letouzé, 1997; Feher et al., 2003; Vieitez, 1995a,b). Maintaining synchrony during the induction of embryogenesis is also difficult in date palm.

EXPLANTS AND TISSUE CULTURE PROTOCOLS

It is very important to select the correct explant for successful propagation, since the successful morphogenesis *in vitro* depends on explants that are meristematic and healthy. Explants can be obtained from the offshoots as the axillary buds, leaf primordia, shoot tip meristems, or sections of young leaf or inflorescence (Bhaskaran and Smith, 1992; Fki et al., 2003). The apical bud meristems with one or two layers of thin sheath and lateral buds can also be used (Gabr and Tisserat, 1985; Rashid and Quraishi, 1994). Schroeder (1970) used shoot tip explants and obtained rhizogenesis. Reuveni et al. (1972) reported differential response of shoot tips to various hormones. Initiation of embryogenic callus and subsequently plantlets through somatic embryogenesis from lateral buds and shoot tip explants has been reported (El-Hennawy and Wally, 1978; Tisserat, 1979; Tisserat and DeMason, 1980). Leaf explants were used by Schroeder (1970), Reuveni (1972), Sharma et al. (1980), and Zaid (1984). Explants from roots have been used by Schroeder (1970), Sharma et al. (1980), and Zaid and Tisserat (1983).

Seedling can be used as the source for shoot-tip, juvenile leaves, while immature inflorescences are obtained from the spathe (protective sheath) of adult plants. Shoot-tip explants can be obtained from 2- to 4-year-old seedlings and offshoots. Shoot tips are ideal for culturing, since they are devoid of the lateral cambial layer. These can be obtained from the offshoots by removing the fibrous tissue surrounding the leaf base and dissecting parental tissue upwards until the shoot tip is exposed. It is carefully separated and trimmed and surface sterilized. Pieces of shoot-tip meristem are plated on Murashige and Skoog (MS) media with high auxin. To avoid browning due to phenolics, antioxidant ascorbic or citric acid is added to the media.

Explant types varied with different workers. Shoot-tip explants ranged from 1 to 5 mm in length with a few or several leaf primordial. Gabr and Tisserat (1985) noted the best growth response from the explant tips trimmed to 3 mm in length. Smaller shoot-tip explants did not grow well due to excision injury, whereas growth was slow in longer explants and these were quite prone to contamination.

Explants are cultured on a basal nutrient medium composed of MS salts (Murashige and Skoog, 1962), along with vitamins and plant growth regulators (PGRs) and activated charcoal. pH of the medium is adjusted to 5.7 before adding agar and then autoclaved. Shoot-tip cultures were incubated in a temperature-controlled room at 29°C and illuminated for 16 h/day with cool fluorescent lamps. For callus production, cultures are incubated in the dark (Gabr and Tisserat, 1985). Nutrient requirements vary from tissue to tissue even in the same plant. Seedling segments have been successfully cultured to obtain calluses (Reuveni and Lilien-Kipnis, 1974; Eeuwens, 1976, 1978). When the cultures are initiated, usually the explant and the media turn to brown color and this results in premature death of the explant (Jones, 1974; Oppenheimer and Reuveni, 1972; Tisserat, 1979). Browning is due to the oxidation of polyphenols and the formation of quinones that are toxic to the tissues (Reuveni and Lilien-Kipnis, 1974). To overcome this, antioxidants are used; cysteine has a favorable effect, whereas ascorbic acid is ineffective. Addition of activated charcoal and incubation of cultures in the dark would reduce phenolic secretions from the explant. The use of activated charcoal is preferred because cysteine at higher concentrations could be toxic to the plant tissues (Sharma et al., 1980). Other adsorbents such as polyvinylpyrrolidine (PVP) have also been successfully employed to combat explant browning in date palms (Rhiss et al., 1979). Nevertheless, after the formation of callus from the explant, browning is negligible.

Date palm callus is highly totipotent (Tisserat, 1979) and is capable of regenerating numerous embryoids and plantlets even after 2 years of serial subculturing (Tisserat, 1982). Direct regeneration was reported by Sudhersan et al. (1993). Le Thi et al. (1999) developed an efficient method for plant propagation of *Phoenix canariensis* through somatic embryogenesis by using two types of explants, zygotic embryo and shoot tip and at the same time finding the factors affecting embryo induction, development, and maturation. They have collected the zygotic embryos excised from mature seeds and shoot tips excised from dissected shoots, surface sterilized with 2.6% sodium hypochlorite incubated in the dark in MS basal medium supplemented with different concentrations of 2,4-dichlorophenoxyacetic (2,4-D) combined with 2-iP or of Picloram combined with kinetin. The zygotic embryos showed a greater potential to produce embryogenic callus than shoot tip.

In a study carried out by Gabr and Tisserat (1985), naphthalene acetic acid (NAA) enhanced shoot tip growth. Changes in the composition of either hormones or other additives affect the formation of axillary shoots; the addition of cytokinins stimulated lateral bud growth. The extent of budding varied considerably among cultures (Gabr and Tisserat, 1985). The best growth responses were obtained from tips that were initially cultured on agar medium for 2 weeks and then transferred to liquid medium, or from tips cultured on agar medium only throughout the initial culture passage. Tips from various palm species and cultivars were found to grow and develop into seedlings like palms with well-formed root and shoot systems. Adventitious roots were produced from established tips when cultured on basal medium containing low NAA. The plantlets were successfully transferred to soil where they developed into free-living palms (Gabr and Tisserat, 1985). Similar findings were also quoted by Le Thi et al. (1999) on solid medium, but the development of somatic embryos was very slow, taking 1 or 2 months *in vitro*. The liquid medium was found to stimulate the development of somatic embryos. Similar results were reported by Pierik (1987). According to him, the agitated liquid medium significantly increased the number of buds, which could be due to the increase of nutrient availability and uptake in the liquid medium compared to the solid medium, as well as the movement of explants which led to gas exchange, and removal of some mineral deficiency symptoms that occur in solid medium.

Khierallah and Bader (2007) have divided whole process of date palm micropropagation into six stages: (1) explant preparation and sterilization, (2) initiation, (3) multiplication stage, (4) elongation stage, (5) rooting stage, and (6) acclimatization stage. They have indicated the superiority of benzyl adenine (BA) over other cytokinins (kinetin and isopentenyladenine [2ip]) in the initiation stage for the development of buds, whereas the addition of 2ip rather than BA was used in the multiplication stage. Gibberellic acid (GA3) was found to have a positive effect on the elongation of shoots, while the addition of NAA helps in root development. The transfer of rooted shoots to the MS salt solution and a gradual increase of light intensity and reduction of humidity are recommended for acclimatization.

EFFECT OF PGRs

Exogenously supplied PGRs are essential for somatic embryogenesis. Auxin is required initially for the induction and proliferation of cells that later differentiate into somatic embryos. Schroeder (1970) and Reuveni (1972) suggested two methods of propagation: the first was direct organogenesis in which buds eventually form plantlets without passing through callusing and the second was somatic embryogenesis through embryogenic callus produced from explants. Comparatively more success was reported in direct organogenesis (Tisserat, 1984; Omar et al. 1992; Al-Maari and Al-Ghamdi 1997; Al-Khateeb et al. 2002). In the second method, the production of somatic embryos from embryogenic callus has been reported by many researchers (Al-Khalifah, 2000; Al-Musawi, 2001; Al-Khayri, 2003). Despite the fact that the second method is most commonly used in commercial plant tissue culture labs, there is a possibility of inducing undesirable genetic variability during the culture and regeneration of plants, which is not detected until the fruiting stage. The first method results in the production of genetically identical plants and true-to-type and hence can be employed in large-scale micropropagation of date palm (Khierallah and Bader, 2007).

After a month in culture, apical meristems are removed and cut longitudinally into four equal segments and cultured on MS media supplemented with BA, kinetin, 2ip, NAA, and naphthoxyacetic acid. The activated charcoal was replaced with PVP (Al-Marri and Al-Ghamdi, 1997; Bekheet and Saker, 1998; Al-Khateeb et al., 2002).

In the same study, the effect of GA3 was also tested on shoot length and it was found that shoot length increased with increasing GA3 concentration in the medium, which directs our attention to the well-known role of gibberellins in the elongation of plant cells. After rooting was checked, this was found to be greater with the addition of NAA, with the concentration 0.5 mg/L resulting in the best-rooting percentage (90%) and average root length (5.4 cm), which were significantly different from other treatments. No reduction in root length was observed with an increase in the concentration of auxin. It is known that auxin plays an active role in root formation by the induction of root initials (IPGSA, 1998), although contradictory results were also reported by El-Hammady (1999) using NAA. He noted that the average root length decreased with increasing auxin concentration. Many researchers have mentioned the importance of NAA in the rooting of date palm shoots *in vitro* (Al-Maari and Al-Gamdi, 1997; Tisserat, 1984; Mater, 1990).

SOMATIC EMBRYOGENESIS

Fki et al. (2003) reported regeneration of plants from somatic embryos derived from suspension cultures. Embryogenic calli were initiated from leaf and inflorescence explants. The inflorescences have a higher capability of forming callus than the leaves. Suspension cultures were established from calli showing a high competency for somatic embryogenesis containing proembryonic masses. The medium in suspension culture that had a low amount of auxin 2,4-D differentiates somatic embryos in large numbers. Partial desiccation of the mature somatic embryos, with a decrease in water content, improved germination. The cutting of the cotyledonary leaf stimulated the germination of embryo as it is known to inhibit the development of apical meristem. Such a stimulation may be attributed to an inhibitory substance released by the surrounding embryogenic cells; consequently, embryos enter the maturation phase. Also, it is possible that the physical stress caused by fine chopping of callus stimulates *de novo* protein synthesis, allowing a rapid division and differentiation of embryogenic cells. In addition, chopping may minimize the intercellular connections between cells and contribute to a partial isolation of cells (Othmani et al., 2009). Alromaihi and Elmeer (2009) studied the influence of different media on the development of roots and leaves from somatic embryos of Kapkap and Tharlaj cultivars, and noted that embryos cultured on hormone-free medium produced only shoot and require NAA for further shoot elongation and rooting.

Addition of silver nitrate ($AgNO_3$) to the media has a positive effect on callus formation and somatic embryogenesis. Callus was cultured on MS medium containing NAA, 2ip, and $AgNO_3$. Callus fresh weight increased in different $AgNO_3$ concentrations (Al-Khayri and Al-Bahrany, 2004). Callus growth gradually increased as the $AgNO_3$ concentration increased; the optimum concentration depended upon the cultivar and ranged from 50 to 62.5 M. Variations were observed in callus proliferation and regeneration efficiency with $AgNO_3$, which either stimulated or inhibited growth and morphogenesis depending upon the concentration and genotype. The concentration for optimal callus growth may not give a similar response for somatic embryogenesis. This information provides insight into improving micropropagation of recalcitrant date palm genotypes. Exploration of the effect of other ethylene antagonists may be worthy of future investigation.

Othmani et al. (2009) tried different concentrations and combinations of 2,4-D, picloram, 6-benzylaminopurine (BAP), and NAA to induce somatic embryogenesis from juvenile leaf explants of date palm. Of these, 2,4-D (10 mg/L) was best in enhancing the induction of embryogenic callus and somatic embryos. This effect is probably mediated by a signal cascade trigged by exogenous auxin. In contrast, Bhaskaran and Smith (1992) reported that, in certain date palm cultivars, embryogenesis occurred only in the absence of 2,4-D. According to Othmani et al. (2009), the highest induction frequency of embryogenic calli occurred after 6–7 months on MS medium supplemented with 2,4-D and activated charcoal. Subculture of these calli resulted in the formation of proembryos. According to Zouine and El-Hadrami (2007), 2,4-D stimulated secondary embryogenesis due to the stimulation of cell division on the apical part of primary somatic embryos and on the shoot apex of plantlets, leading to the development of the globular embryos on the surface of the primary somatic embryos suppressing their normal germination. Fine chopping and partial desiccation of embryogenic calli with proembryos that were transferred to MS medium supplemented with abscisic acid (ABA) stimulated the rapid maturation of somatic embryos. Subsequent substitution of ABA with NAA resulted in the germination and conversion of the somatic embryos into plantlets with normal roots and shoots. According to Rance et al. (1994), the desiccation treatment might trigger rapid biochemical changes in the calli and under-water stress-specific enzymes or polypeptides probably appear in callus culture. Attree and Fowke (1993) reported that desiccation treatment stimulates the accumulation of storage reserves and triglycerides that have a positive effect on somatic embryo maturation. Al-Utbi and Al-Husaibi (2008) successfully initiated procallus and embryogenic callus from shoot tips of date palm in MS medium with high levels of auxin. Procallus appeared as lobed, whereas embryogenic callus was friable (Tisserat and Demason, 1980; Tisserat, 1979, 1982; Mater, 1986). Depending on the number of embryonic centers, the callus response was different in the production of somatic embryos. The use of low levels of auxin NAA in the presence of cytokinins induced the formation of somatic embryos.

Sugar is the principal carbon source normally used, but other carbon sources like ground corn, banana fruit pith, and onion root tips were also tried as a source of vitamins (banana), cytokinin zeatin (corn), and auxin (actively dividing root tips). Results indicate that banana juice is superior to corn and onion. All the sugars, namely, glucose, arabinose, and sucrose, were found to be significantly superior to the control. Al-Khateeb (2008) investigated the effect of different concentrations of date palm syrup and noticed that the effect was second to sucrose. Similar results were reported by Kinnersley and Henderson (1988) and El-Assar et al. (2004). Similarly studies were performed adding plant extracts such as juice of coconut, tomato, potato, onion, banana, orange, apple, pineapple, and yeast to the culture medium (Chen et al., 2005; Amo-Marco and Picazo, 1994; El-Assar et al., 2004).

Al-Khayri (2001) has shown that vitamin content in the culture medium is an important factor in date palm tissue culture. In addition to thiamine, the importance of biotin was highlighted. Callus growth and embryo formation were best achieved on a medium containing 0.5 mg of thiamine combined with 2 mg/L biotin. This combination was effective in embryo elongation.

HISTOLOGY OF DATE PALM CALLUS

Histological analysis of callogenesis and somatic embryogenesis from cell suspensions of date palm was studied by Sane et al. (2006), starting from the initial callusing to the establishment of the cellular embryogenic cell suspensions, the formation of somatic embryos and their development into rooted plants. Foliar segments of seedlings were cultured in the presence of 2,4-D produced primary calli that were cut to produce fine friable granular calli and subsequently used in cell suspensions in liquid medium. The somatic proembryos developed after the removal of 2,4-D were shifted to agar medium, where they developed into complete plants. Thin sections of tissue fragments that were stained with periodic acid, Schiff, and Naphthol Blue-Black reveal that the first cellular divisions were localized closer to the vascular tissue in the leaf. The primary calli were obtained within 2 months. Fine friable granular calli grew quickly after the primary calli were chopped. Individual embryogenic cells were identified that rapidly started to divide and developed into globular proembryos. In the microcalli, breaking zones appeared in the thick pectocellulosic walls, which delimited the pluricellular proembryos. The anatomy of somatic embryos is similar to that of zygotic embryos despite a deficit in the accumulation of intracellular proteins. When rooted with NAA, the vitro plants developed a strong elongated taproot.

Gueye et al. (2009a,b) worked on callogenesis and rhizogenesis in date palm and found similarities between the two auxin-induced pathways. As per this report, in immature leaf explants, callogenesis is a prerequisite for the initiation of somatic embryogenesis, and requires the presence of auxin (NAA) in the medium to induce either rhizogenesis or callogenesis. Histological analysis established that precocious cell reactivation is similar in both morphogenetic pathways. Cytological modifications are associated with cell reactivation and are observed in the cells of perivascular sheaths. Divergence between the callogenesis and rhizogenesis pathways is observed later, during the subsequent determination and morphological differentiation phases. They found that, in date palm, the rhizogenesis and callogenesis pathways are initiated from the same cell type and the ultimate developmental fate depends on auxin concentration. They determined three phases in date palm: (a) the acquisition phase—characterized by the early cytological modifications in response to NAA; (b) the determination phase—for one specific morphogenesis pathway, depending on the concentration of NAA; and (c) the morphological differentiation phase, where organ morphogenesis/callogenesis occurs.

Gueye et al. (2009a,b) also worked on acquisition of callogenic capacity in date palm leaf tissues in response to 2,4-D treatment and noted that callogenesis is difficult in certain genotypes, and to understand the process they studied the developmental events at cellular level. The callogenic capacity of leaf segments depends on the state of cell differentiation. The most callogenic segments were within the leaf elongation zone that required polar auxin transport to initiate callogenesis and contained the highest quantities of free endogenous indole-3-acetic acid (IAA). At the cellular level, callus induction involves two spatially and temporally separated events. The first event involves cells from the fascicular parenchyma in an equatorial plane perpendicular to the vascular axis, which in the presence of 2,4-D acquire structural features typical of the meristematic cells. The second event occurs 3 days later and is characterized by the modification of adjacent perivascular sheath cells, which become callogenic and undergo cell division, leading to callus formation. Their experiments provide evidence that callus initiation in leaf vascular tissue of date palm involves a sequential response of two distinct cell types and requires polar auxin transport.

The major bottleneck is the acclimatization of somatic embryo-derived plantlets, since the success rate is low. Successful monitoring of the acclimatization step requires in-depth studies of photosynthetic activity in the plantlets with reference to seed-derived plants (Rival et al., 1998; Masmoudi et al., 1999). In spite of this problem, thousands of rooted plantlets have already been produced commercially with an average success rate of over 50%.

Female inflorescences exhibit a high ability to undergo either *in vitro* somatic embryogenesis or organogenesis (Drira and Benbadis, 1985; Fki et al., 2003). Siljak-Yakovlev et al. (1996) reported that date palm has heteromorphic X/Y sex chromosomes. De Mason and Tisserat (1980) reported that bisexual flowers can occur naturally in some cases, in female date palm trees. These authors also reported that male flowers can be transformed into bisexual by 2,4-D treatment. Masmoudi-Allouche et al. (2009) cultured inflorescence pedicels on MS medium supplemented with indole-3-butyric acid and BAP, and observed initiation of anther development within the female flowers. Recently, Masmoudi-Allouche et al. (2010) also reported the *in vitro* reversion of a shoot meristem from a vegetative to reproductive state. *In vitro* flowering induction was obtained from apical vegetative meristem that brought about a change in its anatomy without changing its phyllotaxy and the leaf shape.

PHYSIOLOGICAL CHANGES DURING CALLUSING AND EMBRYOGENESIS

Ibrahim et al. (2005) worked on biochemical changes in date palm cultivars and found differences in isozyme-banding patterns that would characterize the cultivars and developmental stages. Hormonal and chemical changes during embryogenesis and germination were also studied by Ibrahim et al. (2005). Samples were analyzed for IAA, GA3, zeatin, ABA, total soluble sugars, free amino acids, and proline. The amount of IAA and GA3 increased in shoot tip explants, but decreased at the embryo stage and then increased again from the embryo stage to the germinated embryo stage. The zeatin content was high at the embryogenic callus stage. The ABA content was at a relatively higher level at the callus stage, but the contents of IAA, GA3, and ABA decreased at the embryo stage. The callus stage recorded high total soluble sugar content, while the free amino acid content was high in the germinated embryo and shoot tip stages. The proline content was high in the shoot tip, embryo, and germinated embryo stages. El-Din et al. (2007) investigated the biochemical and morphological changes at different stages in tissue culture propagation. The morphological studies indicated that shoot tip explants of Bartamuda cv. produced a higher degree of callus and a lower degree of browning than the other cultivars. An increase in indole content was noted in different developmental stages except phenol content of plantlets. SDS–PAGE of water-soluble protein fraction and isozymes of peroxidase and polyphenol oxidase were used to identify and characterize the effect of genotypes and their developmental stages on embryogenesis. SDS–PAGE of water protein fraction was used to determine the molecular weights of protein subunits. The electrophoregrams revealed that a number of polypeptides increased or decreased in intensity. Isozyme analyses by electrophoresis detected the differences, at gene levels, corresponding to genotypes and developmental stages. The peroxidase patterns showed different peroxidase isozyme bands, whereas polyphenol oxidase profiles showed three isozyme bands with different intensities. The increment of these isozymes is correlated with the ratio of phenol and phenol-indole (auxin) at plantlet stage.

Growth and physiological responses of callus to salinity stress was studied by Al-Khayri (2002). Callus induced by shoot tips of offshoots was cultured on MS medium supplemented with NaCl at concentrations ranging from 0 to 225 mM, with an increase of 25 mM. Data obtained after 6 weeks of exposure to salt have shown a significant increase in callus proliferation in response to 25 mM NaCl, the lowest level tested, beyond which callus weight decreased. At 125 mM NaCl and higher, callus growth was nearly completely inhibited. Physiological studies on callus exposed to salt stress have shown an increase in proline accumulation in response to increased salinity, which leads to inhibition of callus growth. Furthermore, increasing the concentration of NaCl in the culture medium generally resulted in a steady increase in Na^+ and a reduction in K^+ concentrations. However, 25 mM NaCl the only level at which callus growth was significantly enhanced and an increase in K^+ content was noted in comparison to the NaCl-free control. In response to increasing external NaCl level, the Na^+/K^+ ratio increased. The Na^+/K^+ ratio was positively correlated to proline accumulation and hence callus growth inhibition.

Al-Khayri and Al-Bahrany (2004) conducted a study to examine the response of date palm (cvs. Barhee and Hillali) calli to water stress. Callus derived from shoot tip explants was grown in liquid MS medium containing 10 mg dm^3 α-NAA, 1.5 mg dm^3 2ip, and 0%–30% (m/v) polyethylene glycol (PEG 8000) to examine the effect of water stress. After 2 weeks, callus growth, water content, and proline accumulation were checked. Increasing water stress caused a progressive reduction in growth as expressed in callus fresh mass, relative growth rate, and index of tolerance. Cultivar Barhee was found to be more tolerant to drought stress than cv. Hillali. Increasing PEG concentration was also associated with a progressive reduction in water content and increased content of endogenous free proline.

ABA, arginine, and sucrose were tested for their effects on the morphology, germination rates, and protein content of date palm somatic embryos by Sghaier-Hammamia et al. (2009a,b). Culture medium was supplemented with different concentrations of these substances. There was no difference in length and thickness of somatic embryos either treated or untreated, except for ABA (20 μM), which increased the thickness. A decrease of water content increases the dry weight of treated somatic embryos and causes increased proliferation rate than that of the control. All the tested compounds increased protein content but ABA (20 μM) was more effective in protein enrichment than arginine and sucrose treatments. The SDS–PAGE protein profiles showed a significant difference between treated and untreated somatic embryos. A protein band of 22 kDa, identified as glutenin, was accumulated after treatment with 20 μM ABA or 3 mM arginine.

Two-dimensional gel electrophoresis along with mass spectrometry was used to compare the proteome of zygotic and somatic embryos (Sghaier-Hammamia et al., 2009a,b). Proteins were trichloroacetic acid–acetone–phenol extracted, quantified, and resolved by 2-DE in the pH range 5–8. Total protein content and the number of resolved spots were higher in zygotic embryos than in somatic embryos. The 2-DE map of both systems showed qualitative and quantitative differences. Variable spots were subjected to mass spectrometry analysis, and the identified proteins were classified into different functional categories. Most of the somatic embryo-specific proteins belonged to glycolysis pathways, whereas those of the zygotic embryo belonged to storage and stress-related proteins. In another study, Sghaier-Hammamia et al. (2010) supplemented the initial embryo culture medium with (ABA or sucrose) to improve the vigor of *in vitro* plants derived from somatic embryogenesis. They recorded increased embryo thickness with no difference in length. ABA increased the embryo proliferation rate and caused a 1.5–1.7-fold increase in the amount of proteins. The 1-D and 2-DE protein profiles showed qualitative and quantitative differences between the somatic embryos cultured in the initial medium and in the modified medium. Using mass spectrometry analysis, proteins were identified and classified. While "up-regulation" of stress-related proteins and "down-regulation" of energy metabolism proteins were observed in somatic embryos matured in medium supplemented with ABA, storage proteins (legumin) were "up-regulated" in somatic embryos matured in medium supplemented with increased sucrose.

Masmoudi et al. (1999) studied the carbon metabolism and the role of carboxylases, phosphoenolpyruvate carboxylase (PEPC), and ribulose 1,5-bisphosphate carboxylase/oxygenase (RubisCO) in embryogenic and nonembryogenic cultures. The presence of three active isoforms of PEPC was detected after electrophoresis from the embryogenic callus, whereas only a single band was present in the nonembryogenic one. Further changes in carboxylase (PEPC and RubisCO) activities during the growth and development of somatic embryo-derived plantlets were also studied. The PEPC/

RubisCO ratio progressively decreased throughout the development of plantlets; RubisCO assumed greater importance and became the main route for inorganic carbon fixation.

SOMACLONAL VARIATION

Thousands of years of selection have produced over 3000 varieties in the date palm (Al-Khayri, 2007). Recently, efforts were made to develop phenotypic, biochemical, and DNA-based markers to characterize the genetic diversity in date palm populations and to establish the relationships between different cultivars.

In micropropagation, it is critical to select proper genotypes and grow mother plants under a controlled environment, and to determine the maximum number of subcultures before initiating new cultures. By failing to do so, *in vitro* grown plants will show somaclonal variation in the field. For mutation induction, fine "embryogenic cell suspension cultures" are treated with γ irradiation and the cells are cultured onto fresh medium for the development of mutated somatic embryos. This approach produces mutated somatic seedlings in a short period of time and also prevents occurrence of chimera (Mohan Jain, 2007). A survey of embryogenesis-derived date palms in the field reveal various morphological abnormalities including abnormal leaves and inflorescences, dwarfing, leaf bleaching, deformed offshoots, delayed flowering time, pollination failure, abnormal fruiting, and so on. All of these are negative characters from the grower's point of view. One of the major weaknesses of mass tissue culture propagation is the appearance of undesired off-type plants, such as poor fertility and delayed flowering (Awad, 2007). More research work is needed on tissue culture-derived date palms to understand more clearly the nature of abnormalities that appear in some cultivars.

There are many reports on the variability in the morphological traits or isozyme makers (Rhouma, 1994; Reynes et al., 1994; Salem et al., 2001). The data based on molecular markers such as restriction fragment length polymorphisms (RFLPs) and random-amplified polymorphic DNAs (RAPDs) have been used to characterize date palm genotypes (Sedra et al., 1998; Ben Abdallah et al., 2000; Trifi et al., 2000; Trifi, 2001). Lately, Karim et al. (2010) have used inter simple sequence repeat (ISSR) markers, which are regarded as superior to other markers (isozymes, RFLPs, and RAPDs), since they permit the detection of polymorphisms in microsatellites and intermicrosatellites loci without previous knowledge of the DNA sequence (Gupta et al., 1994). Microsatellites regions are abundant throughout the eukaryotic genome, are highly polymorphic in length (Tautz, 1989), and are interspersed (Zietkiewicz et al., 1994). ISSR gives information about many loci and can discriminate closely related genotype variants (Roose and Fang, 1997). Microsatellites molecular markers were used to detect the polymorphisms in Tunisian date palm varieties. Karim et al. (2010) have observed the divergence between Tunisian varieties that are organized in clusters, but were unable to identify all tested genotypes as mono-varietal groups. Among 40 regenerated plants, aberrant plants were 12.8%, whereas the total numbers of variant DNA fragments showed a much higher level of variability (0.6%).

Genetic diversity in date palm germplasm from Sudan was investigated using 16 loci of microsatellite (simple sequence repeat [SSR]) primers (Elshibli and Korpelainen, 2007). The tested SSR markers showed a high level of polymorphism. A total of 343 alleles were detected at the 16 loci. The number of alleles per marker ranged from 14 to 44 with an average of 21.4 per locus. A high level of expected heterozygosity was observed among Sudan cultivars (0.841), Morocco cultivars (0.820), and male accessions (0.799). The results indicate that the genetic groups of the Sudan cultivars and/or males do not follow a clear geographic pattern. However, the Morocco group showed significant differentiation in relation to the Sudan groups, as measured by F_{ST} values and genetic distances. The effect of the methods of pollination and cultivar selection on the genetic structure was clearly detected by the weak clustering association that was observed for the majority of accessions originating from Sudan and Morocco as well. This suggests the need for further investigation on the genetic diversity of Sudanese date palm germplasm (Elshibli and Korpelainen, 2007). High genetic diversity was reported by Rhouma et al. (2010) in Tunisian date palm varieties.

GENETIC POLYMORPHISM

Genetic variation was studied using RAPD markers among 43 date palm accessions that included 37 accessions from Morocco and 6 from Iraq and Tunisia. The prescreening was done with 123 primers on four genotypes of which 19 primers were selected that revealed polymorphism. All the analyzed genotypes were distinguishable by their band patterns. RAPD-based genetic distance was used to determine the relationships between the accessions. The grouping-association identified by cluster analysis was weak. So they have clustered together morphologically similar varieties. The results indicate a relatively low polymorphism and lack of organization among the date palm varieties grown in Morocco. This could be related to the mode of introduction and maintenance of the Moroccan date palm germplasm involving limited foundation germplasm, exchange of cultivars between plantations, and periodic development of new recombinant cultivars following sexual reproduction.

Propagation by either offshoots or tissue culture generally results in true-to-type plants, but some off-types with abnormal phenotypes have developed in tissue-cultured plants, which may be due to somaclonal variation. RAPD analysis was used to detect the genetic variation among "Khalas," "Barhy," and "Sukkary" cultivars (Al-Khalifah and Askari, 2007). The cultivars showed unique RAPD-banding patterns with 18 primers. The cluster analysis by the unweighted-paired group method of arithmetic mean showed two clusters. Offshoots and tissue cultured plants of "Khalas" showed a relatively low level of polymorphism. However, tissue-cultured "Barhy" showed a high level of genetic variation. Ali et al. (2007) also studied the RAPD polymorphism in Iraqi date palms obtained from tissue culture. Al-Kaabi et al. (2007) reported somaclonal variation in amplified fragment-length polymorphism (AFLP) analysis in tissue culture-derived date palm in the United Arab Emirates. The incidence of somaclonal variation in plants regenerated through organogenesis was low in contrast to the plants derived from somatic embryogenesis and the organogenesis-derived plants were genetically closer to the mother plant. Some cultivar specificity was noted in AFLP which could be used as a fingerprinting tool to identify somaclonal variation in plants derived from tissue culture.

In spite of several advantages with tissue culture and micropropagation in date palm, there are several points to be addressed. The low survival of regenerated plants at the acclimatization stage and the failure to get true-to-type plants due to the occurrence of abnormalities are the most serious problems. Nevertheless, the early detection and identification of variability in tissue culture would be advantageous.

REFERENCES

Al-Baker, A. 1972. *The Date Palm.* Al-Ani Press, Baghdad, Iraq.

Al-Kaabi, H.H., Zaid, A., Shephard, H., and Ainsworth, C. 2007. AFLP variation in tissue culture-derived date palm (*Phoenix dactylifera* L.) plants. *Acta Hort.* (ISHS) 736:135–160.

Al-Khalifah, N.S. 2000. *In vitro* culture studies in date palm. *Plant Tissue Cult.* 10:1–8.

Al-Khalifah, N.S. and Askari, E. 2007. Early detection of genetic variation in date palms propagated from tissue culture and offshoots by DNA fingerprinting. *Acta Hort.* (ISHS) 736:105–112.

Al-Khateeb, A. 2008. Comparison effects of sucrose and date palm syrup on somatic embryogenesis of date palm (*Phoenix dactylifera* L). *Am. J. Biotechnol. Biochem.* 4:19–23.

Al-Khateeb, A., Abdalla, G.R., Alt-Dinar, H.M., Abdalhameed, A., Al-Khateeb, and Abugulia, K.A. 2002. Auxin: Cytokinin interactions in the *in vitro* micropropagation of date palm (*Phoenix dactylifera* L.) *Egypt. J. Appl. Sci.* 17:409–415.

Al-Khayri, J.M. 2001. Optimization of biotin and thiamine requirements for somatic embryogenesis of date palm (*Phoenix dactylifera* L.) *In Vitro* Cell. *Dev. Biol. Plant* 37:453–456.

Al-Khayri, J.M. 2002. Growth, proline accumulation, and ion content in sodium chloride-stressed callus of date palm cell. *Dev. Biol. Plant* 38:79–82.

Al-Khayri, J.M. 2003. *In vitro* germination of somatic embryos in date palm: Effect of auxin concentration and strength of MS salts. *Curr. Sci.* 84:680–682.

Al-Khayri, J.M. 2007. Date Palm *Phoenix dactylifera* L. In: *Micropropagation Protocols for Woody Trees and Fruits*, Mohan Jain, S. and Haggman, H. (eds), Springer, the Netherlands, pp. 509–526.

Al-Khayri, J.M. and Al-Bahrany, A.M. 2004. Growth, water content, and proline accumulation in drought-stressed callus of date palm. *Biol. Plant.* 48:105–108.

Al-Maari, K.W. and Al-Ghamdi, A.S. 1997. Micro-propagation of five date palm cultivars through *in vitro* axillary buds proliferation. *Du. J. Agri. Sci.* 13:55–71.

Al-Musawi, A.H. 2001. Effect of auxin, cytokinine, and activated charcoal on embryo development and germination of date palm (*Phoenix dactylifera* L.) cultured *in vitro. Basrah Date Palm Res. J.* 1:8–18.

Alromaihi, K.I.K.B. and Elmeer, K.M.S. 2009. Influence of different media on *in vitro* roots and leaves of date palm somatic embryos cvs. *Kapkap and Tharlaj: American-Eurasian J. Agric. Environ. Sci.* 6:100–113.

Al-Utbi, F.M. and Al-Husaibi. 2008. *In vitro* culture of date palm *Phoenix dactylifera* L. and the effect of the addition of some materials on shoot tip and callus growth. *Basrah J. Agric. Sci.* 21:2008.

Al-Yahyai, R. 2007. Improvement of date palm production in the sultanate of Oman. *Acta Hort.* (ISHS) 736:337–343.

Ali, T.A., Jubrail, J.M., and Jasim, A.M. 2007. The use of RAPDs for the detection of genetic stability of regenerated plantlets of Barhee palm in Iraq. *Acta Hort.* (ISHS) 736:127–134.

Amo-Marco, J.B. and Picazo, I. 1994. *In vitro* culture of albedo tissue from fruits of *Citrus sinensis* cv. Washington Navel: Effect of fruit age and orange juice. *J. Hortic. Sci.* 69:929–993.

Anonymous, 2008. http://www.newworldencyclopedia.org/entry/Date_palm.

Attree, S.M. and Fowke, L.C. 1993. Somatic embryogenesis and synthetic seeds of conifers. *Plant Cell Tissue Organ Cult.* 35:1–35.

Awad, M.A. 2007. Fruit set failure in tissue culture-derived date palm trees (*Phoenix dactylifera* L.) cv. 'Nabt Saif' as affected by pollinator type and pollination density. *Acta Hort.* (ISHS) 736:441–448.

Bekheet, S.A. and Saker, M.M. 1998. *In vitro* propagation of Egyptian date palm: II Direct and indirect shoot proliferation from shoot tip explant of (*Phoenix dactylifera* L.) cv. Zaghlool. In: *1st International Conference Date Palms*, Mar. 1998, Al-Ain, UAE, p. 28.

Bekheet, S.A., Taha, H.S., Saker, M.M., and Moursy, H.A. 2002. A synthetic seed system of date palm through somatic embryogenesis encapsulation. *Annals Agric. Sci. Ain Shams Univ. (Cairo)* 47:325–337.

Ben Abdallah, A., Stiti, K., Lepoivre, P., and du Jardin, P. 2000. Identification de cufiivars de palmier dattier (*Phoenix dactylifera* L.) par lamplification aleatoire dADN (RAPD). *Cahiers Agric.* 9:103–107.

Benbadis, B.K. 1992. Coconut and date palm. In: *Biotechnology of Perennial Fruit Crops.* Hammerschlag, F.A. Litz, R.W. (eds), CAB International, UK, pp. 383–400.

Bhaskaran, S. and Smith, R. 1992. Somatic embryogenesis from shoot tip and immature inflorescence of *Phoenix dactylifera* cv. *Barhee. Plant Cell Rep..* 12:22–25.

Branton, R.L. and Blake, J. 1989. Date palm (*Phoenix dactylifera* L.). In: *Biotechnology in Agriculture and Forestry*, Bajaj, Y.P.S. (ed.), Springer-Verlag, Heidelberg, Vol. 5, pp. 161–175.

Brown, T.W., Perkins, R.M., and Vis, E.G. 1969. Temperature and heat unit occurrences during date pollination in Coachella Valley of California. *Date Grower's Inst. Rep.* 46:21–24.

Chen, T., Qing, Y.S., and Wei, L. 2005. The orthogonal test of induction and proliferation of *Dendrobium nobile* protocrom-like bodies (PLBs). *J. South China Agric. University* 26:60–63.

Daguin, F. and Letouze, R. 1988. Régéneration du palmier dattier (*Phoenix dactylifera* L.) par embryogenèse somatique: Amélioration de l'efficacité par passage en milieu agité. *Fruits* 43:191–194.

Daguin, F. and Letouzé, R. 1997. Le clonage végétal. *Biofutur* 163: 1–12 (Le Technoscope, Cahier 88).

De Mason, D.A. and Tisserat, B. 1980. The occurrence and structure of apparently bisexual flowers in the date palm, *Phoenix dactylifera* L. (Arecaceae). *Bot. J. Linn. Soc.* 81:283–292.

Drira, N. 1983. *In Vitro* flowering induction in date palm (*Phoenix dactylifera* L.). *J. Plant Growth Regul.* 29:35–43.

Drira, N. and Benbadis, A. 1985. Multiplication végétative du palmier dattier (*Phoenix dactylifera* L.) par reversion, en culture *in vitro*, d'ébauches florales de pieds femelles adultes. *J. Plant Physiol.* 119:227–235.

Eenwens, C.J. 1978. Effect of organic nutrients and hormones on growth and development of tissue explants from coconut (*Cocos nucifera* L.) and date (*Phoenix dactylifera* L.) palms cultured *in vitro. Physiol. Plant* 42:173–178.

Eeuwens, J. 1976. Mineral requirements for growth and callus initiation of tissue explants excised from mature coconut palms (*Cocos nucifera*) and cultured *in vitro. Physiol. Plantarum* 36:23–28.

El-Mardi, M.O., Labiad, S.N., Consolacion, E., and Abdelbasit, K.M. 1995. Effect of pollination methods and pollen dilution on some chemical constituents of Fard dates at different stages of fruit development. *Emr. J. Agric. Sci.* 7:1–19.

El Hadrami, I., Cheikh, R., and Baaziz, M. 1995. Somatic embryogenesis and plant regeneration from shoot-tip explants in *Phoenix dactylifera* L. *Biol. Plant.* 37:205–211.

El Hadrami, I., El Bellaj, M., El Idrissi, A., J'Aiti, F., El Jaafari, S., and Daayf, F. 1998. Biotechnologie végétales et amelioration du palmier dattier (*Phoenix dactylifera* L.) Pivot de l'agriculture oasienne Marocaine. *Cah. Agric.* 7:463–468.
El-Assar, A.M., El-Messeih, W.M., and El- Shenawi, M.R. 2004. Applying of some natural extracts and growth regulators to culture media their effects on Sewi cv. date palm tissues grown *in vitro. Assuit J. Agric. Sci.* 35:155–168.
El-Din, Z., Amal, F.M., AbdEl-Rasoul, M., Ibrahim, I.S., Aly, A.S., and Sharaf Eldeen, H.A.M. 2007. Micropropagation of some date palm cultivars: Changes of some chemical constituents related to embryogenesis. *Acta Hort. (ISHS)* 736:233–241.
El-Hammady, A.A. 1999. Regeneration of date palm "Sewy" cv. plantlets by somatic embryogenesis through callus with reference to the genetic stability. In *The International Conference Date Palm*, Nov. 1999, Assiut University, Egypt, pp. 117–131.
El-Hennawy, H.M. and Wally, Y.A. 1978. Date palm (*Phoenix dactylifera* L.) bud differentiation *in vitro. Egypt. J. Hort.* 5:81.
Elshibli, S. and Korpelainen, H. 2007. Microsatellite markers reveal high genetic diversity in date palm (*Phoenix dactylifera* L.) germplasm from Sudan. *Genetica* 134:251–260.
Feher, A., Pasternak, T., and Dudits, D. 2003. Transition of somatic plant cells to an embryogenic state. *Plant Cell Tissue Organ. Cult.* 74:201–228.
Fki, L., Masmoudi, R., Drira, N., and Rival, A. 2003. An optimized protocol for plant regeneration from embryogenic suspension cultures of date palm (*Phoenix dactylifera* L.) cv. *Deglet Nour. Plant Cell Rep.* 21:517–524.
Gabr, M.F. and Tisserat, B. 1985. Propagating palms *in vitro* with special emphasis on the date palm (*Phoenix dactylifera* L.). *Sci. Hort.* 25:255–262.
Gueye, B., Morcillo, F., Collin M., Gargani, D., Overvoorde, P., Aberlenc-Bertossi, F., Tranbarger, T.J., Sane, D., Tregear, J.W., Borgel, A., and Verdeil, J. 2009a. Acquisition of callogenic capacity in date palm leaf tissues in response to 2,4-D treatment. *Plant Cell Tissue Organ Cult.* 99:35–45.
Gueye, B., Saıd-Ahmed, H., Morcillo, F., Borgel, A., Sane, D., Hilbert, J.-L., Verdeil, J.-L., and Blervacq, A.-S. 2009b. Callogenesis and rhizogenesis in date palm leaf segments: Are there similarities between the two auxin-induced pathways? *Plant Cell Tissue Organ Cult.* 98:47–58.
Gupta, M., Chyi, Y.S., Romero-Severson, J., and Owen, L. 1994. Amplification of DNA markers from evolutionarily diverse genomes using single primers of simple-sequence repeats. *Theor. Appl. Genet.* 89:998–1006.
Hussein, F. 1982. Pollination of date palm and its effect on production and quality of fruits. *Proceedings of the First Symposium on Date Palm in Saudi Arabia.* King Faisal University, Al-Hassa, Saudi Arabia, Vol. 1, pp. 14–24.
Hussein, F., Al-Kahtani, M.S., and Wally, Y.A. 1979. *Date Cultivation and Production in both Arab and Islamic Nations.* Egyptian Horticulture Society, Ain Shams University Press, Cairo.
Ibrahim, I.S., Sharaf, H.A.M., Amal, F.M.Z.E., and El-Rasoul, M.A. 2005. Physiological studies on micropropagation of some date palm cultivars: 1. Changes of some biochemical constituents related to embryogenesis. *Annals of Agricultural Science Cairo: Faculty of Agriculture, Ain Shams University (Cairo)* 50:381–395.
IPGSA. 1998. *International Conference on Plant Growth Substances*, Aug. 13–17, 1998, Makuhari Masse, Chiba, Japan.
Jones, L.H. 1974. Propagation of clonal oil palm by tissue culture. *Oil Palm News* 17:1–9.
Karim, K., Chokri, B., Amel, S., Wafa, H., Richid, H., and Nouredine, D. 2010. Genetic diversity of Tunisian date palm germplasm using ISSR markers. *Int. J. Bot.* 6:182–186.
Khierallah, H.S.M. and Bader, S.M. 2007. Micropropagation of date palm (*Phoenix dactylifera* L.) var. maktoom through direct organogenesis. *Acta Hort. (ISHS)* 736:213–224.
Kinnersley, A.M. and Henderson, W.E. 1988. Alternative carbohydrates promote differentiation of plant cells. *Plant Cell Tissue Organ Cult.* 15:3–16.
Kwaasi, A.A.A. 2003. *Date Palms 2003*, Elsevier Science Ltd. Elsevier Science Publishers B.V., Amsterdam, Printed in The Netherlands.
Le Thi L.H., Baiocco, M., Phan Huy, B., Mezzetti, B., Santilocchi, R., and Rosati, P. 1999. Somatic embryogenesis in Canary Island date palm plant cell. *Tissue Organ Cult.* 56:1–7.
Levee, V., Garin, E., Klimaszewska, K., and Seguin, A. 1999. Stable genetic transformation of white pine (*Pinus strobus* L.) after cocultivation of embryogenic tissues with *Agrobacterium tumefaciens. Mol. Breeding* 5:429–440.
Masmoudi-Allouche, F., Chaâri-Rkhis, A., Kriaâ, W., Gargouri-Bouzid, R., Jain, S.M, and Drira, N. 2009. *In vitro* hermaphrodism induction in date palm female flower. *Plant Cell Rep.* 28:1–10.

Masmoudi-Allouche, F., Meziou, B., Kriaâ, W., Gargouri-Bouzid, R., and Drira, N. 2010. *In vitro* Flowering induction in date palm *Phoenix dactylifera* L. *J. Plant Growth Regul.* 29:35–43.
Masmoudi, R.A., Rival, A., Nato, D., Lavergne, N., Drira, G., and Ducreux, G. 1999. Carbon metabolism in *in vitro* cultures of date palm: The role of carboxylases (PEPC and RubisCO). *Plant Cell Tissue Organ Cult.* 57:139–143.
Mater, A. 1986. *In vitro* propagation of *Phoenix dactylifera* L. *Date Palm J.* 4:137–152.
Mater, A.A. 1990. Effect of auxin–cytokinin interaction on micropropagation of date palm. *King Faisal Univ. Saudi Arabia J. Agric. Sci.* 2:211–223.
Mohan Jain, S. 2007. Recent advances in date palm tissue culture and mutagenesis. *Acta Hort. (ISHS)* 736:205–211.
Murashige, T. and Skoog, F. 1962. A revised medium for rapid growth and bioassays with tobacco tissue cultures. *Physiol. Plant* 15:473–479.
Omar, M.S., Hameed, M.K., and Al-Rawi, M.S. 1992. Micropropagation of date palm (*Phoenix dactylifera*). In: *Biotechnology in Agriculture and Forestry, Protoplast and Genetic Engineering*. Vol. 18. Bajaj, Y.P.S. (ed.), Springer-Verlag, Berlin, pp. 471–492.
Oppenheimer, C. and Reuveni, O. 1972. Development of a method for quick propagation of new and superior date varieties. Agriculture Research Organization, The Volcani Centre, Bet Dagan, Israel, pp. 1–92.
Othmani, A., Bayoudh, C., Drira, N., Marrakchi, M., and Trifi, M. 2009. Somatic embryogenesis and plant regeneration in date palm *Phoenix dactylifera* L., cv. Boufeggous is significantly improved by fine chopping and partial desiccation of embryogenic callus. *Plant Cell Tissue Organ Cult.* 97:71–79.
Pierik, R.L.M. 1987. *In vitro Culture of Higher Plants*. Martinus Nizhoff. Publishers, Boston.
Raj Bhansali, R., Kaul, R., and Dass, H. 1988. Mass cloning of date palm plantlets through repetitive somatic embryogenesis. *J. Plant. Anat. Morphol.* 5:73–79.
Rance, I.M., Tian, W., Mathews, H., Kochko, A., Beachy, R.N., and Fauquet, C. 1994. Partial desiccation of mature embryo-derived calli, a simple treatment that dramatically enhances the regeneration ability of Indica rice. *Plant Cell Rep.* 13:647–651.
Rashid, H. and Quraishi, A. 1994. Micropropagation of date palm through tissue culture Pakistan. *J. Agric. Res.* 15:1.
Reuveni, O. 1970. Pistil receptivity of "Khadrawi", "Zahdi" and "Deglet Noor" date flowers. *Date Grower's Inst. Rep.* 47:3–4.
Reuveni, O. 1979. Embryogenesis and plantlets growth of date palm (*Phoenix dactylifera* L.) derived from callus tissues. *Plant Physiol.* 63:138.
Reuveni, O., Adato, Y., and Lilien-Kipnis, H. 1972. A study of new and rapid methods for vegetative propagation of date palms. *Date Grower's Inst. Rep.* 49:17–24.
Reuveni, O. and Lilien-Kipnis, H. 1974. Studies on the *in vitro* culture of date palm (*Phoenix dactylifera* L.) tissues and organs. Pamphlet No. 145, Volcani Inst. Agric. Res., Israel.
Reynes, M., Bouabidi, H., Piombo, G., and Rirterrucci, A.M. 1994. Caracterisatbn des principales varietes de dattes cuttivees dans la region du djerid en tunisie. *Fruits* 49:289–298.
Reynolds, J.F. and Murashige, T. 1979. Asexual embryogenesis in callus cultures of palms. *In vitro* 15:383–387.
Rhiss, A., Poulain, C., and Beauchesne, G. 1979. La culture *in vitro* appligee a la multiplication vegetative du palmier-dattier (*Phoenix dactylifera* L.). *Fruits* 34:551–554.
Rhouma, A. 1994. *Le Palmier Dattier en Tunisie. I: Le Patrimoine Genetique*. Arabesque, Tunis, Tunisia, pp. 127.
Rhouma, S., Zehdi-Azouzi, S., Dakhlaoui-Dkhil, S., Ould Mohamed Salem, A., Othmani, A., Cherif, E., Marrakchi, M., and Trifi, M. 2010. Genetic variation in the Tunisian date palm (*Phoenix dactylifera* L.). *Desert Plants* 4:355–370.
Rival, A., Tregear, J., Verdeil, J.L., Richaud, F., Beule, T., Duval, Y., Saker, M., Moursy, H., and Bekheet, S. 1998. *In vitro* propagation of Egyptian date palm: I—Morphogenic responses of immature embryos. *Bull. Fac. Agric., Univ. Cairo* 49:203–214.
Roose, M.L. and Fang, D.Q. 1997. Identification of closely related citrus cultivars with inter-simple sequence repeat markers. *Theor. Appl. Genet.* 95:408–417.
Salem, A.O.M., Trifi, M., Sakka, H., Rhouma, A., and Marrakchi, M. 2001. Genetic inheritance analyses of four enzymes in date palm (*Phoenix dactylifera* L.). *Genet. Resour. Crop Evol.* 48:361–368.
Sane, D., Aberlenc-bertossi, F., Gassama-dia, Y.K., Sagnal, M., Trouslot, M.F., Duval, Y., and Borgel, A. 2006. Histocytological analysis of callogenesis and somatic embryogenesis from cell suspensions of date palm (*Phoenix dactylifera*). *Ann. Bot.* 98:301–308.
Schroeder, C.A. 1970. Tissue culture of date shoots and seedling. *Date Grower's Inst. Rep.* 47:25–47.

Sedra, M.H., Lashermes, P., Trouslot, P., Combes, M.C., and Hamon, S. 1998. Identification and genetic diversity analysis of date palm (*Phoenix dactylifera* L.) varieties from Morocco using RAPD markers. *Euphytica* 103:75–82.

Sghaier-Hammamia, B., Drira, N., and Jorrín-Novob, J.V. 2009a. Comparative 2-DE proteomic analysis of date palm (*Phoenix dactylifera* L.) somatic and zygotic embryos. *J. Proteomics* 73:161–177.

Sghaier-Hammamia, B., Jorrín-Novob, J.V., Gargouri-Bouzida, R., and Drira, N. 2010. Abscisic acid and sucrose increase the protein content in date palm somatic embryos, causing changes in 2-DE profile. *Phytochemistry* 71:1223–1236.

Sghaier-Hammamia, B., Kriaaa, W., Bahloula, M., Novob, J.V.J., and Drira, N. 2009b. Effect of ABA, arginine and sucrose on protein content of date palm somatic embryos. *Sci. Hort.* 120:379–385.

Shaheen, M.R. 1986. Pistil receptivity in three cultivars of date palm (*Phoenix dactylifera* L.). *Proceedings of the 1st Hort. Sci. Conference*. Tanta University, Egypt, September 1986 (H), pp. 489–499.

Sharma, D.R., Deepak, S., and Chowdury, J.B. 1986. Regeneration of plantlets from somatic tissues of the date palm (*Phoenix dactylifera* L.). *Indian J. Exp. Biol.* 24:763–766.

Sharma, D.R., Kumari, R., and Chowdhury, J.B. 1980. *In vitro* culture of female date palm (*Phoenix dactylifera* L.) tissues. *Euphytica* 29:169–174.

Siljak-Yakovlev, S., Cerbah, M., Sarr, A., Benmalek, S., Bounaga, N., Coba de la Pena, T., and Brown, S. 1996. Chromosomal sex determination and heterochromatin structure in date palm. *Sex Plant Reprod.* 9:127–132.

Smith, R.J. and Aynsley, J.S. 1995. Field performance of tissue cultured date palm (*Phoenix dactylifera*) clonally produced by somatic embryogenesis. *Principes* 39:47–52.

Sudhersan, C., AboEl-Nil, M.M., and Al-Baiz, A. 1993. Direct somatic embryogenesis and plantlet formation from the leaf explants of *Phoenix dactylifera* L. cultivar Barhi. *J. Swamy Bot. Cl.* 10:37–43.

Tautz, D., 1989. Hypervariability of simple sequences as a general source of polymorphic DNA markers. *Nucleic Acids Res.* 17:6463–6471.

Tisserat, B. 1979. Propagation of date palm (*Phoenix dactylifera* L.) *in vitro*. *J. Exp. Bot.* 30:1275–1283.

Tisserat, B. 1982. Factors involved in production of plantlets from date palm callus cultures. *Euphytica* 31:201–214.

Tisserat, B. 1984. Propagation of date palm by shoot tip culture. *Hortic. Sci.* 19:230–231.

Tisserat, B. and Demason, D. 1980. A histological study of development of adventive embryos organ cultures of *Phoenix dactylifera* L. *Ann. Bot.* 46:465–472.

Trifi, M. 2001. Polymorphisme et Typage Moleculaire de Varietes Tunisiennes de Palmier Dattier (*Phoenix Dactylifera* L.): Relation Avec la Resistance au Bayoud. These Doctorat d'Etat, Universite Tunis-El Manar, Tunis, Tunisia, p. 141.

Trifi, M., Rhouma, A., and Marrakchi, M. 2000. Phylogenetic relationships in Tunisian date palm (*Phoenix dactylifera* L.) germplasm collection using DNA amplification fingerprinting. *Agronomie* 20:665–671.

Veramendi, J. and Navarro, L. 1997. Influence of explant sources of adult date palm (*Phoenix dactylifera* L.) on embryogenic callus formation. *J. Hortic. Sci.* 72:665–671.

Vieitez, A.M. 1995a. Somatic embryogenesis in Camellia spp. In: *Somatic Embryogenesis in Woody Plants*. Jain, S.M., Gupta, P.K. and Newton, R. (eds). Angiosperm Vol. 2. Kluwer Academic Publishers, The Netherlands, pp. 235–276.

Vieitez, A.M. 1995b. Somatic embryogenesis in Camellia spp. In: *Somatic Embryogenesis in Woody Plants*. Jain, S.M., Gupta, P.K. and Newton, R.J. (eds.). Angiosperms Vol. 2, Kluwer Academic Publishers, Dordrecht, pp. 235–276.

Zaid, A. 1984. *In vitro* browning of tissues and media with special emphasis to date palm cultures. *Date Palm J.* 3:269–275.

Zaid, A. and Tisserat, B.H. 1983. Morphogenetic responses obtained from a variety of somatic explant tissues. *Bot. Mag. Tokyo* 96:67–73.

Zietkiewicz, E., Rafalski, A., and Labuda, D. 1994. Genomic fingerprinting by simple sequence repeat (SSR)-anchored polymerase chain reaction amplification. *Genomics* 20:176–183.

Zouine, J., El Bellaj, M., Meddich, A., Verdeil, J.-L., and El Hadrami, I. 2005. Proliferation and germination of somatic embryos from embryogenic suspension cultures in *Phoenix dactylifera*. *Plant Cell Tissue Organ Cult.* 82:83–92.

Zouine, J. and El-Hadrami, I. 2007. Effect of 2,4-D, glutamine and BAP on embryogenic suspension culture of date palm (*Phoenix dactylifera* L.). *Sci. Hortic.* 112:221–226.

3 The Date Palm Genome Project in the Kingdom of Saudi Arabia

Xiaowei Zhang, Jun Tan, Meng Yang, Yuxin Yin, Ibrahim S. Al-Mssallem, and Jun Yu

CONTENTS

INTRODUCTION

Genomics studies the genome of an organism. A genome is the sum total of DNA molecules that harbor all genes of an organism. Thus, genomics studies all the genes of a given cell, tissue, and organism; it studies not only DNA (genome) but also RNA (transcriptome), and protein (proteome) in a context of regulatory network as well as across taxa (evolution). The field includes intensive efforts to determine the entire DNA sequence of various organisms and genetic map, using large-scale sequencing technology to generate massive, adequate, and high-quality data, by using bioinformatic tools for assembly, annotation, and in-depth analysis. Genomics was initiated in principle by Fred Sanger when he first sequenced the complete genomes of a virus and a mitochondrion. A major branch of genomics is still working on sequencing the genomes of various species, but the knowledge of full genomes has created the possibility for the field of transcriptomics, proteomics, bioinformatics, function genomics, metagenomics, and system biology. Following is a brief history of genomics.

1. In 1972, Walter Fiers and his team first determined the sequence of a coat protein gene of bacteriophage MS2. And then in 1976, they sequenced the complete genome of bacteriophage MS2-RNA.
2. In 1977, Frederick Sanger sequenced the first DNA-based genome phage Φ-X174.
3. In 1995, the first free-living organism, *Haemophilus influenzae*, was sequenced.

4. In 2001, the draft genome of human was completed and the fine map was finished in 2007.
5. In 2002, the draft map of rice genome was sequenced by Chinese scientists.
6. In 2005, the first genome and metagenomics were sequenced using the next-generation sequencer (Illumina GA, Roche 454 GS FLX).
7. Up to June 2010, there were 6414 complete genomic sequences, organelle genomes, plasmids as well as draft genomes, including 2385 viruses, 41 viroids, 95 archaea, 39 plasmids, 1482 bacteria, and 2372 eukaryota.

According to our knowledge, 10 plant genomes had been sequenced and published; most of them are <1 Gb in genome size (Table 3.1). With the development of genome sequencing technology, more and more plant genomes will be sequenced in the next few years. Table 3.2 lists the major plant genome sequencing projects, completed or in progress. There is no doubt that most important economic plant species will be sequenced in time and with high quality.

DNA Sequencing Technology

DNA sequencing refers to determining the order of the nucleotide (adenine, guanine, cytosine, and thymine) in a DNA molecule. The advancement of DNA-sequencing technology has significantly accelerated biological research and discovery. The first DNA sequencing was performed using manual experimental methods based on two-dimensional chromatography. Frederick Sanger used dideoxynucleotide triphosphates (ddNTPs) as DNA chain terminators. The dye-based sequencing method was automated and high throughput. Since then, DNA sequencing has become one of the essential laboratory methods for molecular biology, together with DNA cloning, DNA microarray, polymerase-chain-reaction (PCR), electrophoresis, chromatography, and centrifugation. It has become not only a fundamental method for determining the genome sequences, but also a basic tool for identifying the DNA methylation sites and sequence polymorphisms. In both scale and scope, DNA sequencing dwarfs all other molecular technologies in the acquisition of basic genomic information, paving a way for other genome-scale, technology-oriented disciplines or, in contemporary terminology, the numerous "-omics" (transcriptomics and proteomics being well-accepted

TABLE 3.1
List of Completed Plant Genome

Species	Chromosomes	Genome Size (Mb)	References
Arabidopsis thaliana (thale cress)	5	157	Mozo et al. (1999)
Cucumis sativus (cucumber)	7	350	Sanwen et al. (2009)
Brachypodium distachyon (purple false brome)	5	355	International Brachypodium Initiative (2010)
Carica papaya (papaya)	9	372	Ming et al. (2008)
Oryza sativa (rice)	12	450	Yu et al. (2005)
Lotus japonicus (lotus)	6	470	Sato et al. (2008)
Vitis vinifera (wine grape)	19	500	Zharkikh et al. (2008)
Populus trichocarpa (black cottonwood)	19	550	Tuskan et al. (2006)
Sorghum bicolor (sorghum)	10	760	Paterson et al. (2009)
Glycine max (soybean)	20	1115	Schmutz et al. (2010)

TABLE 3.2
List of In-Progress Plant Genome

Species	Chromosomes	Genome Size	Available Resources
Medicago truncatula (barrel medic)	8	500	Transcriptome, plastid genome, genetic maps
Manihot esculenta (cassava)	20	770	Transcriptome, genetic map
Phoenix dactylifera (date palm)	18	800	Transcriptome, plastid genome, sequence map
Solanum tuberosum (potato)	12	840	Transcriptome, plastid genome, genetic map
Solanum lycopersicum (tomato)	12	950	Plastid genome, genetic map
Zea mays (corn)	10	2400	Transcriptome, plastid genome, mitochondrial genome, genetic map

examples) leading biologists into the future decades, providing new data, ample information, fresh knowledge, and innovative insights.

In the mid-1980s, the first-generation sequencer was developed in Leroy Hood's laboratory at Cal Tech. Then ABI introduced its first semi-automated DNA sequencer ABI 370. Toward the end of the last century, the second version of the first-generation technology appeared (3730xl, MegaBACE3000, and so on), which can sequence up to 384 DNA samples in a single load and up to 24 runs a day. It carries out capillary electrophoresis for size separation, detection and recording of dye fluorescence, and data output as fluorescent peak trace chromatograms. Through decades of gradual improvement, the first-generation sequencer can be applied to achieve sequencing length up to 1000 bp, with raw accuracy higher than 99.999%, at a cost as little as $0.50/kilobase and throughput close to 1.2 mbp/day (Table 3.3). However impressive those numbers may seem, the first-generation technology has reached its pinnacle both in terms of speed and in terms of cost.

With the rapid development of genomics research, demands for low-cost and high-throughput DNA sequencing have driven the invention of new sequencing technologies. Second-generation

TABLE 3.3
Specification and Cost of First-, Second-, and Third-Generation Sequencers

Sequencer	3730Xl	GS FLX	SOLiD 4	Hiseq 2000	Hilicos	Complete Genomics	PacBio[a]
Company	AB	Roche	AB	Illumina	BioScience	Complete Genomics	Pacific Biotechnology
Method	Sanger	Pyro sequence	Sequence by ligation	Sequence by synthesis	tSMS technology	cPAL DNA sequencing	SMRT DNA sequencing
Output per day	1.3 Mb	1 Gb	7 Gb	25 Gb	4 Gb	6 Gb	96 Gb
Run time	1 h	8 h	14 days	8 days	8 days	/	15 min
Rd length (bp)	1,000	500	50	100	32	35	1,250
Cost per run ($)	50	7,000	6,000	10,000	12,000	10,000/200 Gbp	/
Human genome coverage per run	5× (300,000 run)	5× (30 run)	30× (1 run)	30× (0.5 run)	40× (4 run)	45×	/
Cost per genome ($)	15,000,000	200,000	6,000	5,000	48,000	1,726	1,000

[a] No paper published, so all the data are from the websites, news reports, or estimates based on experience.

sequencers, parallelizing the sequence process and producing millions or billions of reads at once, were produced by Roche, Illumina, and Applied Biosystem. We would like to briefly describe the technical parameters of the three major sequencers.

1. Genome Sequencer FLX (Roche 454), based on sequencing-by-synthesis, is the first second-generation sequencer. It offers more than 1 million reads per run and read lengths of 400 bp, and the system is ideally suitable for *de novo* sequencing of whole genomes and transcriptomes.
2. Hiseq 2000 (Illunina) relies on DNA fragment extended and bridge amplified on a slide to create an ultrahigh-density sequencing flow cell with hundreds of millions of clusters it can generate up to 200 Gb per run in a read-length of 2×100 bp, equivalent to up to 25 Gb per day.
3. SOLiD4 system is a product of Applied Biosystem (also known as Life Technologies as part of Invitrogen). It employs sequencing by ligation method and delivers 100 Gb of mappable reads per run, and the accuracy is about 99.94%.

The second-generation sequencing platforms based on sequencing by synthesis or ligation have dramatically increased the speed of data production and reduced sequencing cost per base over the first-generation machines. It is common for these platforms to generate Giga-base level of sequence data per day at a cost only $500 per Gbp (Table 3.3). However, their short sequence reads are a serious limitation, which is mainly attributable to the dephasing problem of optical signal in DNA sequencing cluster. One solution to remedy this would be to eliminate the ensemble effect all together—sequencing on single a molecule.

To overcome one of the major drawbacks of second-generation sequencing technology, namely the relatively short read, efforts have been made to develop single molecule sequencing platforms—which is called third-generation sequencing technology. Single molecule also helps increase the number of DNA fragments that can be independently analyzed in a given surface area and, therefore, achieves a much higher level of throughput.

Pacific Bioscience Inc. is developing a third-generation sequencer based on a new technology: the SMRT (Single Molecule Real Time) technology. This DNA sequencer uses single DNA molecules as templates. The main characteristic of this sequencer is real-time monitoring of nucleotide incorporation with DNA polymerase. Unlike the second-generation sequencer, the sequencer can obtain reads of more than 1 kb from a single template, and generate about 1 mbp sequence data per second. Other than PacBio, Oxford Nanopore, VisiGen Biotechnologies and Mobious Biosystems will present their third-generation sequencers separately. Till date, there are already two types of third-generation sequencing platforms available in the market from Complete Genomics and Helicos. The third-generation sequencers promise researchers to finish the human genome within a day and at less than one thousand dollars. The specification and cost of selected sequencers from the first- to the third generations are summarized in Table 3.3.

The Date Palm Genome Project

Date palm cultivars are giant woody plants within the genus *Phoenix*. There are over 340 domesticated *Phoenix* cultivars and their genetic diversity is relatively high, indicating multiple origins from different wild hybrids between two principal ancestral species. The total number of date palm trees in the world is over 100 million with a production of 15 million metric tons. However, there are major challenges encountered in date palm production and marketing from virulent diseases, dangerous pests, abiotic stresses, and new demands for sustainability, quality, transport, and yield.

Date palm trees are mostly diploid ($2n = 36$) and propagated sexually by seeds and asexually by offshoots. The genome size is estimated to be 1.2 billion base pairs, based on flow cytometry. Although it is an economically important plant species, very few DNA sequences can be found in GenBank databases. With the development of genome sequencing technology, date palm genome sequencing became a target of two research teams.

A team from Weil Cornell Medical College in Qatar tried to sequence the whole genome of date palm using Solexa (illumine) sequencer based on a shot-gun method. They announced the completed draft map in 2009 and released the sequence data subsequently (http://qatar-weill.cornell.edu/research/datepalmGenome/index.html). According to their analyses, the genome assembly has a predicted genome size of ~550 Mbp. The following are genome parameters of their draft sequence assembly:

1. 45,000 scaffolds greater than 2 kb
2. Scaffold N50 is 4250 bp
3. 850,000 novel high-quality SNPs between parental alleles
4. GC content of the nuclear genome is 37%
5. 302 Mb of assembled sequence with 18.5 Mb of ordered gaps
6. Unique sequence is 292 Mb at the 24-mer level

As a collaborative effort between KACST and BIG (CAS), the Date Palm Genome Project (DPGP) was initiated. The Project goals are manifold and comprise several components at genomics, bioinformatics, genetics, and biochemistry levels. The first phase of the Project focuses on genomics and bioinformatics that pave the way for genetic and biochemical studies.

The specific aims of the Project are

1. Working draft with sequence coverage: 10× from 454 and 50× from SOLiD
2. Complete map will be built with end-sequences from BACs and Fosmids
3. Genome diversity map built with shotgun sequencing of 30 varieties, each having 30× of SOLiD reads
4. The Date Palm Transcriptomes
   a. Full-length cDNA: over 30,000 unigenes
   b. Expression profiles for leaves, roots, and flowers (~50 tissue samples)

In the following sections, some of the preliminary data in genome sequencing and assembly, chloroplast genome sequencing, and transcriptomics are presented.

DATE PALM GENOME ASSEMBLY

Assembly Software Package

Both Roche 454 and ABI SOLiD are used for generating sequencing data. 454 usually produces sequences in a length of 400 bp; sometimes, paired-end information for a better assembly is generated. SOLiD produces short reads <50 bp in length and paired-end information in a range of 3–8 kb.

De nova assembly is carried out by using a hybrid strategy where contigs with unique reads are built and paired end information is added to the contigs to obtain supercontigs. To deal with short reads in tens of millions, three approaches are used De Bruijn graph, traditional overlap-layout-consensus, and Prefix tree-based (Trie) by using several assembly tools. These tools are readily available and have been used for similar data types from different or similar sequencing strategies.

TABLE 3.4
Examples of Sequence Assembly Software Packages for Data Produced by Next-Generation Sequencers

Algorithm	Software Package	Author's Name	Institute
De Bruijn graph	ABySS	Jared T. Simpson et al.	British Columbia Cancer Agency
	ALLPATHS	Jonathan Butler et al.	Broad Institute of MIT and Harvard
	Contrail	Michael Schatz et al.	University of Maryland
	Euler-SR	Mark J. Chaisson et al.	University of California San Diego
	Velvet	Daniel R. Zerbino, Ewan Birney	Wellcome Trust Genome Campus
	Celera/WGS	Gennady Denisov et al.	J. Craig Venter Institute
Traditional overlap-layout-consensus	Edena	David Hernandez et al.	Geneva University Hospitals and the University of Geneva
	Newbler		F. Hoffmann-La Roche Ltd.
	SeqCon	Tobias Rausch et al.	International Max Planck Research School for Computational Biology and Scientific Computing
	SHORTY	Mohammad Sajjad Hossain et al.	Stony Brook University
	QSRA	Douglas W. Bryant Jr et al.	Oregon State University
Prefix tree-based (Trie) approach	SHARCGS	Juliane C. Dohm et al.	Max-Planck-Institute for Molecular Genetics
	SSAKE	René L. Warren et al.	Genome Sciences Centre
	VCAKE	William R. Jeck et al.	University of Carolina—Chapel Hill
Smith-Waterman	MOSAIK	Michael Strömberg	Boston College—Higgins Hall
Simple Greedy	MIRA	Bastien Chevreux et al.	German Cancer Research Centre Heidelberg
Burrow Wheeler Transformation	SOAP	Ruiqiang Li et al.	Beijing Genomics Institute at Shenzhen

Scaffolds are built and the sequence assembly will be completed by using various tools for large plant genome assemblies, such as those used for the date palm and tomato genomes. Optimistically new algorithms and tools will be developed by research groups around the globe to deal better with large quantity of data and repeat-rich genomes, such as large plant genomes. In scaffold building, paired-end information from large-insert clones is also used such as BACs (120 kb in size) and fosmids (30 kb in size). Some of the major software packages for next-generation sequence assembly are listed in Table 3.4.

Date Palm Genome Assembly

Date palm genome consists of 18 chromosome pairs and the estimated size is 250 M (Abdelali et al. 1999), 550–680 Mb. At the same time, both oil palm and coconut have 16 chromosome pairs and around 2 G genome sizes. Based on 30 M 454 reads (average length is 383 bp, near 15× coverage) produced by KACST, 20-mer filtering was done to pick out unique reads, which resulted in about half of them being qualified to be assembled by newbler runAssembly tool. A quantity of 226,501 contigs were obtained, with a total length of 416,498,895 bps. That is to say, the estimated genome size of date palm is more than 800 M considering 454 sequencing missed regions.

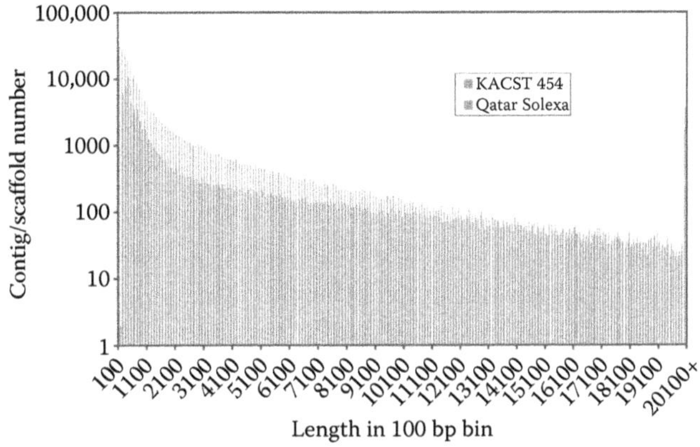

FIGURE 3.1 Comparison of contig/scaffold length distribution of KACST 454 and Qatar Solexa assemblies. A cutoff is shown at 500 bp in Qatar assembly, whereas more than 100,000 contigs of KACST 454 are <500 bps. So, the numbers and total lengths are similar for contigs/scaffolds of both researches. But, there are more than 30 M "N/Ns" in Qatar scaffold sequences.

Comparison with the Qatar Draft Map

Since the majority of the data have been acquired for the Project, the data were compared to the Qatar scaffolding result acquired by using Solexa alone. The Qatar assembly has 18,024 scaffolds with a length of more than 4000 bp (the estimated average gene length of date palm is about 4000 bp) (Figure 3.1); it sums up to ~252 Mb in total length. This assembly is similar to the assembled length of 258 Mb in 27,389 contigs (Table 3.5). The next step is to add more than 50× SOLiD data in mate-pairs to obtain supercontigs as well as to improve the sequence quality.

Date Palm Chloroplast Genome

Chloroplasts are one of many different types of organelles in the cell. In general, they are considered to have originated from cyanobacteria through endosymbiosis. They are found in plant cells and other eukaryotic organisms that conduct photosynthesis. All chloroplasts are thought to derive directly or indirectly from a single endosymbiotic event. The chloroplast genome has long been a focus of research in plant molecular evolution and systematics because of its small size, high copy number, conservation, and extensive characterization at the molecular level. In angiosperm, most cp genomes are circular DNAs with the length ranging from 120 to 160 kb and have a quadripartite

TABLE 3.5
Comparison of Assembly Results of KACST 454 Contigs and Qatar Solexa Scaffolds

	KACST 454	Qatar Solexa
Total length (bps)	416,498,895	302,873,018
Contig/scaffold number	226,501	73,747
Average length (bps)	1,839	4,107
Max length (bps)	89,547	169,087
Min length (bps)	100	222
GC content (%)	39	35
N50 length (bps)	6,075	14,840

organization consisting of two copies of inverted repeats (IRs) of ~20–28 kb in size, dividing the rest into an 80–90 kb large-single-copy region (LSC) and a 16–27 kb small-single-copy (SSC) region. Usually, the gene content of angiosperm cp genome is rather conserved, encoding 4 rRNAs, ~30 tRNAs, and ~80 unique proteins.

Since the publication of the first angiosperm chloroplast genome sequence of *Nicotiana* in 1986, researchers have been keen to use complete chloroplast genome sequences to infer a wide range of information about plants, in which the phylogenetic relationships are most concerned in three forms: restriction fragment/site comparisons, structural rearrangements, and DNA sequencing. In the last 10 years, with the widespread application of new DNA-sequencing technology, the complete chloroplast genome available in public database has rapidly increased. Until now, there are 132 complete chloroplast genomes of land plants deposited GenBank Organelle Genome Resources and most of them are angiosperms (Table 3.6).

TABLE 3.6
Completely Sequenced Land Plant Chloroplast Genomes (as of July 1, 2010)

Higher Classification	Family	Species	Accession Number	Length
Mosses	Pottiaceae	*Syntrichia ruralis*	NC_012052	122,630
	Funariaceae	*Physcomitrella patens*	NC_005087	122,890
Liverworts	Marchantiaceae	*Marchantia polymorpha*	NC_001319	121,024
	Aneuraceae	*Aneura mirabilis*	NC_010359	108,007
Ferns	Selaginellaceae	*Selaginella uncinata*	AB197035	144,170
	Selaginellaceae	*Selaginella moellendorffii*	NC_013086	143,780
	Lycopodiaceae	*Huperzia lucidula*	NC_006861	154,373
	Pteridaceae	*Adiantum capillus-veneris*	NC_004766	150,568
	Psilotaceae	*Psilotum nudum*	NC_003386	138,829
	Marattiaceae	*Angiopteris evecta*	NC_008829	153,901
	Cyatheaceae	*Alsophila spinulosa*	NC_012818	156,661
Gymnosperms	Cupressaceae	*Cryptomeria japonica*	NC_010548	131,810
	Cycadaceae	*Cycas taitungensis*	NC_009618	163,403
	Ephedraceae	*Ephedra equisetina*	NC_011954	109,518
	Gnetaceae	*Gnetum parvifolium*	NC_011942	114,914
	Pinaceae	*Keteleeria davidiana*	NC_011930	117,720
	Pinaceae	*Picea sitchensis*	NC_011152	109,798
	Pinaceae	*Pinus contorta*	NC_011153	115,615
	Pinaceae	*Pinus gerardiana*	NC_011154	116,997
	Pinaceae	*Pinus koraiensis*	NC_004677	117,190
	Pinaceae	*Pinus krempfii*	NC_011155	115,555
	Pinaceae	*Pinus lambertiana*	NC_011156	112,521
	Pinaceae	*Pinus longaeva*	NC_011157	115,896
	Pinaceae	*Pinus monophylla*	NC_011158	114,607
	Pinaceae	*Pinus nelsonii*	NC_011159	110,575
	Pinaceae	*Pinus thunbergii*	NC_001631	119,707
	Welwitschiaceae	*Welwitschia mirabilis*	NC_010654	119,726
Angiosperms				
Monocots	Acoraceae	*Acorus americanus*	NC_010093	153,819
	Acoraceae	*Acorus calamus*	NC_007407	153,821
	Araceae	*Lemna minor*	NC_010109	165,955
	Arecaceae	*Phoenix dactylifera* L.	GU811709	158,462
	Dioscoreaceae	*Dioscorea elephantipes*	NC_009601	152,609
	Orchidaceae	*Phalaenopsis aphrodite*	NC_007499	148,964
	Orchidaceae	*Oncidium Gower Ramsey*	NC_014056	146,484

TABLE 3.6 (continued)
Completely Sequenced Land Plant Chloroplast Genomes (as of July 1, 2010)

Higher Classification	Family	Species	Accession Number	Length
	Poaceae	*Agrostis stolonifera*	NC_008591	136,584
	Poaceae	*Anomochloa marantoidea*	NC_014062	138,412
	Poaceae	*Bambusa oldhamii*	NC_012927	139,350
	Poaceae	*Brachypodium distachyon*	NC_011032	135,199
	Poaceae	*Dendrocalamus latiflorus*	NC_013088	139,394
	Poaceae	*Festuca arundinacea*	NC_011713	136,040
	Poaceae	*Hordeum vulgare* subsp. *vulgare*	NC_008590	136,462
	Poaceae	*Lolium perenne*	NC_009950	135,282
	Poaceae	*Oryza nivara*	NC_005973	134,494
	Poaceae	*Oryza sativa* Indica Group	NC_008155	134,496
	Poaceae	*Oryza sativa* Japonica Group	NC_001320	134,525
	Poaceae	*Saccharum* hybrid cultivar SP-80-3280	NC_005878	141,182
	Poaceae	*Saccharum officinarum*	NC_006084	141,182
	Poaceae	*Sorghum bicolor*	NC_008602	140,754
	Poaceae	*Triticum aestivum*	NC_002762	134,545
	Poaceae	*Coix lacryma-jobi*	NC_013273	140,745
	Poaceae	*Zea mays*	NC_001666	140,384
	Typhaceae	*Typha latifolia*	NC_013823	161,572
Dicots	Amaranthaceae	*Spinacia oleracea*	NC_002202	150,725
	Amborellaceae	*Amborella trichopoda*	NC_005086	162,686
	Apiaceae	*Daucus carota*	NC_008325	155,911
	Araliaceae	*Panax ginseng*	NC_006290	156,318
	Asteraceae	*Helianthus annuus*	NC_007977	151,104
	Asteraceae	*Guizotia abyssinica*	NC_010601	151,762
	Asteraceae	*Lactuca sativa*	NC_007578	152,765
	Asteraceae	*Parthenium argentatum*	NC_013553	152,803
	Berberidaceae	*Nandina domestica*	NC_008336	156,599
	Brassicaceae	*Lobularia maritima*	NC_009274	152,659
	Brassicaceae	*Draba nemorosa*	NC_009272	153,289
	Brassicaceae	*Arabis hirsuta*	NC_009268	153,689
	Brassicaceae	*Aethionema cordifolium*	NC_009265	154,168
	Brassicaceae	*Aethionema grandiflorum*	NC_009266	154,243
	Brassicaceae	*Arabidopsis thaliana*	NC_000932	154,478
	Brassicaceae	*Capsella bursa-pastoris*	NC_009270	154,490
	Brassicaceae	*Barbarea verna*	NC_009269	154,532
	Brassicaceae	*Olimarabidopsis pumila*	NC_009267	154,737
	Brassicaceae	*Lepidium virginicum*	NC_009273	154,743
	Brassicaceae	*Nasturtium officinale*	NC_009275	155,105
	Brassicaceae	*Crucihimalaya wallichii*	NC_009271	155,199
	Buxaceae	*Buxus microphylla*	NC_009599	159,010
	Calycanthaceae	*Calycanthus floridus*	NC_004993	153,337
	Campanulaceae	*Trachelium caeruleum*	NC_010442	162,321
	Caricaceae	*Carica papaya*	NC_010323	160,100
	Ceratophyllaceae	*Ceratophyllum demersum*	NC_009962	156,252
	Chaetosphaeridiaceae	*Chaetosphaeridium globosum*	NC_004115	131,183
	Chloranthaceae	*Chloranthus spicatus*	NC_009598	157,772
	Convolvulaceae	*Cuscuta obtusiflora*	NC_009949	85,286
	Convolvulaceae	*Cuscuta gronovii*	NC_009765	86,744
	Convolvulaceae	*Cuscuta reflexa*	NC_009766	121,521

continued

TABLE 3.6 (continued)
Completely Sequenced Land Plant Chloroplast Genomes (as of July 1, 2010)

Higher Classification	Family	Species	Accession Number	Length
	Convolvulaceae	*Cuscuta exaltata*	NC_009963	125,373
	Convolvulaceae	*Ipomoea purpurea*	NC_009808	162,046
	Cucurbitaceae	*Cucumis sativus*	NC_007144	155,293
	Euphorbiaceae	*Manihot esculenta*	NC_010433	161,453
	Euphorbiaceae	*Jatropha curcas*	NC_012224	163,856
	Fabaceae	*Lathyrus sativus*	NC_014063	121,020
	Fabaceae	*Pisum sativum*	NC_014057	122,169
	Fabaceae	*Medicago truncatula*	NC_003119	124,033
	Fabaceae	*Cicer arietinum*	NC_011163	125,319
	Fabaceae	*Trifolium subterraneum*	NC_011828	144,763
	Fabaceae	*Phaseolus vulgaris*	NC_009259	150,285
	Fabaceae	*Lotus japonicus*	NC_002694	150,519
	Fabaceae	*Vigna radiata*	NC_013843	151,271
	Fabaceae	*Glycine max*	NC_007942	152,218
	Geraniaceae	*Pelargonium* × *hortorum*	NC_008454	217,942
	Magnoliaceae	*Liriodendron tulipifera*	NC_008326	159,886
	Malvaceae	*Gossypium hirsutum*	NC_007944	160,301
	Malvaceae	*Gossypium barbadense*	NC_008641	160,317
	Moraceae	*Morus indica*	NC_008359	158,484
	Myrtaceae	*Eucalyptus globulus*	NC_008115	160,286
	Nymphaeaceae	*Nymphaea alba*	NC_006050	159,930
	Nymphaeaceae	*Nuphar advena*	NC_008788	160,866
	Oleaceae	*Olea europaea*	NC_013707	155,872
	Oleaceae	*Jasminum nudiflorum*	NC_008407	165,121
	Onagraceae	*Oenothera parviflora*	NC_010362	163,365
	Onagraceae	*Oenothera biennis*	NC_010361	164,807
	Onagraceae	*Oenothera argillicola*	NC_010358	165,055
	Onagraceae	*Oenothera glazioviana*	NC_010360	165,225
	Onagraceae	*Oenothera elata*	NC_002693	165,728
	Orobanchaceae	*Epifagus virginiana*	NC_001568	70,028
	Piperaceae	*Piper cenocladum*	NC_008457	160624
	Platanaceae	*Platanus occidentalis*	NC_008335	161,791
	Polygonaceae	*Fagopyrum esculentum*	NC_010776	159,599
	Ranunculaceae	*Ranunculus macranthus*	NC_008796	155,129
	Ranunculaceae	*Megaleranthis saniculifolia*	NC_012615	159,924
	Rubiaceae	*Coffea arabica*	NC_008535	155,189
	Rutaceae	*Citrus sinensis*	NC_008334	160,129
	Salicaceae	*Populus alba*	NC_008235	156,505
	Salicaceae	*Populus trichocarpa*	NC_009143	157,033
	Schisandraceae	*Illicium oligandrum*	NC_009600	148,553
	Solanaceae	*Solanum tuberosum*	NC_008096	155,298
	Solanaceae	*Solanum bulbocastanum*	NC_007943	155,371
	Solanaceae	*Solanum lycopersicum*	NC_007898	155,461
	Solanaceae	*Nicotiana tomentosiformis*	NC_007602	155,745
	Solanaceae	*Nicotiana sylvestris*	NC_007500	155,941
	Solanaceae	*Nicotiana tabacum*	NC_001879	155,943
	Solanaceae	*Atropa belladonna*	NC_004561	156,687
	Vitaceae	*Vitis vinifera*	NC_007957	160,928
	Winteraceae	*Drimys granadensis*	NC_008456	160,604

The variations of land plants brought challenges to biologists interested in the origin and evolution of these species, and resolving these issues critically depends on having a well-resolved and strongly supported phylogenetic framework. Over the past 20 years, numerous phylogenetic studies have used both morphological and molecular data to assess the relationships among the major clades. For nearly two decades, most phylogenetic analyses of land plants have relied on DNA sequences of one to several genes from the plastid, mitochondrial, and nuclear genomes; however, the support is often low due to insufficient data. Coding sequences from the completely sequenced genomes provide more facilities to classify the relationship of land plants and find the earliest lineage of angiosperm. Recent phylogenetic study has used 61 protein-coding genes from completely sequenced plastid genomes to estimate the relationships among the major clades of angiosperms. Another famous study used 81 genes from 64 plastid genome to identify the genome-scale evolutionary patterns to provide strong support for the position of *Amborella* as the earliest diverging lineage of flowering plants.

As a subject of Date palm genome project, studying its cp genome will provide great facility to interpret the evolutionary changes among angiosperms. The sequencing and assembling of date palm chloroplast genome has been completed using the pyrosequencing technology. The genome is a 158,462 bp double-stranded circular DNA molecule, with a typical quadripartite structure of the large (LSC, 86,198 bp) and small single copy (SSC, 17,712 bp) regions separated by a pair of inverted repeats (IRs, 27,276 bp). Similar to other cp genomes, the date palm cp genome is also rich in *A* and *T* (62.77%). The value is decreasing in regions of noncoding, protein-coding, tRNA, and rRNA, which is 66.60%, 61.03%, 57.94%, and 52.19%, respectively. As a regular cp genome, the gene content and structure are more like those in tobacco without rearrangements and gene gain or loss (Figure 3.2). Inverted repeats have expanded into LSC, but not the SSC region. The number of repeat sequence (>= 30 bp with 90% identity but not including the IRs) is lower than that in Poaceae family and some dicots. Intravarietal SNP also exists not only in intergenic region, but also in coding region of genes with essential function, and some SNPs can also be detected in oil palm. Transcriptome analyses showed that three genes (*atpF*, *trnA-UGC*, and *rrn23*) have a dominant number of transcripts and most protein-coding genes belong to polycistronic transcripts. A rectangular phylogenetic tree was built to estimate the phylogenetic position using 76

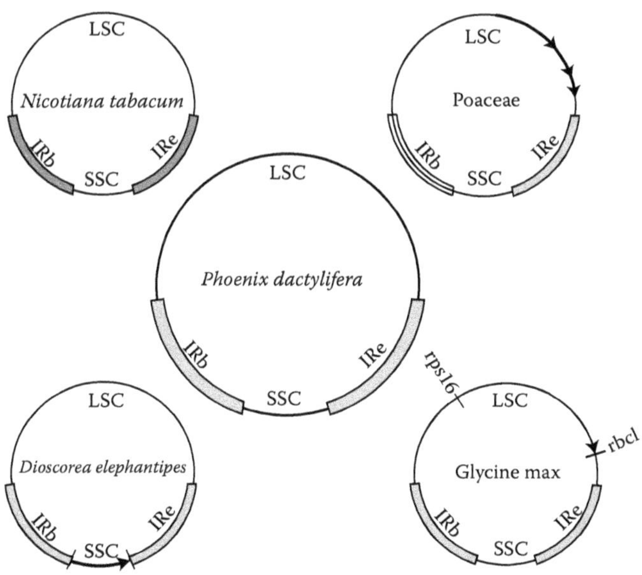

FIGURE 3.2 General description of chloroplast genome organization in representative angiosperms. The arcs with arrows indicate the inversion fragments.

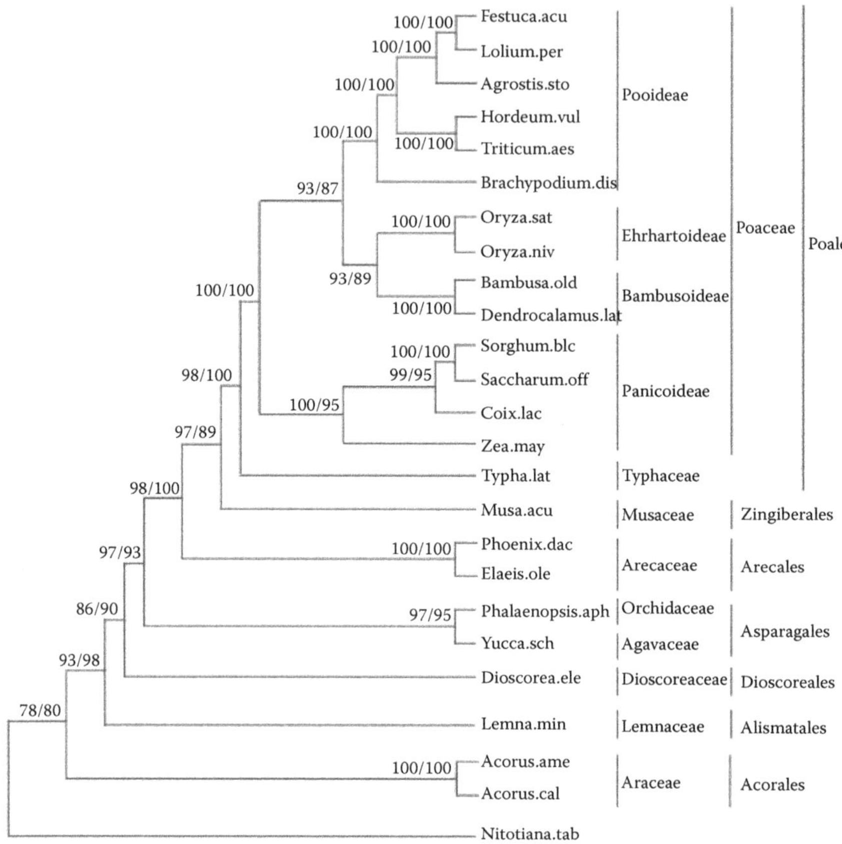

FIGURE 3.3 A phylogenetic tree of 76 concatenated cp genes from 25 taxa, derived from both MP and ML methods. Branch-swapping algorithm of both methods is based on TBR. Numbers at nodes indicate the bootstrap support.

concatenated genes from 26 taxa with *Nicotiana* as outgroup (Figure 3.3). Date palm shares extremely high homology with oil palm according to the bootstrap sampling test. It is also closer to Poales than other monocot families.

In addition to chloroplast genome, work is in progress to assemble the date palm mitochondrial genome. Up to now, 25 mitochondrial genomes in land plants (different species or cultivars) have been completely sequenced and the genome sizes range from 200 to 1000 kb (Table 3.7). The major contigs of date palm mitochondrial genome have been assembled and the genome size is estimated to be 600 kb. The essential genes accepted in the mitochondrial genome of other land plants have been predicted in date palm. The assembling process is also based on pyrosequencing technology, and mitochondrion-related reads were first extracted from the whole genome sequencing data to assemble the preliminary contigs. Subsequent wok aims to extend these contigs, filter the contamination from nuclear and chloroplast reads, and organize the order of all contigs until the final genome sequences are obtained.

Date Palm Transcriptome Analysis

Recently, next-generation sequencer developed a scalable, highly parallel DNA-sequencing system that is faster than standard-sequencing methods. This increase in throughput comes at the expense of read length. In contrast, the read-length limitation associated with next-generation sequencers is

TABLE 3.7
Completely Sequenced Land Plant Mitochondrial Genomes (as of July 1, 2010)

Higher Classification	Family	Species	Accession Number	Length
Angiosperms				
Monocots	Poaceae	*Bambusa oldhamii*	EU365401	509,941
	Poaceae	*Oryza rufipogon*	NC_013816	559,045
	Poaceae	*Tripsacum dactyloides*	NC_008362	704,100
	Poaceae	*Oryza sativa* Indica Group	NC_007886	491,515
	Poaceae	*Oryza sativa* Japonica Group	NC_011033	490,520
	Poaceae	*Sorghum bicolor*	NC_008360	468,628
	Poaceae	*Triticum aestivum*	NC_007579	452,528
	Poaceae	*Zea luxurians*	NC_008333	539,368
	Poaceae	*Zea mays* subsp. *mays*	NC_007982	569,630
	Poaceae	*Zea mays* subsp. *parviglumis*	NC_008332	680,603
	Poaceae	*Zea perennis*	NC_008331	570,354
Dicots	Solanaceae	*Nicotiana tabacum*	NC_006581	430,597
	Brassicaceae	*Arabidopsis thaliana*	NC_001284	366,924
	Vitaceae	*Vitis vinifera*	NC_012119	773,279
	Cucurbitaceae	*Citrullus lanatus*	NC_014043	379,236
	Cucurbitaceae	*Cucurbita pepo*	NC_014050	982,833
	Caricaceae	*Carica papaya*	NC_012116	476,890
	Brassicaceae	*Brassica napus*	NC_008285	221,853
	Amaranthaceae	*Beta vulgaris* subsp. *vulgaris*	NC_002511	368,801
Gymnosperms	Cycadaceae	*Cycas taitungensis*	NC_010303	414,903
Liverworts	Marchantiaceae	*Marchantia polymorpha*	NC_001660	186,609
	Pleuroziaceae	*Pleurozia purpurea*	NC_013444	168,526
Hornworts	Dendrocerotaceae	*Megaceros aenigmaticus*	NC_012651	184,908
	Anthocerotaceae	*Phaeoceros laevis*	NC_013765	209,482
Mosses	Funariaceae	*Physcomitrella patens*	NC_007945	105,340

less of a concern for transcriptome sequencing and analysis. They also avoid expensive cloning-based library construction, and it is feasible to sequence a wide variety of cDNA samples, thereby increasing the recovery of highly specialized transcripts (Jarvie and Harkins 2008). Moreover, the new technologies are particularly well suited for EST-based gene discovery, because it generates hundreds of thousands of tags per run, greatly increasing the chances of capturing rare transcripts. Table 3.8 summarizes the top 20 plant Unigene entries' collections from public databases.

Full-length cDNAs are necessary to identify exon–intron boundaries and gene-coding regions within genomic sequences and for comprehensive gene function analysis at the transcriptional and translational levels. More than 5,000,000 data palm cDNAs have been sequenced, and these sequences have not only advanced the functional analysis of the date palm genes, but have also played an important role in date palm genome annotation (Figure 3.4). To identify and characterize all the transcribed regions in a genome is one of the greatest challenges. In order to provide genomic

TABLE 3.8
List of Top 20 Plant Unigene Entries' Collections from Selected Species[a]

Species	# of UniGenes Entries	# of ESTs
Zea mays	97,486	2,021,364
Artemisia annua	88,127	94,734
Oryza sativa	40,978	1,272,399
Triticum aestivum	40,870	1,099,713
Glycine max	33,001	1,465,708
Arabidopsis thaliana	30,579	1,788,991
Brassica napus	27,139	656,152
Theobroma cacao	24,958	164,435
Nicotiana tabacum	24,069	326,135
Malus × domestica	23,731	329,920
Hordeum vulgare	23,595	533,726
Picea glauca	22,472	341,325
Vitis vinifera	22,083	364,469
Panicum virgatum	20,973	442,269
Gossypium hirsutum	20,671	270,813
Picea sitchensis	19,828	290,364
Solanum tuberosum	18,825	245,308
Raphanus raphanistrum	18,794	164,120
Medicago truncatula	18,785	285,262
Solanum lycopersicum	18,346	300,688

[a] This list is ordered by the total number of Unigene entries at http://www.ncbi.nlm.nih.gov/, accessed on June 10, 2010.

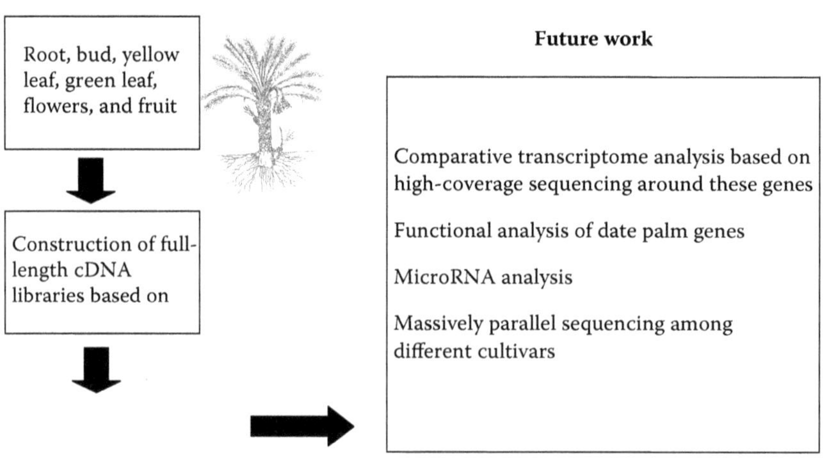

FIGURE 3.4 Schematic diagram of large-scale cDNA collection, sequencing, and future works.

resources for the analysis of date palm, and for the analysis of genes differentially expressed among different tissues, the objectives of this study were to create cDNA libraries (Beldade et al. 2006) suitable for the analysis of expressed sequence tags (ESTs) and to generate an EST resource for date palm to annotate the whole genome. This technique of sequencing also costs just a fraction of the cost of the whole genome.

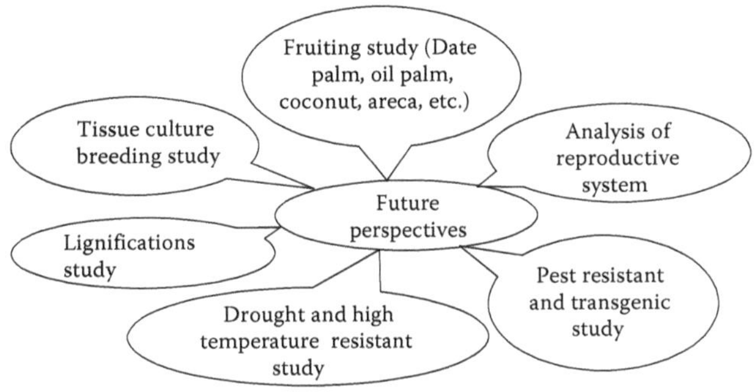

FIGURE 3.5 Schematic diagram of future perspectives.

Transcriptome sequencing (mRNA transcript-expression analysis, novel gene discovery, gene space identification in novel genomes, assembly of full-length genes, etc.) becomes more encompassing, as it is now possible to detect and identify nearly every class of molecules. Naturally, it is one of the most efficient ways for the identification of differentially expressed genes. Furthermore, a certain number of scientific propositions are being solved and our future perspectives are illustrated in Figure 3.5. Transcriptomes are very important for both understanding the physiology of date palm and the annotation of genome. Till now, significant progress has been made in transcriptome analysis of the genomes based on new sequencers. Recent advances in new methods (SOLiD total RNA-Seq; cDNA rapid library preparation method—Roche 454) for a deep and comprehensive transcriptome analysis of the date palm genome have been reviewed.

The alignment of full-length cDNA sequences to the genome sequence has provided direct experimental evidence for many of the gene models, which were predicted by computer programs. But a large number of the gene models should be confirmed by even more experimental data (Kikuchi et al. 2007). Current computer programs are still not enough for gene structural annotation, and the problem of exon–intron structure is still a major challenge for date palm genome biologists. In addition, further collection and complete sequencing of cDNA are needed, and these data will improve current date palm genome annotation.

ACKNOWLEDGMENT

The Date Palm Genome Project is supported by a research grant from KACST and institutional support from BIG and CAS. It also received support from the Chinese Ministry of Sciences and Technology for administration and traveling.

FURTHER READING

Abdelali B, Tran Han D, Benslimane A-A, Rode A, and Bernardi G. 1999. The gene distribution in the genomes of pea, tomato and date palm. *FEBS Letters* 463(1):139–142.

Al-Bakr A. 1972. *The Date Palm: Past, Present and Future*, 2nd ed. Al-Ani, Baghdad, Iraq.

Al-Salih AA, Hussain N, and Al-Jarrah A. 1987. Chromosome number of a date palm male Cultivar's Ghanname Akhdar. *Date Palm J* 5:128–133.

Beldade P, Rudd S, Gruber JD, and Long AD. 2006. A wing expressed sequence tag resource for *Bicyclus anynana* butterflies, an evo–devo model. *BMC Genomics* 7:130.

Brandham PE and Dorothy M-J. 1998. Genome size in the Aloaceae an angiosperm family displaying karyotypic orthoselection. *Ann Bot* (Suppl) 82(6):67–73 (7).

Chumley TW, Palmer JD, Mower JP, Fourcade HM, Calie PJ et al. 2006. The complete chloroplast genome sequence of Pelargonium × hortorum: Organization and evolution of the largest and most highly rearranged chloroplast genome of land plants. *Mol Biol Evol* 23:2175–2190.

Fiers W, Contreras R, Duerinck F, Haegeman G, Iserentant D, Merregaert J, Min Jou W et al. 1976. Complete nucleotide sequence of bacteriophage MS2 RNA: Primary and secondary structure of the replicase gene. *Nature* 260(5551):500–507.

Gao L, Yi X, Yang YX, Su YJ, and Wang T. 2009. Complete chloroplast genome sequence of a tree fern *Alsophila spinulosa*: Insights into evolutionary changes in fern chloroplast genomes. *BMC Evol Biol* 9:130.

International Brachypodium Initiative. 2010. Genome sequencing and analysis of the model grass *Brachypodium distachyon*. *Nature* 11(463):763–768.

Jansen RK, Cai Z, Raubeson LA, Daniell H, Depamphilis CW et al. 2007. Analysis of 81 genes from 64 plastid genomes resolves relationships in angiosperms and identifies genome-scale evolutionary patterns. *Proc Natl Acad Sci USA* 104:19369–19374.

Jarvie T and Harkins T. 2008. Transcriptome sequencing with the genome sequencer FLX system. *Nature Methods* 5:vi–viii.

Jim Leebens-Mack LAR, Cui L, Kuehl JV, Fourcade MH, Chumley TW, Boore JL, Jansen RK, and dePamphilis CW. 2005. Identifying the basal angiosperm node in chloroplast genome phylogenies: Sampling one's way out of the Felsenstein Zone. *Mol Biol Evol* 22:1948–1963.

Kikuchi S, Wang GL, and Li L. 2007. Genome-wide RNA expression profiling in rice. In: *Rice Functional Genomics—Challenges, Progress and Prospects*, Upadhyaya NM (ed). Springer, New York, NY, pp. 31–59.

Kikuya K. 2009. Impact of the next generation DNA sequencers. *Int J Clin Exp Med* 2:193–202.

Margulies M, Egholm M, Altman WE, Attiya S, Bader JS, Bemben LA, Berka J, Braverman MS et al. 2005. Genome sequencing in microfabricated high-density picolitre reactors. *Nature* 437:326–327.

Min Jou W, Haegeman G, Ysebaert M, and Fiers W. 1972. Nucleotide sequence of the gene coding for the bacteriophage MS2 coat protein. *Nature* 237(5350):82–88.

Ming R et al. 2008. The draft genome of the transgenic tropical fruit tree papaya (*Carica papaya* Linnaeus). *Nature* 24;452(7190):991–996.

Mozo T et al. 1999. A complete BAC-based physical map of the *Arabidopsis thaliana* genome. *Nat Genet* 22(3):271–275.

Poinar HN, Schwarz C, Qi J, Shapiro B, Macphee RD, Buigues B, Tikhonov A, Huson DH, Tomsho LP, Auch A, Rampp M, Miller W, and Schuster SC. 2005. Metagenomics to paleogenomics: Large-scale sequencing of mammoth DNA. *Science* 311:392–394.

Radoje D, Sparks AB et al. 2009. Human genome sequencing using unchained base reads on self-assembling DNA nanoarrays. *Science* 327:78–81.

Raubeson LA JR. 2005. Chloroplast genomes of plants. In *Plant Diversity and Evolution: Genotypic and Phenotypic Variation in Higher Plants*. CABI Publishing, London, UK, pp. 45–68.

Raubeson LA, Peery R, Chumley TW, Dziubek C, Fourcade HM et al. 2007. Comparative chloroplast genomics: Analyses including new sequences from the angiosperms *Nuphar advena* and *Ranunculus macranthus*. *BMC Genomics* 8:174.

Sanger F, Air GM, Barrell BG, Brown NL, Coulson AR, Fiddes CA, Hutchison CA, Slocombe PM, and Smith M. 1977. Nucleotide sequence of bacteriophage phi X174 DNA. *Nature* 265(5596):687–695.

Sanwen H et al. 2009. The genome of the cucumber, *Cucumis sativus* L. *Nature Genet* 41(12):1275–1283.

Sato S et al. 2008. Genome structure of the legume, *Lotus japonicus*. *DNA Res* 15(4):227–239.

Shendure J, Mitra R, Varma C et al. 2004. Advanced sequencing technologies: Methods and goals. *Nat Rev Genet* 5:335–344.

Shinozaki K, Ohme M, Tanaka M, Wakasugi T, Hayashida N et al. 1986. The complete nucleotide sequence of the tobacco chloroplast genome: Its gene organization and expression. *EMBO J* 5:2043–2049.

Tanaka H and Partial KT. 2009. Sequencing of a single DNA molecule with a scanning tunnelling microscope. *Nat Nanotechnol* 4:518–522.

Tuskan GA et al. 2006. The genome of black cottonwood, *Populus trichocarpa* (Torr. & Gray). *Science* 15;313(5793):1596–1604.

Yu J et al. 2005. The genomes of Oryza sativa: A history of duplications. *PLoS Biol* 3(2):e38.

Zharkikh A et al. 2008. Sequencing and assembly of highly heterozygous genome of *Vitis vinifera* L. cv Pinot Noir: Problems and solutions. *J Biotechnol* 31;136(1–2):38–43.

Zhou X, Ren L, Li Y, Zhang M, Yu Y, and Yu J. 2010. The next-generation sequencing technology: A technology review and future perspective. *Sci China Life Sci* 53(1):44–57.

Zonneveld BJM, Leitch IJ, and Bennett MD. 2005. First nuclear DNA amounts in more than 300 angiosperms. *Ann. Bot.* 96:229–244.

4 Water Management in Date Palm Groves

M. Mumtaz Khan and S. A. Prathapar

CONTENTS

INTRODUCTION

Water is a significant factor for socio-economic growth of the world. The availability of fresh water is limited, while the demand for this precious natural resource is continuously on the rise. It is anticipated that nearly two-third population of the world will be living in water-stressed areas by the year 2025. It is, therefore, a discrete requirement of today to design strategies to economize the uses of the available water: in particular, for agriculture, which utilizes as much as 70%–90% of it.

In the current scenario of immense pressure on land and water resources in terms of availability and accessibility, food security to the fastest growing population is a big challenge, which has received attention worldwide. Among all the inputs of crop production, water is vital as it is the carrier of all nutrient constituents. This precious resource has competition from industry, domestic use, and crop production. Therefore, harnessing the available potential of surface and ground water resources and its judicious use is the most important agenda to face future challenges of the world. In the majority of countries, 70%–90% of water is consumed in the agriculture sector with efficiency as low as 40% in many countries. The ratio of gross irrigated area to gross areas under agricultural production is very low at many places. Increasing areas under irrigation will largely depend upon the increase in water use efficiency. Otherwise, the misuse will not only lead to shortage of irrigation water, but also impact the soil health and crop production. In this context, modern irrigation methods including pressurized irrigation systems, for example, sprinkler, micro and bubble irrigation, may play a significant role in achieving higher crop productivity, water use efficiency, and sustainability in agriculture.

Date palm-producing regions are largely centered in the Near East and Africa (Figure 4.1), which are largely water-scant regions of the world. Water scarcity is becoming more acute in areas where the increased water use is coupled with overgrowing human population. The agriculture sector contributes 14% to the economy of the Arab region but uses 88% of water resources. Industrialization and service sector expansions are likely to increasingly compete with the agricultural sector for this scarce commodity. Thus growing strain on limited water wealth from nonagricultural sectors pose important questions on the trade-off between agriculture, industrialization, and residential uses and the possibility of increasing water use efficiency.

Innovated technologies will likely play an important role in future water supply, including desalination plants, wastewater recycling, and agricultural research focused on drought stress and salinity tolerance. Global per capita availability of water is 7130 m^3, which is almost 10 times higher than the available water in the Middle East and North Africa (MENA) region (757 m^3). A projected curve of water availability in this region for 2025 and 2050 shows a further significant decrease (Breisinger et al., 2010). All countries in MENA regions have water availability per capita <1700 m^3/year, which is generally considered as a threshold between water abundance and water stress.

There is a common misconception among farmers as to the date palm's relationship with aridity that palms can be cultivated productively with little or no water. No doubt, the date palm is a resilient plant with an ability to survive prolonged dry periods but it does need enough amounts of water to sustain growth through keeping all metabolic processes intact, and produce marketable fruit. This misunderstanding originated from the experience that most of the date groves, in general, are found near oases or alongside field boundaries, where, in fact, plants fulfill their consumptive use through seepage from neighboring canals or capillary rise from shallow groundwater table. Plant metabolism can only occur when a sufficient amount of water is present in the plant system. In the Near East, where date palm is grown abundantly under short water supply and water poverty, the quantity of water shared out to grow date palm carries strategic importance for the nationwide water reserves management. For growers whose livelihood is dependent on date palm cultivation, it is

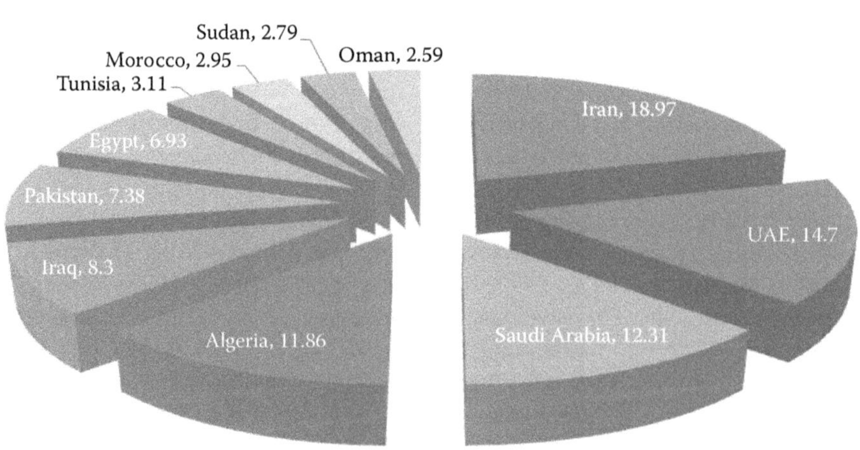

FIGURE 4.1 Date palm-producing countries (% area harvested).

imperative to implement the best water management practices that make best use of each drop of water for sustainable crop growth and enhanced productivity and/or fruit quality.

Irrigation scheduling, that is, the application of the right amount of water to the crops/plants at the right time, is crucial for crop establishment and productivity. Water-saving technologies like micro and sprinkler irrigation systems are being adopted in many countries to reduce the loss of water through overirrigation. Microirrigation is relatively a better system of irrigation water management for all crops, in particular, for orchards. This system has gained widespread popularity in areas of acute water scarcity and in areas where horticultural and commercial crops are the backbone of the local as well as regional economies.

DATE PALM-GROWING REGIONS AROUND THE GLOBE

In the Near East and African regions, date palm has been a major agricultural crop for centuries with a significant role in the socio-economic livelihoods of their residents. Because of the most favored dry subtropical and high-temperature climate, the world's largest date palm cultivation is concentrated in the Near East and Africa. The total world date production during 2008 was 7.05 million tons, of which two-third (about 4.40 million tons) came from Asia and the remaining (2.61 million tons) from Africa (FAO, 2010). Table 4.1 shows the top 20 date palm-producing countries with 74.4% of its production concentrated in Egypt, Saudi Arabia, Iran, the United Arab Emirates, Pakistan, and Algeria, 20.8% from Iraq, Sudan, Oman, Libya, China, Tunisia, and Morocco and the remaining 5% from 22 other countries.

TABLE 4.1
Major Date Palm-Producing Countries of the World

Rank	Countries/Regions	Production (MT)	Area Harvested (Ha)	World Production (%)
1	Egypt	1,326,133	87,685	18.82
2	Iran	1,000,000	240,000	14.19
3	Saudi Arabia	982,546	155,734	13.94
4	UAE	755,000	186,000	10.71
5	Pakistan	680,107	93,336	9.65
6	Algeria	500,000	150,000	7.094
7	Iraq	440,000	105,000	6.24
8	Sudan	336,000	35,295	4.77
9	Oman	255,871	32,746	3.63
10	Libya	175,000	30,000	2.48
11	China	135,000	9,000	1.91
12	Tunisia	127,000	39,300	1.80
13	Morocco	72,700	37,300	1.03
14	Yemen	55,204	14,465	0.78
15	Israel	30,191	5,500	0.43
16	Mauritania	22,000	8,200	0.31
17	Qatar	21,564	1,444	0.31
18	Chad	18,300	7,700	0.26
19	USA	16,511	2,064	0.23
20	Bahrain	13,293	2,102	0.19
	World Production	*7,048,089*	*1,264,611*	*100*

Source: FAO, Stat. 2010. *Food and Agriculture Organization of the United Nations*. http://faostat.fao.org. This table is constructed from various FAO sources.

FACTORS INFLUENCING IRRIGATION REQUIREMENTS

For the estimation of water requirement for date palm cultivation, it is important to understand the factors that have a direct or indirect influence on its production. The nature of climate, soil, water quality, and cultivar has a significant impact on water requirement calculation for date palm. The available information shows spatial and temporal variations in the date palm water requirements (Table 4.2).

Climate

The climatic factors that directly influence crop water requirements include air temperature, rainfall, wind, humidity, cloud cover, and radiation.

Air temperature: Prolonged dry and hot summers with a modest winter temperature are essential for successful date palm cultivation. However, the evapotranspiration (ET) rate (the rate of water transpiration from soil to atmosphere through plant roots, stem, and leaves) increases significantly when dates are grown in hotter climates due to higher levels of radiation. Therefore, water consumptive use of date palm is greater during the hottest months of the year.

Rainfall: Date palm is a crop of the arid and subtropical zone. Annual rainfall in such regions is low to very low. During insufficient precipitation, the date palm tree requires supplemental watering to keep its growth and development throughout its growing season.

Wind: Continuous high wind coupled with high air temperature in date palm-growing areas causes significant losses of available water through ET. In such environments, the ET rate from the soil surface and trees is much higher. Therefore, extra water in persistent windy regions is required for successful cultivation.

Humidity: Humidity has a role in determining the date palm water requirements. It is known that if the humidity is lower, plants will need more water. Relative humidity with air temperature and solar radiation intensity affects the ET of the plant. Under high evaporative demand and water deficit, plants cannot deliver enough water to the leaves and close their stomata partially or completely. Under high relative humidity, there is lesser evaporative water loss and irrigation demand is reduced to some extent. Rainfall and high relative humidity at the time of flowering and ripening are injurious and spoil the palm fruit badly due to splitting and rotting.

Cloud cover and radiation: Cloud cover can be split into low, middle, and upper level clouds, which are averaged to yield total cloud cover. Low clouds tend to be an extremely beneficial means of cooling the climate (by reflecting sunlight), whereas high clouds are more effective in warming the climate (by trapping infrared radiation). Thus, knowledge of cloud cover can help in explaining the temperature variations. Furthermore, cloud cover also has an impact on vegetation, because it is a major factor in controlling photosynthetically active radiation. ET is dependent primarily on the solar energy available to vaporize the water. Because of the importance of solar energy, ET also varies with latitude, season of year, time of day, and cloud cover.

TABLE 4.2
Water Requirements of Date Palm Trees in Algeria, UAE, and Yemen

	Water Requirement (L/tree/day)											
Country	**Jan**	**Feb**	**March**	**April**	**May**	**June**	**July**	**Aug**	**Sept**	**Oct**	**Nov**	**Dec**
Algeria	198	280	388	513	630	783	831	771	570	407	247	180
UAE	155	155	250	250	250	310	310	310	310	200	200	155
Yemen	350	378	362	532	546	574	574	504	483	420	399	371

Source: FAO. 2008. *Proceedings, Workshop on "Irrigation of Date Palm and Associated Crops"*. Damascus, Syrian Arab Republic, May 27–30, 2007.

ESTIMATING ET REQUIREMENTS AND SCHEDULING

During the process of evaporation from the soil and open water surface, water is released in the form of vapors to the atmosphere. The transformation of water from liquid to vapor form is caused by the energy movement in the ecosystem. The energy movement can be measured through climatic conditions, which include sunshine, humidity, temperature, and wind speed.

Transpiration is the process of removal of liquid water contained in plant tissues to the atmosphere. Plants lose water from their leaves and stem through transpiration that happens mainly during day time. ET results due to loss of water through evaporation and transpiration. The climate, crop/plant growth stage and plant species determine the ET rate.

Different methods have been used, over time, to calculate the ET requirements of different crops. These methods vary in accuracy and ease of use due to the availability of the required information. Some of these methods are discussed in subsequent sections.

Different approaches are employed for establishing irrigation schedules. The water balance approach includes the amount of water input into the soil and the amount of water output from the soil on a daily basis (Doorenbos, 1976). Once reference crop ET is known, the actual evapotranspiration/crop water requirement (ETc) of a crop is calculated after considering the crop factor and crop growth stages. Therefore, the reference crop ET (ETo) of a specific area is a very useful parameter for developing irrigation schedules for a crop.

ET/Class A pan method: Evaporation pans provides accurate information about the ETo. There are two types of evaporation pans that are used commonly. The class A evaporation pan is a circular pan with a measurement of 120.7 × 25 cm. The Class A evaporation pan is placed 5 cm above the surface. It measures combined effects such as humidity, temperature, sunshine, and wind speed. The pan is filled with water and the amount of water evaporated after 24 h is recorded.

$$\mathrm{ETo} = \mathrm{Epan} \times \mathrm{Kpan}$$

$$\mathrm{ETc} = \mathrm{ETo} \times \mathrm{Kc}$$

where

ETc = Crop evapotranspiration or crop water requirement (mm/day)
Kc = Crop factor
ETo = Reference crop evapotranspiration (mm/day)
Kpan = Pan coefficient (0.80)

Penman–Monteith equation: Penman (1948) derived an equation by combining the energy balance with the mass transfer method (also known as the Penman method). This equation is used to calculate evaporation from the open water surface from the standard climatic data on temperature, humidity, wind speed, and sunshine. The Penman–Monteith equation to estimate ETo is given below:

$$\mathrm{ETo} = \frac{0.408\Delta(Rn - G) + (\gamma 900/T + 273)\, u2\, (es - ea)}{\Delta + \gamma(1 + 0.34u2)}$$

where

ETo = Reference evapotranspiration (mm/day)
Rn = Net radiation at the crop surface (MJ/m^2/day)
G = Soil heat flux density (MJ/m^2/day)

T = Mean daily air temperature at 2 m height (°C)
$u2$ = Wind speed at 2 m height (m/s)
es = Saturation vapor pressure (kPa)
ea = Actual vapor pressure (kPa)
$es–ea$ = Saturation vapor pressure deficit (kPa)
Δ = Slope vapor pressure curve (kPa/°C)
γ = Psychrometric constant (kPa/°C)

Due to the complexity of the formula, it is suggested to use the equation through a computer program, Cropwat 7, based on the Penman–Monteith method (Smith et al., 1992). The Eto is determined first and then the water requirement is worked out using the following formula:

$$ET_{crop} = Kc*ETo \text{ (mm/day)}$$

where

Kc = Crop factor
ETo = Reference evaporation (mm/day)
ET_{crop} = Crop evapotranspiration (mm/day)

Date palm water requirement estimated through the Penman–Monteith mathematical model showed that field observations and theoretical estimates of palm water requirements have generally good agreement, particularly during the periods of mid- and end seasons (Alazba, 2004).

Scheduling based on soil data: Soil water conditions strongly influence the irrigation scheduling. Knowledge of the actual water content (soil moisture) and the soil water storage (SWS) capacity allows measuring the exact amount of supplemental irrigation water to be applied at one time and at different intervals.

SWS capacity: Soil has three main water condition levels. The water saturation capacity (SC) is the moisture content of the soil when soil pore spaces (all) are fully saturated with water. The field capacity (FC) is the amount of water that a soil holds after 48 h of irrigation or rain events to SC. Finally, the permanent wilting point (PWP) is the water content of the soil at which no water can be extracted from the soil, meaning that the soil water at this point, that is, PWP is no longer available for plant growth. The available water capacity (AWC) is the amount of water used by the plant between FC and PWP. The following steps should be considered before scheduling irrigation for any crop.

First, estimate the soil available water storage capacity (AWSC), also known as available water capacity, which is the amount of moisture that the soil can hold between PWP and FC as

$$AWSCw = FCw - PWPw \text{ established on weight basis}$$

where

AWSCw = Available water storage capacity on weight basis
FCw = Field capacity on weight basis
PWPw = Permanent wilting point on weight basis

$$AWSCv = FCv - PWPv \text{ established on volume basis}$$

where

AWSCv = Available water storage capacity on volume basis

FCv = Field capacity on volume basis
PWPv = Permanent wilting point on volume basis

$$\text{AWSC in mm/1 m soil depth} = \text{AWSC} \times (\text{hs/As}) \times 10$$

where

AWSC = Available water storage capacity in cm^3 or mL
hs = Soil depth unit (1 m = 100 cm)
As = Soil surface unit (1 m^2 = 10,000 cm^2)

Second, estimate the SWS capacity, which can easily be calculated from the AWSC if the plant effective root depth (RD) is known. It indicates the amount of water that should be applied at one time and the frequency with which the crop should be irrigated. AWSC can be calculated as

$$\text{SWS} = \text{RD} \times \text{AWSC}$$

where

SWS = Soil water storage
RD = Rooting depth
AWSC = Available water storage capacity

Third, estimate the maximum soil water deficit (MSWD), also known as readily available water (RAW) or plant available water capacity (PAWC). This is the amount of water in soil that is allowed to be removed from the soil prior to irrigation. Van der Gulik (2005) described it as the amount of water that can be stored in the soil that is readily available to the plant. It is the maximum amount of water that can be applied at one time before the risk of deep percolation occurs.

To calculate MSWD from SWS capacity, a factor called the availability coefficient (AC) or management allowed depletion, should be applied. The AC is expressed in percentage and it refers to the amount of water that may be extracted by the plant without putting unnecessary stress on the plant itself. The AC varies with plant species. For high-value crops, the AC is 40% or less, while for low-value crops, the AC is 60% or more. Usually, an average value of 50% for AC represents a reasonable value for most row crops. MSWD is calculated as

$$\text{MSWD} = \text{SWC} \times \text{AC}$$

where

MSWD = Maximum soil water deficit (MSWD = RAW = PAWC)
SWS = Soil water storage
AC = Availability coefficient

Soil types: Date palm is grown in a range of diverse soil types including light, medium, and heavy soils that can be used for sustainable crop production but require good drainage, aeration, and fertility. It is a peculiar tree crop which is tolerant to high pH, saline, and alkaline environment conditions (Chundawat, 1990; Chandra and Gupta, 1992). Having large capital investment on date palm plantation to keep its sustained growth and profitable production, it is most desirable to grow date orchards on sandy loam soils with good water-holding capacity and drainage. Date palm can survive well on soils having salt concentration up to 4%, provided that its root system should not come into contact with a soil stratum where the sodicity is >1%. Date palm requires frequent watering in sandy and sandy loam soils as compared to clay or clay loam soils. Less water is needed in soils of high water-holding capacity. Studies conducted in Kuwait have shown that date palm required 2685 mm net irrigation requirements with 2553 mm on loamy sand soil. It was argued that these values seemed higher and further field experiments are needed for validation (Salam and Al Mazrooei, 2007).

In saline areas, due to accumulation of salts on soil surface through evaporation, frequent watering is needed to leach down the salts. Although the date palm tree shows high tolerance of salinity, increased soil salinity can cause significant reduction in tree growth, yield, and fruit quality (Al-Rawi and Al-Mohemdy 2001; Furr and Armstrong, 1962). A detailed survey in Algeria showed that date palm potential yield was far below when it was grown under saline and waterlogged soil conditions (Bohoun et al., 2011).

Water quality: Water quality is an important factor in successful crop production. Date palm can be grown successfully with moderate saline water compared to many other fruit crops; however, crop yield is affected adversely if the salt concentration is high (Table 4.3). It has been observed that the date palm trees can bear full crop load under 2000 ppm saline water, but the yield decreases considerably when irrigated with saline water from 5000 to 8000 ppm.

Quality of irrigation water has an impact on crop production and quality. Depending on the sources of water, all irrigation water contains varied amounts of dissolved salts. The commonly dissolved salts in irrigation water include NaCl, $MgSO_4$, $CaSO_4$, and $NaHCO_3$. In general, two water quality parameters are used to access the salinity in irrigation water. One parameter is the total dissolved salts, which is usually expressed in milligram of salt per liter of water (mg/L). The other commonly used parameter is the electric conductivity (EC) expressed in deciSiemens per meter (dS/m) or millimhos per centimeter (mmhos/cm).

Accumulation of salts in the root zone of the tree creates a physiological stress on the plant, resulting in restricted nutrient uptake. The main criteria for evaluating the quality of water for irrigation are salinity (EC), sodicity (sodium adsorption ratio [SAR] and residual sodium carbonate [RSC]), specific ion effects, and infiltration problems. The guidelines for interpreting water quality are described in Table 4.4.

Irrigation time: Calculating the irrigation interval (IInt) is one of the most critical steps to establish an efficient irrigation system. The simplest way to establish the periods between irrigations is to collect information about the MSWD and the crop water use (ETc).

According to the climate and stage of plant development, the ET and plant water need values change during the growing period; therefore, the IInt is calculated on a monthly basis and expressed

TABLE 4.3
Relative Salt Tolerance of Date Palm to Other Fruit Crops

High Salt Tolerance ($EC_e \times 10^3 = 18$[a])	Medium Salt Tolerance ($EC_e \times 10^3 = 10$)	Low Salt Tolerance ($EC_e \times 10^3 = 5$)
Date palm	Fig	Apple, apricot
	Pomegranate	Pear, almond
	Grape	Grapefruit, strawberry
	Olive	Orange, peach

Source: Ayers, R.S. and D.W. Westcot. 1994. *Water Quality for Agriculture.* FAO Irrigation and Drainage paper No. 29 Rev. 1, Rome.

[a] The numbers following $EC_e \times 10^3$ are the electrical conductivity values of the saturation extracts in mmhos/cm at 25°C associated with a 50% decrease in yield.

TABLE 4.4
Guidelines for Interpreting Water Quality for Irrigation

	Degree of Problem		
Irrigation Water Quality Index	**None**	**Increasing**	**Severe**
Salinity (water availability) EC_{iw}, dS/m	<0.75	0.75–3	>3
Permeability (infiltration rate)			
EC_{iw}, dS/m	>0.5	0.5–0.2	<0.2
SAR_{adj}			
Montmorillonite[a]	<6	6–9[a]	>9
Illite-vermiculite	<8	8–16[a]	>16
Kaolinite-sesquioxide	<16	16–25[a]	>24
Specific ion toxicity (affects sensitive crops)			
Sodium[b] (adj SAR)	<3	3–9	>9
Chloride,[b] $mmol_c$/L	<4	4–10	>10
Boron, $mmol_c$/L	<0.75	0.75–2	>2
Miscellaneous			
NO_3^-–N or NH_4–N, mg/L	<5	5–30	>30
HCO_3^-, $mmol_c$/L (overhead spray)	<1.5	1.5–8.5	>8.5
Cl^-, $mmol_c$/L	<1	1–2	>2
pH		Normal range is 6.5–8.4	

Source: Gupta, I.P. and S.P. Yadav. 1986. *J. Indian Soc. Soil Sci.* 34:279–286 and Ghafoor, A., M. Qadir, and G. Murtaza. 2004. *Salt-affected Soils: Principles of Management.* Meraj Printers, Allied Bank Centre, Urdu bazaar, Lahore, Pakistan, Table 12.11.

[a] Dominant type of clays minerals.

[b] Sodium absorption ratio.

[c] Charge.

in minutes or hours. The IInt is expressed in days. The volume of irrigation water at one time is equal to the amount of water of MSWD.

$$\text{IInt} = \text{MSWD/ETc}$$

where

IInt = Irrigation interval
MSWD = Maximum soil water deficit (MSWD = RAW = PAWC)
ETc = Crop daily water need

Cultivars: Date palm cultivar response is independent and varies with water quality and quantity. A classical study on irrigation with saline water for a longer period (40 years) revealed that all four cultivars, namely Zahdi, Khestawi, Maktoom, and Barban, behaved differently (Al-Rawi and Al-Mohemdy, 2001). Another study on three date palm cultivars (Khalas, Khunaizy, and Abunarenjeh) irrigated with varied saline water showed 50% reduction in growth, while "Khunaizy" exhibited a little better performance over others (Al-Rasbi et al., 2010). The Barban cultivar excelled among all other cultivars in terms of growth and fruit quality under various salinity levels. A study on date palm varietal response to different IInt showed an independent behavior toward irrigation regimes (Dawoud, 2001; Walid and Abdelhadi, 1990).

IRRIGATION METHODS IN COMMERCIAL GROVES

Water scarcity is a worldwide problem, particularly in arid climates where water resources have always been a limiting factor for the production (Goosen and Shayya, 1999). At present, several irrigation systems are being used depending on water availability, the types of cultivated crops and the capital investment. A range of irrigation methods are practiced in many date-growing regions as briefed below.

Flood or Surface Irrigation

Largely, date palm groves are irrigated through the surface irrigation system as it comes from the historical tradition. Furrow and basin is the modified form of flood irrigation in which water is directed to flow through narrow channels dug between the rows of plants, instead of distributing throughout the whole field evenly. Furrows must have equal dimensions in order to guarantee even distribution of water. However, its suitability for dates grown in sandy soils is low (Zaid and Arias-Jimenez, 1999). The method offers high water use efficiency, uniform application, less expensive if the land is flat and leveled, and fairly easy in managing the irrigation operations. The weaker side of this method is its laborious nature and heavy losses of water through leaching and surface evaporation.

Bubbler Irrigation

The bubbler irrigation system was initially designed to irrigate fruit tree crops (Hull, 1981). It is a localized, low-pressure, solid permanent installation system used in tree groves. Each tree has a round or square basin which is flooded with water during irrigation. Water creeps into soil and wets the root zone of the plant. Water is applied through bubblers and there are small emitters placed in the basins which discharge water at flow rates of 60–240 L/h (Nakayama and Bucks, 1986). Each basin can have one or two bubblers as required. The technology is simple and no highly sophisticated equipment is used. The system can be operated by unskilled farmers and laborers. No filters or fertilizer injectors are needed. Furthermore, the entire piping network is buried, so there are no field operation problems. It also gives high water use efficiency of up to 75%, resulting in considerable water savings, with absolute control of the irrigation water from the source to the tree basin.

But there is high initial purchase cost and in sandy soils with high infiltration rates, it is difficult to achieve a uniform water distribution over the tree basins. Another disadvantage is that the small water flows cannot be used as in other micro-irrigation systems.

Drip Irrigation

The concept of drip irrigation is based on the best use of available water resources and better plant performance comes from preventing rather than relieving moisture stress (Morris, 1999). Comparison among various irrigation methods (drip, basin, and bubbler) on date palms has shown that the drip system has the highest water use efficiency, followed by the basin system and the bubbler irrigation system, respectively (Al-Amoud et al., 2000). However, a recent study on mature date trees revealed that there is an increase in yield and considerable water saving when the subsurface irrigation system is compared to the conventional drip irrigation method. A high increase in water use efficiency was also recoded in date palm through the subsurface system of irrigation (Al-Amoud, 2010).

A comparative study on drip and bubbler irrigation in date palm showed higher water distribution in the drip system (97%) as compared to the bubbler system (62%) (Al-Amoud, 2008).

The main limitation of profitable farming is poor water management techniques in the arid areas (Baxter, 2007). In the preceding study, it was found that water use efficiency through drip irrigation was certainly much better. There was significant water saving (up to 45%) when compared to traditional watering practices and labor engagement was also decreased by 71% in the arid areas.

However, under traditional farming systems, drip irrigation alone may not be very cost effective. This may be because of the lack of technical competence and better inputs availability to achieve higher crop yields. It was observed that farmer's financial return was higher when the drip irrigation system was practiced in combination with improved soil fertility, high-value crops, and trained horticultural work force. The majority of farmers are convinced that labor, time, and capital can be saved through the drip irrigation system. Growers may get higher financial benefits through improved water productivity (WP) and its role in enhancing farmer's participation in commercial horticultural farming is significant.

Irrigation in Traditional Farms

In addition, to modern irrigation systems, old traditional irrigation systems are still intact and working in many areas of the date palm-producing countries. Aflaj is one of the 2700-year-old irrigation methods which existed in Oman in the regions of dry and desert environments (Wilkinson, 1983). However, some Aflaj were constructed in the past, ~150 years ago (Bonnefant and Al-Harthy, 1977); the estimated number of Aflaj in Oman is 4300 (Al-Rawas et al., 1999).

Aflaj Irrigation System

In arid regions, indigenous community-managed systems access ground water by gravity flow from the underground aquifers or surface springs on neighboring mountain slopes. An example of such traditional systems of water mining in Oman is the use of the Falaj (Figure 4.2), a structure resembling a large horizontal well, which skims the upper part of the aquifer and imparts a reliable and reasonably regular supply to domestic and agriculture users. A large area under date palm is irrigated by the Falaj system. Basically, the Falaj system is a flooding system of irrigation which depends on gravity. Date palm growers irrigate their palm fields for a couple of hours. The farm is usually irrigated once a week in the summer and after every 2 to 3 weeks during the winter season. Falaj water is inherited or brought from the village Falaj committee that is responsible for water distribution (Al-Marshudi, 2002).

In many parts of the world, including members of the Gulf Countries Council, the assumption is often made that high water losses are inherent in all forms of surface irrigation, with efficiencies often estimated at 50% or less (Ahmad, 1996). However, little quantitative work has been reported to evaluate actual, on-farm water use among traditional systems in Oman. Among Aflaj in particular, performance levels of on-farm water management have remained unknown. This form of baseline information is crucial in the search for (and evaluation of) solutions to the crisis facing many of Oman's Aflaj and in the attempt to ameliorate water conservation within irrigated systems in the Sultanate. Norman et al. (1998) reported that on-farm ratios of crop water demand to irrigation supply (*D/S*) were relatively high in Falaj. Mean "actual" demand/supply ratios (*Da/S*) which reflect farmer practices are in excess of 0.80, for both wheat and date palm crops. When "design" demand/supply ratios (*Dd/S*) are considered, the values are even higher (considerably higher than the often-quoted efficiencies of 50% or less for traditional surface irrigation methods). It is likely that an important portion of surface-irrigated farms in the region does operate at low irrigation *D/S* values, particularly those which have been developed in more recent years. But a case study at Falaj indicates that rural Falaj systems, in which traditional Omani management methods continue to be employed, tend to use less water since the supply is limited and may be operated at *D/S* values considerably higher than presumed. The data indicate that farmers understand how to manage water carefully when using traditional surface irrigation methods, in so long as flow rates (in particular, base flow rates) can be anticipated and remain within their control.

The rapid social and economic changes that have occurred in Oman during the past 25 years have had a profound impact on the viability of Falaj systems. Measures are currently being sought

FIGURE 4.2 Traditional Aflaj irrigation system in Oman.

by which irrigation water can be used more efficiently. Particular attention has been given to the improvement of "traditional," on-farm surface irrigation, since this method is employed in the majority of farms with wells and within all Falaj systems. Keeping in view its role in traditional agriculture and cultural heritage, Aflaj needs to be preserved in a more systematic way in Oman (Siebert et al., 2007).

LEACHING AND DRAINAGE

Leaching has an environmental concern as it refers to loss of water-soluble nutrients from the soil due to irrigation or precipitation to the fresh groundwater. Leaching depends upon many factors including soil type, structure, rate of fertilization, weather conditions, and crop type. Sometimes, leaching is applied deliberately to control excess salt buildup in soil from high-salt-content water. When water from rain, flood or any other source seeps into the ground by carrying hazardous wastes, fertilizers, biological wastes, and biocides (e.g., fungicides, herbicides, pesticides, and insecticides), it contributes to ground water contamination (Ayers and Westcot, 1985). Drainage is the natural or artificial phenomenon of surface and subsurface water removal from an area. Drainage is required to improve the productivity of soils as well as water management (Mass and Grattan, 1999; Hanson et al., 1999).

ARTIFICIAL/DELIBERATE LEACHING OF ACCUMULATED SALTS

A substantial amount of salt is detected in many irrigation waters (Sleeman, 1963; Sharma et al., 1997). Salts accumulate in the root zone due to insufficient leaching of salts and by upward movement of shallow saline water table. Leaching is very useful for salt removal from plants' root zone and installation of drains is necessary in the field to overcome salinity from high saline water tables (Ayers and Westcot, 1985). Date palm needs sufficient water (estimated at ≈150–200 L) of acceptable quality to reach its potential yield, like any other fruit tree. Rain-fed cultivation of date palm is not possible. The average EC of the soil (EC_e) should not exceed 4 dS/m (Ayers and Westcot, 1985), and that of water (EC_w) should not be >2.7 dS/m. In saline soils, clearing of salt from the soil can be carried out using a sufficient quantity of water (Sehgal and Stoops, 1972; Jaynes and Rice, 1993). For mature date palms, the radius around the trunk is 3 m and the depth is about 5 m. Depending on plant size, this radius could vary from 10 to 30 cm and depth from 25 to 50 cm for young date plantlets. For dates, almost 40% of all water is extracted from the first 50 cm, 70% is from the first 100 cm, 90% is from the top 150 cm, and only 10% is from the last layer or 150–200 cm and deeper. Thus, the irrigation water should be applied by considering these boundaries. The irrigation water application should be in such a way that it must not drain to the deeper levels of soil to ensure proper root development in date palms (Soil Survey Manual, 1993). In Table 4.5, EC_w and EC_e values in correlation to date palm percentage yield are provided (Ayers and Westcot, 1994).

The quantity of water required for leaching can be calculated from the following:

$$LR = EC_w/[5(EC_e - EC_w)]$$

TABLE 4.5
EC_w and EC_e Values and Percentage Yield of Date Palm

Yield %	EC_w (dS/m)	EC_e (dS/m)
100	2.7	4.0
90	4.5	6.8
75	7.3	11.0
50	12.0	18.0
0	21.0	32.0

Source: Ayers, R.S. and D.W. Westcot. 1994. *Water Quality for Agriculture*. FAO Irrigation and Drainage paper No. 29 Rev. 1, Rome.

where

LR = Leaching requirement (fraction)
EC_w = Electric conductivity of water (dS/m)
EC_e = Electric conductivity of soil at percentage yield to be obtained (dS/m)

This quantity of water is in excess of irrigation water required for growing crop in a particular season. The total annual requirement can be determined as follows:

$$AW = ET/[1 - LR]$$

where

AW = Depth of water supply (mm/year)
ET = Total annual water demand (mm/year)

Overwatering must be avoided because it can kill plants and encourage pathogen development, which can result in specific disease problems. It promotes root diseases due to anaerobic conditions in the area of the root system. Overwatering can also cause wood rotting on the exterior trunk of date palm. It should be kept in mind that the date palm trunk is covered with dead wood; therefore, it will start decaying and become a source fungi food if allowed to remain wet (Daniels et al., 1967; Khalifa et al., 1989).

WP OF DATE PALM

Productivity is a measure of performance expressed in the ratio between output and input. Productivity of water could be for a whole system or for one of its component (Molden et al., 1998). WP, like land productivity, is a partial factor of productivity that measures how water is converted into services and goods (Molden et al., 2003). It is generally represented as follows:

$$WP = \text{Output deviced from water use/water input}$$

WP was introduced to compare the efficiency of classic and modern irrigation systems (Keller et al., 1996). This parameter can be used to get the highest benefits from irrigation water as well as the extent of water saving. WP is a very useful measure and can use various scales to determine the needs of different stakeholders. WP can be defined in one or more of the following ways: (i) physical output: could be harvestable product and/or total biomass, (ii) economic output: either net or gross benefit.

The water input can be determined as volume in m^3 or as the value of water used as the highest opportunity cost in alternative uses of the water. WP assessment can be done by using the following parameters: (i) crop scale, (ii) field scale beneficial use, (iii) field/farm scale beneficial and nonbeneficial use, (iv) irrigation system scale, (v) subbasin scale, and (vi) basin scale.

Keeping future needs in mind and to overcome problems such as water shortage, it is important to increase WP that varies greatly in relation to the specific conditions under which the crop is grown. Palms require water for growth and production of dates whether they are being irrigated or not. They can uptake shallow groundwater from oases, deeper groundwater or from surface streams. The very high average yield of date palm in Egypt in comparison to other countries is due to the better access to water for palms growing along the Nile valley.

Recently, a number of innovations were made by small-scale farmers to manage water effectively during the dry season and to increase on-farm water efficiency by using simple, affordable micro-irrigation systems. These are affordable to many small farmers in Asia and Africa, even to farmers with small plots who earn <$1 a day (Polak and Yoder, 2006). The studies have shown that drip irrigation can significantly increase water and labor productivity, and can marginally increase land productivity.

It is better to promote drip irrigation among farmers for whom water is a key constraint: scarce, costly to pump, and so on. When integrated with improved crop management practices, drip irrigation can be cost effective when compared with drip irrigation using traditional cultural practices. On-farm water management can be improved further with more research into appropriate watering regimes.

The existing irrigation systems have serious restrictions. Drastic measures need to be taken in the agriculture sector, particularly to save deterioration of WP in rain-fed and noncanal areas. Recent developments to improve WP and control water losses such as the use of polyvinyl chloride (PVC) pipes for water conveyance and the introduction of high-efficiency irrigation systems (HEIS) are very important. The HEIS methodology optimizes water utilization by dripping water in the plant root zone through a network of emitting devices, pipes, and valves. HEIS is a demand-based system in farm for crops and is found very useful all over the globe, particularly in water-scare areas, in increasing the WP by avoiding evaporation losses and seepage. The quantity of water to be dispensed in the root zone using IHES depends on the crop's water requirement, climatic conditions, condition of water and soil, and other factors such as farm size. A wide variety of emitting devices with varied discharge rates are available, for example, microtubes, bubblers, drippers, and sprinklers. HEIS offers unlimited benefits by increasing water efficiency, uniformity, and consistency in water supply, and ease and effectiveness in fertilizer application. Mostly PVC pipes are used in HEIS systems in tube well casing, distribution and conveyance, and micro-irrigation systems due to low cost, durability, noncorrosive nature, and ease in installation. The results of a comparative study between drip and sprinkler irrigation on the date palm growth and yield clearly demonstrated that drip irrigation was more effective for an increase in leaves, flowers, and fruits (Reuveni, 1971, 1974). In another comparison between drip and bubbler irrigation systems for date palms, it was concluded that salts accumulation on the surface layer was significantly higher for the drip compared to the bubbler system (Nimah, 1985).

Date palm can withstand salinity up to 4 dS/m without any significant reduction in yield (Ayers and Wescot, 1985). Date palm trees could uptake 65%–80% of water within 1.2 m of the root zone although the depth of the root zone ranges from 1.5 to 2.5 m (Doorenbos and Pruitt, 1977; Yaaqoob, 1996). Usually, date palm is irrigated through the surface irrigation method, mainly the flooding or basin system in many countries. Date palm water requirements are comparatively higher than other plants (Salter and Goode, 1967). Water consumption varies with conditions and the annual water requirements of a mature date palm may vary from 115 to 306 m^3 (Al-Baker, 1972). Water use and its productivity in date palm vary greatly in palm-growing regions (Table 4.6).

TABLE 4.6
Water Use and Productivity of Date Palm from Different Countries

Countries	Annual Gross Water Use (m^3/palm)	Annual Gross Water Use (m^3/ha)	Productivity (kg/m^3 of Water)
Egypt	86–124	10,280–14,880	2.28–3.31
Saudi Arabia	150–350	18,180–42,600	0.15–0.37
Iran	102–164	12,270–19,720	0.21–0.34
Algeria	43–210	5200–25,400	0.14–0.67
Oman	183–240	21,950–29,320	0.15–0.21
Libya	183–240	7,200–29,700	0.15–0.21
Tunisia	100	12,000	0.28
Morocco	105–200	12,600–23,900	0.21–0.4
UAE	130–173	15,500–20,740	0.20–0.26
Yemen	130–173	15,500–20,740	0.13–0.16

Source: FAO. 2008. *Proceedings, Workshop on "Irrigation of Date Palm and Associated Crops."* Damascus, Syrian Arab Republic, 27–30 May, 2007.

Experiments conducted on date palm water consumption in the Riyadh area of Saudi Arabia have indicated that a water consumption of 108 m^3/tree (1.08 m/ha), 216 m^3/tree (2.16 m/ha), and 324 m^3/tree (3.24 m/ha) per year showed an evaporation rate of 50%, 100%, and 150%, respectively (Al-Amoud et al., 2000). A comparison of various irrigation systems has shown date yield in the order of the drip, basin, and bubbler, respectively. To obtain the highest efficiency of irrigation water for date palm, an amount of 108 m^3/year/tree was found to be enough (Al-Amoud et al., 2000). The optimum date palm response to drip irrigation was due to slow delivery of irrigation water for a relatively long period of time, which provides better distribution and control for water through soil profile; therefore, the date palm tree used almost all the water delivered.

EFFECT OF WATER ON DATE QUALITY

Date palm contains a high percentage of carbohydrates (total sugars, 44%–88%), dietary fiber (6.4%–11.5%), pectin (0.5%–3.9%), proteins (2.3%–5.6%), fats (0.2%–0.5%), 15 minerals/salts (percentage of each mineral varies from 0.1 to 916 mg/100 g of dry dates), and vitamins (vitamin A, vitamin C, vitamins B1 thiamine, B2 riboflavin, and nicotinic acid [niacin]). The percentage of oil in flesh and seed is 0.2%–0.5% and 7.7%–9.7%, respectively. The seed weight is 5.6%–14.2% of the date total weight. The date flesh and seed contain 8 and 14 types of fatty acids, respectively. The percentage of fatty acids is very low in the date flesh. Unsaturated fatty acids found in date include oleic (41.1%–58.8% in seed), linoleic, palmitoleic, and linolenic acids. The various minerals and metals present in various proportions include B, Ca, Co, Cu, F, Fe, Mg, Mn, K, P, Na, Zn, and Al. Additionally, the seeds contain Al, Cd, Cl^-, Pb, and S (Ahmed and Ahmed, 1995; Al-Mukhtar, 1994; Al-Hooti et al., 1998; Al-Shahib and Marshal, 2002, 2003).

The quality and quantity of irrigation water determines the chemical constituents and physical appearance of the date flesh and seed. In general, date quality is graded on the basis of its physical and chemical characteristics as both play a vital role in its quality assessment. Physical factors include color, uniformity in color and size, shape, texture, and seed/flash weight ratio. Chemical constitutions important for the determination of date quality are moisture, fibers, sugar, enzymes, acidity, and tannins. Fungal attack, pest infestation, remains of pesticides, mold, sand, and dust are the external factors responsible for the determination of date quality.

Water and soil salinity has been recognized as a major agriculture problem in arid and semiarid regions. Saline water not only affects plant physiology through changes of water and ionic balance in the cell due to excessive accumulation of Na and Cl, but also reduces the uptake of other minerals (Alabdulhadi et al., 2005). Furthermore, irrigation with saline water causes salts accumulation in soil.

The effects of water quality on the growth and yield as well as on fruit physical properties have been well documented in the literature. The average fruit yield was highly affected since it was reduced by about 85%. It can be concluded from previous studies that irrigation with saline water for a long period causes a severe reduction in the growth and yield as well as a reduction in fruit quality of date palm trees. Certain management practices including the addition of fertilizers to maintain nutrient balance as well as watching soil salinity are needed when irrigating with saline water (Bernstein and Hayward, 1958; Furr and Ream, 1968; Al-Baker, 1972; Al-Rawi and Khalaf-Allah, 1980).

OPPORTUNITIES TO IMPROVE WP OF DATE PALM GROOVES

With the above challenges in mind, there is a potential for sustainable date palm production through modern innovations in irrigation and agronomic practices.

AGRONOMY

Layout and Planting Density of Date Palm: The following are popular methods of orchard layout opted for fruit trees:

- Square system
- Rectangular system
- Quincunx system

Each method has its own advantages and disadvantages. Date palm growers are not practicing any standard method of layout in general; row-to-row and plant-to-plant distance varies considerably throughout the date palm world. This could be due to a farmer's personal choice, differences in varietal behavior and agro-climatic conditions. For example, in Namibia, normal practice of spacing is 10×8 m^2, 10 m between rows, and 8 m in the rows. Some farmers elsewhere also practice 8×8 m^2 spacing; however, close planting in date palm groves is not advised.

The denser the planting, the more water is needed. Date palm is mostly grown in sandy soils and micro-irrigation is recommended because this kind of irrigation system has proven high efficiency. Water should not be sprayed into the crown of the small palms because it may cause fungal growth. As the young palm tree keeps the root system shallow, a frequent irrigation is suggested in the early years of its development. A 4-year-old tree normally covers an area of about 12 m^2 with the flow rate of 96 L/h/palm, a 5–10-year-old tree covers an area of about 18 m^2 with the flow rate of 104 L/h/palm, while after 10 years, the area covered by the tree is 28 m^2 with the flow rate of 156 L/h/palm. This huge area covered in the initial years (0–3 and 5–8) causes waste of water but, on the other hand, it serves as a leaching operation that benefits date palm as a whole.

Cultivars: Date palm genetic resources are broader and a large number of cultivated varieties exist around the date palm world. However, there are a few cultivars being grown commercially for the world market (Table 4.7). Palm cultivars' response to water is varied and independent. Earlier studies showed that some can perform better in drought, while others cannot. In the scenario of water poverty and hunt for drought-resistant cultivars, screening of wild date palm biotypes is needed. Modern tools, for example, genetic engineering and biotechnology, can help in exploiting hidden water-resistant genes and their use in variety development.

Intercropping: Intercropping is a practice of concurrently growing more than one crop in the same field. Farmers exploit space and time through intensifying more than one crop at a time in the same field. However, there is a considerable competition for resource utilization among multiple crops during their growing period. Intercropping carries some benefits; for example, each pattern of intercropping is commonly opted to lessen a particular limitation in resources. Intercropping is practiced mostly on a small scale and it generates extra income to farmers. It promotes interdisciplinary

TABLE 4.7
Date Palm Cultivars Grown Commercially

Cultivar	Description
Medhjoul	Dried date: Large fruit size, soft and sweet, color light brown to dark brown, appearance attractive
Deglet Nour	Dried date: Semisoft and pleasant flavor, light to brown color, unique taste and popular in the Middle East
Barhi	Fresh consumption: Consumed at Khala stage (partially ripe), juicy and sweet, color yellow, available in harvest season only
Hayani	Fresh consumption: Black and shiny color, long fruit size, and less sweet
Amari	Dried stage: Sweet, soft, and medium in size
Hadrawi	Dried stage: Sweet and fleshy, color dark brown
Zahidi	Dried stage: Round shape, golden in color size medium, and less sweet

activities and avoids dependence on one crop for economic return of the farms. It favors wildlife growth and reduces the outlay for fertilizers. The farmers having little economic resources can grow a range of useful products, which may constitute a type of savings for the future.

There are disadvantages of intercropping systems too, for example, competition between plants for light, soil nutrients, and water. There is a possibility of having allelopathic impact on the main crop. Sowing and harvesting of a crop become difficult because of mechanization issues. Loss of water is expected to be higher because of increased ET. Multiple cropping environments may help in proliferating certain harmful diseases and pests.

Traditionally, intercropping with alfalfa, barley, and beans is practiced in many date-producing areas. Intercropping with alfalfa and other leguminous crops provides nitrogen enrichment and livestock fodder production as well. Date palm plantation can be practiced at three cropping levels: palm trees, fruit trees, and low crops, for example, legumes, cereals, forages, and vegetables. This is known as three-storey cropping. Water requirement of date palm may change when intercropping or cover crops are grown in date groves. It is difficult to determine water requirement of date palm when intercropping is in practice.

Mulching: Mulching is practiced to minimize the evaporation from the soil surface, lower the soil surface temperature, and control weed and salt accumulation. Terasaki et al. (2009) compared the effects of gravel and date palm mulching on heat, moisture, and salt movement through soil columns in the United Arab Emirates. Results showed that elapsed time when 50% of irrigation water was lost due to evaporation was greatly prolonged by mulching with 2.3 times for gravel mulch and 3.4 times for date palm mulch compared to bare soil. The daily maximum soil surface temperature was 5°C lower for date mulch compared to bare soil; however, there was a very little difference between the both types of mulches. The gravels and date palm chips used as mulch were also effective in reducing the salt accumulation.

CONCLUSION

Water use is increasing day by day for industrial, domestic, environmental, recreational, and agricultural purposes. More water allocation is needed for nonagricultural purposes such as nature and environment with socio-economic development. Consequently, improvements are needed in the management of the water used for agricultural purposes, energy, and soil conservation while satisfying society's increasing demand for food crops and fiber, aquatic products, forest products, and livestock as well. The emphasis of the last decade was on "improved agricultural water management" but more recently emphasis is on "producing more with relatively less water—increasing water productivity." WP per unit of water used can also be increased by improving biological and economic output in both rain-fed and irrigated agricultural systems. It was estimated that several hundreds of kilograms of water per 1 kg of total or economic biomass produced is evaporated through crop plants.

In arid and semiarid regions, most of the countries are dependent on desalination of sea water to meet their daily water needs. Fresh water demand for agriculture and domestic use is continuously on the rise in this region. Escalating water supply from aquifers and desalination entails higher costs and poses disastrous environmental consequences. Agricultural strategies under extreme water scarcity should be based on sustainable maximization of economic and environmental returns from the unit volume of water. Crops should be carefully selected to satisfy this criterion together with water and agricultural management practices that ensure minimum water losses with maximum WP. Indigenous high-value crops and plants tolerant to drought and salinity are highly valuable in this respect. Date palm cultivation is still a viable option in dry regions; however, current irrigation practices are inefficient and adaptive research at the local level is required to develop options to minimize irrigation water losses and improve plant productivity.

Water requirements and/or consumption for date palm production vary from country to country depending upon site-specific agro-climatic conditions. Advance irrigation technology has not yet reached many of the date palm growers. Centuries-old irrigation methods are still in practice in many parts of the world. Irrigation scheduling is not practiced as a water management tool for date palm cultivation. Furthermore, irrigation management practices need to be tested under different environmental conditions before disseminating them to the farmers. Investigations should concentrate on the impact of adequate or deficit irrigation on date's quality and yield.

REFERENCES

Ahmad, M. 1996. Sustainable water policies in the Arab region. *Paper presented at the 14th International Symposium, Water and Arab Gulf Development: Problems and Policies*, University of Exeter, UK, pp. 10–12, September.

Ahmed, I.A. and A-WK. Ahmed. 1995. Chemical composition of date varieties as influenced by the stage of ripening. *Food Chem.* 54:305–309.

Alabdulhadi, I., H. Ali-Dinar, and G. Ebert. 2005. Ion distribution and partitioning of date palm (*Phoenix dactylifera* L.) under salinity conditions. In: Eric Tielkes, Christian Hülsebusch, Inga Häuser, Andreas Deininger, and Klaus Becker, (eds.). *The Global Food and Product Chain—Dynamics, Innovations, Conflicts, Strategies*. Deutscher Tropentag, Hohenheim, October 11–13, 2005.

Alazba, A. 2004. Estimating palm water requirements using Pennman-Monteith mathematical model. *J. King Saud Univ. Vol.16, Agric. Sci.* 2:137–152.

Al-Amoud, A.I. 2008. Performance of bubbler irrigation system as compared to trickle for large size date palm tree farm. *The Canadian Society for Bioengineering*. Paper No. CSBE 08-172.

Al-Amoud, A.I. 2010. Subsurface drip irrigation for date palm trees to conserve water. *Acta Hort. (ISHS)* 882:103–114.

Al-Amoud, A.I., M.A. Bacha, and A.M. Al-Darby. 2000. Seasonal water use of date palms in central region of Saudi Arabia. *Agric. Eng. J.* 9(2):51–62.

Al-Baker, A. 1972. *Date Palm Trees*. Ministry of Higher Education, Baghdad, Iraq, p. 225 (in Arabic).

Al-Hooti, S., J.S. Sidhu, and H. Qabazard. 1998. Chemical composition of seeds of date fruit cultivars of United Arab Emirates. *J. Food Sci. Technol.* 35:44–46.

Al-Marshudi, A.S. 2002. Oman traditional date palms: Production and improvement of date palms in Oman. *Tropicultura* 4:203–209.

Al-Mukhtar, M.M. 1994. *Nutritional Aspects of Date with Special Reference to the Tryptophan, Lysine and Dietary Fibre Content*. M.Sc. Thesis, Strathclyde University.

Al-Rasbi, S.A.R., N. Hussain, and H. Schmeisky. 2010. Evaluation of the growth of date palm seedlings irrigated with saline water in the Sultanate of Oman. *Acta Hort. (ISHS)* 882:233–246.

Al-Rawas, A., A.W. Hago, A. Al-Harthy, and S. Al-Shaqsi. 1999. Construction problems in traditional irrigation systems: A case study from the Aflaj of Oman. In: Goosen, M.F.A. and W.H. Shayya (eds.). *Water Management (Vol 1): Water Management, Purification and Conservation in Arid Climates*. Technomic Publishing Co., Inc., USA, pp. 351–366.

Al-Rawi, A.A.H. and A.F. Al-Mohemdy. 2001. Effect of water quality on growth and yield of date palm *Phoenix dactylifera* L. *Proceedings of the 2nd International Conference on Date Palm*, Al-Ain, UAE, pp. 128–136.

Al-Rawi, K.M. and A.M. Khalaf-Allah, 1980. *Design and Analysis of Agricultural Experiments*. University of Mosul Press, Iraq.

Al-Shahib, W. and R.J. Marshall. 2002. Dietary fibre content of 13 varieties of datepalm (*Phoenix dactylifera* L.). *J. Food Sci. Technol.* 37:719–721.

Al-Shahib, W. and R.J. Marshall. 2003. The fruit of the date palm: Its possible use as the best food for the future. *Int. J. Food Sci. Nutr.* 54(4):247–259.

Ayers, R.S. and D.W. Westcot. 1985. *Water Quality for Agriculture*. FAO Irrigation and Drainage paper No. 29, Rome.

Ayers, R.S. and D.W. Westcot. 1994. *Water Quality for Agriculture*. FAO Irrigation and Drainage paper No. 29 Rev. 1, Rome.

Baxter, N. 2007. *Market Strategies for Water Productivity*. CGIAR Challenge Program on Water and Food and Canadian International Development Agency. International Development Enterprises, Canada.

Bonnefant, P.G. and S. Al-Harthy. 1977. Architecture and social history at Mudhairib. *J. Oman Stud.* 3:107–135.

Bouhoun, M.D., S. Marlet, L. Brinis, M.L. Saker, J. Rabier, and M. Côte. 2011. A survey of the combined effects of waterlogging and salinity on fruit yield in the date palm groves of the Wargla basin, Algeria. *Fruits* 66:11–24.

Breisinger, C., M.H. Collion, X. Diao, and P. Rondot. 2011. The impacts of the triple crises on growth and poverty in Yemen. *Development Policy Review* 29(2):155–184.

Bernstein, L. and H.E. Hayward. 1958. Physiology of Salt tolerance. *Ann. Rev. Plant Physiol.* 9:25–46.
Chandra, A. and I.C. Gupta. 1992. *Date Palm Research in Thar Desert.* Scientific, Jodhpur.
Chundawat, B.S. 1990. *Arid Fruit Culture.* Oxford and IBH, New Delhi.
Daniels, R.B., E.E. Gamble, and L.A. Nelson. 1967. Relation between A2 horizon characteristics and drainage in some fine loamy ultisols. *Soil Sci.* 104:364–369.
Dawoud, D.H. 2001. Effect of different irrigation intervals on the growth rate, yield and fruit quality of MWL and MWK date palm cultivars. *Proceedings of the 2nd International Conference on Date Palm*, Al-Ain, UAE, p. 138. http://www.pubhort.org/datepalm/datepalm2/datepalm2-14.pdf
Doorenbos, J. 1976. Agro-meteorological field stations (FAO irrigation and drainage paper 27). Food and Agriculture Organization of the United Nations (FAO), Rome.
Doorenbos, J. and W.O. Pruitt. 1977. *Guidelines for Predicting Crop Water Requirements.* FAO Irrigation and Drainage paper No. 24, Rome, p. 144.
FAO. 2008. *Proceedings, Workshop on "Irrigation of Date Palm and Associated Crops".* Damascus, Syrian Arab Republic, May 27–30, 2007.
FAO. Stat. 2010. *Food and Agriculture Organization of the United Nations.* http://faostat.fao.org
Furr, J.R. and C.L. Ream. 1968. Salinity effects on growth on salt uptake of seedlings of the date *Phoenix dactylifera* L. *Proc. Am. Soc. Hort. Sci.* 92:268–273.
Furr, J.R. and W.W. Armstrong. 1962. A test of mature Hallawyi and Medjool date palms for salt tolerance. *Date Growers Inst. Report No.* 39:11–13.
Ghafoor, A., M. Qadir, and G. Murtaza. 2004. *Salt-affected Soils: Principles of Management.* Meraj Printers, Allied Bank Centre, Urdu Bazaar, Lahore, Pakistan.
Goosen, M.F.A. and W.H. Shayya (eds.). 1999. *Water Management, Purification and Conservation in Arid Climates.* Technomic Publishing Co., Lancaster, PA, p. 36.
Gupta, I.P. and S.P. Yadav. 1986. Crop tolerance to saline irrigation waters. *J. Indian Soc. Soil Sci.* 34:279–286.
Hanson, B., S.R. Grattan, and A. Fulton. 1999. *Agricultural Salinity and Drainage.* University of California Division of Agriculture and Natural Resources Publication, Oakland, 160 pp.
Hull, P.J. 1981. A low pressure irrigation system for orchard tree and plantation crops. *Agric. Eng.*, Summer:55–58.
Jaynes, D.B. and R.C. Rice. 1993. Transport of solute as affected by irrigation method. *Soil Sci. Soc. Am. J.* 57:1348–1353.
Keller, A., J. Keller, and D. Seckler. 1996. *Integrated Water Resource Systems: Theory and Policy Implications.* Research Report. International Water Management Institute, Colombo, Sri Lanka.
Khalifa, E.M., M. Reda, and M.H. Al-Awajy. 1989. Changes in soil fabric of torripsamments under irrigated date palms Saudi Arabia. *Geoderma* 44:307–317.
Mass, E.V. and S.R. Grattan. 1999. Crop yields as affected by salinity. In: R.W. Skaggs and J. Van Schilfgaarde (eds.). *Agricultural Drainage Agronomy Monograph 38.* ASA, CSSA, SSSA. Madison, WI, pp. 55–110.
Molden, D., H. Murray-Rust, R. Sakthivadivel, and I. Makin. 2003. A water-productivity framework for understanding and action. In: Kijne, W., R. Barker and D. Molden (eds.). *Water Productivity in Agriculture: Limits and Opportunities for Improvements.* CAB International, Wallingford, UK, 1018 pp.
Molden, D., R. Sakthivadivel, C.J. Perry, C. Fraiture, and W.H. Kloezen. 1998. *Indicators for Comparing Performance Indicators of Irrigated Agricultural Systems.* Research Report 20. International Water Management Institute, Colombo, Sri Lanka.
Morris, J.R. 1999. Evaluating drip irrigation in eastern vineyards. *Water Effeciency* 24(4):16–21.
Nakayama, F.S. and D.A. Bucks. 1986. *Trickle Irrigation for Crop Production: Design, Operation and Management.* Elsevier Science Publishers B.V., Amsterdam, The Netherlands, p. 383.
Nimah, M. 1985. Localized versus drip irrigation system. *Proceedings of the Third Drip Irrigation Congress*, Fresno, California, USA. pp. 552–554.
Norman, W.R., W.H. Shayya, A. Al-Ghafri, and I.R. McCann. 1998. Aflaj irrigation and on-farm water management in northern Oman. *Irrigation Drainage Systems* 12:35–38.
Penman, H.L. 1948. Natural evaporation from open water, bare soil, and grass. *Proc. Roy. Soc. London* A193:120–146.
Polak, P. and R. Yoder. 2006. Creating wealth from groundwater for dollar-a-day farmers: Where the silent revolution and the four revolutions to end rural poverty meet. *Hydrogeol. J.* 14:424–432.
Reuveni, O. 1971. Trickle irrigation of date palms. *Ann. Date Growers Inst. Rep.* 48:16–19.
Reuveni, O. 1974. Drip versus sprinkler irrigation of date palms. *Ann. Date Growers Inst. Rep.* 51:3–5.
Salam, M.A. and S. Al Mazrooei. 2007. Crop growth and irrigation water requirment of datepalm (*Phoenix dactylifera*) in the loamy sands of Kuwait. *Acta Hort. (ISHS)* 736:309–315.

Salter, P.J. and J.E. Goode. 1967. *Crop Responses to Water at Different Stages of Growth.* Commonwealth Agricultural Bureaux, Farnham Royal, Bucks, England, pp. 177–179.

Sehgal, J.L. and G. Stoops. 1972. Pedogenic calcic accumulation in arid and semi-arid regions of the Indo-Gangetic alluvial plain of the Erstwhile Punjab (India). Their morphology and origin. *Geoderma* 8:59–72.

Sharma, B.D., S.S. Mukhopadhyay, and P.S. Sidhu. 1997. Microtopographic controls on soil formation in the Punjab Region, India. *Geoderma* 81:357–368.

Sleeman, J.R., 1963. Cracks, peds and their surfaces in some soils of the Riverine Plain, N.S.W. *Aust. J. Soil Res.* 1:91–102.

Siebert, S., M. Nagieb, and A. Buerkert. 2007. Climate and irrigation water use of a mountain oasis in northern Oman. *Agric. Water Manage.* 89:1–14.

Smith, M., R.G. Allen, J.L. Monteith, L.S. Pereira, A. Perrier, and W.O. Pruitt. 1992. Report on the expert consultation on procedures for revision of FAO guidelines for prediction of crop water requirements. Land and Water Development Division, United Nations Food and Agriculture Service, Rome, Italy.

Soil Survey Manual. 1993. *Soil Survey Manual.* United States Department of Agriculture, Natural Resources and Conservation Service (NRCS), p. 315.

Terasaki, H., T. Fukuhara., M. Ito, and C. He. 2009. Effect of gravel and date palm mulch on heat moisture and salt movement in a desert soil. *Adv. Water Resour. Hydraulic Eng.* 1:320–325.

Van der Gulik, T.W. 2005. *Landscape Irrigation Scheduling Calculator*, IIABC, Abbotsford, Canada.

Walid, A.A. and Y.M. Abdelhadi. 1990. Effect of salt tolerancy levels on farming of date palm varieties: A case study. *Arid Land Res. Manage.* 4(4):269–272.

Wilkinson, J.C. 1983. The origins of the Aflaj of Oman. *J. Oman Stud.* 6:177–194.

Yaaqoob, A. 1996. Date palm irrigation; abstracted review article by: Achtrich, W. (translation from German language). Bassel Al-Assad. *J. Agric. Eng. Sci.* 1(1):85–88 (in Arabic).

Zaid, A. and E.J. Arias-Jimenez. 1999. *Date Palm Cultivation.* FAO Irrigation and Drainage paper No. 156, FAO, Rome, 287 pp.

5 Fertilizer Application in Date Palm Cultivation

Ahmed Al-Busaidi

CONTENTS

INTRODUCTION

Date palm (*Phoenix dactylifera* L.) has a wide ecological range, including desert oases, mangrove margins, steep limestone cliffs, and cultivated lands. It can thrive on a wide range of soil types, from light sandy to heavy clay soils, from waterlogged soils to extremely arid soils, and it can tolerate up to 1.5% salinity (Amer 2000).

There is lack of information on date palm fertilization in spite of the fact that date palm is grown in the desert soil that is very poor in fertility. However, date palm fertilization produces variable results according to location, type of fertilizer, cultivar, and time of application. The chemical composition of dates can vary depending on cultivar, soil conditions, agronomic practices, as well as the ripening stage (Al-Hooti et al. 1997; Myhara et al. 1999; Al-Farsi et al. 2005; Ismail et al. 2006). Such huge reductions in soil mineral contents must be compensated annually by adding fertilizers to maintain high yield and good quality of dates. Organic manuring can play a valuable role, as organic materials supply all plant nutrients. However, their content may be low if the organic material is derived from a deficient soil. Furthermore, organic supplies are often limited, and this may be exacerbated as farm size diminishes, and organic manuring is often costly in terms of labor. Many researchers have reported that organic manure is not sufficient to provide the palm its requirements to produce good yield (Al-Bakr 1982; Bacha and Abo-Hassen 1983). Most of the nutrients required for future high-yielding agriculture will have to come from mineral fertilizers, providing a range of nutrients, not merely one or two (FAO 1995). This requires availability, understanding by farmers and extension workers, availability of reliable information, etc. Therefore, alternative sources of nutrients to the commercial fertilizers will continue to be a subject of interest.

ROLE OF FERTILIZERS

Date growers in many parts of the world believe that the date palm remains productive as long as it keeps receiving water and organic manure. The production of good-quality dates, however, depends on adequate irrigation, fertilization, disease protection, pollination, and harvesting, as well as post-harvest handling techniques (Osman 1995; Iqbal et al. 2004; MAF 2005).

Organic or animal manures are widely used in most date gardens of the Arabian Peninsula and North Africa. The quantities of animal manure used naturally vary to a great extent. The application is in the range of 5–15 tons per acre. Sheep manure is preferred because of its relatively higher nitrogen content (about 2%) than that of cow or horse manure (0.5%–1%). Chicken manure (nitrogen content—3%–5%) is also being used when available. Inorganic chemical fertilizers have been widely accepted as a major source of improving and maintaining soil fertility. Palm trees need sufficient nutrients in proper balance for normal growth and development. Sixteen chemical elements are known to be essential for date palms. Except carbon, hydrogen, and oxygen, the remaining 13 elements are taken up by plants during development and crop production (Hussein 2004). The amount applied should be determined on the basis of whether it is used alone or in combination with manure or cover crops. Results of some studies clearly show the vital role of microelements, such as Zn, Cu, Fe, B, and Mn, as well as their combined effect, whether used in organic or inorganic form in improving the total yield and fruit quality of the date palm. Fertigation has been developed through the addition of fertilizers to irrigation water. Although fertilization and fertigation have proved to improve yield and fruit quality, more investigations are needed to study the nutrient concentration in soil solution, labor and energy saving, and flexibility in application time in relation to crop demand, regardless of the growth stages of palm and fruits (Hussein 2004).

ELEMENTAL DEPLETION

It has been reported that large amounts of macro- and micronutrients are depleted annually from the soil as a result of heavy crop production/yields and pruning of leaves (Hass and Bliss 1935; El-Shurafa 1984). Attalla and Harraz (1996) analyzed (dry weight basis) elemental composition and organic constituents of pits of 11 selected date palm cultivars grown in the Qassim region, Saudi Arabia. Regarding the elemental composition, their concentration ranges in the pits of the studied cultivars were as follows: nitrogen (N) 0.81%–1.20%, phosphorus (P) 0.19%–0.26%, potassium (K) 0.36%–0.40%, calcium (Ca) 0.36%–0.42%, magnesium (Mg) 0.10%–0.15%, chloride (Cl) 0.27%–0.51%, sodium (Na) 0.03%–0.04%, iron (Fe) 124.8–172.0 $\mu g\ g^{-1}$, manganese (Mn) 17.0–24.8 $\mu g\ g^{-1}$, zinc (Zn) 8.3–18.3 $\mu g\ g^{-1}$, and copper (Cu) 8.8–17.3 $\mu g\ g^{-1}$, while the concentrations of organic constituents were: total carbohydrates 57.65%–68.92%, total sugars 37.73%–58.30%, total proteins 5.11%–7.52%, crude fats 8.67%–12.31%, and tannins 1.59%–3.48%. Significant differences were also observed among the studied cultivars in most of their mineral and organic constituents. This study was supported by El-Shurafa (1984) when the annual loss of dry matter from a date palm carrying 10 bunches of 100 kg weight was calculated as 82.4 kg. The dry matter contained 472 g N, 48 g P, 427 g K, 219 g Ca, 36.4 g Na, 5.8 g Fe, 1.2 g Mn, and 1.3 g Zn. It was found that most of the nitrogen, phosphorus, and potassium (NPK) was removed by fruits, whereas most of the Ca, Na, and Fe was removed by the leaves.

Generally, the composition of date fruits keeps on changing during the various fruit development stages, internationally known by their Arabic names: *Kimri* (unripe), *Khalal* and *Bisir* (full size, crunchy), *Rutab* (ripe, soft), and *Tamr* (ripe, sun-dried). Each stage usually consumed many nutrients and such losses in the nutrient elements need to be substituted annually. The study of Sawaya et al. (1982) indicated a rapid reduction in the concentration of Ca, Mg, K, and Na, as the fruit develops from Kimri to Bisir. This reduction was slower between other stages. Sodium and potassium increased between Bisir and Rutab, and decreased between Rutab and Tamr. However, Ca and Mg showed a slight increase between Bisir and Tamr. Zn, Cu, Fe, and Mn were higher in the Kimri stage, then decreased afterwards. A recent study (El-Mardi et al. 2005) showed similar trend in the mineral concentration. Hass and Bliss (1935) analyzed six cultivars of dates for mineral content. They found that the potassium content of these cultivars was approximately three times that of sodium and more than 10 times the amount of phosphorus. They also calculated the amount of nutrient elements removed by one Deglet Noor palm bearing 9000 fruits as 242, 41, and 583 g of N, P, and K, respectively.

Chemical Fertilizer

Limited studies have reported the effect of fertilizer applications on the chemical characteristics of date fruits (Bacha and Abo-Hassan 1983; Al-Juburi 1995). However, it has been proved that fertilization is very important for the date palm to improve the growth and yield. Usually, fertilizer regime does not show its effect on date palm tree in the short run. Al-Ghamidi et al. (1999) reported that date palm seedlings did not indicate any significant differences in growth parameters in response to urea application in the nutrient culture. However, root elongation was significantly affected. A clear effect of N sources on leaf N, and Ca and chlorophyll content was not observed. Mekki et al. (1998) recommended an increase in N, P, K, and Mg from the 1st year of planting till the 8th year at which the increase can go to 10 times. In order to prove the effect of fertilizer on date palm, long-term treatments are needed to influence the palm at the productive stage. Thus, type, quantity, and time of fertilizer application can vary according to the type of soil and the development stage of the palm. The fruits remove most of the N, P, and K, whereas the pruned leaves removed most of the Na, Ca, Fe, and Zn.

Kolsi-Benzina and Zougari (2008) studied the mineral composition of date palm leaflets. At pollination, three palms per palm tree were taken at the median position, the lower green palm, and the last open palm. Leaflets were analyzed on the entire palm, split into four zones. The median zone leaflet nitrogen and magnesium contents were the highest for all the palm positions. A similar nitrogen and magnesium determination was noticed along each palm. The highest calcium content was at the base of the youngest palm and at the apex of the oldest palm; phosphorus content was the reverse. The highest potassium content was determined in the base palm leaflets. For foliar diagnosis, median zone leaflet sampling of median palm at pollination is recommended. Moreover, and in order to adjust the level of the lost nutrients, source fertilizers must be applied annually. Therefore, application of fertilizer should be based on the physical characteristics and chemical composition of the soil as well as the analysis of leaves and fruits.

Nitrogen, Phosphorus, and Potassium Fertilizers

Uptake of nitrogen by date palm trees did not exceed 60% of that added in fast-release N fertilizer. These drawbacks of the fast-release N fertilizers have promoted a search for a fertilizer that is characterized by lower loss of their own nutrients via drainage water. Using a slow-release fertilizer was proved to be very favorable in improving growth, nutritional status of the trees, fruit set, yield, and fruit quality of date palms. Little work has been done on such slow-release N fertilizers. Much more information is still needed for assessing the appropriate source, the actual and optimum rates, and application time for gaining maximum productivity in date palm trees (Hussein 2004). Al-Dekaili and Al-Dejaili (1989) indicated that date palm is sensitive to fertilizer addition, especially to nitrogen fertilizer. They reported that the average amount of nitrogen needed by date palm ranges from 1.8 to 2.7 kg yearly. This amount of nitrogen should be added in three doses during the year.

An increase in nitrogen was found to increase the yield of date palm (Al-Juburi 1993; Kassem et al. 1997; Soliman and Osman 2003) like potassium element (Abdullah et al. 1987; El-Hamaady et al. 1994; Kassem et al. 1997). It has been found that addition of both nitrogen and potassium to the fertilizer regime increased yield and bunch weight, whereas application of small amounts of nitrogen and a copious amount of potassium increased the weight of the fruit, length, and diameter (Soliman and Osman 2003).

Bacha and Abo-Hassan (1983) investigated the effect of seven fertilizer treatments consisting of NPK fertilizers and organic manure on yield and fruit quality of Khudari date palm variety. They reported an increase in yield of Khudari palm with an application of chemical fertilizer as compared with organic manure. They found that an increase in nitrogen level increased the yield of Khudari in Saudi Arabia. The percentage increase in yield over the control can be seen when N was added with organic matter alone or in combination with organic matter, P_2O_5, and K_2O. They also reported that the addition of chemical fertilizers increased the weight, volume, diameter, and length of Khudari fruits as compared to the addition of organic manure only (Table 5.1).

Furr (1951) studied the influence of fertilization treatments on yield and rate of gain in the palm trunk of Deglet Noor during a 7-year period. He reported that during the first 3 years of the experiment, nitrogen fertilization had no effect on growth or yield, but during the last 4 years of the experiment the fertilized trees had about 20% greater rate of trunk growth and yield of fruit than the unfertilized trees. Furr (1951) also showed that the addition of 2.7–3.6 kg of nitrogen to date palm Deglet Noor resulted in 20% increase in date production as compared with unfertilized trees through 7 years of experiment. The trunk growth as well as leaves and number of bunches/tree and number of fruits/bunch but not fruit weight were also increased, while the quality of fruits did not change or was even lowered (Table 5.2).

In another study, Furr (1957) introduced a comparison on the effect of equal amounts of nitrogen from manure and from chemical fertilizer on Deglet Noor. Differences in average yield were not significant because of the long-term effect of manure alone versus chemical fertilizer alone. Furr (1957) suggested that heavy fertilization improved growth production and yields of the palm trees. The reduction in crop may improve the quality of fruit produced by the fertilized trees. Application of nitrogen fertilizers in excess of the need of the trees is largely wasted because of the relatively rapid decomposition of organic matter and conversion of nitrogen into other forms.

Al-Juburi (1995) reported that the application of nitrogen fertilization did not show significant difference in N contents of Khassab fruit among various treatments during the first and second growing seasons, except at the fourth stage in the second growing season, where fruit nitrogen increased significantly for all treatments as compared with the control. Nitrogen levels in date palm fruit decreased significantly with age for all treatments in the first growing season, whereas *N* decreased

TABLE 5.1
Effect of Chemical Fertilizer on Khudari Cultivar Yield in 1980 and 1981

	Year 1980		Year 1981	
Treatment/Tree[a]	Yield (kg/palm)	Increase over Control (%)	Yield (kg/palm)	Increase over Control (%)
25 kg Org (control)	75.00		73.50	
500 g N + 25 kg Org	78.50	4.67	80.80	9.93
1000 g N + 25 kg Org	84.50	12.67	85.00	15.70
1500 g N + 25 kg Org	87.00	16.00	87.50	19.05
500 g N + 500 g P_2O_5 + 500 g K_2O + 25 kg Org	79.50	6.00	81.00	10.20
1000 g N + 500 g P_2O_5 + 500 g K_2O + 25 kg Org	85.00	13.30	87.00	18.40
1500 g N + 500 g P_2O_5 + 500 g K_2O + 25 kg Org	88.00	17.30	88.80	20.80

Source: Bacha, M. and A. A. Abo-Hassan. 1983. *Proceedings of the 1st Symposium on the Date Palm in Saudi Arabia*, Al-Hassa, Saudi Arabia, pp. 174–179.

[a] Nitrogen (N), phosphorus (P_2O_5), and potassium (K_2O) fertilizers, Org = organic manure.

TABLE 5.2
Effect of Fertilization on Yield and Trunk Growth of Deglet Noor Dates during Seven-Year Period

	Crop Year								
Treatments	**1944**	**1945**	**1946**	**Mean**	**1947**	**1948**	**1949**	**1950**	**Mean**
				Mean yield per tree (pounds)					
With nitrogen	228	114	238	193	179	260	172	231	210
Without nitrogen	239	94	243	192	158	202	146	193	175
F value				NS					*
				Mean annual gain in trunk height (feet)					
With nitrogen	2.90	2.20	1.80	2.28	1.90	1.20	1.60	1.70	1.62
Without nitrogen	2.90	2.10	1.70	2.23	1.60	0.80	1.20	1.70	1.33
F value				NS					**

Source: Furr, J. R. 1951. *Date Growers Institute* 20:17–20.
Note: * and ** = significant at the 0.05 and 0.01 probability level, respectively; NS = nonsignificant.

in the second and third stage and increased slightly in the fourth stage. Application of nitrogen fertilization did not show significant effect on P concentration. The means of fruit P concentration at the first stage for all treatments were significantly higher than the second and third stage of fruit development during the first growing season, and second, third, and fourth stages of development during the second growing season. Khassab fruit K at all four stages of development did not show any response to the various concentrations of nitrogen applications of either four or six equal doses a year compared with fruit of untreated date palm trees during the first and second growing season. The means of fruit K level tended to be significantly high in the first stage of development, and then progressively and significantly decreased with fruit age throughout the two successive seasons. Application of 600 g nitrogen/tree/year in four equal doses produced significantly higher levels of Khassab fruit Mn at the first stage of development compared with control, whereas the application of 300 or 600 g N/tree/year in four or six equal doses respectively reduced significantly fruit Mn levels as compared with control during the first growing season. Fruit Mn date palm trees did not respond to application of various concentrations of nitrogen fertilizers during the second growing season. Application of different concentrations of nitrogen fertilizers in four or six equal doses/tree/year did not show significant increase of fruit Zn levels in Khassab, whereas fruit Zn levels reduced significantly and progressively with fruit age for all treatments during the first growing season. Application of 600 g N either in four or six equal doses/tree/year increased significantly the Fe level at the first stage of fruit development during the first growing season compared with control (Al-Juburi 1995).

Abdullah et al. (1987) found that an increase in potash fertilization (K_2O) increased palm yield (Hayani cultivar) in Egypt when applied in equal doses during March, May, and July than when applied in March and July. They also found that the application of 1000 g of K_2O per palm/year in three doses in March, May, and July resulted in higher yield, heavier fruit, higher dry matter, and total soluble solids (TSS) percentages than the application in two equal doses during March and July. These results indicate the need for studying the response of the palm to the time of fertilizer application.

Saleh (2008) showed that injection of 25 g $FeSO_4$ per tree into the trunk caused the best results. Other injection treatments were placed in second class. Localized placement of iron fertilizer was the next treatment from the viewpoint of desirable effects. Finally, soil surface application appeared as an inept method because it was statistically similar to control treatment, in most cases.

DATE PALM BEHAVIOR UNDER SALINE ENVIRONMENT

The impact of salinity on nutrient uptake is a secondary negative effect on growth of plants. The passive nutrient uptake is related to water uptake and any decrease in water availability (osmotic effects of salinity) should be reflected in reduced nutrient uptake. The imbalance in composition of saline soil solution can affect the absorption of nutrients, for example, excess of Cl, Na, or Mg ions decreases the uptake of NO_3, K, Zn, and Ca ions. It has been further suggested that the ability of a plant to retain K in the presence of excess Na ion could be an important mechanism in salt tolerance (Ghafoor et al. 2004). Ion imbalances and nutrient deficiency, particularly for K^+ nutrition, can also occur (Tejera et al. 2006). According to criteria of Maas (1986), date palm plant is included in the group that is tolerant to salinity. Their salinity threshold for reduction of yield is 4.0 dS m^{-1} and has 50% reduction in yield at 17.9 dS m^{-1} (Figure 5.1). Good yield has been observed where the electrical conductivity (EC) of ground water is around 10.0 dS m^{-1}. However, salt tolerance level can vary a lot depending upon the texture and drainage of soil, climate, and the genetic potential of the variety. Ayers and Westcot (1985) proposed the threshold of date palm as 4.0 and 2.7 dS m^{-1} for soil and water, respectively, whereas 50% yield reduction limits were found by them as 18.0 and 12.0 dS m^{-1}, respectively. Marcar et al. (1995) reported that date palm, depending upon the variety, can tolerate up to at least 18,000 ppm salts or even more. However, fruit production usually stops at about 10,000 ppm salts. The salt marsh *Phoenix reclinata* of tropical Africa can tolerate full-strength ocean water, but it does not produce fruit.

Al-Mansoori (2001) conducted research to screen date palm cultivars for salt tolerance and to investigate the physiological mechanisms involved in this tolerance. Different levels of tissue organization and stages of plant life cycle, namely, callus induction, callus fast-growing phase, seed germination, and 1-month-, 3-month-, and 2-year-old plants, were investigated under *in vitro* and *in vivo* conditions. Different cultivar performances were evident at the cellular and the whole-plant levels, without counting the 2-year-old plants. The results suggest that *in vitro* screening for salinity tolerance is possible in date palm. However, this cannot be conclusive, as the 2-year-old plants did not show any differential cultivator performance. The characters showing variation were inconsistent and insufficient to assort the cultivars according to their degree of tolerance. Nevertheless, callus induction using immature embryos may offer the potential of selection from the preexisting date palm gene pool for enhanced salt-tolerant cultivars or individuals.

The results also indicate that date palms may rely on inorganic ion accumulation to maintain osmotic adjustment. Proline accumulation in response to salinity was detected but its role in osmotic adjustment was not clear. The differential cultivar performance was not related to proline. Proline cannot, therefore, be used as an indicator for salinity tolerance in date palm. Whole-plant tolerance may be a result of Na^+ compartmentalization in the root and Na^+ exclusion from the

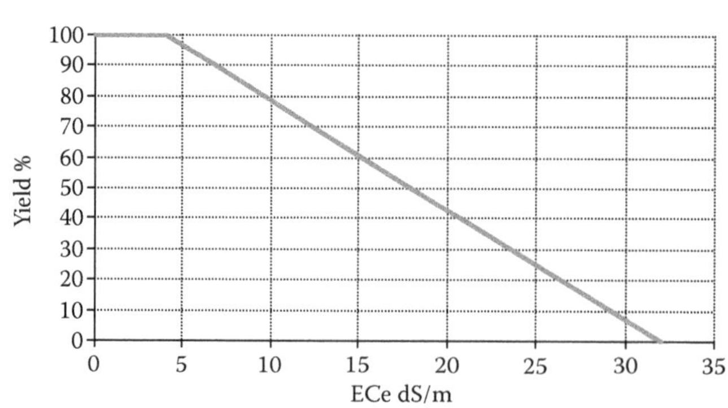

FIGURE 5.1 Date palm productivity with different soil salinity. (From Maas, E. V. 1986. *Appl. Agric. Res.* 1:12–26. With permission.)

shoot. However, the roots have specific holding capacity and exceeding this capability enhances Na^+ release into the shoot.

In another study conducted by Ahmed and Al-Mohemdy (2001) on four varieties of date palm trees in Kubaisa-Hit area, North West of Mesopotamian plain, the trees were irrigated with saline and fresh water for a long period. The study includes vegetative growth measurements as well as fruit yield. The results showed that irrigation with saline water (5–6 dS m^{-1}) caused salt accumulation in soil, which caused a reduction in tree growth and low yield as well as low quality of fruits. Different varieties were affected differently in this respect.

Physiological Responses of Date Palm to Salinity

Growth and physiological responses of date palm (variety: Barhee) callus to salinity stress were examined by Al-Khayri (2002) who found that, at 125 mM or higher level of NaCl, callus growth was completely inhibited. However, the lowest level (25 mM) increased the growth of date palm callus. Proline accumulation was correlated to callus growth inhibition. In response to increasing external NaCl level, the Na/K ratio increased. This ratio was positively correlated to proline accumulation and hence callus growth. Shani and Dudley (2001) related the yield loss to reduced photosynthesis, high energy, and carbohydrate expenses in osmoregulation, and interference with cell functions under saline conditions. Ghadiri et al. (2005) reported restricted water uptake by salinity due to the high osmotic potential in the soil and high concentrations of specific ions that may cause physiological disorders in the plant tissues and reduce yields.

Djibril et al. (2005) conducted a study to evaluate date palm tolerance to osmotic stress induced by polyethylene glycol (PEG) or NaCl during the early stages of plant development. Two varieties Nakhla Hamra (NHH) and Tijib widely cultivated in Mauritania were tested. Nakhla Hamra showed an increase in epicotyl length, primary root length, secondary root number, and proline content when water deficit was induced by PEG. In contrast, for the same developmental and biochemical characters, the Tijib cultivar was more tolerant in salinity stress. It has been reported that environmental stresses, especially water stress and salinity, increase the generation of toxic active oxygen species (O_2^-, H_2O_2, and OH) causing cellular injury or damage, which frequently results from impaired or perturbed metabolism (Sairam and Saxena 2000; Sreenivasulu et al. 2000). In another study, it was found that application of antistress substances, such as gamma aminobutyric acid, vitamin E, and salicylic and acetyl salicylic acids, increased the survival percentage of "Khalas" date palm plants during the acclimatization program (Awad et al. 2006).

Al-Khateeb et al. (2001) carried out a study on date palm trees at Al-Ojair in the coast of Arabian Gulf during 1999/2000. Ion and water relationships were investigated using trees of Khalas variety grown at Al-Hassa under field conditions as a control. The results revealed that high external EC in Al-Ojair resulted in greater internal Na^+ and Cl^- concentrations, compared with nonsaline media, which contributed to a decrease in leaf and root osmotic potentials. Concentrations of macronutrients represented levels adequate for growth even under saline conditions. All the trees showed no reduction in chlorophyll content even under saline conditions but the chlorophyll content decreased as the leaf aged. All date palm trees generated sufficiently larger water potential for import gradients of water with no significant differences in turgor pressure.

The study by Al-Hammadi (2006) aimed to address the current status of the United Arab Emirates date palms. In a greenhouse experiment, optimal growth was found in control and 3000 ppm of NaCl. Relative growth rate (RGR), biomass, and NL decreased significantly by increasing salinity to 6000, and to 12,000 ppm. However, no significant differences were observed in the average RGR for any cultivars. Increased NaCl led to significant decreases in K^+, Mg^{2+}, and Ca^{2+} contents of plants. The Na:K ratios were lower in shoots than in roots. Lulu, Fard, Khunaizy, Nabtat Safi, and Razez cultivars showed higher RGR and biomasses, whereas Khunaizy, Mesally, and Safri had higher Na:K ratios than other cultivars in the control, indicating higher Na^+ discriminations from plant parts.

A study was conducted by El-Sharabasy et al. (2008) in 2005–2006 to evaluate an *in vitro* technique for salt tolerance of three date palm cultivars, namely, Samani, Sewy, and Bartamuda, cultured on MS medium with 0.0, 4000, 8000, and 12,000 ppm NaCl after 12 weeks for three subcultures. The effect of salinity on shoot length did not show significant differences under 0.0 ppm, but it increased significantly for the three cultivars, especially Bartamuda cv., to 4000 ppm, then decreased significantly at 8000 and 12,000 ppm NaCl. The number of shoots showed high significant value at 4000 ppm for the Bartamuda cv.; however, at 8000 and 12,000 ppm, it decreased and the differences between Sewy and Bartamda cvs. were insignificant. The number of burned leaves was the highest at 12,000 ppm NaCl for the Sewy and Samani cvs. as compared with Bartmuda cv. Significant differences were recorded between Bartamuda and the other two cultivars. The growth vigor of Bartmuda was still the highest even at 12,000 ppm NaCl.

Youssef and Awad (2008) examined photosynthetic gas exchange characteristics, salt uptake, pigment contents, and electrolyte leakage in the date palm seedlings subjected to seawater treatments at 1, 15, and 30 dS m^{-1} salinity levels in the presence or absence of 0.08% ALA-based (5-aminolevulinic acid-based) functional fertilizer commercially known as Pentakeep-v. Date palm seedlings accumulated significant amounts of Na^+ in the foliage with increasing salinity, about a threefold increase in the accumulated Na^+ between the control and 30 dS m^{-1} salinity treatment. Electrolyte leakage indicated a significant reduction in membrane integrity as salinity increased. ALA-based fertilizer Pentakeep-v improved salt tolerance and growth of date palm seedlings by increasing photosynthetic assimilation. The latter was mediated via boosting light-harvesting capabilities of the treated plants by increasing chlorophyll content and by reducing stomatal limitation to photosynthetic gas exchange.

Yield and transpiration of juvenile date palms (*Phoenix dactylifera* L., cv. Medjool) were studied by Tripler et al. (2007) under conditions of increasing salinity and boron (B) in lysimeters. Twenty seedlings were planted and grown in 20 lysimeters, and irrigated with combinations of four salinity and five B irrigation concentrations. A linear decrease was found for both yield and transpiration in response to increased soil saturated paste salinity for the treatments having lower B concentrations (0.0278, 0.185, and 0.4625 mmol L^{-1}). Yield and transpiration also decreased with increased B concentration. While increases in soil saturated paste B from 0.3 to 1.5 mmol L^{-1} caused substantial declines in yield and transpiration, subsequent increase of B to 3 mmol L^{-1} caused only minor reductions. Response to salinity and to excess B was witnessed from the lowest tested levels when each of the variables was isolated. Growth response to combined conditions of salinity and B behaved according to the dominance of the two stress-causing factors and did not show additive effects. The dynamics of plant water uptake and tree growth observed for salinity and boron occurring independently and together were found to decrease water uptake but not ion accumulation for NaCl and CaCl salts and by boron that was accumulated in leaves and subsequently was associated with reduced tree size. It is suggested that, while mechanisms for plant response to salinity are dominated by lowered soil water potential (osmotic stress), boron becomes toxic as it accumulates to a threshold level in plant tissue.

The impact of straw incorporation (6 Mg ha^{-1} $year^{-1}$) into agricultural soils compared with straw removal on organic matter mineralization and salinity was studied by Badi (2000). The mineralization coefficient (evolved CO_2/organic C ratio) was obtained to evaluate organic matter mineralization. Soil salinity was measured as means of electrolytic conductivity of saturation paste extract. Both parameters were measured seasonally during 2 years in two salt-affected soils of the semiarid Central Ebro Valley (northeast Spain), a saline soil and a saline sodic soil.

Electrolytic conductivity (ranging from 2.5 to 24.3 dS m^{-1}) and mineralization coefficient (ranging from 5.9×10^{-4} to 37.9×10^{-4} day^{-1}) varied widely during seasonal samplings of both soils. The lowest electrolytic conductivity values coincided with the highest mineralization coefficient values. Straw mulching and burying decreased significantly the average seasonal electrolytic conductivity of both soils: 2.5 times in the saline soil and 1.9 times in the saline sodic soil. The EC reduction only increased significantly ($p < 0.05$) the mineralization coefficient on saline soil (1.6 times). Straw amendment, followed by rainy periods, allowed soluble salts leaching but did not modify the sodium content significantly. Infiltration, water aggregate stability, and CO_2 were improved with straw amendment, but only in saline soil. Soil differences showed the existence of osmotic and specific ion effects.

FERTILIZATION FOR DATE PALMS IN OMAN

Date palm is the primary crop in Oman covering 49% of the cultivated areas and 82% of all fruit crops grown in the country. It is considered an integral part of Oman's national heritage and social life. Oman is currently the eighth largest world producer of dates. According to the Ministry of Agriculture and Fisheries (MAF), the production of dates during the year 2007 was estimated to be about 255,871 metric tons (MAF 2007). Oman has diverse topographical and climatic ecoregions that allow for cultivation of various types of date palm cultivars, particularly in the northern coastal and the interior regions. There are approximately 180 female and 48 male cultivated varieties of the 7.8 million trees of date palm. Despite the great diversity of the cultivars, over 78% of the total production is only from 10 commercial cultivars. These cultivars are dominant because of their marketable high fruit quality, or early and late season production. Physical phenotypic diversity index of selected date palm cultivars indicated large biodiversity among the population (Al-Yahyai and Al-Khanjari 2008).

There are few studies on the fertilization rates for date palm in the Sultanate. The rate usually varies with soil type and size of the palms (Al-Yahyai 2007). However, there is lack of information on date palm fertilization in spite of the fact that date palm is grown in the desert soil, which is very poor in fertility. Poor soil fertility in these areas is due to high temperature and dry weather, besides its tendency to be sandy with rapid leaching of the fertilizer applied. Organic matter is also easily lost in the desert and semidesert areas.

Fertilization in the past was only based on green and animal manure. Date growers in Oman used to apply animal manure and mix it with straw at the rate of two donkey-loads per tree twice a year. Intercropping of alfalfa, which is common in most farms, contributed a lot in providing nitrogen to the soil and, subsequently, to the trees. The fibrous root system of the date palm allows them to uptake nutrients from a wide range of soil layers. The Omanis have practiced date palm orchard operations for a long time and have gained valuable experience on pollination, irrigation, fertilization, and postharvest techniques. Such operations are still carried out in the traditional way. However, modern agricultural technologies should be able to bring changes to these traditional operations. Date cultivars grown in the Sultanate, such as Khalas, Khuneizi, Khassab, and Zabad, are of high quality that can compete with the best cultivars in the world.

Most of the mineral fertilizers have been used very little in date palm-growing regions. Results of the 1994 Agricultural Census (MAF 1994) showed that 74% of date growers in Batinah region in the Sultanate of Oman used only manure as fertilizer and 13.7% used mineral fertilizers. Moreover, it was recorded in MAF reports that around 86% of date growers in the Batinah area did not use mineral fertilizer (MAF 1994). In this region, soil fertility is crucial to obtain and sustain high crop yields. Serious depletion of nutrients has already taken place, involving not only the macronutrients but also micronutrients. Soil type in Batinah is small and sandy gravelly; it therefore requires frequent irrigation and fertilization with organic matter and minerals because of rapid leaching.

RESPONSE OF DATE PALM TO DIFFERENT FERTILIZER APPLICATIONS: A CASE STUDY

A study was conducted during 2002–2004 to increase yield, improve quality, and optimize fertilization (El-Mardi et al. 2005). Thirty date palms from each of Khalas and Khassab cultivars at 6–7 years of age were selected randomly in Al-Fairooz farm in Khabourah in Batinah. Experiment treatments applied in factorial design included: (1) control, (2) organic peat, (3) 1000 g urea divided into five and four doses mixed with 500 g triple superphosphate and 800 g K_2SO_4, and (4) with or without micronutrients. The control included no organic or chemical fertilizer. Applied treatments will be abbreviated in the text as follows: NPK = urea, superphosphate, potassium sulfate; (4-N) and (5-N) = 4 and 5 nitrogen doses; Mic = micronutrients; Org = organic peat.

The results showed that mineral fertilizer tended to increase Khassab strand weight and decrease fruit number. The highest number and yield associated with smallest length and diameter were produced by NPK + micronutrients and organic peat, whereas no significant differences were observed in Khalas in fruit number, weight, and measurements; however, NPK 5- or 4-N doses with micronutrients and organic peat produced the highest yield (Table 5.3). Four nitrogen doses with phosphate and K_2SO_4 resulted in an increase of Khalas yield by 22% over the control and 24.7% over organic peat. The same treatment increased Khassab yield by 77.5% over the control and 130% over organic peat (Table 5.4). The 5-N doses increased the yield of Khalas by 17.8% over the control and 20.4% over organic peat. On Khassab, it increased the yield by 63% over the control and 112.4% over peat. Five nitrogen doses in Khalas produced higher weight of strand, number of fruits per strand, weight and length of fruit, and yield than 4-N doses, whereas they produced lower strand weight, fruit length, and diameter, and higher yield and number of fruits per strand than the 4-N doses on Khassab. In both cultivars, fruits/strand, number, fruit weight, and diameter decreased from Bisir to Rutab, but in contrast to Khalas, yield in Rutab was higher than in Bisir. In 2004, both cultivars produced higher yield, fruit number, and weight/strand but lower fruit weight than in 2002 and 2003 (Table 5.5). Five nitrogen doses increased tannins, pectins, and pH over 4-N doses, but reduced percentage of titratable acidity (TA) in both cultivars.

There were no significant differences in the effect of fertilizers on the pectin, glucose, and fructose content of the fruit. Pectin content increased when the peat and mineral fertilizers were applied together. An increase in acidity was associated with a decrease in tannin and vice versa. There was a gradual increase in fructose when 5-N doses were added with K and P and organic peat, but not with 4-N doses, an indication of sweeter dates. Chemical fertilizers resulted in gradual increase in the dry matter percentage. It was the highest with four doses of urea + K + P and lower with the control and organic peat and five doses of urea. In both cultivars, tannin and moisture decreased

TABLE 5.3
Effect of Fertilizer on the Physical Characteristics of Khalas Fruits

	Date Strand		Individual Fruit			
Fertilizer	Fruit Weight (g)	Fruit Number	Weight (g)	Length (cm)	Diameter (cm)	Yield (kg/palm)
Control	103.10[a]	7.50[a,b]	13.46[a]	3.83[a]	2.40[a]	38.31[a,b]
Organic	84.39[a]	6.50[a,b]	14.65[a]	3.92[a]	2.45[a]	37.48[a,b]
NPK (5-N)	101.65[a]	7.40[a,b]	14.14[a]	3.85[a]	2.43[a]	40.15[a,b]
NPK (5-N) + Org	93.63[a]	6.80[a,b]	15.18[a]	4.00[a]	2.48[a]	34.57[a,b]
NPK (5-N) + Mic	97.06[a]	8.7[a]	16.44[a]	3.89[a]	2.43[a]	34.42[a,b]
NPK (5-N) + Org + Mic	85.07[a]	6.00[b]	14.38[a]	3.86[a]	2.37[a]	45.15[a,b]
NPK (4-N)	92.10[a]	6.60[a,b]	15.14[a]	3.73[a]	2.41[a]	31.2[b]
NPK (4-N) + Org	98.35[a]	7.40[a,b]	14.44[a]	3.77[a]	2.43[a]	31.97[a,b]
NPK (4-N) + Mic	96.44[a]	6.80[a,b]	14.14[a]	3.92[a]	2.44[a]	34.23[a,b]
NPK (4-N) + Org + Mic	88.53[a]	6.40[a,b]	14.78[a]	3.86[a]	2.44[a]	46.75[a,b]
±5-N dose from 4-N dose	+0.53	+6.25	+2.80	+2.09	−0.10	+7.03

Note: Numbers followed by the same letter in the same column are not significantly different at 0.05 levels. NPK = urea, superphosphate, potassium sulfate; (4-N) and (5-N) = 4 and 5 nitrogen doses; Mic = micronutrients; Org = organic peat.

TABLE 5.4
Effect of Fertilizer on the Physical Characteristics of Khassab Fruits

	Date Strand		Individual Fruit			
Fertilizer	Fruit Weight (g)	Fruit Number	Weight (g)	Length (cm)	Diameter (cm)	Yield (kg/palm)
Control	187.80[d]	18.39[c]	10.72[c,d]	3.35[a,b]	2.28[b,c]	55.86[b,c]
Organic	202.76[c,d]	19.84[d,c]	11.05[c]	3.34[a,b]	2.28[b,c]	44.29[c]
NPK (5-N)	244.82[a,b]	23.81[a,b]	10.89[c]	3.54[a]	2.28[b,c]	96.15[a,b]
NPK (5-N) + Org	216.11[b,c,d]	21.13[a,b,c,d,e]	10.95[c]	3.36[a,b]	2.28[b,c]	72.41[a,b,c]
NPK (5-N) + Mic	222.99[a,b,c]	22.79[a,b,c,d,e]	10.22[cd]	3.30[a,b]	2.25[c,d]	87.84[a,b]
NPK (5-N) + Org + Mic	208.42[c,d]	23.19[a,b,c,d]	9.40[d]	3.26[b]	2.18[d]	91.10[a,b]
NPK (4-N)	251.54[a]	20.38[c,d,e]	13.95[a]	3.53[a]	2.34[b,c]	61.58[a,b,c]
NPK (4-N) + Org	223.27[a,b,c]	23.31[a,b,c]	10.33[c,d]	3.36[a,b]	2.27[b,c]	62.50[a,b,c]
NPK (4-N) + Mic	228.52[a,b,c]	20.77[b,c,d,e]	12.29[b]	3.44[a,b]	2.43[a]	79.00[a,b,c]
NPK (4-N) + Org + Mic	245.67[a,b]	24.43[a]	10.02[c,d]	3.27[b]	2.23[c,d]	99.14[a]
±5-N dose from 4-N	−5.97	+2.28	−12.37	−1.03	−3.11	+14.98

Note: Numbers followed by the same letter in the same column are not significantly different at 0.05 levels. NPK = urea, superphosphate, potassium sulfate; (4-N) and (5-N) = 4 and 5 nitrogen doses; Mic = micronutrients; Org = organic peat.

TABLE 5.5
Plant Response to Fertilizer Availability with Time on Khalas Fruits

	Date Strand		Individual Fruit			
Year	Fruit Weight (g)	Fruit Number	Weight (g)	Length (cm)	Diameter (cm)	Yield (kg/palm)
2002	80.72[b]	6.42[b]	13.82[b]	3.89[a]	2.52[a,b]	39.91[b]
2003	92.17[a]	5.24[c]	16.94[a]	3.96[a]	2.65[a,b]	20.40[c]
2004	96.73[a]	7.67[a]	12.21[c]	3.64[b]	2.34[b]	51.31[a]

Note: Numbers followed by the same letter in the same column are not significantly different at 0.05 levels.

with mineral fertilizer. In contrast, TA, dry matter increased, whereas pectin and sugars followed a different trend. Pectins, tannins, and TA were lower in Rutab than in Bisir. In 2004, pectins and tannins were lower than in 2002 and 2003, but TA reduced in Khalas and increased in Khassab with time (SQU 1997; El-Mardi et al. 2005; Al-Kharusi et al. 2009).

Application of micronutrients with four or five doses of NPK with or without peat produced lower tannin than when they were excluded. On the other hand, micronutrients increased tannin in Khasab fruits when applied with NPK four doses with or without peat and it reduced tannin with the five doses NPK. In fruits of both cultivars, mineral fertilizer with or without organic peat tended to increase TA. Organic peat with NPK produced a more pronounced effect on increasing TA than micronutrients with NPK. Khalas showed no significant difference in fruits N, P, K, Na, Ca, Mg, Fe, Zn, Cu, and Mn content as influenced by fertilizer application in five and four doses compared to the control and organic peat. Similar results were obtained from Khasab except peat treatment, which produced significantly higher Ca than the other treatments, and Mg significantly higher than fertilizer in four doses. Likewise, it produced the highest Fe and Zn (SQU 1997; El-Mardi et al. 2005; Al-Kharusi et al. 2009).

It can be concluded that date palm in the Sultanate of Oman does not receive great attention with regard to fertilization, which has resulted in a reduction of date palm yield. Application of only organic peat or manure is not sufficient to optimize the date palm production for both Khalas and Khassab cultivars in Batinah region. However, a combination of both organic and inorganic fertilizers is recommended for higher yields and better physical and chemical fruit characteristics. Application of 1 kg of nitrogen, 500 g of phosphorus, and 800 g of potassium sulfate with trace elements per palm per year in four doses is recommended to maintain good fruit growth and development. It is recommended to apply nitrogenous and potash fertilizers in March, May, July, and October, and phosphorus in March and double the amounts of each in May, July, and October. Organic peat is needed in winter during flower bud transformation in November and January. Long trials on the use of mineral fertilizers in Batinah are also required. Nitrogenous fertilizers applied in great excess of the needs of the date palm are largely wasted because of leaching and the tendency to delay fruiting and enhance vegetative growth, so optimized application is necessary. Several years may be required for the effect of fertilizers to increase yield, because the growth of leaves and production of inflorescences must be increased first. It would be a good practice to provide each year increased nitrogen, phosphorus, and potassium supply adequate to sustain good yield and quality (SQU 1997; El-Mardi et al. 2005; Al-Kharusi et al. 2009).

CONCLUSIONS

Date palm is one of the most important fruit trees in Oman as well as in the Arab world. It can be affected by improper mineral nutrition, whether due to insufficient or incorrect fertilization. They also may exhibit certain nutritional disorders when they grow under saline conditions. Therefore, fertilizers are recommended in these situations. Little research exists on fertilizer application and management. Fruit growers are using some artificial fertilizers (such as ammonium nitrate or urea) for their higher nitrogen content and solubility, whereas other farmers are applying organic fertilizers to supply date palms with the essential elements. Moreover, using organic manure reduces some environmental risks caused by mineral fertilizers and minimizes the loss of nutrients by leaching. Application of only organic manure is not sufficient to optimize date palm production. A combination of both organic and inorganic fertilizers is recommended for higher yields, and better physical and chemical fruit characteristics.

In order to show the effect of fertilizers on date palm, long-term treatments are needed to influence the palm at the productive stage. Thus, type, quantity, and time of fertilizer application can vary according to the type of soil and the development stage of the palm. Moreover, to adjust the level of the lost nutrients, source fertilizers must be applied annually. Thus, application of fertilizers should be based on the physical characteristics and chemical composition of the soil, as well as the analysis of leaves and fruits.

ACKNOWLEDGMENT

The author would like to thank Dr. Salim Al-Rasbi (Directorate General of Agriculture & Livestock Research, Rumais, Ministry of Agriculture, Oman) and Ms Latifa Al-Kharusi (Department of Crop Sciences, College of Agricultural and Marine Sciences, Sultan Qaboos University, Oman) for their valuable contribution and support in completing this chapter.

REFERENCES

Abdullah, K. M., S. I. Gaafar, A. S. Khalifa, and A. M. El-Hammady. 1987. Influence of fertilization with potash on Hayany dates grown on sandy soil. *Ann Agric. Sci. Fac. Agric.* 32:649–656.

Ahmed, A. H. and A. F. Al-Mohemdy. 2001. Effect of water quality on the growth and yield of date palm (*Phoenix dactylifera* L.), *Proceedings of the Second International Conference on Date Palms*, Al-Ain, UAE, March 25–27, pp. 128–137.

Al-Bakr, A. 1982. *The Date Palm, Its Past and Present Status*, 2nd ed., Al-Satan Print Shop, Beirut, Lebanon, p. 1085.

Al-Dekaili, A. A. and J. A. Al-Dejaili. 1989. *Fruits Production*. Ministry of Higher Education Press, Mousil.

Al-Farsi, M., C. Alasalavar, A. Morris, M. Baron, and F. Shahid. 2005. Compositional and sensory characteristics of three native sun-dried date (*Phoenix dactylifera* L.) varieties grown in Oman. *J. Agric. Food Chem.* 53:7586–7591.

Al-Ghamidi, A. A., A. A. Hlal, and M. H. ALWhaibi. 1999. Nitrogen source and growth of date palm seedlings. *Pak. J. Biol. Sci.* 2:1207–1210.

Al-Hammadi, M. S. 2006. *Salt Tolerance and Current Status of the Date Palms in the United Arab Emirates*. Ph.D. dissertation, The University of Arizona, Arizona, USA, Publication number: AAT 3207858.

Al-Hooti, S., J. S. Sidhu, and H. Qabazard. 1997. Physiochemical characteristics of five date fruit cultivars grown in the United Arab Emirates. *Plant Foods Hum. Nutr.* 50:101–113.

Al-Juburi, H. J. 1993. *The Date Palm*, Al Ain Press, Faculty of Agricultural Sciences, UAE University, United Arab Emirates, pp. 65–84.

Al-Juburi, H. J. 1995. Analysis of minerals in date palm fruit under different nitrogen fertilizations. *Fruits* 50:153–158.

Al-Kharusi, L. M., M. O. Elmardi, A. Ali, F. A. J. Al-Said, K. M. Abdelbasit, and S. Al-Rawahi. 2009. Effect of mineral and organic fertilizers on the chemical characteristics and quality of date fruits. *Int. J. Agric. Biol.* 11:290–296.

Al-Khateeb, S. A., I. A. Abdulhadi, K. M. Al-Barrak, and A. A. Al-Khateeb. 2001. Ion and water relations of date palm trees grown in Al-Hass coast of Arabian Gulf of Saudi, *Proceedings of the Second International Conference on Date Palms*, Al-Ain, UAE, March 25–27, pp. 139–149.

Al-Khayri, J. M. 2002. Growth, proline accumulation and ion content in sodium chloride-stressed callus of date palm. *Vitro Cell Dev. Biol. Plant.* 38:79–82.

Al-Mansoori, T. A. 2001. *Salt Tolerance in Date Palm* (*Phoenix dactylifera* L.). Ph.D. dissertation, University of Reading, Reading, pp. 52–174 (BL: DXN061960).

Al-Yahyai, R. 2007. Improvement of date palm production in the Sultanate of Oman. *ISHS Acta Hort.* 736:337–343.

Al-Yahyai, R. and S. Al-Khanjari. 2008. Biodiversity of date palm in the Sultanate of Oman. *African J. Agric. Res.* 3:389–395.

Amer, W. M. 2000. Date palm (*Phoenix dactylifera* L.) cultivars in Egypt. *El Minia Sci. Bull.* 13:1–15.

Attalla, A. M. and F. M. Harraz. 1996. Chemical composition of the pits of selected date palm cultivars grown in the Qassim region, Saudi Arabia Arab. *Gulf J. Sci. Res.* 14:629–639.

Awad, M. A., A. A. Soaud, and S. M. El-Konaissi. 2006. Effect of exogenous application of anti-stress substances and elemental sulfur on growth and stress tolerance of tissue culture derived plantlets of date palm (*Phoenix dactylifera* L.) c.v. 'Khalas' during acclimatization, *J. Appl. Hort.* 8:129–134.

Ayers, R. S. and D. W. Westcot. 1985. *Water Quality for Agriculture*. Irrig. Drainage Paper 29, FAO, Rome, Italy.

Bacha, M. and A. A. Abo-Hassan. 1983. Effects of soil fertilization on yield, fruit quality and mineral content of Khudairi date palm variety. *Proceedings of the 1st Symposium on the Date Palm in Saudi Arabia*, Al-Hassa, Saudi Arabia, pp. 174–179.

Badi, D. A. 2000. Straw management effects on organic matter mineralization and salinity in semiarid agricultural soils. *Arid Soil Res. Rehabil.* 14:193–203.

Djibril, S., O. K. Mohamed, D. Diaga, D. Diegane, B. F. Abaye, S. Maurice, and B. Alain. 2005. Growth and development of date palm (*Phoenix dactylifera* L.) seedlings under drought and salinity stresses. *African J. Biotech.* 4:968–972.

El-Hamaady, A. M., A. S. Khalifa, and A. S. Montasser. 1994. Effect of potash fertilization on "Seewy" date palm. II. Effect on yield and fruit quality. *Hort. Abst.* 64:4035.

El-Mardi, M. O., A. E. Pillay, J. R. Williams, C. S. Bakheit, S. M. Hassan, M. Al-Hadabbi, and A. Al-Hamdi. 2005. Effect of pollination method on changes in physical and chemical characteristics of date fruit during development. *SQU J. Agric. Sci.* 10:5–12.

El-Sharabasy, S. F., W. H. Wanas, and A. Y. Al-Kerdany. 2008. Effect of salinity stress on some date palm cultivars during proliferation stage *in vitro*. *Arab J. Biotech.* 11:273–280.

El-Shurafa, M. Y. 1984. Annual loss of minerals from date palm. *Date Palm J.* 3:278–290.

FAO. 1995. *Integrated Plant Nutrition System* 12:20–178.

Furr, J. R. 1951. Irrigation and fertilization experiments with Deglet Noor dates. *Date Growers Institute* 20:17–20.

Furr, J. R. 1957. Nitrogen fertilization of dates: A review and progress report. *Date Growers Institute* 34:6–9.

Ghadiri, H., I. Dordipour, M. Bybordi, and M. J. Malakourti. 2005. Potential use of Caspian Sea for supplementary irrigation in North Iran. *Agric. Water Manag.* 79:209–224.

Ghafoor, A. G., M. Qadir, and G. Murtaza. 2004. *Salt Affected Soils: Principles of Management*. Institute of Soil and Environmental Sciences, University of Agriculture, Faisalabad, Pakistan, pp. 110–111.

Hass, D. E. and A. Bliss. 1935. Mineral nutrition of date palm. *Date Growers Institution Rep.* 7:19–21.

Hussein, F. 2004. Nutrients requirements of date palm and fertilizer use. *Proceedings of the Regional Workshop on Date Palm Development in the Arabian Peninsula*, Abu Dhabi, UAE, 29–31 May, pp. 168–172.

Iqbal, M., A. Ghafoor, and S. Rehman. 2004. Effect of pollination times on fruit characteristics and yield of date palm cv. Dhakki. *Int. J. Agric. Biol.* 6:96–99.

Ismail, B., I. Haffar, R. Baalbaki, Y. Mechref, and J. Henry. 2006. Physicochemical characteristics and total quality of five date varieties grown in the United Arab Emirates. *Int. J. Food Sci. Technol.* 41:919–926.

Kassem, H. A., M. B. El-Shabrout, and M. M. Attia. 1997. Effect of nitrogen and potassium fertilization on yield, fruit quality and leaf mineral content in some Egyptian soft date varieties. *Alex. J. Agric. Res.* 42:137–157.

Kolsi-Benzina, N. and B. Zougari. 2008. Mineral composition of the palms leaflets of the date palm. *J. Plant Nut.* 31:583–591.

Maas, E. V. 1986. Salt tolerance of plants. *Appl. Agric. Res.* 1:12–26.

Marcar, N., D. Crawford, P. Leppart, T. Jovanvic, R. Floyl, and R. Farrow. 1995. *Trees for Saltland: A Guide to Select Native Species for Australia*. CSRIO, East Melbourne, Victoria, p. 72.

Mekki, M. I., A. M. Hamouda, and A. S. Al Abri. 1998. *Date Palm*. 1st ed., The Royal Sultani Deewan, Muscat, Oman, pp. 70–350.

Ministry of Agriculture and Fisheries (MAF). 1994. *Agriculture Census*. Ministry of Agriculture and Fisheries, Muscat, Oman.

Ministry of Agriculture and Fisheries (MAF). 2005. *Agriculture and Fisheries Research Report*. Ministry of Agriculture and Fisheries, Muscat, Sultanate of Oman.

Ministry of Agriculture and Fisheries (MAF). 2007. *Agricultural and Fisheries Research Report*. Ministry of Agriculture and Fisheries, Muscat, Oman.

Myhara, R. M., J. Karkalas, and M. S. Taylor. 1999. The composition of maturing Omani dates. *J. Sci. Food Agric.* 79:1345–1350.

Osman, A. M. 1995. Date-palm production and protection in the Arab countries. *Expert Consultation on Date Palm Pest Problems and Their Control in the Near East*, April 22–26, Al-Ain, United Arab Emirates.

Sairam, R. K. and D. C. Saxena. 2000. Oxidative stress and antioxidants in wheat genotypes: Possible mechanism of water stress tolerance. *J. Agron. Crop Sci.* 184:55–61.

Saleh, J. 2008. Yield and chemical composition of 'Piarom' date palm as affected by levels and methods of iron fertilization. *Int J. Plant Prod.* 2:1735–6814.

Sawaya, W. N., W. M. Safi, A. A. Al- Shalhat, and H. Al-Mahmmod. 1982. Growth and development of Khudari, Sillaj and Sifri date cultivars grown in Saudi Arabia. *First Symposium on Date Palm*, March 23–25, King Faisal University, Al-Hssa.

Shani, U. and L. M. Dudley. 2001. Field studies of crop response to water and salt stress. *Soil Sci. Soc. Am. J.* 65:1522–1528.

Soliman, S. S. and S. M. Osman. 2003. Effect of nitrogen and potassium fertilization on yield, fruit quality and some nutrients content of Samany date palm. *Ann. Agric. Sci.* 48:283–296.

Sreenivasulu, N., B. Grimm, U. Wobus, and W. Weschke. 2000. Differential response of antioxidant compounds to salinity stress in salt-tolerant and salt-sensitive seedlings of foxtail millet (*Setaria italica*). *Physiol. Plant* 109:435–444.

Sultan Qaboos University (SQU). 1997. *Dates Quality Improvement in the Sultanate of Oman: A Comprehensive Document*. Sultan Qaboos University Press, Oman, October, pp. 01–19.

Tejera, N. A., M. Soussi, and C. Lluch. 2006. Physiological and nutritional indicators of tolerance to salinity in chickpea plants growing under symbiotic conditions. *Env. Exptal. Bot.* 58:17–24.

Tripler, E., A. Ben-Gal, and U. Shani. 2007. Consequence of salinity and excess boron on growth, evapotranspiration and ion uptake in date palm (*Phoenix dactylifera* L., cv. Medjool). *Plant Soil.* 297:147–155.

Youssef, T. and M. A. Awad. 2008. Mechanisms of enhancing photosynthetic gas exchange in date palm seedlings (*Phoenix dactylifera* L.) under salinity stress by a 5-aminolevulinic acid-based fertilizer. *J. Plant Growth Regul.* 27:1–9.

6 Environmental Concern in Date Palm Production from Atmospheric Trace Metal Contamination

Salim H. Al-Rawahy and Hameed Sulaiman

CONTENTS

INTRODUCTION

Trace metals are currently of much environmental concern in human consumption. They are harmful to humans and animals since they tend to bioaccumulate in the food chain (Yoon et al., 2006). As always the increasing anthropogenic activities intensify the emission of various pollutants into the environment and introduce different types of harmful substances into the atmosphere including metals. Their sources could be from fertilizers, polluted air, polluted water, and pesticides. Atmospheric trace metals can reach leaves by dry and wet deposition, resulting in changes in their metal concentrations. Although metals are naturally present in soil, contamination comes from local sources, mostly industry, agriculture, sewage sludge, waste incineration, and road traffic (Celik et al., 2005). Since most of the date palm production in the farms nowadays practices sustainable farming inputs, these trace metals enter through external sources other than farm inputs. In the arid environments of the Arabian peninsula, the date palm (*Phoenix dactylifera* L.) is commonly grown over large areas, even where a high level of urban pollution exists (e.g., on road shoulders, in industrial and residential areas). Trace metals tend to build up in flora and most vegetation can accumulate metals so that their metal levels are much higher than those found in the air (Onder and Dursun, 2006).

METALS IN DATE PALM LEAVES

Gasim et al. (1986) studied windblown dust accumulation on date palm leaves and its effect on leaves as a function of distance from the roadside in Saudi Arabia. They found that the dust consisted of very fine grains of sand particles which land on the leaves and settle in such a manner as to form an impervious layer and that its accumulation on the leaves decreased significantly with increasing distance from a farm road than the leaves of palms away from the road.

These dust deposits are primarily responsible for carrying the metal pollutants in the atmosphere in contact with the leaves. Later, another study by Al-Shayeb et al. (1995) tried to identify the sources of metal based on zoning areas. They determined the metal content (Pb, Zn, Cu, Ni, Cr, and Li) for washed and unwashed leaflets collected from a wide range of sites with different degrees of metal pollution (urban, suburban, highway, and industrial areas) and from a rural (control) site. When trying to see the differences between washed and unwashed samples, they revealed that metal pollutants exist as superficial contaminants, especially the levels of Pb and Zn. The other metals in the leaf varied widely according to the nearby contribution from the source. On the contrary, a similar type of study in Kuwait by Bu-Olayan and Thomas (2002) found a significant difference between unwashed and washed samples wherein only 20% of Pb level in the leaves leached by the washing procedure and thus 80% of Pb was found to be accumulated by the leaves.

Although uptake from the soil is one of the possible ways to raise metal concentrations in leaves, another work has shown that although roots do absorb metal, there is only limited translocation to shoots. Wilson and Cline (1966) concluded that soil lead is largely unavailable for plant uptake and that only 0.003%–0.005% of the total lead in the soil was available. Similarly, Motto et al. (1970) have shown that the uptake of lead by plants in the greenhouse is through the root system but in the field it is mostly through the leaves.

El-Shobokshy et al. (1990) have shown that copper has a relatively high mean level of 6.4 pg m^{-3} in Riyadh environments and concluded that surface soil contributes more than industrial processes to Cu particulate pollution. A subsequent study by Al-Shayeb et al. (1995) showed that the amounts of Cu removed by washing the leaflets were higher than that of the control area, indicating that atmospheric deposition of these metals differs greatly in these areas. In the same study, unwashed leaflets from the industrial area in Riyadh city show significantly high zinc concentrations compared with the control area. Elevated zinc levels in urban, suburban, and highway areas show the effects of other pollution sources, such as traffic volume and tyre wear from the vehicles. The researchers attribute chromium levels in the unwashed leaflets to the sources such as erosion from automobile engine and body parts as well as extensive road marking by yellow lead chromate. Certain industrial activities have also been shown to be other sources of this metal in the study area.

In a previous zonation study in Riyadh city, Al-Shayeb et al. (1995) is not conclusive with respect to different locations with different degrees of pollution. But samples of unwashed leaves collected from different locations with different degrees of metal pollution (urban, suburban, industrial, highway, and rural sites) in Jordan by Abdelaziz and Al-Khashman (2007) showed that leaf samples collected from industrial areas have high contents of all metals except nickel, copper, and lead, which were found at high concentrations in the samples collected from highway sites. The mean metal concentration values in this study were found in the following order: Fe > Pb > Zn > Ni > Cu > Cr. The lowest mean concentrations of metals were observed at the control site. Thus, the study concludes that the degree of metals in the leaves was found to be proportional to industrial and human activities as reported by Celik et al. (2005), where he found high metal concentrations in plants located in industrial sites and urban and highway roadsides due to the anthropogenic activities in addition to the traffic density.

METALS IN DATE PALM FRUITS

There seems to be a very few exclusively studies on metal level in fruits. But there are studies conducted with the primary objective of metal contribution from location with different degree of pollution in all the plant parts of the date palm including fruits. One such study was conducted in Kuwait in 2000 on lead accumulation alone to determine the lead levels in industrial, express highway and city areas (Bu-Olayan and Thomas, 2002). The authors came out with the general finding summarized in Table 6.1.

TABLE 6.1
Concentration of Lead (μg g^{-1}) in Fruits Compared to the Other Parts of Date Palm and Soil

	Industrial Area	Express Highway	City Area
Sample	**Mean ± SD**	**Mean ± SD**	**Mean ± SD**
Unwashed leaves	5.06 ± 3.23	3.39 ± 1.88	1.99 ± 1.32
Washed leaves	3.52 ± 2.19	2.70 ± 0.77	1.80 ± 1.19
Barks	1.47 ± 0.14	1.08 ± 0.70	0.74 ± 0.92
Fruits	2.88 ± 2.54	1.87 ± 1.61	0.94 ± 0.85
Soil	6.86 ± 3.10	4.69 ± 2.13	3.90 ± 1.83

The highest concentration of Pb among the plant parts was noted in the unwashed leaves (5.06 μg g^{-1} fresh weight) followed by washed leaves (3.52 μg g^{-1} fresh weight), fruit (2.88 μg g^{-1} fresh weight), and bark (1.47 μg g^{-1}). In actual terms, however, the concentration in the fruit is less than that in the leaves; the difference is minor and a transformation from the vegetative to reproductive parts is evident. Interestingly, the Pb level in the surface soil was the highest compared to all plant parts. It is obvious to expect that the soil can be a potential source of Pb available for plant uptake. They also found that Pb levels in all the samples were found to be higher around industrial areas followed by express highways, followed by the samples taken from city limits.

When it comes to human health risk, the concern is more about metal in edible parts. Heavy metals are known to cause deleterious effects on human health (Davydova, 2005; Bo et al., 2009). One of the main routes of exposure to toxic elements is through dietary intake (Fu et al., 2008) and date palm fruit is an important part of the human diet in the Arabian Gulf. From similar studies cited above with respect to different locations with different degrees of pollution, this study was conducted in Muscat city in Oman to estimate the metal concentration in the fruits of date palm grown along highways. The experiment was designed to collect date palm fruits of the type "Khuneizi" at two sites: on the highway curb side in Muscat and in a rural date palm farm during fruiting season in 2006. A total of 100 matured Khuneizi fruits obtained from 10 plants/site were used for analysis of cadmiun (Cd), chromium (Cr), copper (Cu), mercury (Hg), manganese (Mn), lead (Pb), and zinc (Zn) using an inductively coupled plasma optical emission spectrometer of type Perkin Elmar 3300 DV ICP (USA). As shown in Table 6.2, the concentration levels of metal elements were as follows, in descending order: In rural sites Zn > Cu > Pb > Mn > Cr > Hg > Cd; and in highways: Zn > Pb > Cu > Mn > Cr > Hg > Cd. Metals in fruits from date palm grown in highways exceeded the corresponding values grown in rural farms by 1.9 (Cd), 4.6 (Cr), 2.1 (Cu), 1.9 (Hg), 5.7 (Mn), 5.3 (Pb), and 1.8 (Zn) folds.

The estimated daily intake (EDI) of heavy metals was calculated according to Fu et al. (2008). Based on our date consumption survey in 2008, during the harvest season of Khuneizi date fruit (between July and August), human consumption of this date variety was the essential part of food and the adult (average weight 66 kg) residence of Muscat had an average daily intake of seven fruits and children (average weight 30 kg) had an average daily intake of four fruits. For all metal elements in this study, the EDIs for fruits grown in highway were higher at least by 1.6-fold compared with the EDIs for fruits grown in rural areas. The amounts of intake for all metals except Pb in Khuneizi date fruit grown along highways in Muscat were within the provisional tolerable daily intake (PTDI) suggested by World Health Organization (WHO, 1993) and FAO/WHO (2001) for human consumption. Lead is a nonessential element and the PTDI suggested was 3.57 μg/kg bw. It is a cumulative poison that is built up in tissues and organs and eventually reaches a point at which symptoms and disabilities occur (Ayres and Hellier, 1998). It should be noted that EDI here was only evaluated for metals in Khuneizi date fruit, which may account for only a fraction of the contamination through daily food intake. If considering the total contamination from other food types, such as rice, vegetables, and

TABLE 6.2
Concentration of Heavy Metals in Khuneizi Date Palm Fruits Grown in Rural and Highway Sites and the EDI for Children and Adult

	Concentration ($\mu g\ g^{-1}$ Wet Weight)			EDI (μg/kg bw)			
				Rural		Highway	
Metal	Rural	Highway	Ratio[a]	Adult	Children	Adult	Children
Cd	0.05	0.25	1.9	0.31	0.42	1.63	2.13
Cr	0.32	1.46	4.6	1.95	2.72	9.62	18.11
Cu	1.03	2.82	2.1	6.28	8.81	17.48	23.92
Hg	0.27	0.46	1.7	1.65	2.30	2.86	3.92
Mn	0.44	2.51	5.7	2.69	3.77	15.47	21.38
Pb	0.13	3.64	6.2	0.79	1.07	22.21	30.58
Zn	15.57	48.93	1.9	94.98	130.79	293.42	408.96

[a] Ratio: Highway/Rural.

other fruits, the Pb intake would probably reach unsound levels. It is known that serious systemic health problems can develop as a result of excessive accumulation of dietary Pb in the human body (Oliver, 1997). The extent of Khuneizi date fruit contamination in highways was evaluated by the health risk associated with metals ingested through the consumption of fruits using the target hazard quotient (THQ) as explained by Bo et al. (2009). If THQ is <1, there is no obvious risk from the element over a lifetime of exposure, while if THQ is >1, the toxicant may produce an adverse effect (Bo et al., 2009). The THQs for children are normally higher than that for adults because the exposure risk of heavy metals to children through food consumption is higher (Bo et al., 2009). Children are very sensitive to metals since the dietary intake of food that contains these metals may constitute a major source of a long-term low-level body accumulation (Bahemuka and Mubofu, 1999). In this study, the THQs of all metal elements in Khuneizi date fruits for the adult and children in Muscat were equal to or below 0.35, indicating that these metals pose no serious probability of causing risk for those consuming average amount of dates grown along the highways. However, the authors are of the opinion that exposure to two or more heavy metals through food consumption may result in additive or synergism effects. Moreover, consumption of dates in Oman is mostly accompanied by consumption of other foods such as other fruits, rice, coffee, or milk, which may have some amount of heavy metals. The mathematical sum of each individual THQ from each type of food consumed daily may add up to dangerous levels and the detrimental impact of these metals may become apparent only after several years of exposure.

CONCLUSION

Studies conducted on the accumulation of metals in leaves shows that the difference between washed and unwashed samples is conspicuous and reveals that metals exist as superficial contaminants. Works with the primary objective to understand the sources of metal pollution from different locations with different degrees of metal contribution vary widely from one study to another. This is very obvious since they are specific to the site and prevailing conditions in the site such as metal-contributing sources and subsequent exposure to the plants. The level of accumulation in different parts such as leaves, barks, and fruits needs to be investigated thoroughly to understand the metal transformation from vegetative to reproductive plant parts, which may be expected to vary between species and varieties within the same species. These investigations might help us identify those species that accumulate lesser concentration of metal in edible parts such as fruits. As the study in

Oman clearly indicates, the evidence of higher metal accumulation in date fruits harvested from date palm grown closer to the highways compared to the rural site also shows that this concentration will not pose any health risk to the populations consuming the average amount of dates grown along the highways. Regular monitoring of metals in fruits grown along the highways or industrial areas is essential to prevent buildup of any serious metal pollutants in the food chain.

REFERENCES

Abdelaziz, L. Al-Khlaifat, and O. A. Al-Khashman, 2007. Atmospheric heavy metal pollution in Aqaba city, Jordan, using *Phoenix dactylifera* L. leaves. *Atmospheric Environment*, 41: 8891–8897.

Al-Shayeb, S. M., M. A. Al-Rajhi, and M. R. D. Seaward, 1995. The date palm (*Phoenix dactylifera* L.) as a biomonitor of lead and other elements in arid environments. *The Science of the Total Environment*, 168: 1–10.

Ayres, D. C. and D. G. Hellier, 1998. *Dictionary of Environmentally Important Chemicals*. Blackie Academic and Professional, London, UK.

Bahemuka, T. E. and E. B. Mubofu, 1999. Heavy metals in edible green vegetables grown along the sites of the Sinza and Msimbazi rivers in Dar es Salaam, Tanzania. *Food Chemistry*, 66: 63–66.

Bo, S., L. Mei, C. Tongbin, Z. Yuanming, X. Yunfeng, L. Xiaoyan, and G. Ding, 2009. Assessing the health risk of heavy metals in vegetables to general population in Beijing, China. *Journal of Environmental Science*, 2: 1702–1709.

Bu-Olayan, A. H. and B. V. Thomas, 2002. Bio-monitoring studies on the effect of lead in date palm (*Phoenix dactylifera*) in the arid ecosystem of Kuwait. *Journal of Arid Environments*, 51: 133–139.

Celik, A., A. Kartal, A. Akdogan, and Y. Kaska, 2005. Determination of heavy metal pollution in Denizli (Turkey) by using *Robinio Pseudo-acacia* L. *Environmental International*, 31: 105–112.

Davydova, S., 2005. Heavy metals as toxicants in big cities. *Microchemical Journal*, 79: 133–136.

El-Shobokshy, M. S., S. A. Al-Tamrah, and F. M. Hussein, 1990. Inhalable particulates and meteorological characteristics of the city of Riyadh, Saudi Arabia. *Atmospheric Environment*, 24B: 261–265.

FAO/WHO, 2001. Report on the 32nd Session of the Codex Committee on Food Additives and Contaminants. Joint FAO/WHO Food Standard Programme, Codex Alimentarius Commission, Geneva, Switzerland.

Fu, J., Q. Zhou, J. Liu, W. Liu, T. Wang, Q. Zhang, and G. Jiang, 2008. High levels of heavy metals in rice (*Oryza sativa* L.) from a typical E-waste recycling area in southern China and its potential risk to human health. *Chemosphere*, 71: 1269–1275.

Gasim, A. A., M. I. Asif, and O. A. Altahir, 1986. Effect of dust on the leaves and fruits of date palm (*Phoenix dactylifera*). *Proceedings of the 2nd Symposium on Date Palm in Saudi Arabia*, 2: 619–625.

Motto, H. L., R. P. Daines, D. M. Chilko, and C. K. Motto, 1970. Lead in soils and plants: Its relationship to traffic volume and proximity to highways. *Environmental Science and Technology*, 4: 231–237.

Oliver, M. A. 1997. Soil and human health: A review. *European Journal of Soil Science*, 48: 573–592.

Onder, S. and S. Dursun, 2006. Airborne heavy metal pollution of *Cedrus libani* (A. Rich.) in the city center of Konya (Turkey). *Atmospheric Environment*, 40: 1122–1133.

WHO, World Health Organization, 1993. Evaluation of certain food additives and contaminants (41st report of the joint FAO/WHO expert committee on food additives). WHO Technical Report Series No. 837.

Wilson, D. O. and J. F. Cline, 1966. Removal of plutonium-239, tungsten-185 and lead-210 from soils. *Nature*, 209: 941.

Yoon, J., X. Cao, Q. Zhou, and L. Ma, 2006. Accumulation of Pb, Cu, and Zn in native plants growing on a contaminated Florida site. *Science of the Total Environment*, 368 (2–3): 456–464.

7 Salinity Problems and Their Management in Date Palm Production

Nazir Hussain, Salim Al-Rasbi, Nasser Salim Al-Wahaibi, Ghanum Abdul Rehman Al-Ghanum, and Osman El-Sharief Abdalla

CONTENTS

INTRODUCTION

Date palm (*Phoenix dactylifera*) is considered as the oldest fruit tree in the Arabian Peninsula and perhaps all over the world. The earliest evidence of date palm cultivation has been reported in 4000 BC in Ur, lower Mesopotamia (now Iraq), while in the Nile Valley, date palm cultivation goes back to 3000 BC (Erskine et al., 2004). This plant is deeply rooted in the traditions, history, and culture of the Arabian Peninsula region since ancient times and has become an integral part of economy and Arab culture. Worldwide date production has been found to concentrate at four main regions: the Arabian Peninsula (Saudi Arabia, the UAE, Oman, Yemen, Bahrain, Qatar, and Kuwait), Al-Maghrab Arabian countries (Algeria, Morocco, Mauritania, Libya, and Tunisia), other Arab countries (Egypt, Iraq, and Sudan), and non-Arab Asian countries (Iran, Pakistan, and China). A few other countries of the world such as Israel, the United States, and Chad are included in the list. According to 2008 data (FAOSTAT), Arabian countries contributed 72.3% to the world production of dates (Gulf countries = 29.8%, Al-Maghrab Arab countries = 12.7%, and other Arab countries = 29.8%). Non-Arab Asian countries shared 25.7%, whereas the remaining 2.1% was produced in the rest of the world. When data of the last 48 years (1961–2008) is considered, it can be noticed that total world production of dates continued to increase, but merely due to an increase in the harvested area, because there has been no increase in yield ha^{-1} in any of the decades or years. The last 8 years (the beginning of the twenty-first century that is claimed to bring innovative technologies) have been the most discouraging.

The hydrological cyclones, floods, excessive or meager rainfall and droughts, increase of the mean sea level, and melting of ice are expected under global climatic changes. Date palm-growing regions are included among the areas where a significant increase in surface temperature and evapotranspiration may occur that will strongly influence the water cycles and could ultimately result in decreased rainfall, lesser moisture availability, desiccation, and continuous droughts. The dryness of these dry regions of the world would increase further, and it may become even more difficult for such water-scarce countries to cope with natural conditions with lesser water than what is now available (Bino, 2008).

Surface water scarcity will increase pressure on the utilization of more and more groundwater for growing date palm. The major part of groundwater in the date palm-growing areas is not of safe and usable quality and its extended utilization will affect the crop negatively. All these expected changes are closely related to surface salinity and sodicity, because high temperature, low rainfall, and excessive evaporation favor salt concentration and salinity build-up. Thus, secondary salinity can be enhanced tremendously that will affect date palm growth and yield negatively. Although date palm is generally considered to be salt tolerant, at higher values of EC, the growth and fruit productivity may be affected negatively (Qureshi and Barrett-Lennard, 1998).

There has been a tremendous increase in world date production, but it was all due to an increase in harvested area and not because of an increase in per ha yields, as was the case for almost all the food crops such as wheat, rice, maize, sorghum potato, and so on. The harvested area cannot be increased endlessly because there is competing demand for other commodities as well. There had been a tremendous decrease in average yields at the world level as well as the leading countries individually. Only a few countries newly entering this field could show up positive performance in this regard. The consistent decrease of date palm yields over the past 48 years (Table 7.1) is a great challenge for the highly developed world laden with modern technology of the twenty-first century. The real causes of low yields are

1. Low-yielding cultivars
2. Deficiency of irrigation water of good quality and appropriate irrigation systems
3. Soil and water salinity
4. Insect and disease attack
5. Poor agronomic techniques operative in the field

TABLE 7.1
Pattern of Average Yield (T ha^{-1}) of Date Palm of the Top 21 Countries during the Last Five Decades

Countries	1961	1970	1980	1990	2000	2005	2008
World	7.720	7.137	7.444	5.496	6.200	5.850	5.573
Egypt	23.950	16.950	22.300	24.842	34.736	33.428	15.124
Iraq	–	–	–	4.412	8.473	8.080	4.190
Saudi Arabia	7.273	7.742	5.671	7.293	5.158	6.438	6.309
UAE	12.000	12.500	9.194	6.385	4.088	4.054	4.059
Pakistan	8.270	8.503	8.044	6.838	7.793	6.056	7.287
Algeria	2.500	2.477	2.996	2.618	3.652	3.491	3.333
Iran	3.841	3.875	3.049	4.339	4.707	4.173	4.167
Sudan	6.023	6.000	10.004	7.0097	9.495	9.764	9.520
Oman	3.077	3.462	3.482	4.800	7.886	7.886	7.813
Libya	–	–	–	4.933	5.000	5.357	5.833
China	4.000	4.349	3.722	6.000	20.833	19.480	15.000
Tunisia	3.630	3.714	3.222	3.867	3.322	2.827	3.231
Morocco	4.778	4.500	4.237	5.454	2.434	1.369	1.949
Yemen	3.644	2.871	1.226	1.351	1.311	2.177	3.816
Israel	8.571	11.500	12.250	10.335	5.668	5.509	5.849
Mauritania	2.889	3.000	3.000	1.786	4.400	2.750	2.683
Qatar	–	–	3.801	5.723	8.346	13.742	14.933
Chad	–	–	–	3.289	2.368	2.368	2.377
USA	11.420	10.192	12.543	10.777	8.124	6.880	7.999
Bahrain	9.375	9.375	9.500	5.729	20.008	8.823	6.324
Kuwait	–	–	–	4.000	7.522	11.286	10.000

Note: Compiled from FAOSTAT data by the authors.

6. Harvesting losses and postharvest care
7. Abandoning/degradation of date palm orchards due to miscellaneous factors
8. Economic considerations
9. Poor awareness of the date palm growers, the need for training and hampered technology transfer to the end users
10. Lack of advanced packing and processing techniques as well as industry
11. Lack of coordination/collaboration
12. Trade and marketing

Although salinity is one of the conspicuous date palm problems at present, it will aggravate further under the expected climatic changes. Therefore, soil and water salinity is presented in more detail in this chapter.

WHAT IS SOIL SALINITY AND SODICITY?

Soil salinity and sodicity are two separate global problems but they often coexist and are identified as major threats to the present world agriculture. Based on the FAO/UNESCO Soil Map of the World, the total area of saline soils is 397 million hectares, while that of sodic soils is 434 million ha. Out of the current 230 million ha of irrigated land, 45 million ha are salt-affected soils (19.5%), and of the almost 1500 million ha of dry land agriculture, 32 million ha are salt-affected (2.1%), to varying degrees by human-induced processes (Oldeman et al., 1991). Salinity may be primary or secondary. Primary salinity resulted from natural geological cycles such as migration and redistribution of salts. Also deposition of salts during processes of weathering and soil formation by surface and groundwater resulted in salinization. Secondary salinity emerges from soil and water mismanagement through utilization of saline water, improper irrigation, un-leveling, leaving the lands uncultivated, and poor drainage.

The term "soil salinity" is used in a dual sense. In the true sense, "salinity" denotes a particular condition of a soil when its total soluble salts exceed a specified limit (4 dS m^{-1}). In a broader meaning, "soil salinity" indicates all problems and conditions of soil related to salts, including sodicity as well. However, technically, salinity and sodicity are two separately distinguishable abiotic stresses with well-defined definitions and limits. Soil sodicity indicates excess of Na on soil exchange sites. Under sodic conditions, the physical properties (bulk density, porosity, permeability, water infiltration rate, and hydraulic conductivity) of the soil are deteriorating. Problems like hardening of soil, closure of macropores, restricted penetration of air, less infiltration of water, more runoff, more evaporation, lesser availability of water to plants, standing of water on the soil surface for longer periods, less root development and proliferation, and decreased workability emerge as a consequence of sodicity and, finally, very low yields are obtained from a soil in degraded physical conditions (Ghafoor et al., 2004). Thus, soil and water salinity, sodicity, physical properties, and crop yields are interrelated and interdependent. These relationships are important for understanding the appropriate management.

Salt-affected soils can be classified in many ways such as natural occurrence, genesis, original soil taxonomical classes, or manmade grouping. However, identification on the basis of dominating salts (soluble salts and exchangeable Na) present in a soil is now globally accepted.

SALINE SOILS

These soils have a concentration of soluble salts that adversely affects the growth of most plants. The general limit specified for classification is $EC_e > 4$ dS m^{-1} (Soil Science Society of America, 2008), while the pH and SAR are <8.5 and 13 ($mmol_c$ $L^{-1})^{1/2}$, respectively. Soil salinity is usually measured through electrical conductivity of the extract of saturated paste (EC_e), although suspensions of higher soil-to-water ratios (1:1, 1:2, or 1:5) are also used. However, the critical limit of $EC = 4.0$ dS m^{-1} is just arbitrary and wide variations can be observed among different plants and crop varieties, as well as under variable soil and climatic conditions. The structure and other physical conditions of these soils are not impaired. Therefore, these soils remain permeable.

SODIC SOILS

Sodic or black alkali soils have EC_e lower than 4 dS m^{-1} but ESP (exchangeable sodium percentage) more than 15 or SAR (sodium adsorption ratio) more than 13 (mol_c $L^{-1})^{1/2}$, while the pH is more than 8.5. Soil sodicity can be measured by the determination of exchangeable cations and by calculating ESP, but SAR can also denote sodicity because a very close relationship was found between the two parameters (US Salinity Lab. Staff, 1954). The parameter of SAR can be determined from an analysis of saturation paste extract using Na and Ca + Mg values. The organic matter present in the soil gets dissolved and gives a dark brown or black color to these soils. The soil structure deteriorates and physical characteristics are impaired. Rehabilitation of such soils becomes difficult and needs an amendment application (Gupta and Sharma, 1990).

SALINE SODIC SOILS

Values of EC_e and SAR of these soils are more than 4 dS m^{-1} and 13 ($mmol_c$ $L^{-1})^{1/2}$, respectively, while pH may or may not be more than 8.5 depending upon relative magnitudes of soluble salts (EC_e) and exchangeable sodium (measured through ESP or SAR). Thus, these soils have the characteristics of both saline and sodic soils. Initially, these soils have good permeability, but physical conditions may deteriorate gradually when rainfall or irrigation water leaches down soluble salts and sodium becomes dominant and consequently physical characteristics become closer to a sodic soil.

WATER SALINITY AND SODICITY

Although irrigation of crops can achieve higher yields but is also coupled with a variety of associated problems, such as waterlogging from seepage, soil salinization and sodication, specific ion effects; especially from boron, carbonates and bicarbonates, sodium and chlorides, impairment of soil physical properties, and related health hazards. When the irrigation water is coming from rivers and canals, it is mostly problem-free but not always, as indicated in the Australian Murray region (Tee et al., 2003), but groundwater and untreated industrial effluents cause manifold problems to soils and plants. Enhanced utilization of such waters has become a major source of salt addition and accumulation even in fertile soils during recent decades. Like soils, water can be saline, sodic, or saline sodic as well. Irrigation water can be classified in many ways. For more details, consult the FAO paper by Ayers and Westcot (1985).

At least four hazards are associated with use of saline. These are: loss in soil productivity due to salinity, sodicity or waterlogging, pollution of associated water resources with salts and toxicants by drainage, damage to the prevailing ecosystems, and increased threat to public health resulting from water pollution and waterlogging (FAO, 1992). Consistent usage of salty water for irrigation purposes has emerged as the biggest cause of secondary salinization in arid and semiarid regions. In studies by Hussain (1997), EC and SAR of the soil profile (60 cm) were tremendously increased (Tables 7.2 and 7.3) converting normal soil into saline sodic in 3 years (after the harvesting of six crops) when saline sodic water (EC_{iw} = 1.43 dS m^{-1}, SAR = 6.5, and RSC = 5.7 $mmol_c$ L^{-1}) was used continuously. The hydraulic conductivity was also reduced significantly (Table 7.4).

SALINITY AND GLOBAL CLIMATIC CHANGES

Soil and water salinity are the problems of arid and semiarid regions because these are directly associated with levels of precipitation and temperature. More rainfall causes dissolution of salts that are leached down into the soil profile and subsequently drained off horizontally in the natural drainage

TABLE 7.2
EC (dS m^{-1}) of Soil Profile after Irrigation with Saline Sodic Water

Quality of Irrigation Water	First Crop Rice	Second Crop Wheat	Third Crop Rice	Fourth Crop Wheat	Fifth Crop Rice	Sixth Crop Wheat
			0–15 cm			
Canal water	1.01 NS	1.12 B	1.21 B	1.29 B	1.36 B	1.49 B
Tube well water	1.63	1.81 A	2.49 A	2.89 A	3.69 A	4.15 A
			15–30 cm			
Canal water	1.36 NS	1.47 NS	1.39 B	1.40 B	1.56 B	1.48 B
Tube well water	1.40	1.69	2.22 A	2.75 A	2.89 A	3.87 A
			30–60 cm			
Canal water	1.42 NS	1.57 NS	1.63 B	1.66 B	1.58 B	1.72 B
Tube well water	1.56	1.66	2.04 A	2.60 A	2.89 A	3.67 A

Source: Hussain N. 1997. *Use of Brackish Water for Sustained Crop Production While Maintaining Soil Health.* PhD Thesis, Department of Soil Science, University of Agriculture, Faisalabad, Pakistan, 148 pp.

TABLE 7.3
SAR of Soil Profile during Irrigation with Saline Sodic Water

Quality of Irrigation Water	First Crop Rice	Second Crop Wheat	Third Crop Rice	Fourth Crop Wheat	Fifth Crop Rice	Sixth Crop Wheat
			0–15 cm			
Canal water	6.82 B	7.80 B	7.95 B	8.02 B	7.87 B	8.15 B
Tube well water	9.25 A	13.47 A	16.89 A	20.74 A	26.45 A	32.76 A
			15–30 cm			
Canal water	6.47 B	7.24 B	7.37 B	7.40 B	7.65 B	7.80 B
Tube well water	7.20 A	10.29 A	12.57 A	12.69 A	21.52 A	30.40 A
			30–60 cm			
Canal water	6.69 NS	7.15 NS	7.20 B	7.35 B	7.47 B	7.52 B
Tube well water	7.05	7.15	9.24 A	14.38 A	19.62 A	29.59 A

Source: Hussain N. 1997. *Use of Brackish Water for Sustained Crop Production While Maintaining Soil Health.* PhD Thesis, Department of Soil Science, University of Agriculture, Faisalabad, Pakistan, 148 pp.

TABLE 7.4
Hydraulic Conductivity (cm h^{-1}) of Soil Profile during Irrigation with Saline Sodic Water

Quality of Irrigation Water	First Crop (Rice)	Second Crop (Wheat)	Third Crop (Rice)	Fourth Crop (Wheat)	Fifth Crop (Rice)	Sixth Crop (Wheat)
			0–15 cm			
Canal water	1.86	1.85	1.83 A	1.82 A	1.82 A	1.76 A
Tube well water	1.88	1.77	1.42 B	1.00 B	0.65 B	0.23 B
			15–30 cm			
Canal water	1.63	1.58	1.56 A	1.50 A	1.50 A	1.45 A
Tube well water	1.65	1.50	1.24 B	1.08 B	0.85 B	0.64 B
			30–60 cm			
Canal water	1.25	1.26	1.26 A	1.24 A	1.20 A	1.16 A
Tube well water	1.25	1.24	1.14 B	1.05 B	0.94 B	0.78 B

Source: Hussain N. 1997. *Use of Brackish Water for Sustained Crop Production While Maintaining Soil Health.* PhD Thesis, Department of Soil Science, University of Agriculture, Faisalabad, Pakistan, 148 pp.

system. These processes occur during weathering of minerals of the parent material and soil genesis. But dissolution and leaching processes remain incomplete when the rainfall is insufficient, as is the case in arid and semiarid regions. The salts generated during soil formation or subsequent weathering or added through irrigations or runoff remain in the profile and come to the surface or subsurface with the capillary movement of soil water when dry conditions coupled with high temperature occur. The water evaporates and leaves dissolved salts on the surface or just underneath. When such a process continues for longer periods, the consequence is salinization (accumulation of salts) followed by sodication (substitution of Ca + Mg by sodium on the clay complex). Irrigation or rainfall temporarily reverts these processes, but prolonged dry periods and high temperatures persisting in arid climates prove favorable for salinization. Thus, relatively small mismanagement of soil and water as well as cultural practices like fallowing (leaving the lands uncultivated) or unleveled conditions may become a big factor in creating secondary salinity. Data of Hussain et al. (2002b) presented in Table

TABLE 7.5
Relationship of Rainfall, Soil EC, and SAR

Seasons	Rainfall (mm)	Soil EC (dS m^{-1})	Soil SAR ($(mol_c\ L^{-1})^{1/2}$)
Premonsoon 2000	11	6.8	75
Post monsoon 2000	123	4.9	72
Premonsoon 2001	121	17.8	200
Post monsoon 2001	500	6.1	103

Source: Hussain N, Mujeeb F, Sarwar G, Hussan G, and Ullah MK. 2002b. *Proc. International Workshop on Conjunctive Water Management for Sustainable Irrigated Agriculture in South Asia*, April 16–17, 2002, Islamabad, Pakistan, pp. 91–99.

7.5 indicate a close relationship between salinity/sodicity parameters (soil EC and SAR) and rainfall. There was salt accumulation during premonsoon periods in the case of lesser rainfalls but the accumulated salts were washed away during intensive monsoon rainfall, especially that of the year 2001.

Climatologists claim that arid and semiarid regions may become drier and environmental temperatures of hot regions may increase due to global climatic changes (IPCC 4th Report, 2007). These two important factors are closely related to salinity and sodicity. Evaporation may tremendously increase under high temperature and less relative humidity. Consequently, net capillary movement of water will remain upwards during a major part of the year. Thus, soil and water salinity/sodicity problems are expected to be enhanced under changing environments of the future. The data and observed evidence of the last two decades that had been relatively drier and hot support this hypothesis. Despite the best possible efforts and improved technologies, it has not been possible by the global community to reduce the sanity and sodicity menaces.

CAUSES, ORIGIN, AND GENESIS OF SOIL SALINITY IN OMAN AND OTHER GULF COUNTRIES

The main causes of salinity are the climatic conditions (scanty rainfall and high temperature), nearness to sea, salty parent material, and development of secondary salinity due to consistent usage of very saline water. Many soils have a surface accumulation of secondary lime, which may also be Aeolian. The salinization of irrigated soils by groundwater has become a major process of soil formation in many areas of the Sultanate, in particular, in coastal areas. Salt pans in many soils can be attributed to Aeolian addition originating from the nearby sea through wind flows toward the inland. The genesis and classification studies indicated that salt-affected soils belong to only two orders, the Aridisols and Entisols. The suborders are Salids, Psamments, Fluvents, and Orthents, while the major group is Typic. The subgroups are mostly Typic Aquisalids, Typic Haplogypsids, Typic Calcigypsids, Typic Haplocalcids, and Typic Torrifluvents (Hussain et al., 2006).

WATER SALINITY IN THE ARAB WORLD (THE MAJOR DATE PALM AREAS)

Owing to scarcity of good-quality water, the usage of saline water is very high in date palm irrigation because of the general notion that the date palm is a salt-tolerant tree. However, this statement cannot be blindly accepted as truth, because studies have indicated that growth and yield losses start occurring at the water EC of 4–6 dS m^{-1} depending upon varying potential, soil texture, drainage, and climatic factors (temperature and rainfall; total quantity and intensity). The water quality assessment studies indicated that a major part of the groundwater in date palm-growing countries is saline, which can pose negative effects on date palm growth and yields. As an example, the groundwater quality categorization of Oman is presented here:

1. EC > 3 dS m^{-1} = 48.3%
2. EC = 3–5 dS m^{-1} = 12.1%
3. EC = 5–10 dS m^{-1} = 9.5%
4. EC < 10 dS m^{-1} = 4.2%
5. Uncovered or mixed quality = 25.9%

When saline water is applied to a field, the major part of irrigation water evaporates and leaves the salt load on the surface or the subsurface. Thus, the soil salinity slowly and gradually keeps on increasing and starts affecting the date palm plants negatively when it goes beyond its tolerance potential. This phenomenon is now becoming common in many date palm-growing countries. A continuous concentration of salts is highly damaging for date palm orchards because groundwater is reaching almost the EC of sea water, especially in coastal areas (Figure 7.1).

FIGURE 7.1 **(See color insert.)** Damaged date palm orchards due to salinity in coastal area of Batinah, the Sultanate of Oman.

SOIL SALINITY IN THE MAJOR DATE PALM-GROWING AREAS

The gravity of the salinity problem in the Arab world can be noticed from the following data (Haechicha and Abdelgawed, 2003):

1. Arab land occupies 1402 mha area
2. Arable area is 197 mha (14.1%)
3. Actually cultivated 70 mha (5%)
4. Irrigated land 9.5 mha (0.68%)
5. Salt affected area 82.7 mha (42% of arable)

The salt-affected area in different Arab countries growing date palm presents an alarming picture. There are vast salt-affected areas in different Arab countries such as KSA (3.64 mha), Oman (9.442 mha), Mauritania (86.3 mha), Tunisia (1.5 mha), Qatar (70,000 ha), Bahrain (41,000 ha), Kuwait (8500 mha), Yemen (0.483 mha), Egypt (1.2 mha), Sudan (4.8 mha), Iraq (8.0 mha), Syria (0.53 mha), Jordan (6500 ha), Libya (0.70 mha), Algeria (0.35 mha), and Morocco (0.35 mha).

Besides concentration of salts, excess Na (sodicity) poses manifold negative effects on many plants including date palm when its values increase beyond a certain limit, which is particular for a plant. This problem is common in countries such as Pakistan, Iran, and China where soil and water contain higher quantities of carbonates and bicarbonates of Na instead of chlorides and sulfates, as is found in the Arabian Peninsula. The area affected by salinity/sodicity in Pakistan covers 6.5 mha (Khan, 1998). Thus, water and soil salinity is a huge problem that is affecting date palm yields tremendously and is expected to even increase in future if very serious management practices are not adopted.

SALINITY EFFECTS ON PLANTS

Reduction in water potential, causing ion imbalance, and disturbing availability of nutrient ions are the threefold effects on plants caused by soil and water salinity stress. Terms like physiological unavailability of water or physiological drought are used when the plants are unable to take water from the soil due to its increased osmotic potential, although water may physically be present in sufficient quantities. Such an alteration in water status leads to loss of turgor pressure, closure of stomata, significant reductions in transpiration, photosynthesis (Brugnoli and Lauteri, 1991), protein synthesis, and energy and lipid metabolism. Consequently, many physiological and morphological disorders in plants may emerge. Ion uptake and transportation by vascular tissues (xylem and phloem) can be retarded due to impaired water flow within the plants. The energy requirements are increased due to more utilization in osmoregulation for saving the plants from stress damage. Plants may suffer due to restricted long-distance transport of nutrients and metabolic processes like transpiration, photosynthesis, respiration, energy transformations, and excretion of toxicants under excessive salt conditions.

The final consequences are growth reduction and limitation of plant productivity. Growth suppression is directly related to total concentration of soluble salts or osmotic potential of soil water. The detrimental effects are observed at the whole plant level as well as the tissue level. Suppression/reduction of growth occurs in all plants, but their tolerance levels and rates of growth reduction at lethal concentrations of salt vary widely among plants (Alia et al., 1995; Yoshiba et al., 1997; Hayashi and Murata, 1998; Yeo, 1999; Sanchez et al., 1998; Munns, 2002; Ramoliya et al., 2004; Flowers, 2004; Sánchez-Blanco et al., 2004; Parida and Das, 2005; Benlloch-González et al., 2005; Rabie and Almadini, 2005; Ghadiri et al., 2005).

The dominant ions in saline water and soils are Na^+ and Cl^-. Excess Na^+ often harms nonhalophytes by displacing Ca^{2+} from root membranes and thus changes their integrity and normal functioning

(Cramer et al., 1985). Excessive concentration of salts under saline conditions and drastic shifts in soil pH may affect the conversion of ions into soluble form. Consequently, ions like Na, Cl, B, Li, Se, Cd, Cr, and Pb may have toxic effects when their concentration within the plants exceeds a specific level. These effects may appear as improper functioning or death of cells and tissues or the suppression of the uptake of other essential ions.

SALINITY EFFECTS ON DATE PALM

An increase in salinity level of soils due to the prolonged use of saline water for irrigation or other salt accumulation factors lead to decreased productivity of most crop plants and marked changes in the growth pattern. Large differences are found between plant species under a saline environment. For example, with 200 mmol NaCl a salt-tolerant species such as sugar beet might have a reduction of only 20% in dry weight, a moderately tolerant species such as cotton might indicate 60% reduction, and sensitive leguminous species such as soybean might be dead at the same level (Greenway and Munns, 1980; Cordovilla, 1994; Delgado et al., 1994; Tate, 1995).

The date palm plant originates from arid and semiarid climates, where the salt concentration in soils is generally high; therefore, due to its natural evolutionary characteristics, it has good tolerance against salinity and is affected to a lesser extent at moderate salinity. As all the plants are finally affected negatively by salinity, date palm is no exception to this phenomenon. Physiological and biological processes described in the section like physiological unavailability of water, reductions in transpiration, photosynthesis, protein synthesis, and supply of energy and lipid metabolism, reduced ion uptake and transportation by vascular tissues are also operative in date palm. The ultimate consequences are disruption of growth, bearing of fruit, decrease in yield and quality of fruit as well as its keeping quality. However, the quantum of negative effect will depend upon the date palm variety/cultivar itself in addition to prevailing soil and climatic factors coupled with growing techniques. Therefore, a wide variation in behavior toward salinity is frequently observed within date palm varieties. The nutritional imbalance may occur due to unavailability of essential macro and micronutrients under higher salinity stresses. The specific ion effect from toxic ions such as B, Cl, CO_3, HCO_3, and Na can also appear if the concentration of these ions exceeds the tolerance level of date palm plants. This subject is presented in more detail in subsequent sections.

ROOT DEVELOPMENT OF DATE PALM UNDER SALINITY STRESS

The first plant part faced and affected by increased salt concentration in the rhizosphere are the plant roots. These are in direct contact with salts and the root radical starts suffering right from germination. Roots remain exposed to the salt stress throughout the life cycle of the plant. The roots may lose the growth, proliferation, and bearing of root hairs. There may be root mortality in acute cases for want of water and nutrients, uptake of toxic ions, and extreme hardening of soil (Ferreria et al., 2001). Roots may also suffer due to restricted respiration under less aeration, a consequence of closure of macropores, soil dispersion, and temporary or prolonged waterlogging. Djibril et al. (2005) conducted salinity and drought studies on two varieties of date palm: Nakhla hamra (NHH) and Tijib widely cultivated in Mauritania. Increasing levels of salinity were developed using NaCl (up to 16 g L^{-1}), while osmotic stress was induced by polyethylene glycol (PEG) during the early stages of plant development. NHH showed increase of secondary root length, secondary root number (Table 7.6), and proline content when water deficit was induced by PEG. In contrast, on the basis of the same developmental and biochemical characters, the Tijib cultivar proved to be more tolerant to salinity, indicating a wide variation of genetic variability in root behavior of date palm plant under salinity stress.

TABLE 7.6
Influence of PEG and NaCl on Number, Elongation, and Biomass of Secondary Roots in NHH and Tijib Cultivar Seedlings 3 Months after Sowing

	Mean Number of Secondary Roots/Plant		Mean Elongation of Secondary Roots/Plant (cm)		Mean Weight of Fresh Roots' Matter (mg)	
Media (g L^{-1})	NHH	Tijib	NHH	Tijib	NHH	Tijib
Control (no PEG or NaCl)	36 a	40 a	05 a	07 a	426 a	567 a
PEG 6.9	45 a	20 b	12 a	02 b	430 a	364 a
PEG 13.75	60 a	19 b	10 a	2.5 b	528 a	359 a
PEG 27.5	89 a	07 b	08 a	1.5 b	766 a	214 b
NaCl 4	30 b	70 a	02 b	07 a	392 b	1,048 a
NaCl 8	20 b	175 a	02 b	10 a	383 b	1,208 a
NaCl 16	08 b	09 a	1.5 b	3.8 a	314 b	288 a

Source: Djibril S, Mohamed OKD, Diaga D, Diégane BF, Abaye SM, and Alain B 2005. *Afr. J. Biotech.* 4(9): 968–972.

NUTRITIONAL IMBALANCES AND ION TOXICITIES IN DATE PALM UNDER SALINITY

The presence of large quantities of salts in arid zones often causes soil infertility. Imbalanced plant nutrition is an indirect effect of salinity and sodicity in contrast to direct osmotic and specific ion effects. Soil and water salinity and sodicity bring drastic changes in availability and uptake of many plant nutrients. The absorption and passive uptake of ions such as N, K, Ca, Mg, P, and Zn may decrease and plants can suffer scarcity of the required nutrients, while concentrations of Na, B, CO_3, HCO_3, Cl, and SO_4 may increase to excessive limits. Thus, the plants may suffer simultaneously for want of certain essential nutrients or toxicities of others due to higher concentrations. However, with regard to date palm, fewer specific data are available on ionic behavior. Al-Khateeb et al. (1993) carried out a study on date palm trees grown at Al-Ojair in the coast of the Arabian Gulf during 1999/2000. Ion and water relations were investigated using trees of the Khalas variety grown at Al-Hassa under field conditions as the control. The result of ion relations revealed that high external EC in Al-Ojair resulted in greater internal Na^+ and Cl^- concentrations, compared with the nonsaline media, which contributed to the decrease in leaf and root osmotic potentials. Concentration of macronutrients represented levels regarded adequate for growth even under saline conditions. All the trees showed no reduction in chlorophyll content even under saline conditions but chlorophyll content decreased as leaf aged. All date palm trees generated sufficiently larger water potential for import gradients of water with no significant differences in turgor pressure. A study conducted by Aljuburi and Maroff (2007) at Qatar University indicated that salinity in irrigation water reduced leaf and root Mn, Zu, Fe and leaf K, Ca, Mg concentrations and the ratio of K/Na and Mg/Ca, but increased leaf N and root and leaf P, Cu, Na, and Cl concentrations. A recent study by Al-Rasbi (2010) indicated that some of the ions in date palm leaf P, Na, Mg, and Mn are increased due to an increase in water salinity while Ca, Fe, and Zn decrease. The correlation between water salinity and concentration was significant for leaf P, Na, Mg, Mn, and Zn, while for a few other ions, namely N, K, Ca, Cl, Fe, and Cu, the correlation was found to be insignificant (Table 7.7). He reported that date palm exercises the phenomenon of ion selectivity to control the entry of unwanted ions into plant parts and thus protects itself from toxicities of specific ions such as Cl and Na and keeps the

TABLE 7.7
Correlation of Water Salinity (EC_{iw}) with Different Parameters

Sr. No.	Parameter	Coefficient of Corr. (r^2)	Slope	Significance
1	Leaf N	+0.090	+0.007	NS
2	Leaf P	+0.305	+0.003	*
3	Leaf K	−0.090	+0.008	NS
4	Leaf Na	+0.503	+0.012	**
5	Leaf Na/K	+0.532	+0.007	**
6	Leaf Ca	−0.093	−0.004	NS
7	Leaf Mg	+0.386	+0.016	**
8	Leaf Cl	−0.096	−0.005	NS
9	Leaf Mn	+0.717	+0.384	**
10	Leaf Fe	−0.062	−2.289	NS
11	Leaf Zn	−0.503	−1.371	**
12	Leaf Cu	+0.058	+0.93	NS

Source: Al-Rasbi, SAR. 2010. *Evaluation of the Growth of Date Palm Seedlings Irrigated with Saline Water in Sultanate of Oman.* PhD Thesis, Department of Landscape Ecology and Nature Conservation, University of Kassel, Germany.

Na/K ratio narrow to avoid the bad effects of Na. Thus, it preferably undertakes more K compared with Na for keeping the ionic balance favorable.

EFFECT ON GROWTH OF DATE PALM

Salinity stress is known to retard plant growth through restricted water uptake and nutrients, disturbing basic biochemical processes such as photosynthesis, transpiration, and respiration, influencing vital facets of plant metabolism, or disturbing the concentration of endogenous plant hormones. Shani et al. (2001) related the yield loss in date palm to reduced photosynthesis, high energy and carbohydrate expenses in osmoregulation, and interference with cell functions under saline conditions, while Ghadiri et al. (2005) reported restricted water uptake by salinity due to the high osmotic potential in the soil and high concentrations of specific ions that may cause physiological disorders in the plant tissues and reduce the yields. Recent studies of Al-Rasbi (2010) have revealed that increasing levels of water salinity hamper the date palm growth parameters such as plant height, plant girth, the number of fronds and leaf length significantly and that relative values keep on decreasing with the quantum of water salinity (Figure 7.2). These parameters significantly but negatively correlated with water and soil salinity (Tables 7.8 and 7.9). The mechanisms of this negative effect may be: osmotic effects that restricted water and nutrient uptake, the specific ion effect of Na and Cl ions, nutritional imbalance and inhibition of physiological processes such as photosynthesis, transpiration, energy reactions, metabolism of nutrients, and the use of extra energy in osmoregulation.

Studies of El-Sharabasy et al. (2008) on three varieties of date palm, namely Samani (SM), Sewy (SE), and Bartamuda (BA), indicated that growth vigor during the proliferation stage was significantly increased by supplementation of 4000 and 8000 ppm NaCl, while it was noticed similar with the control when 12,000 ppm salts were added (Table 7.10). Growth vigor (GV) was calculated visually in this study by allotting scores: A—negative growth = 1 (the lowest growth), B—below average growth = 2, C—average growth = 3, D—above average growth = 4, and E—good growth = 5 (the highest growth).

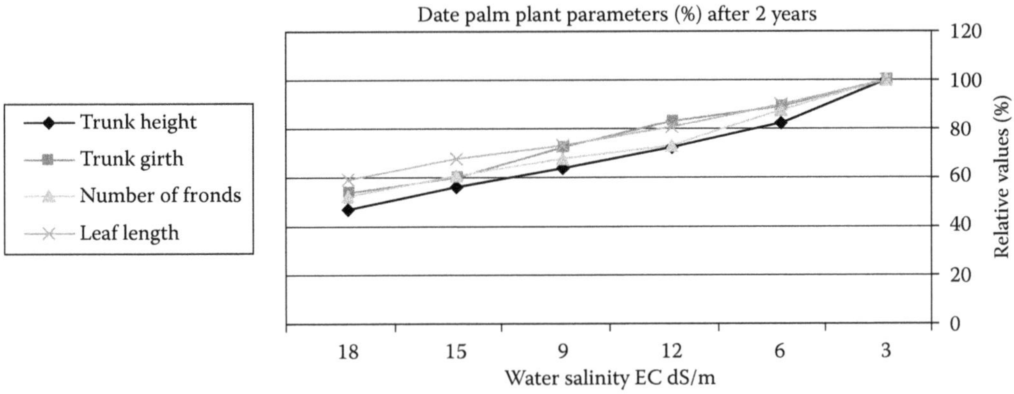

FIGURE 7.2 **(See color insert.)** Effect of water salinity on plant characters of date palm. (Adapted from Al-Rasbi, SAR. 2010. *Evaluation of the Growth of Date Palm Seedlings Irrigated with Saline Water in Sultanate of Oman.* PhD Thesis, Department of Landscape Ecology and Nature Conservation, University of Kassel, Germany.)

TABLE 7.8
Correlation of Water Salinity (EC_{iw}) with Growth Parameters

Sr. No.	Parameter	Coefficient of Corr. (r^2)	Slope	Significance
1	No. of fronds	–0.909	–2.432	**
2	Length of frond	–0.539	–8.020	*
3	Height of trunk	–0.847	–3.587	**
4	Dry wt of leaflets	–0.477	–	**
5	Leaf moisture	–0.560	–	**

Source: Al-Rasbi, SAR. 2010. *Evaluation of the Growth of Date Palm Seedlings Irrigated with Saline Water in Sultanate of Oman.* PhD Thesis, Department of Landscape Ecology and Nature Conservation, University of Kassel, Germany.

TABLE 7.9
Correlation of Soil EC_e with Growth Parameters

Sr. No.	Parameter	Coefficient of Corr. (r^2)	Significance
1	No. of fronds	–0.695	**
2	Length of frond	–0.318	**
3	Height of trunk	–0.655	**
4	No. of leaflets	–0.454	**
5	Trunk girth	–0.682	**

Source: Al-Rasbi, SAR. 2010. *Evaluation of the Growth of Date Palm Seedlings Irrigated with Saline Water in Sultanate of Oman.* PhD Thesis, Department of Landscape Ecology and Nature Conservation, University of Kassel, Germany.

TABLE 7.10
Effect of NaCl Salt Treatments on Growth Vigor of Some Date Palm Cultivars after 12 Weeks *In Vitro* during Proliferation Stage

	NaCl Salt Concentration (ppm)				
Cultivar	**Control**	**4000**	**8000**	**12,000**	**Means of Cultivars**
Samani	2.00 d	4.00 b	3.25 c	2.50 d	2.93 C
Sewy	2.50 d	4.75 a	3.50 bc	2.50 d	3.31 B
Bartamuda	3.25 c	5.00 a	4.00 b	3.25 c	3.87 A
Means of treatments	2.58 C	4.58 A	3.58 B	2.75 C	–

Source: El-Sharabasy SF, Wanas WH, and Al-Kerdany AY. 2008. *Arab J. Biotech.* 11(2): 273–280.

EFFECT ON FRUIT YIELD AND QUALITY OF DATE PALM

Growth hampering is ultimately translated into reduced fruit yield of date palm because many metabolic processes responsible for the contribution toward productivity either slow down or stop totally depending upon the salinity stress level. In acute cases, plants may die down before completing their yearly cycle and fruit yield is not obtained at all. Although no recent data are available on this particular subject because of a limited number of research studies, especially mature plants, the results of Al-Rawi and Al-Mohemdy (1975) have been presented in Table 7.11. These data clearly indicate that the irrigation water EC of 6.0 dS m^{-1} drastically reduces the fruit yield of date palm. However, noticeable differences could be seen in between the varieties Zaidi, Kestwari, Makroom,

TABLE 7.11
Mean Effect of Water Quality on the Fruit Characters of Four Varieties of Date Palm

	Varieties				
Fruit Characters	**Zaidi**	**Kestawi**	**Maktoom**	**Barban**	**Mean**
	Irrigation Water EC of 1.1 dS m^{-1}				
Fruit yield (kg $tree^{-1}$)	132.3	115.0	84.8	148.0	120.0
Fruit girth (cm)	8.1	9.3	9.9	10.5	9.45
Fruit length (cm)	4.2	4.9	4.7	5.0	4.70
Fruit wt. (g)	10.6	13.2	17.6	20.0	15.35
Seed wt. (g)	1.26	1.14	1.18	1.18	1.19
Flesh ratio (%)	88.0	91.3	93.3	94.1	91.67
	Irrigation Water EC of 6.0 dS m^{-1}				
Fruit yield (kg $tree^{-1}$)	11.6	24.6	9.6	28.0	18.45
Fruit girth (cm)	8.1	8.4	8.2	9.3	8.50
Fruit length (cm)	3.5	3.6	3.9	4.5	3.90
Fruit wt (g)	7.9	7.8	9.2	13.5	9.60
Seed wt (g)	1.2	0.8	1.2	1.5	1.17
Flesh ratio (%)	84.8	89.2	86.8	89.2	87.5

Source: Al-Rawi Ahmed AH and Al-Mohemdy AF. 1975. Effect of water quality on the growth and yield of date palm (*Phoenix dactylifera* L.). http://www.pubhort.org/datepalm/datepalm2/datepalm2_13.pdf
LSD 0.05; for salinity = 0.16 0.14 0.26 0.94, for varieties = 0.23 0.20 0.36 0.11 1.33.

and Barban with, respectively, 91%, 79%, 89%, and 81% decrease in yield of dates. Similarly, fruit quality parameters such as fruit length, fruit girth, fruit weight, seed weight, and flesh ratio were also affected negatively in all the four varieties.

INDIRECT EFFECTS OF SODICITY ON DATE PALM

Many indirect effects of sodicity can appear in the case of date palm. Although this problem is mainly limited to Pakistan, Iran, and China among the important date palm-growing countries, it has far-reaching indirect effects in addition to specific ion effects of Na, Cl, CO_3, HCO_3, B, and so on. The indirect effects appear as soil dispersion after breakage of soil granules. Thus, the soil porosity decreases due to closure of macropores. The consequence is hardening of soil and exhibition of lesser water and air penetration. Infiltration of water and hydraulic conductivity decrease significantly (Table 7.4). Thus irrigation water may stand on the soil surface for longer periods unless it ultimately evaporates, leaving salt load on the soil surface. There may be temporary or permanent waterlogging. Plants may suffer because of deficient aeration and roots may die.

SALT TOLERANCE OF DATE PALM

Salt tolerance is a complex; quantitative and genetic character controlled by many genes in plants and is a complex phenomenon that involves morphological and developmental changes as well as physiological and biochemical processes. Plant salt tolerance is generally an inherent ability of the plants to withstand the effects of high salts in the root zone or on the plant's surfaces without a significantly adverse effect. Salt resistance is another term that is often used for this phenomenon. Although some have tried to differentiate the two terms (Levitt, 1972), these are mostly used interchangeably. In an agronomic context, salt tolerance is described as a complex function of yield decline across a range of salt concentrations (Maas and Hoffman, 1977; Van Genuchten and Hoffman, 1984). It can be measured in different ways. Using a simple convention, salt tolerance can be measured on the basis of two parameters: the EC threshold (EC_a), the salinity that is expected to cause the initial significant reduction in the maximum expected yield, and the slope. The slope (S) is simply the percentage of yield expected to be reduced for each unit of added salinity above the threshold value. The relative yield (Y) at any salinity level (EC_e) exceeding EC_a can be calculated as

$$Y = 100 - S\,(EC_e - EC_a), \quad \text{where } EC_e > EC_a$$

As with most cultivated crops, the salinity response of date palm varieties/cultivars varies greatly and depends upon soil and climatic factors and the stage of growth. Tolerance to salinity more specifically depends upon the specific plant species (genetic variability), salinity levels, ionic composition of the salts, soil characteristics (texture, structure, drainage, and organic matter), and climatic factors such as rainfall and temperature (Munnus and Tester, 2008; Yano-Melo et al., 2003). Ayers and Wescott (1985) reported that, in general, the minimum EC_e for 100% yield of date palm was 4.0 dS m^{-1}, while the plant produced no yield at the EC_e of 32 dS m^{-1} (Table 7.12). The respective values for irrigation water EC_{iw} are 2.7 and 21.0 dS m^{-1}. Fifty percent reduction in yield can occur at an EC_e of 18.0 dS m^{-1} and an EC_{iw} of 12.0. Date palm was regarded as tolerant by Maas (1990) having threshold value (salinity effects starting) at 4.0 dS m^{-1} with a slope value of 3.6% per unit dS m^{-1} increase in soil salinity (Table 7.13). Recent studies of Al-Rasbi (2010) indicated that 53%, 48%, 39%, and 46% decrease occurred, respectively, in date palm trunk height, the number of fronds, leaf length, and trunk girth, respectively, at an EC_{iw} of 18 dS m^{-1} (Table 7.14). Leeden et al. (1990) included date palm in the tolerant group to boron in irrigation water (permissible limit <3.00 ppm of boron), while no data were available for Cl and sodicity tolerance.

TABLE 7.12
Salinity Tolerance Levels of Date Palm

Adapted from	EC of Soil/Water	Yield Potential 100%	75%	50%	Maximum EC_e
Ayers and Westcot (1985)	EC_e	4.0	11.0	18.0	32
Ayers and Westcot (1985)	EC_{iw}	2.7	7.3	12.0	21.0

TABLE 7.13
Salt Tolerance of Date Palm

Plant Botanical Name	Salt Tolerance Parameters Threshold (EC_e) in dS m^{-1}	Slope % per dS m^{-1}	Rating	Reference
Phoenix dactylifera	4.0	3.6	Tolerant	Maas (1990)

Note: These data serve only as a guideline to relative tolerances, absolute tolerances vary, depending upon climate, soil conditions, cultural practices, and crop varieties.

TABLE 7.14
Negative Effect of Saline Water on Plant Parameters of Date Palm (% Decrease Compared to Control = 3 dS m^{-1})

Water EC in dS m^{-1}	Plant Height	Number of Fronds	Leaf Length	Trunk Girth
3	100	100	100	100
6	18	12	8	11
9	28	27	11	17
12	36	32	21	28
15	46	39	29	40
18	53	48	39	46

Source: Al-Rasbi, SAR. 2010. *Evaluation of the Growth of Date Palm Seedlings Irrigated with Saline Water in Sultanate of Oman*. PhD Thesis, Department of Landscape Ecology and Nature Conservation, University of Kassel, Germany.

MANAGEMENT TECHNIQUES WITH RESPECT TO SOIL SALINITY AND BRACKISH WATER IRRIGATION

To cope with future consequences such as aggravation of salinity/sodicity, enhanced use of brackish water and deterioration of soil physical properties, special management techniques are the only alternative to avoid worsening of the situation. The management techniques may be of various types.

IMPROVED HYDRAULIC TECHNIQUES

Scarcity of water will remain a permanent major constraint of crop production in arid and semiarid regions. It may be tackled through appropriate, timely, effective, economical, and easily practicable hydraulic management techniques.

LEACHING OF SALTS

Salt accumulation, salinization, and sodication are the ultimate results of aridity, drought, high temperature, and supplemental irrigation with brackish water, but the concentration of salts in the rhizosphere must remain within tolerance potential of the plant to be grown; otherwise a significant reduction in growth and yield may occur. Therefore, accumulated salts must necessarily be leached down through either rainfall or deep and continuous irrigations with good-quality water. When crops are irrigated with saline water but monsoon rains are effectively harvested, the accumulated salts can be leached down fully or partially depending upon the intensity of rainfall. In some of the investigations, the surface salinity build-up (EC_e and ESP) during an irrigation cycle was taken care of by one above average yearly rainfall or two subseasonal rainy seasons, especially monsoon rains (Dhir, 1977; Jain, 1981). Rains could convert or keep 40 cm soil surface as nonsaline. For this purpose, the field boundaries should be kept strong and fallow plots need to be plowed to control water runoff and increase infiltration during the rainy season. The basin of tree plants like date palm should be kept clean from weeds and hoed to catch the rainy water for natural leaching of salts. Another phenomenon has been observed in the Gulf countries. The continuous irrigation with saline water causes high salt accumulation around the plant basin. When a rainfall event occurs, the accumulated salts rush into the plant basin and very high concentration damages the plants. Therefore, scrapping of salts accumulated around the basin of trees should be made a permanent practice before the start of the rainy season.

When irrigation with brackish water is a continuous practice, extra water over consumptive use (that used by the crops) has to be applied for preventing salt accumulation and keeping the flow of water going into lower soil depths so that the rooting zone of the soil remains salt-free or within tolerable limits of the crops under cultivation. The additional water is called leaching water (leaching fraction [LF] or leaching requirement [LR]). This is dependent upon the quality and quantity of irrigation water and crop tolerance of plant for salt stress. It can very easily be calculated by dividing the depth of drainage water (D_{dw}) by depth of irrigation water (D_{iw}) or the EC of irrigation water (EC_{iw}) by EC of drainage water (EC_{dw}) as proposed by US Salinity Laboratory Staff (1954).

$$LR = D_{dw}/D_{iw} = EC_{iw}/EC_{dw}$$

However, when the irrigation water is sodic as well, some amendment like gypsum has also to be added that can be calculated on the basis of soil or water analysis. Data of Rasbi (2010) presented in Table 7.15 indicated that leaching fraction is positively correlated with growth parameters and proves helpful in controlling increases in soil EC and pH when saline irrigation water is being used consistently (Figures 7.3 and 7.4).

TABLE 7.15
Correlation of Leaching Fractions with Different Parameters

Sr. No.	Parameter	Coefficient of Corr. (r^2)	Slope	Probability	Significance
1	No. of fronds	+0.332	+0.708	0.048	*
2	Length of frond	+0.136	+1.013	0.429	NS
3	Height of trunk	+0.398	+1.417	0.016	**
4	No. of leaflets	+0.261	+1.463	0.123	NS
5	Trunk girth	+0.220	+0.567	0.019	*

Source: Al-Rasbi, SAR. 2010. *Evaluation of the Growth of Date Palm Seedlings Irrigated with Saline Water in Sultanate of Oman*. PhD Thesis, Department of Landscape Ecology and Nature Conservation, University of Kassel, Germany.

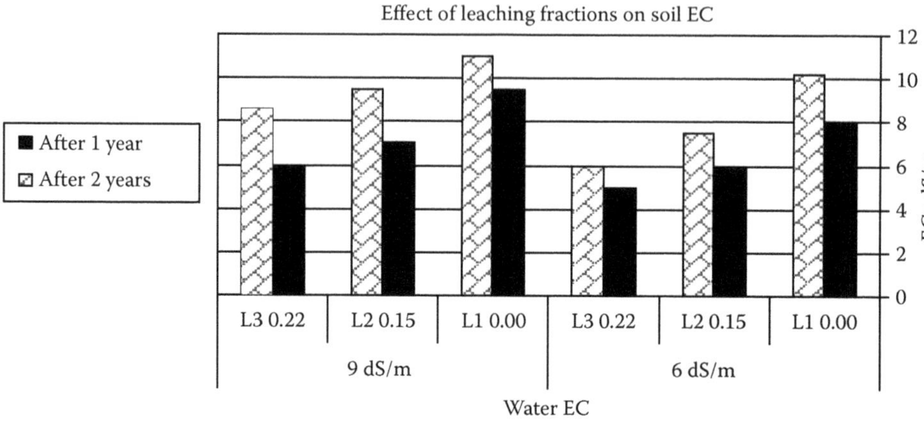

FIGURE 7.3 Effectiveness of leaching fractions in controlling increase in soil EC during irrigation with saline water. (Adapted from Al-Rasbi, SAR. 2010. *Evaluation of the Growth of Date Palm Seedlings Irrigated with Saline Water in Sultanate of Oman*. PhD Thesis, Department of Landscape Ecology and Nature Conservation, University of Kassel, Germany.)

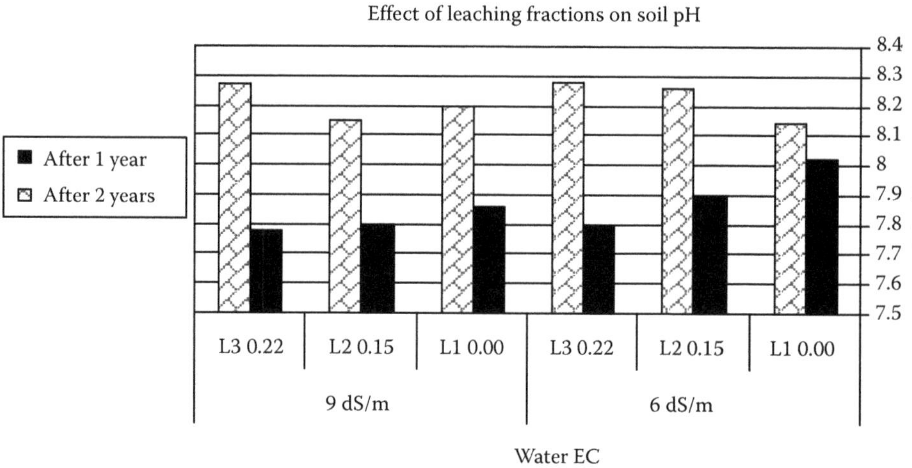

FIGURE 7.4 Effectiveness of leaching fractions in controlling increase in soil pH during irrigation with saline water. (Adapted from Al-Rasbi, SAR. 2010. *Evaluation of the Growth of Date Palm Seedlings Irrigated with Saline Water in Sultanate of Oman*. PhD Thesis, Department of Landscape Ecology and Nature Conservation, University of Kassel, Germany.)

There may also be problems with leaching fractions where subsoil drainage is poor because it may cause a rise of the water table. Thus, a temporary waterlogging may occur. In such a situation, the soil drainage has also to be addressed.

IMPROVEMENT OF SOIL DRAINAGE

Natural drainage is the soil's ability to drain water (rainfall or irrigations) away laterally or horizontally. Excessive or impeded drainage can both cause problems. Rapid drainage of soil will allow less water for plant utilization, whereas poor drainage can cause waterlogging coupled with poor aeration and salt accumulation. If the soil is poorly drained or there is impeded drainage, its drainage must be improved through an artificial drainage system (surface or subsurface). In the case of rapid drainage, the addition of organic matter or a soil conditioner may help us to avoid excessive leaching of water.

IMPROVED IRRIGATION PRACTICES

Irrigation practices for crops, especially date palm, have to be amended under the changing scenes of climate because this plant is growing on very vast areas. Requirements for irrigation will increase due to lesser or scanty rainfall that will demand increased usage of groundwater, the major part of which will be brackish. Therefore, not only water-saving strategies will direly be needed, but also such practices and techniques have to be selected that can control or minimize the deleterious effects of enhanced quantities of saline or sodic water. In the first step the irrigation system itself has to be changed and modified into bubbler, drip, or subsurface instead of surface irrigation. Early warning systems for the prediction of irrigation scheduling will be required so that under- and over-irrigations are avoided. When water is saline then leaching fractions have to be added into consumptive water use of crops. Cyclic use of brackish and good-quality water is also a good and safer strategy (Hussain et al., 2002a), but awareness of sensitive stages to salinity stress must be acquired. For example highly saline water must be avoided at the fruit-bearing stage. Brackish water can wisely be used at the vegetative stage, which is relatively more salt tolerant. However, pertinent research with respect to date palm and intercrops, if practiced, has to be conducted to generate appropriate data.

More frequent irrigations may be required under salinity stress because the available water between the field capacity and the permanent wilting point is lesser under stressed conditions. Treatment of sodic water before its entry into the field with powdered gypsum or gypsum stone (put in the tube well tank) has also proven to be a useful technique for reducing SAR and RSC of water. To increase the partial usefulness of gypsum stone, its supplemental placement in the water channel in a zigzag manner at a distance of 0.5 m on both sides was also found to be useful. The effectiveness increased directly with an increase in length of gypsum stone lining (Hussain et al., 2002a).

APPROPRIATE AGRONOMIC PRACTICES

Appropriate agronomic practices have to be adopted to cope with future climatic changes and the resultant scarcity of water. However, these practices have to be selected according to the prevailing set of conditions of soil and climate. The selection of cultural practices must match the specified conditions. In order to reduce salt accumulation, deep plowing/chiseling has to be practiced before transplanting saplings of date palm, especially when irrigation water is saline.

Mulching is another agronomic technique that helps reduce evaporation, conserve moisture, and thus reduce salt accumulation. Putting a few centimeters of chopped date palm residues on the soil surface significantly cuts down evaporation and keeps the soil with higher moisture and cooler than the black plastic sheet during tomato growing in the Sultanate of Oman (Al-Wahaibi et al., 2007). Brackish water was used for irrigation, but salt concentration (Soil ECe) was less than that found in the control (without any mulch). This technique may prove useful in date palm as well. Hence, mulching can be adopted as the proven technology to face drought and salinity stresses simultaneously. The practice can also be useful in conserving the moisture for longer periods, ultimately helping in early establishment of roots, rapid growth and proliferation, resulting in healthy and stout plants capable of withstanding salinity stress and unfavorable soil physical properties.

Application of organic matter helps a lot in improving water infiltration, keeping the soil porous, and increasing microbial activity, supplying plant nutrients. Plants growing in high organic matter soil are stronger and able to face unfavorable conditions and stresses. Organic matter applications such as farm manures, composts, crop residues, industrial wastes, green manures, etc. also help in fighting against salinity. The release of H ions during decomposition of organic material behaves like Ca ions and keeps the soil aggregated, resulting in larger porosity and less evaporation that ultimately cuts down surface salt concentrations. Application of farm manure and compost helped mitigate deleterious effects of soil and water salinity on soil chemical and physical parameters and controlled yield losses (Hussain et al., 1995, 1998). Application of organic matter proved highly useful to control the increase in soil EC and pH in the study of Rasbi (2010) when saline water was the irrigation source (Figures 7.5 and 7.6, Table 7.16).

Balanced nutrition can also assist plant growth. However, nutrient carriers and the method of application have to be selected according to soil texture, type, and method of irrigation and salinity level. The type and quantity of soil and water salinity must be considered in the selection of fertilizer sources and their application method. When the soil is sodic or sodic water is in use for irrigation, the acidic or acid-forming nutrient sources prove more useful. Split application of nitrogen proves good because date palm is mostly grown in sandy soils. The quantity of fertilizers has to be increased by 15%–20% over recommendations for normal conditions but specific data still have to be produced on this subject.

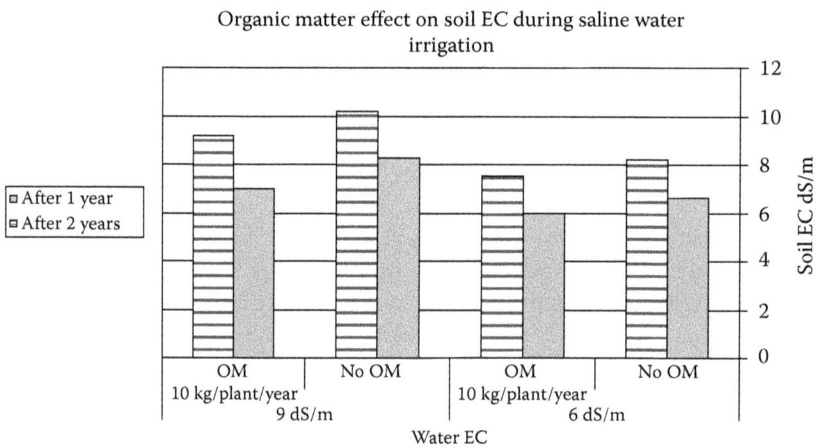

FIGURE 7.5 Effectiveness of organic matter application to control increase in soil EC. (Adapted from Al-Rasbi, SAR. 2010. *Evaluation of the Growth of Date Palm Seedlings Irrigated with Saline Water in Sultanate of Oman.* PhD Thesis, Department of Landscape Ecology and Nature Conservation, University of Kassel, Germany.)

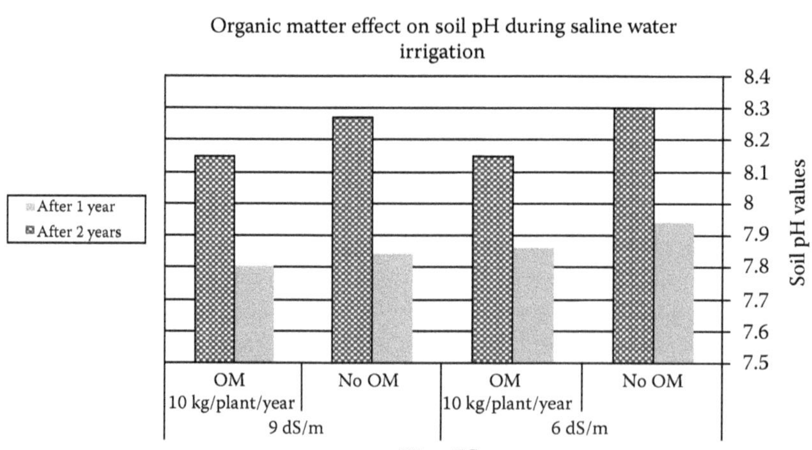

FIGURE 7.6 Effectiveness of organic matter application to control increase in soil pH. (Adapted from Al-Rasbi, SAR. 2010. *Evaluation of the Growth of Date Palm Seedlings Irrigated with Saline Water in Sultanate of Oman.* PhD Thesis, Department of Landscape Ecology and Nature Conservation, University of Kassel, Germany.)

TABLE 7.16
Correlation of Application of Organic Matter with Different Parameters

Sr. No.	Parameter	Coefficient of Corr. (r^2)	Slope	Probability	Significance
1	No. of fronds	+0.447	+1.556	0.006	**
2	Length of frond	+0.483	+5.872	0.003	**
3	Height of trunk	+0.124	+0.722	0.470	NS
4	No. of leaflets	+0.573	+5.239	0.00	**
5	Trunk girth	+0.552	+2.317	0.000	**

Source: Al-Rasbi, SAR. 2010. *Evaluation of the Growth of Date Palm Seedlings Irrigated with Saline Water in Sultanate of Oman.* PhD Thesis, Department of Landscape Ecology and Nature Conservation, University of Kassel, Germany.

APPROPRIATE INTERCROPS

Mono crop of date palm is relatively less profitable when compared with other orchard plants or field cash crops. Therefore, traditionally intercrops are often practiced. However, under the changing scenarios and expected increased use of saline water, the selection of intercrops should be done carefully. Relatively salt-tolerant crops but requiring a lesser amount of water, for example, wheat, barley, sorghum, pearl millet, tomato, spinach, and so on, may be selected. Inclusion of leguminous plants in crop rotations may keep the physical properties of soil favorable for good plant establishment and subsequent growth under severely stressed conditions. This practice also helps in conserving moisture and keeping water available for longer periods.

APPLICATION OF AMENDMENTS

When salts in soil solution are dominated by carbonates and bicarbonates of Na instead of chlorides and sulfates and ESP/SAR exceed critical limits, the stage is set for sodication. To mitigate sodicity and get rid of excessive Na from the exchange complex, the application of amendments like gypsum, sulfuric acid, sulfur, FeS, and CaS becomes necessary (Hussain et al., 1995, 2000, 2002a; Qadir et al., 2001). The divalent ion (Ca) can also be released from native lime ($CaCO_3$) in the soil by the application of inorganic acids. The excessive Na will leach down with subsequent irrigations.

In the case of sodic water irrigations, amendments have also to be added regularly to mitigate the negative effects of Na; otherwise the physical properties of the soil rapidly deteriorate and are expressed in reduced growth and yields. Calculations for quantities of amendments can be made on the basis of soil and water analyses.

GENETIC VARIABILITY AND APPROPRIATE SELECTION OF VARIETIES

A wide variation in salt tolerance potential among various crops and different varieties/cultivars of the same crop exists, which give rise to genetic variability. Therefore, relatively salt-tolerant varieties should be selected when the soil is saline or saline water is to be used continuously or frequently. This practice is also preferable when intercrops are also grown in date palm orchards. The selection of varieties and intercrop sequences is highly important and must necessarily be decided in view of the site-specific conditions of soil, climate, and available water well suited to local conditions of a water-scarce country such as the Sultanate of Oman (Hussain et al., 2006). Under the changing situation, the responsibility of plant breeders also enhances manifold.

FUTURE RESEARCH

A great deal of data and information is lacking regarding the salinity field in the case of date palm so that specific problems can be addressed properly. Hence, concerted efforts in research are direly needed to make up these deficiencies. A few of them are presented here, but still there are many more that demand the attention of scientists.

1. Site-specific suitable hydraulic options (leaching and drainage) for a set of soil, climate, salinity level, and locally prevailing conditions.
2. Appropriate agronomic practices like leveling, deep plowing at the time of transplantation, rainfall harvesting, application of organic matter, balanced nutrients, suitable sowing methods, mulching, selection of varieties that can withstand salinity stresses, and appropriate intercrops.
3. Appropriate irrigation technologies: scheduling, modification of the irrigation system (shifting from surface irrigation to drip, bubbler, or subsurface), and cyclic use of good-quality and brackish water at different crop stages.
4. Starting strong screening studies for genetic variability and breeding programs supported by modern approaches: biotechnology, mutation, and genetic engineering to identify and evolve salt-tolerant cultivars.

CONCLUSIONS

1. Significant increase in temperature, droughts, evaporation, utilization of brackish water, and salinity are expected under predicted global climatic changes in growing areas of date palm (arid and semiarid regions).
2. Date palm plant growth, crop yields, and soil characteristics (chemical and physical) can negatively be affected due to water scarcity, enhanced secondary salinity, and inappropriate management practices.
3. Special management practices have to be adopted for coping with changing scenarios that may include leaching fractions, improvement of drainage, irrigation scheduling, application of organic matter and soil conditioners, selection of salt-tolerant varieties, adjustment of agronomic practices and strong breeding programs for the evolution of highly salt-tolerant varieties using modern approaches of biotechnology and mutation breeding.

ACKNOWLEDGMENTS

The authors of this chapter duly acknowledge the Soil Salinity Research Institute, Pindi Bhattian, Pakistan, the Agricultural Research Center, the Ministry of Agriculture, Sultanate of Oman, the FAO, the US Salinity Lab. staff, and Maas, the authors Al-Rawi and Al-Mohemdy, Ayers and Westcot, Djibril et al., El-Sharabasy et al. (2008), and many other authors for the use of their data or classification system in this chapter. Special thanks are due to Hussain, Al-Rasbi, and coauthors for providing their data, which were used in a major part of this chapter.

REFERENCES

Alia KV, Prasad SK, and Saradhi PP. 1995. Effect of zinc on free radicals and proline in *Brassica* and *Cajanus. Phytochemistry* 39:45–47.

Aljuburi HJ and Maroff A. 2007. The growth and mineral composition of Hatamy date palm seedlings as affected by seawater and growth regulators. *Acta Hort. (ISHS)* 736:161–164. http://www.actahort.org/books/736/736_13.htm

Al-Khateeb SA, Abdulhadi IA, Al-Barrak KM, and Al-Khateeb AA. 1993. Ion and water relations of date palm trees grown in Al-Hassa Coast of Arabian Gulf of Saudi Arabia. *Third Symposium on Date Palm*, Jan. 17–20, 1993, Vol. 1, pp. 293–297. www.pubhort.org/datepalm/datepalm2/datepalm2_15.pdf

Al-Rasbi, SAR. 2010. *Evaluation of the Growth of Date Palm Seedlings Irrigated with Saline Water in Sultanate of Oman*. PhD Thesis, Department of Landscape Ecology and Nature Conservation, University of Kassel, Germany.

Al-Rawi Ahmed AH and Al-Mohemdy AF. 1975. Effect of water quality on the growth and yield of date palm (*Phoenix dactylifera* L.). http://www.pubhort.org/datepalm/datepalm2/datepalm2_13.pdf

Al-Wahaibi NS, Hussain N, and Al-Rawahy SA. 2007. Mulching for sustainable use of saline water to grow tomato in Sultanate of Oman. *Sci. Int.* 19(1):79–81.

Ayers RS and Westcot DW. 1985. *Water Quality for Agriculture*. Irrigation and Drainage paper 29. FAQ, Rome, Italy.

Benlloch-González M, Fournier J, Ramos J, and Benlloch M. 2005. Strategies underlying salt tolerance in halophytes are present in *Cynara cardunculus*. *Plant Sci.* 168(3):653–659.

Bino M. 2008. How can MENA countries respond to long term climate change impacts on water resources. *Abstracts of the First International Conference: Water Resources and Climate Change in the MENA Region*. November 2–4, 2008, Muscat, Sultanate of Oman, 10 pp.

Brugnoli E and Lauteri M. 1991. Effects of salinity on stomatal conductance, photosynthetic capacity, and carbon isotope discrimination of salt-tolerant (*Gossypium hisutum* L.) and salt-sensitive (*Phaseolus vulgaris* L.) C_3 non-halophytes. *Plant Physiol.* 95:628–635.

Cordovilla MP, Ligero F, and Lluch C. 1994. The effect of salinity on N fixation and assimilation in Vicia faba. *J. Expt. Bot.* 279:1483–1488.

Cramer GR, Lauchli' A, and Polito VS. 1985. Displacement of Ca^{2+} by Na^+ from the plasmalemma of root cells. *Plant Physiol.* 79:207–211.

Delgado MJ, Ligero F, and Lluch C. 1994. Effects of salt stress on growth and nitrogen fixation by pea, faba-bean, common bean and soybean plants. *Soil Biol. Biochem.* 26:371–376.

Dhir RP. 1977. Saline waters, their potential as a source of irrigation. In: *Desertification and its Control*. ICAR, New Delhi, pp. 130–148.

Djibril S, Mohamed OKD, Diaga D, Diégane BF, Abaye SM, and Alain B. 2005. Growth and development of date palm (*Phoenix dactylifera* L.) seedlings under drought and salinity stresses. *Afr. J. Biotech.* 4(9): 968–972.

El-Sharabasy SF, Wanas WH, and Al-Kerdany AY. 2008. Effect of salinity stress on some date palm cultivars during proliferation stage *in vitro*. *Arab J. Biotech.* 11(2):273–280.

Erskine W, Moustafa AT, Osman AE, Lashine Z, Nejatian A, Badawi T, and Ragy SM. 2004. *Regional Workshop on Date Palm Development in the Arabian Peninsula*. Abu Dhabi, UAE, May 29–31, 2004. http://www.icarda.org/APRP/Datepalm/Fo&Ac/foreword.htm

FAO. 1992. The use of saline waters for crop production—FAO irrigation and drainage paper 48, 6 pp.

FAOSTAT. 2008. http://faostat.fao.org/default.aspx

Ferreira RG, Tavora FJ, and Hernandez FF. 2001. Dry matter partitioning and mineral composition of roots, stems and leaves of guava grown under salt stress conditions. *Pesqui Agropecu Bras.* 36:79–88.

Flowers TJ. 2004. Improving crop salt tolerance. *J. Exp. Bot.* 55:307–319.

Ghadiri H, Dordipour I, Bybordi M, and Malakourti MJ. 2005 Potential use of Caspan Sea for supplementary irrigation in North Iran. *Agr. Water Manage.* 79:209–224.

Ghafoor A, Qadir M, and Murtaza G. 2004. *Salt Affected Soils: Principles of Management*. Institute of soil and environmental sciences, University of Agriculture Faisalabad, Pakistan, pp. 183–215.

Gupta SK and Sharma SK. 1990. Response of crops to high ESP. *Irrig. Sci.* 11:173–179.

Greenway H and Munns R. 1980. Mechanisms of salt tolerance of the cultivated tomato responses of *Lycopersicon esculentum*, non-halophytes. *Annu. Rev. Plant Physiol.* 31:149–190.

Haechicha M and Abdelgawed G. 2003. *Aspects of Salt-Affected Soils in the Arab World*. SALTMED Workshop, Cairo, December 8–10, 2003.

Hayashi H and Murata N. 1998. Genetically engineered enhancement of salt tolerance in higher plants. In: *Stress Response of Photosynthetic Organisms: Molecular Mechanisms and Molecular Regulation*. Ed. N Murata. Elsevier, Amsterdam, pp. 133–148.

Hussain N. 1997. *Use of Brackish Water for Sustained Crop Production While Maintaining Soil Health*. PhD Thesis, Department of Soil Science, University of Agriculture, Faisalabad, Pakistan, 148 pp.

Hussain N, Ahmad M, Saleem M, and Ali A. 2000. Sodic water management with gypsum application for sustainable crop production. *Pak. J. Bio. Sci.* 3(6):996–997.

Hussain N, Al-Wahaibi NS, Habsi SSA, and Al-Bakri A. 2006. Hydro-economical localization of Oman agriculture. *Proc. Int. Conf: Economic Incentives and Demand Management*. March 18–22, 2006, Muscat, Oman, pp. 188–197.

Hussain N, Hassan G, Arshadullah M, Tahir AG, Naseem AR, and Khan GD. 1998. Bio-amelioration of sandy clay loam saline sodic soil. *Drainage in the 21st Century Food Productions and Environment. Proceedings of the Seventh International Drainage Symposium*. Florida, USA, March 8–10, 1998, pp. 293–300.

Hussain N, Khan GD, Mehdi SM, Sarwar G, Dogar MS, and Hussain T. 1995. Use of brackish water for sustained crop production. *Proc. Fifth Int. Micro Irrig. Cong*. April 2–6, 1995, Orlando, FL, USA, pp. 129–134.

Hussain N, Mujeeb F, Sarwar G, Hussan G, and Ullah MK. 2002b. Soil salinity/sodicity and groundwater quality changes in relation to rainfall and reclamation activities. *Proc. International Workshop on Conjunctive Water Management for Sustainable Irrigated Agriculture in South Asia*, April 16–17, 2002, Islamabad, Pakistan, pp. 91–99.

Hussain N, Sarwar G, Zaka MA, Tahir M, and Mujeeb F. 2002a. Strategies for alleviation of harmful effects of brackish water. *Proc. Second South Asia Water Forum*, 14–16 December 2002, Islamabad, Pakistan, Vol. 2, pp. 631–641.

IPCC 4th Report. 2007. *Climate Change 2007: Impacts, Adaptation and Vulnerability*. Eds. M Parry, O Canziani, J Palutikof, P van der Linden and C Hanson. Cambridge University Press, England, 978-0-521-88010-7.

Jain BL. 1981. Salt balance studies in saline water irrigated soils. *J. Ind. Soc. Soil Sci.* 29:160–166.

Khan GS. 1998. *Soil Salinity/Sodicity Status in Pakistan*. Soil Survey of Pakistan, Lahore, 59 p.

Leeden, van der F, Troise FL, and Todd DK. 1990. *The Water Encyclopedia*. Lewis Publishers, USA, 808 pp.

Levitt J. 1972. *Responses of Plants to Environmental Stresses*. Academic Press, New York, 697 pp.

Maas EV. 1990. Crop salt tolerance. In: *Agriculture Salinity Assessment and management Manual*. Ed. K.K. Tanji, ASCE, New York, pp. 262–304.

Maas EV and Hoffman GJ. 1977. Crop salt tolerance—Current assessment. *J. Irrig. Drai. Div., ASCE* 103(IR2):115–134.

Munns R. 2002. Comparative physiology of salt and water stress. *Plant, Cell & Env*. 25(2):239–250.

Munns R and Tester M. 2008. Mechanism of salinity tolerance. *Annu. Rev. Plant Biol.* 59:651–681 (http://www.plant.annualreviews.org).

Oldeman LR, HakkelIng TA, and Sombroek WG. 1991. Second revised edition. *World Map of the Status of Human-Induced Soil Degradation. An Explanatory Note*. International Soil Reference and Information Center (ISRIC), Wageningen, 35 pp.

Parida SK and Das AB. 2005. Salt tolerance and salinity effects on plants. *Ecotoxicol. Environ. Safety*. 60(3):324–349.

Qadir M, Ghafoor A, and Murtaza G. 2001. Use of saline sodic water through phytomediation of calcareous saline sodic soils. *Agric. Water Manage.* 1647:1–14.

Qureshi RH and Barrett-Lennard EG. 1998. *Saline Agriculture for Irrigated Land in Pakistan: A Handbook*. Australian Center for International Agriculture Research (ACIAR), Canberra, Australia, 30 pp.

Ramoliya P, Patel H, and Pandey AN. 2004. Effect of salinization of soil on growth and macro-and micronutrient accumulation in seedlings of *Salvadora persica* (Salvadoraceae). *Forest Ecol. Manage.* 202(1–3):181–193.

Rabie GH and Almadini AM. 2005. Role of bioinoculants in development of salt-tolerance of *Vicia faba* plants under salinity stress. *Afric. J. Biotech.* 4(3):210–222.

Sanchez FJ, Manzanares M, De Andres EF, Tenorio JL, and Ayerbe L. 1998. Turgor maintenance, osmotic adjustment and soluble sugar and proline accumulation in 49 pea cultivars in response to water stress. *Field Crop Res.* 59:225–235.

Sánchez-Blanco MJ, Ferrández TM, Morales A, Morte A, and José J. 2004. Variations in water status, gas exchange and growth in *Rosmarinus officinalis* plants infected with *Glomus deserticola* under drought conditions. *J. Plant Physiol.* 161(6):675–682.

Shani U and Dudley LM. 2001. Field studies of crop response to water and salt stress. *Soil Sci. Soc. Am. J.* 65:1522–1528.

Soil Science Society of America (SSA). 2008. *Glossary of Soil Science Terms* (https://www.soils.org/sssagloss/ https://www.soils.org/sssagloss/).

Tate RL. 1995. *Soil Microbiology (Symbiotic Nitrogen Fixation)*. John Wiley & Sons, Inc., New York, pp. 307–333.

Tee E, Burrows D, Boland AM, and Putland S. 2003. *Best Irrigation Management Practices for Viticulture in the Murray Darling Basin*. Cooperative Research Centre for Viticulture, Adelaide.

US Salinity Laboratory Staff. 1954. *Diagnosis and Improvement of Saline and Alkali Soils*. USDA Handbook 60. Washington, DC, USA.

Van Genuchten MT and Hoffman GJ. 1984. Analysis of crop salt tolerance data. In: *Soil Salinity Under Irrigation—Process and Management Ecological Studies* 51. Eds. I Shainberg and J Shalhevet. Springer-Verlag, New York, pp. 258–271.

Yoshiba Y, Kiyosue T, Nakashima K, and Yamaguchi-shinozaki K. 1997. Regulation of levels of prolines and osmolyte in plants under water stress. *Plant Cell Physiol.* 38:1095–1102.

Yano-Melo AM, Saggin OJ, and Maia LC. 2003. Tolerance of mycorrhized banana (*Musa* sp. cv. Pacovan) plantlets to saline stress. *Agric. Ecosyst. Environ.* 95(1):343–348.

Yeo A. 1999. Prediction of the interaction between the effects of salinity and climate change on crop plants. *Scientia Hort.* 78:159–174.

8 Date Palm Insect and Mite Pests and Their Management

Mohammad Ali Al-Deeb

CONTENTS

INTRODUCTION

The date palm tree (*Phoenix dactylifera* L.) is attacked by a number of insect and mite pests. Pest problems vary in severity from country to country, which makes it difficult to cover all details of each insect in a way that would apply to every country. In addition, the insect life cycle is affected by environmental conditions and local agricultural practices. Such effects can lead to differences in the timing of adult emergence, mating, egg laying times, and so on. Therefore, this chapter is written in a general format knowing that local details could be easily obtained from local pest control personnel in each country. The objectives of this chapter are (1) to mention the field insect and mite pests of date palms, (2) to describe the importance, appearance, and life history of major pests, (3) to provide a general synopsis of damage done by each pest, and (4) to give guidelines on pest control within an integrated pest management (IPM) program. In talking about insects, I have emulated the writing style used by Larry P. Pedigo in his book *Entomology and Pest Management*, 4th edition. Information on date palm insects and mites in this chapter are derived from the author's field work in the United Arab Emirates (UAE) and from El-Haideri and Al-Hafidh (1986), Zaid et al. (2002), Gassouma (1991), Najeeb (2001), Hussain (1963), and RedPalmWeevil (2009). Other references are mentioned within the text.

INSECT PESTS

Several insects attack date palm trees, although not all of them are serious pests. General information will be given on the pests causing major damage. Other less important insects will be mentioned without details.

Red Palm Weevil

Species

Rhynchophorus ferrugineus Olivier (Coleoptera: Curculionidae).

Importance

This species is one of the most destructive insect pests attacking the date palm tree. Infestation is concealed and usually discovered when damage becomes severe (Abraham et al. 1998). Injury is caused by larvae feeding on soft tissues inside the tree trunk. The red palm weevil is native to southern Asia and Melanesia where it originally fed on coconuts (Ferry and Gómez 1998). It was introduced to several countries in the world through transportation of infested date palm trees and offshoots. At present, populations of this pest are found in most countries where date palm trees are grown. Repeated infestations can occur with this insect due to its highly aggregated spatial distribution (Faleiro et al. 2002). This insect prefers young palms <20 years of age (Abraham et al. 1998).

Appearance

The adult is ~30 mm long with a reddish brown body. It has a long curved distinctive snout (rostrum) that is smooth in the female and with short black bristles in the male. The larva is legless with a creamy body and dark brown, strong head capsule. Its body is broad at the middle and tapered at the ends (conical shaped). It can reach 5 cm in length before pupation. The pupa is creamy to light-brown in color and gets darker over time. Pupation occurs inside a cocoon made from date palm tree fibers and masticated trunk tissues. Eggs are creamy-white (2–3 mm) deposited singly inside trunk soft tissues. The symptoms of red palm weevil damage include: (1) sticky, thick liquid (brown-reddish) with a fermented odor oozing out of holes on the trunk, (2) masticated tissues appearing from tunnel holes on the trunk, (3) sounds of larval mandibles chewing and gnawing tissues inside the trunk, and (4) presence of empty cocoons on the trunk or under infested trees.

Life History

Adults do not overwinter and are found throughout the year. There are two population peaks in some countries (e.g., the UAE and Saudi Arabia). Adults feed (without causing measurable damage) and, after mating, the female lays about 200–300 eggs during her life span. Eggs are laid singly into cavities previously made with the female chewing mouthparts or at frond bases. Eggs can also be laid inside wounds made by tree pruning. Eggs hatch and larvae emerge and start feeding on soft tissues inside the date palm trunk. Larvae grow over time, and the damage grows simultaneously. The last larval instar leaves the trunk through an exit hole to make a cocoon of tree fibers and masticated tissues. Pupation takes place inside the cocoon on the tree trunk between frond bases or inside cavities made by larval feeding. At the end of the pupation period, adults emerge. The life cycle is about 3–4 months and this insect has three to four generations per year.

More details on the red palm weevil can be found in Faleiro (2006), which is an excellent and thorough review that covers through 2006 the literature on the red palm weevil over the last 100 years.

Rhinoceros Beetles

Species

Oryctes spp. (Coleoptera: Scarabaeidae). Three species *O. agamemnon* Burmiester, *O. elegans* Prell, and *O. rhinoceros* L. cause damage to the date palm tree. *Oryctes elegans* is referred to as the fruit stalk borer. The three species can be identified as follows (Brett Ratcliffe, Professor and Curator, University of Nebraska State Museum, personal communication):

Apex of hind tibia with two large, fixed teeth	*O. rhinoceros*
Apex of hind tibia with three large, fixed teeth	*O. agamemnon* and *O. elegans*
Underside of front tibia in both sexes with a strong tooth	*O. elegans*
Underside of front tibia in both sexes without a strong tooth	*O. agamemnon*

Importance

Larvae of *Oryctes* spp. have very strong mandibles and as a result they can cause severe damage. Adult *O. agamemnon* are root and frond (leaf) petiole borers whereas adult *O. elegans* are fruit stalk borers. This species complex needs further study before using pheromone traps, especially in countries where the three species coexist. From my personal field observations I have noticed that many pest control workers in the UAE do not know how to differentiate among these species.

Appearance

Adults are large, 30–50 mm long, and black or brownish black in color, with stout bodies. The cephalic horn of the male is larger than that of the female. Larvae are big (30–60 mm), C-shaped, and white in color with three pairs of thoracic legs. The head capsule is large and dark brown in color with two strong mandibles. *Oryctes* beetle larvae can easily be distinguished from red palm weevil larvae because the former has six thoracic legs and the latter is legless.

Life History

In general, early in the spring, adults emerge and, after mating, females deposit eggs in the soil near the date palm roots or on the trunk and frond petioles. Eggs hatch and larvae bore into plant tissues causing damage. Larvae feed for roughly 7–9 months. Pupation occurs in soil or in the plant. Adults emerge and cause damage to roots, fronds, and petioles, or fruit stalks (depending on the species). *Oryctes agamemnon* females were reported to deposit eggs in different parts of the date palm tree (between the hairy roots, on the stem, and at the base of cut fronds). These three species are univoltine insects with a generation time of 9–10 months. In the UAE, *O. agamemnon* breeds in organic manure, which is used as an organic fertilizer in the fields. *Oryctes* beetle adults are active at night and are attracted to light sources. The full details of the life cycle of this species were studied in the laboratory by Soltani et al. (2008).

Long Horn Date Stem Borer

Species

Jebusaea hammerschmidtii Reiche (Coleoptera: Cerambycidae).

Importance

The damaging stage is the larva. Larvae bore inside the trunk, making cavities. The progress of damage by this insect is slower than that caused by the red palm weevil, although some trunks have been found riddled with several hundred larvae feeding inside.

Appearance

Male (21–32 mm) and female (30–45 mm) adults are light to dark brown in color. They are easily distinguished from other date palm insects because of the presence of long antennae which can be as long as the body. Females are bigger than males. The female elytra does not cover the last abdominal segment whereas the male elytra does. Larva can reach as much as 45 mm in length and is creamy with a small brown head. It is legless; however it can be distinguished from the legless larva of the red palm weevil by its slender, elongated body, which has large segmentation with the first thoracic segments broader than the rest of the body.

Life History

After mating, females deposit eggs on the tree trunk and between frond bases. Eggs hatch and larvae bore inside the trunk causing damage. The larval stage takes several months, and this pest thrives on trees in neglected and water-stressed conditions. Larvae feed for about 10 months. Pupation takes place inside the trunk. Adult beetles leave the trunk through exit holes that are round or oval with smooth edges resembling big "bullet holes." Small amounts of tree sap can be seen oozing from these holes. This insect has one generation per year. Adults are active at night and are attracted to light sources.

Plant Hopper (Dubas Bug)

Species

Ommatissus binotatus var. *lybicus*, De Bergevin (Tropiduchidae: Homoptera).

Importance

Adults and nymphs cause damage by their piercing, sucking mouthparts. Tree sap is removed in large amounts under heavy infestation, which affects tree vigor. Also, leaf hoppers excrete large amounts of honeydew, which is an excellent medium for the growth of the sooty mold fungus. The black fungal growth covers the fronds (leaves) with a dark coating that interferes with photosynthesis. In cases of heavy infestations, the pinnae (leaflets) become very sticky and honeydew drops are noticed on the ground under the infested trees. The presence of sticky liquid (honeydew) on fronds (leaves) is an indication of Dubas infestation.

Appearance

The adult female is a small insect (5.5 mm) with four transparent wings nearly equal to the body length. The adult male is 3.5 mm and the wings are longer than the body (about 1.5 times of body length). The plant hopper is green to pale yellowish in color. Eggs are green and become yellow before hatching. Nymphs are wingless and have long hairs at the end of the abdomen.

Life History

This insect has two generations per year (winter and summer generations). Females lay eggs on all green parts of the palm tree except the fruit, and the majority are deposited on the leaflets, especially on the upper surface. Winter-generation eggs hatch in early spring, producing nymphs that start feeding and causing damage by sucking sap and producing honeydew. Nymphs develop into adults in summer, which also feed and cause damage. Before they die, they lay eggs for the summer generation. Summer-generation eggs hatch and nymphs emerge to feed and cause damage. They become adults which then lay eggs for the winter generation and die. The summer generation is shorter than the winter generation. The Dubas bug thrives mainly in humid areas and on trees that are closely planted. Low humidity and proper spacing of trees can significantly reduce the infestation potential.

Lesser Date Moth (Hemira)

Species

Batrachedra amydraula Meyrick (Cosmopterygidae: Lepidoptera).

Importance

This is a very important insect as it attacks the inflorescence and newly formed fruits. Affected fruits fail to grow and become dry and full of larval frass. They may then fall to the ground or remain on the fruit bunch. The larva is capable of moving from one fruit to another causing more damage. Heavy infestations reduce the yield considerably.

Appearance

The adult is a small moth (1–2 mm) brownish in color. Its abdomen is gray to silver and it has brown eyes. The long antennae are slivery with brown dots. Finding brown-reddish dry fruits on the ground under the tree or on the fruit bunch is an indication of lesser date moth infestation.

Life History

The overwintering stage is the pupa inside its cocoon, which is hidden at frond bases or among tree fibers. In the spring, eggs hatch and larvae emerge and penetrate newly formed fruits. Larvae eat the fruit content and destroy it. Pupation takes place on the tree trunk or in soil. Adults emerge, mate, and females lay eggs and die. This insect has two to three generations per year. Date variety selection is very important when establishing a new date palm farm. A farm with mixed varieties of different flowering times (early, middle, and late) makes the control of this pest difficult.

Scale Insects

Species

Green scale/date palm pit scale *Palmaspis phoenicis* (Asterolecaniidae: Homoptera), white scale/parlatoria date scale *Parlatoria blanchardi* (Trag.) (Diaspididae: Homoptera), and red date scale *Phoenicoccus marlatti* Cockerell (Coccidae: Homoptera).

Importance

Scale insects cause damage by sucking the sap from the plants (fruits and fronds). Their damage can be serious, sometimes causing yellowing or death of the heavily infested fronds, but trees and offshoots do not die. Heavily infested plants appear unhealthy. Soft scales excrete large amounts of honeydew. This secretion is an excellent medium for the growth of a black sooty mold fungus, which interferes with photosynthesis.

Appearance

Adult scale insects are small (0.6–1.5 mm), legless, and spherical/oval in shape. The body of *Phoenicococcus marlatti* is small, about 1.5 mm long, spherical, red or reddish brown, embedded or nested in white wax.

Life History

A general life cycle of an armored scale (e.g., *P. blanchardi*) begins with the eggs being laid underneath the waxy covering. The newly hatched scales (crawlers) move about over the plant and begin feeding using their piercing, sucking mouthparts. Female scales lose their legs and antennae during the first molt. They molt again to reach maturity. Adult males are tiny two-winged insects without mouthparts. The life cycle of soft scales is similar to the armored scales. However, legs and antennae of female soft scale are reduced but not lost. In general, scale insects have several generations per year and the number of generations is affected by temperature and humidity.

Wasps

Species

The Oriental Wasp, *Vespa orientalis* (L.) (Vespidae: Hymenoptera), is the main pest. There are two other species of lesser importance, the Yellow Wasp, *Polistes hebroeus* F., and the Spotted Yellow Wasp, *Polistes gallicus* L.

Importance

Wasps are considered secondary pests. Damage is caused by the feeding of adults on fruits. Damaged fruits tend to be more affected by other insects and diseases in the field and in the store.

Appearance

The adult has a stout body (25–30 mm) and large compound eyes and four strong wings yellowish brown in color. The body has yellow stripes.

Life History

Wasp-nesting sites are not located on date palm trees. Early in the spring, the wasp queen lays eggs. After egg hatching, it supplies the larvae with food until they are ready to pupate. At the end of pupation, adult wasps emerge and start searching for food. They attack date palm fruits and cause damage.

Other Insects

- Date Thrips: *Adiheteothrips jambudvipae* Ramok (Thripidae: Thysanoptera)
- Termite (white ants): *Microcerotermes diversus* Slivestri (Termitidae: Isoptera)
- *Carpophilus obseletus* Erichson (Nitidulidae: Coleoptera)
- *Elaphidion villosum* (Cerambycidae: Coleoptera)
- Date Stone Beetle: *Cocotrypes dactyliperda* (Fabricius) (Scolytidae: Coleoptera)
- Frond Borer: *Phonapate frontalis* Fahraeus (Bostrychidae: Coleoptera)
- Greater Date Moth: *Arenipses sabella* (Hampson) (Pyralidae: Lepidoptera)
- Pomegranate Butterfly: *Virachola livia* Klug (Lycaenidae: Lepidoptera)
- Bark beetle: *Xyleborus perforans* Wollaston (Scolytidae: Coleoptera)
- Mole cricket: *Gryllotalpa gryllotalpa* L. (Gryllotalpidae: Orthoptera)
- Locust: *Schistocerca gregaria* Forskal (Acrididae: Orthoptera)
- Pink Hibiscus mealybug: *Maconellicoccus hirsutus* (Green) (Pseudococcidae: Hemiptera)

MITE PESTS

Old World Date Mite

Species

Oligonychus afrasiaticus Mc Gregor (Acari: Tetranychidae).

Importance

All mite life stages (except eggs) cause damage by their rasping-sucking mouthparts. They mainly remove sap from fruits and produce webs that collect the dust. Affected fruits become dark or reddish and are not suitable for marketing. Webbing is one of the problems that reduces the market value of fresh dates as its presence is objectionable to customers. The dusty appearance of affected fruits gives this spider the common name of "date dust mite" in some countries.

Appearance

The adult mite is tiny (0.3–0.5 mm), greenish-yellow, and oval-shaped. The terminal leg segments are reddish in color. The first immature stage (larva) has three pairs of legs. The following nymphal stages and the adult have four pairs of legs.

Life History

The life cycle is composed of the egg, the larva, nymphal stages, and the adult. The eggs are attached to the fruit and the silk webbing and hatch in few days. Temperature affects the length of time from egg to adult. Under favorable conditions, there are many (10–12) generations per year. The dusty appearance of the fruit and the presence of fine silky webbing on the bunch is an indication of mite infestation.

Other Mites

- Palm false spider mite, *Tenuipalus eriophyides* Baker
- Leaflet false spider mite, *Raoiella indica* Hirst.
- Palm bud mite, *Mackiella phoenicis* K.
- Oriental mite, *Eutetranychus orientalis* (Klein)
- Palm bud mite, *Mackiella phoenicis* K.
- Palm rust mite, *Tumescoptes trachycarpi* K.

PEST DAMAGE

Insect and mite pests that attack the date palm can be categorized into three groups according to the mouthpart type (Table 8.1). The damage caused by each pest can be predicted by knowing the type of its mouthparts.

Chewing Pests

Insects with chewing mouthparts masticate plant tissues with their strong mandibles. Their feeding results in parts of the plant being chewed or torn off. Some of the date pests with such mouthparts cause damage by eating fruit and inflorescence, making tunnels and cavities inside the trunk, and eating roots and leaves.

TABLE 8.1
Date Palm Insect and Mite Damage

Pest	Stage	Mouthparts	Affected Part
Adiheteothrips jambudvipae	A, N	RS	Inflorescence
Arenipses sabella	L	C	Inflorescence
Batrachedra amydraula	L	C	Inflorescence, fruit
Carpophilus obseletus	L	C	Inflorescence
Cocotrypes dactyliperda	A, L	C	Fruit stone
Elaphidion villosum	A	C	Fruit
Gryllotalpa gryllotalpa	A, L	C	Root
Jebusaea hammerschmidtii	L	C	Trunk
Maconellicoccus hirsutus	A, N	PS	Fronds
Microcerotermes diversus	A (worker)	C	Root, trunk
Other mites	A, L, N	RS	Leaf, fruit
Ommatissus binotatus	A, N	PS	Leaf
Oligonychus afrasiaticus	A, L, N	RS	Fruit
Oryctes sp.	L	C	Root, trunk, frond
Oryctes sp.	A	C	Fruit stalk
Phonapate frontalis	A, L	C	Frond
Rhynchophorus ferrugineus	L	C	Trunk
Scale insects	A, N	PS	Frond, fruit
Schistocerca gregaria	A, L	C	Fruit, frond
Virachola livia	L	C	Fruit
Wasps	A	C	Fruit
Xyleborus perforans	A, L	C	Trunk

C = chewing, PS = piercing sucking, RS = rasping sucking, A = adult, L = larva, N = nymph.

Piercing–Sucking Pests

Mouthparts of this group of insects are modified to feed on plant sap. Sap removal causes the affected area to become discolored (yellow or brown), mottled, or necrotic (dead), or results in wilt. Many of these insects produce honeydew on the affected plant part, which may support the growth of a black or gray sooty mold. Mold growth impairs or stops the process of photosynthesis in the affected tissues.

Rasping–Sucking Pests

These pests rasp or scrape the surface of plant tissue and suck up the sap oozing from the damaged area. The damage symptoms of this group of pests resemble the damage produced by piercing-sucking insects.

DATE INSECT AND MITE MANAGEMENT

Insect species composition and damage severity vary from country to country. Also, the extent of pest control work varies among countries depending on the local agricultural practices and weather conditions. There is no doubt that the best solution is to have a good IPM program. IPM is defined as "an effective and environmentally sensitive approach to pest management that relies on a combination of common-sense practices" (EPA 2009). Because there are several insects and mites that affect the date palm, and these have different life cycles, the choices for their control vary as well. In this chapter, I do not provide a single detailed program that would be suitable for controlling date palm pests in every country. Instead, I have given guidelines on how to establish a good IPM program. Many pest control programs exist; however, these programs do not always give good results. The reason is simple; they require good practice and adherence to the program details from the user side. So, just having a good pest control program does not guarantee success, rather it represents the first step toward controlling pests successfully, and more steps remain to be accomplished in order to reach the target. Many date palm growers complain about the severe damage caused by some insect pests, such as the red palm weevil even though they have knowledge about IPM and access to control methods. The failure of pest control efforts lies in the way in which they have been implemented. To make this chapter of a general benefit when talking about IPM, I have selected the "step-by-step procedure for developing an IPM program" written by Flint et al. (2003) to be used as a guideline for establishing an IPM program. The 10 steps are:

1. *Identification of all potential pests (including all life stages) in the system.* As discussed earlier, in some countries, the researchers working on the control of *Oryctes* spp. do not know how to distinguish among them. The same problem applies to the identification of natural enemies. The identification issue can be easily solved by sending insect samples to be identified in a specialized laboratory or museum.
2. *Creating monitoring guidelines for each pest.* A common problem is that most of the time and effort of pest control workers in a given country is spent on one pest. For example, in Arabian Gulf states, this pest is the red palm weevil. There is extensive knowledge about this pest and there are monitoring guidelines for it. At the same time, monitoring guidelines for other pests are poor or nonexistent. The long horn beetle in the UAE has started causing extensive damage, and its population continues to build over time, yet no IPM program has been developed for its control.
3. *Setting up injury levels and action thresholds for each pest species before making any treatment.* It is safe to say that this is one of the biggest knowledge gaps of pest control programs in most of the countries where date palms are grown (with some exceptions, of course). In some countries, control operations are launched in response to growers'

demands, and no thresholds are used. Such control efforts will not be fully successful because they are not based on experimental research. In fact, many natural enemies of date palm pests are killed under ill-planned control operations. As a result, in some countries biological control does not play a significant role in the IPM programs for date palm pests. Setting up injury levels, especially in the Arabian Gulf states (where a large number of date palm trees are grown), is urgently needed in order to have successful control programs.

4. *Setting up a record-keeping system.* Unfortunately, this aspect is not considered important by some pest control authorities, especially in the long term. In fact, some pest control researchers have told me that keeping very detailed records or even publishing the results of research work in peer-reviewed journals are pure academic practices. With this mentality, data about pests will always be used in the short term, but long-term data and knowledge get lost. Good records give pest control researchers needed background for designing effective IPM programs. Good records also prevent the repetition of mistakes or experiments that have been done in other parts of the same country or in other countries.
5. *Development of acceptable management strategies for each pest.* Pest management strategies vary from simple preventive measures to approaches that use pesticides. They include legal control, cultural control, physical and mechanical control, biological control, and chemical control. Prevention of pest problems is always better than treatment. Quarantine, for example, is very important in preventing the spread of pests, such as the red palm weevil, into new counties or into new areas within one country. Biological control should be considered because it is a safe management strategy that can play both preventive and control roles. A list of pesticides that can be used to control each pest should be developed based on laboratory and field tests in each country. The selected pesticides should be effective and environmentally sound.
6. *Development of specific criteria for selection of pest management methods.* Pest management methods should be cost-effective, least hazardous, least disruptive to the environment, and least toxic to nontarget organism. Selection criteria of pest management methods should provide the maximum effectiveness and minimum negative side effects. A pest management manual should be prepared for each country.
7. *Development of guidelines to be followed each time a pesticide is used.* Reading the label is a general rule that should be followed before using any pesticide. A comprehensive list of safety and precautionary measures should be prepared and made available to all people using or dealing, directly or indirectly, with pesticide applications. A pesticide application and safe handling manual needs to be developed for each country.
8. *Assigning a person to be responsible for each step of the IPM program.* An organized group of people should be in charge of all IPM program segments from the planning to the postharvest stages. Each team member should be able to track the development and progress of one aspect of the IPM program.
9. *Development of a list of resources.* Having information resources is very important; however, this is useless if people do not know about the existence of such resources. The resource list should be published and made available through multiple knowledge dissemination methods.
10. *Knowing that IPM policy is a living document that changes over time as a result of experience growth and availability of new information.* This is crucial for decision makers because if the IPM program is treated as a fixed policy that should be followed verbatim without change, then the IPM program will not achieve its goals. An IPM program is a dynamic product of teamwork, and it should receive input from all relevant parties and should be periodically updated and revised.

We need to remember that the existing gaps in data about several date palm insect pests is one of the major obstacles against having successful control programs in many countries. A good understanding of the life cycle of each pest and the knowledge of where and how it develops are required to have successful long-term control programs. Knowledge of local natural enemies is also essential. Additionally, sophistication and success of an IPM program depend on the experience, skill, education, and enthusiasm of people working on the program (Flint et al. 2003). IPM programs should take a comprehensive, multidisciplinary approach toward dealing with pest problems.

In general, no matter how simple or complicated the IPM program, it should include the identification of the pest, the quantitative measurement of its population density by means of sampling, a comparison of population density with economic threshold levels, a means of choosing and implementing proper control options, and a means of evaluating the efficacy of pest control measures by retaking samples.

Several control methods can be used to control date palm pests. The methods include legal control (quarantine), biological and microbial control, control using attractants and traps, mechanical control, and chemical control.

LEGAL CONTROL (QUARANTINE)

Prevention is always better than treatment. Insect pests move accidentally to new uninfested areas by several anthropogenic means. Not all pests travel for long distances. Quarantine can play a major role in preventing the spread of pests and should be practiced at inter- and intracountry levels. The red palm weevil is expanding its infestation zone every year, and a strong quarantine program is needed to stop or at least delay its invasion. *Oryctes* beetles use organic fertilizer as a breeding medium and transportation. Once an infested fertilizer is used, the *Oryctes* damage starts to appear in the region and the beetles spread from there to new areas.

BIOLOGICAL AND MICROBIAL CONTROL

This approach uses living organisms to suppress pest populations. Predators, parasites, and pathogens are the three categories of natural enemies of insects. For some insect pests, there is more than one natural enemy (Table 8.2). However, few or no natural enemies exist for some insect pests in some countries. When insects move to new geographical areas, they usually have few or no natural enemies in the new habitat. This applies to the red palm weevil, which invaded many countries and established itself as an alien species. However, the arrival of new pests should not discourage attempts to use biological controls against them. There is a good chance that natural enemies move into the new habitat by the same method through which the invasive species was introduced. Also, indigenous natural enemies may attack invasive pests, especially if the natural enemy is a generalist. There are three primary methods of using biological control agents (Gullan and Cranston 2004): importation, conservation, and augmentation.

Importation

Natural enemies of the country from which an exotic pest has been introduced are collected. They are released to control the exotic pest in the new habitat with the hope that they will establish a permanent population. This is usually called classical biological control. However, this approach may lead to negative ecological impacts if the host range of the new natural enemy has not been extensively evaluated.

Conservation

Indigenous natural enemies can be protected by reducing or eliminating insecticide applications or by providing a shelter. Selecting insecticides with the least damage to nontarget organisms is one of the major requirements of an effective and successful IPM program.

TABLE 8.2
Some Date Palm Pests and Their Biological Control Agents

Biological Control Agent	Biological Control Name	Target Pest	Source
Bacteria	*Pseudomonas aeruginosa*, *Serratia* sp., *Bacillus sphaericus*, *B. megaterium*, *B. laterosporus*, and *B. thuringiensis*	*R. ferrugineus*	Faleiro (2006)
Virus	Cytoplasmic polyhedrosis		
Nematode	*Heterorhabditis* spp., *Teratorhabditis palmarum*, *Steinernema* sp., and *Rhabditis* sp.		
Fungi	*Beauveria bassiana* and *Metarhizium anisopliae*		
Fly (insect)	*Sarcophaga fuscicauda*		
Wasp (insect)	*Scolia erratica*		
Fungi	*Beauveria bassiana* and *Cordyceps* sp.	*J. hammerschmidtii*	El-Haideri and Al-Hafidh (1986), Hussain (1963)
Predatory mites	*Hypoaspis* sp. and *Ameroseius* sp.		
Predatory mites	*Stethorus gilvifrons*	*O. afrasiaticus*	
Wasps	*Parasierola* sp., *Habrocytus* sp., *Pediobius* sp., *Bracon hebetor*, and *Phanerotoma* sp.	*B. amydraula*	
Green lacewing	*Chrysopa carnea*		
Predatory mites	*Bdella* sp.	*O. binotatus*	
Green lacewing	*Chrysopa carnea*		
Lady bugs	*Coccinella septempunctata*, *C. undecimpunctata*		
Virus	Baculovirus and *Bacnlovinis oryctes*	*O. rhinoceros*	Lacy et al. (2001), Gopal et al. (2001)

Augmentation

Mass rearing and periodic release of natural enemies can be used to control insect pests. An inoculative release is done when small numbers of natural enemies are released in certain areas to suppress outbreaks of pests. An inundative release is done by placing large numbers of natural enemies in one location to flood the pest population.

In many countries, numerous pesticides are used on date palm trees without considering the presence of natural enemies. The balance between the pest and its natural enemies gets disrupted by this negative intervention. Also, when using chemical control, selective application should be practiced to save natural enemies in some parts of the field. Pest infestation levels should be evaluated and several zones should be identified based on the gradient of severity of infestation. Undoubtedly, this approach takes more time and effort, but its negative impact on natural enemies is much lower. Traditionally, date palm growers like to see the entire date palm plantation treated with pesticides. However, with this mass application method, the damage to natural enemies is high. Exploration for natural enemies needs to be conducted in each country, especially in areas with no or minimal pesticide applications. This is one of the important areas that deserves more attention, particularly at the governmental level in countries where date palm trees are grown. Many growers are tempted to support the importation approach; however, conservation and local augmentation are more ecologically sound and sustainable. Indigenous natural enemies are part of the local environment and do not need acclimatization. The risks of a new imported natural enemy turning out to be an invasive species and the unsuitability of weather conditions make the importation of natural enemies a questionable approach for the biological control of date palm pests.

CONTROL USING ATTRACTANTS AND TRAPS

A variety of insect-attracting chemical and visual lures can be used to monitor or to directly reduce insect populations (Weinzierl et al. 2005). Currently, traps are used to monitor and control certain insects on date palm trees. Traps are excellent tools which are compatible with IPM programs as they have little environmental impact compared to chemical control using conventional insecticides.

Pheromone Traps (Chemical Attractants)

Some of the date palm insects can be monitored and mass-captured using pheromone traps. The trap should contain the right pheromone for the targeted species. This step requires proper identification of the pest species, which is the first step of the IPM program. In the following paragraphs the pheromone traps of two important pests will be briefly discussed as examples of the use of pheromones in date palm pest management.

Red Palm Weevil Trap

The trap is a bucket containing the synthesized red palm weevil male aggregation pheromone along with water and food bait. The male aggregation pheromone is the "ferrugineol" (4-methyl-5-nonanol) which was identified and synthesized by Hallett et al. (1993). The bucket size, color, number, and shape of holes/windows on its sides and lid, and the condition of its surface (smooth vs. rough) vary between countries. Better catches have been achieved using dark-colored traps such as black (Hallett et al. 1999), green (Ajlan and Abdulsalam 2000), or red (Al-Saoud et al. 2010) compared to white traps. Low-quality dates are mixed with water and used as food bait in the trap. The addition of ethyl acetate to the trap can significantly increase the catch. The pheromone trap is an excellent monitoring and sampling tool in determining the presence of the pest and in assessing its population density. The pheromone food-baited trap has been widely and successfully used in date palm plantations in many countries. It is also used as a mass trapping control device to suppress the red palm weevil population. In fact, it is the most significant tool used in the IPM program for this insect.

Rhinoceros Beetle Trap

As discussed earlier in this chapter, there are three *Oryctes* species that attack the date palm tree. The pheromones of *O. elegans* (Hallett et al. 1995) and *O. rhinoceros* (Rochat et al. 2004) are known, but pheromones of *O. agamemnon* are not. In order to have successful trapping, the right pheromone should be used for each of these species. In the UAE, I have seen a trial in which the *O. rhinoceros* pheromone was used in an attempt to collect *O. agamemnon* but the traps did not catch any beetles.

Light Traps (Visual Lures)

Rhinoceros Beetle and Long Horn Borer

Adults of both insects are attracted to light at night. A regular light trap (light bulb, inverted funnel, bucket to collect trapped insects) is successfully used to collect these two insects. However, the trap needs an electrical source, and in some farms this could be a problem. Also, the cost of the electric bill and trap maintenance are other factors to be considered, especially on big farms. Using light traps may pose some risk to nonpest insects and useful natural enemies, which are attracted to the light and are trapped and killed along with the intended pests.

MECHANICAL CONTROL

Discarding heavily infested trees, offshoots, or damaged tree parts (leaves and fruits) by burying or burning can stop the spread of infestation in the field. This is very important with the red palm weevil infestation. Broken infested trunks that are left in the field serve as an infestation source from which many adults emerge to attack healthy trees.

CHEMICAL CONTROL

Conventional insecticides and acaricides have been used by date palm growers because they act rapidly and are readily available. Such pesticides are available in different formulations: granules or pellets, dusts, soluble or wettable powders, emulsifiable concentrates, aerosols, and ultravolume concentrates (ULV). Pesticide selection should be made based on efficacy and safety. Despite the advantages of conventional pesticides, their damage to nontarget organisms and to the environment in general can be extensive. A good pesticide should be IPM compatible and environmentally sound. These two conditions may seem difficult to find in a pesticide, but at the very least, a thorough review of the pesticide properties should be made before using it in the field. Sound scientific and environmental factors should be considered before adopting insecticides/acaricides to be used in control programs. Business and money should not be made at the expense of efficacy and environmental safety.

Insecticide names, registration status, and application rates vary among countries. Therefore, no mention of such information was included in this chapter.

Several methods of pesticide application, including spray, injection, soil treatment, and fumigation, are used to control date palm insect pests:

Spray

High-pressure and high-volume pesticide sprayers are used on date farms. The pressure should allow the chemical to reach treetops. This is important especially in controlling mites, plant hoppers, and lesser date moth.

Injection

This method is used when controlling trunk borers. It has advantages, including the ability to being used during rainy or windy weather, uses less volume of material, safety to people, natural enemies, and wildlife. Injection holes on the date palm trunk should be plugged after application in order not to attract females of the red palm weevil.

Soil Treatment

Some pesticides are mixed with the soil near the root system. This method is good against the larvae of *Oryctes* beetles. However, the effect on soil fauna should be taken in consideration. In addition, the efficacy of pesticide movement inside the tree should be tested.

Fumigation

The larval and pupal stages of date palm trunk borers can be mechanically removed from the trunk. This mechanical treatment is usually coupled with chemical control using a fumigant such as aluminum phosphide in the form of pellets. This is done manually using special knives and equipment.

After removing the damaged trunk tissues, the pellets are placed inside the hole, which is then filled with cement. The pellets release a toxic gas which kills the immature stages that were not reached by the mechanical cleaning treatment. This method is successfully used with the red palm weevil in several countries.

CONCLUSION

Insects and mites that attack date palm trees can cause severe damage. However, well-planned and well-implemented IPM programs can stop or minimize the damage. It is never too late for countries that do not have well-established IPM programs to start building the infrastructure for such programs. IPM program dynamism is a factor without which control tactics and programs do not give desirable results. Pest eradication should not be the target of our IPM efforts; the focus should be on good management programs.

Local research is needed in order to obtain precise information on pests since insect ecology varies among countries. Additionally, information exchange among affected countries is vital in order to improve control efforts. It is disturbing to see many complaints about insect problems in some countries while there is research being done on these pests. The unfortunate reason is that, too often, the research was done incorrectly. Researchers should build their work on a sound basis of experimentation and not solely on observation. Observation is the first step of scientific methodology, but is not itself sufficient. Low-quality research and extension publication do not lead to good control results.

Insects and mites can develop resistance to pesticides; therefore, resistance monitoring and management are two essential items that need to be studied in any control program that uses pesticides.

REFERENCES

Abraham, V. A., M. A. Al Shuaibi, J. R. Faleiro, R. A. Abozuhairah, and P. S. P. V. Vidyasagar. 1998. An integrated management approach for red palm weevil, *Rhynchophorus ferrugineus* Oliv. a key pest of date palm in the Middle East. *Agric Sci* 3:77–83.

Ajlan, A. M. and K. S. Abdulsalam. 2000. Efficiency of pheromone traps for controlling the red palm weevil, *Rhynchophorus ferrugineus* (Olivier) (Coleoptera: Curculionidae) under Saudi Arabia conditions. *Bull Entomol Soc Egypt (Economic Series)* 27:109–120.

Al-Saoud, A. H., M. A. Al-Deeb, and A. K. Murchie. 2010. Effect of color on the trapping effectiveness of red palm weevil pheromone traps. *J Entomol* 7:54–59.

El-Haideri, H. S. and E. M. T. Al-Hafidh. 1986. *Palm and Date Arthropod Pests in the Near East and North Africa*. Food and Agriculture Organization, Iraq.

EPA. 2009. Integrated Pest Management (IPM) Principles. http://www.epa.gov/pesticides/factsheets/ipm.htm (accessed May 10, 2010).

Faleiro, J. R. 2006. A review of the issues and management of the red palm weevil *Rhynchophorus ferrugineus* (Coleoptera: Rhynchophoridae) in coconut and date palm during the last one hundred years. *Int J Trop Insect Sci* 26:135–154.

Faleiro, J. R., J. Ashok Kumar, and P. A. Rangnekar. 2002. Spatial distribution of red palm weevil, *Rhynchophorus ferrugineus* Oliv. (Coleoptera: Curculionidae). *Crop Prot* 21:171–176.

Ferry, M. and S. Gómez. 1998. The red palm weevil in the Mediterranean area. *J Int Palm Soc* 46:172–178.

Flint, M. L., S. Daar, and R. Molinar. 2003. Establishing integrated pest management policies and programs: A guide for public agencies. University of California ANR Publication 8093.

Gassouma, M. S. S. 1991. *Agricultural Pests in the United Arab Emirates*. Part 1. Fruit trees (in Arabic). UAE Ministry of Agriculture and Fisheries.

Gopal, M., A. Gupta, B. Sathiamma, and C. P. Radhakrishnan Nair. 2001. Control of the coconut pest *Oryctes rhinoceros* L. using the Oryctes virus. *Insect Sci Appl* 21:93–101.

Gullan, P. J. and P. S. Cranston. 2004. *The Insects: An Outline of Entomology*. Wiley-Blackwell, Australia.

Hallett, R. H., B. Gries, R. Gries, J. H. Borden, E. Czyzewska, A. C. Oehlschlager, H. D. Pierce Jr., N. P. D. Ngerilli, and A. Rauf. 1993. Aggregation pheromones of two Asian palm weevils, *Rhynchophorus ferrugineus* and *R. vulneratus*. *Naturwissensechaften* 80:328–331.

Hallett, R. H., A. C. Oehlschlager, and J. H. Borden. 1999. Pheromone trapping protocols for the Asian palm weevil, *Rhynchophorus ferrugineus* (Coleoptera: Curculionidae). *Int J Pest Manag* 45:231–237.

Hallett, R. H., A. L. Perez, G. Gries, R. Gries, H. D. Pierce Jr., J. Yue, A. C. Oehlschlager, L. M. Gonzalez, and J. H. Borden. 1995. Aggregation pheromone of the coconut rhinoceros beetles, *Oryctes rhinoceros* (L.) (Coleoptera: Scarabaeidae). *J Chem Ecol* 21:1549–1570.

Hussain, A. A. 1963. Biology and control of the dubas bug, *Ommatissus binotatus lybicus* de Berg. (Homoptera, Tropiduchidae), infecting date palms in Iraq. *Bull Entomol Res* 53:737–745.

Lacey, L. A., R. Frutos, H. K. Kaya, and P. Vail. 2001. Insect pathogens as biological control agents: Do they have a future? *Biol Control* 21:230–248.

Najeeb, M. A. A. 2001. *Date Palm Diseases and Pests: Diagnostic and Identifying Field Guide (in Arabic).* Kingdom of Saudi Arabia Ministry of Agriculture and Water.

RedPalmWeevil. 2009. *Insects Attack Date Palm.* http://www.redpalmweevil.com/Diseases/insleaves.htm (accessed May 15, 2010).

Rochat, D., K. Mohammadpoor, C. Malosse, A. Avand-Faghih, M. Lettere, J. Beauhaire, A. Pezier, M. Renou, and G. A. Abdollahi. 2004. Male aggregation pheromone of date palm fruit stalk borer *Oryctes elegans. J Chem Ecol* 30:387–407.

Soltani R., I. Chaieb, and M. Ben Hamouda. 2008. The life cycle of the root borer, *Oryctes agamemnon*, under laboratory conditions. *J Insect Sci* 8:61. http://www. insectscience.org/8.61

Weinzierl, R., T. Henn, P. G. Koehler, and C. L. Tucker. 2005. *Insect Attractants and Traps.* University of Florida, IFAS, ENY277.

Zaid, A., P. F. de Wet, M. Djerbi, and A. Oihabi. 2002. Diseases and pests of date palm. In *Date Palm Cultivation*, ed. A. Zaid. FAO, Rome. http://www.fao.org/docrep/006/y4360e/y4360e0g.htm#bm16 (accessed May 10, 2010).

9 Mechanization in Date Palm Pollination

Ahmad Mostaan

CONTENTS

INTRODUCTION

Mechanization in its general meaning is based on the concept of using the maximum capacity of available technologies for higher productivity in agriculture processes (Almasi et al., 2001). Mechanization of cultural practices in the date palm industry is an emerging issue in recent years. In fact, high costs, work intensity, high operation time, and shortage of skilled and willing workers have increased the need for development of mechanized methods for the production of dates (Albozahar, 2004). One season of the date production process includes a chain of operations, each with its own important role in product quality and quantity (BenSalah, 2000). However, the importance of some of these operations becomes a determining factor in maintaining good yield and product quality. Pollination is one such important operation in date production and has specific characteristics and needs that make it the frontier of mechanization of date production. Date growers look at this operation as the investment phase of production, which must be done accurately to ensure good economic return (Mostaan, 2006).

POLLINATION PROCESS IN DATE PALM

Being a dioecious species in character, date palm sexes are borne by separate individuals. The unisexual flowers are pistillate (female) and staminate (male) in character. The male palm produces the pollen and the female palm produces the fruit. The flower stalks are produced from the axils of the leaves in similar positions to those in which offshoots are produced. The inflorescence consists of a long stout spathe, which, on bursting, exposes many thickly crowded floral branchlets that are stout and short in male and long and slender in female (Zaid and DeWet, 1999).

Female palm flowers and male palm pollen grain are considered as two major members in the pollination process and the production of dates. This means that effective transfer of pollen grain within female palm flowers is necessary for successful fruit set and proper yield. The efficiency of natural pollen transfer that is done by wind and insects is so little, and influenced by several

FIGURE 9.1 **(See color insert.)** Traditional pollination in Iran; left: cv. Barhee, Ahwaz; right: cv. Mozafti, Jiroft.

environmental conditions, that it cannot be considered as a reliable method for the commercial production of dates. Therefore, artificial pollination for commercial production is necessary (Eeta, 1985).

In the initial artificial methods, which are known as traditional methods of pollination, the palm worker is the only pollen transfer mechanism to the female palm flowers. Traditional pollination is done by various methods such as (a) the insertion of fresh or dried flowers' male flower strands within female cluster (Figure 9.1), (b) putting cotton balls dipped in pollen into female clusters, and (c) manual dispersion of powder of pollen grains on female flowers (Figure 9.2). The most common method of traditional pollination includes insertion of three to five male flower strands within female clusters after some pollen has been shaken over the female inflorescence. The operation must be repeated on at least three occasions during the pollination season due to unsynchronized

FIGURE 9.2 **(See color insert.)** Manual dry pollen dispersion method: A fine net bag filled with dry pollen was fastened on a stick and shaken remotely on open female inflorescences by the worker.

emergence and opening of the female palm spathes (Eeta, 1986; Dowson, 1982). In general, climbing palms taller than 1.5 m becomes necessary in this method.

In recent years, traditional pollinating has been facing significant problems and challenges such as decline in skilled workers, high production costs and establishment of large-scale and modern date palm plantations that make them unsuitable for a new production system. Therefore, there is a great tendency to develop mechanized pollinating methods (Mostaan, 2006).

MECHANIZATION IN DATE PALM POLLINATION

Alexander (1952) reported the first design of the date palm pollinator. In his short report, he states that by means of this tool he was able to pollinate his 0.8 ha date palm grove within 1.5 h, in contrast with days of pollination work with the manual method of putting cotton balls dipped in the pollen into the female inflorescence. Alexander has made this device using an empty peanut butter jar, aluminum rod as a pole and some copper tube and plastic hose. In this pollinator, the pollen tank was located at the bottom of the pole and a mixture of pollen and air was puffed on the open female clusters after passing through the copper tube of ~8 m length.

Yost (1957) patented a pollinator for fruit trees similar to Alexander's described pollinator in the U.S. Patent Office. The pollen tank was placed on top of the pollinator pole and the pole was made of variable diameter tubes in the new design.

Research on mechanized pollination initiated in 1966 following a severe reduction of date palm labor in the United States in 1965 (Brown and Perkins, 1969). In this regard, several studies on aerial pollination with aircraft and helicopters have been conducted in both the United States and Iraq. Results indicated that aircraft pollination is not reliable for economic date palm pollination (Brown and Perkins, 1972; Eeta, 1985). Furthermore, this method requires at least four to five times the amount of pollen used in the traditional method. Although, aerial pollination had been advised in Iraq to find a more suitable method due to time and cost reduction benefits (Eeta, 1985), recent experiments with ground-level mechanized pollination with different types of mechanical and electrical pollinators show more reliability, time, and economic benefits (Zaid and DeWet, 1999).

Ground-level mechanized pollination, which is also called "Mechanical Pollination," consists of pollinating freshly opened female spathes from the ground level with the use of a variety of handheld, dorsal, cart-type, or tractor-mounted devices with dominant mechanical, electrical, or electrostatic design.

Handheld mechanical pollinators were used in Iraq in 1972 and 2 years later in Iran in 1974. This device is consisted of an air pump, an air tank, pollen repository, two leather belts, and an aluminum pollen delivery (Eeta, 1986) (Figure 9.3). Loghavi (1993) also designed a pollinator of this type. The tank had been placed on top of the aluminum pole. Field test results of this device in pollinating "Shahani" cultivar show that the fruit set is equal to or even better than the traditional method (Loghavi, 1993).

Haffar (1999a) showed that the length and diameter of the duster pollen delivery tube have a great effect on the pollen viability. Results showed that the pollen viability diminishes with a longer and narrower pollen delivery tube. This effect is likely to occur in pollinators with the pollen tank placed at the bottom of the pollinator tube. This effect also highlighted the necessity for improvements in pollination equipment design.

Haffar (1999b) developed and tested a 12 V battery-operated microduster for date palm pollination. The duster is composed of three major systems of dusting, power and control, and extension and reach. The duster delivered pure pollen at various throughputs that ranged from 0.15 to 0.35 g/s depending on the operation time and inclination angle. The maximum and minimum outreaches of the pollen were 2.39 and 0.73 m, respectively, depending on operation condition. Results indicated that the operation of the duster was very gentle and caused no significant damage to the pollen grains.

FIGURE 9.3 **(See color insert.)** Mechanical pollinator at work in date palm pollination, Ahwaz, Iran.

Obaidi (2001) has also developed a lightweight pollinator. The pollinator has two small 12 V electric motors that make the device capable of delivering pollen flow with a speed of 15 m/s in an outreach of 1 m. The pollen tank and overall system equipment are located at the top of the pole, and hence eliminated the long pollen delivery tube. The device is controlled via a button at the bottom of the pole from the ground by the worker. Using this device, a worker is able to pollinate each palm in <5 min.

Another pollinator was also developed by Al-Rawi (2001). The device uses two 1.5 V dry batteries to power the blower motor in order to blow the pollen mixture from the pollen tank at bottom of the pole to the open spathes at the palm crown. Results of 4 years of field tests showed that this pollinator increased the fruit weight and fruit set by 10% and 15%, respectively.

A new electric device for pollinating date palms has been designed and developed in the Date Palm and Tropical Fruits Research Institute of Iran. The device has been designed based on a review of almost of the past pollinators. The dispersing system used is of peripheral dispersion type and is completely optimized in comparison with previous designs. Thus the pollination feasibility and controllability have been enhanced, while the operation time and tool size, weight, and cost diminished in the new tool in comparison with current mechanical pollinators (Mostaan et al., 2010).

Figure 9.4 represents the developed device sketch. This device consists of an extending pole, disperser, pollen pipe, nozzle, and remote controller. The pole is made of aluminum pipes of 1.5 m length steeply increasing in diameter from the top to the bottom. The main objective of this part is to bring the disperser close to the spadixes without the need to climb the palm.

The disperser consists of a double-wall pollen tank, blower, and top cap. The pollen tank has a cylindrical body, an adjustable pollen tank, and an air distributor. The adjustable pollen tank is fixed inside the cylindrical body right above the air distributor. The pollen tube is small in diameter; it is extended from the top cap and bent 45° to make a targeted pollination. Some special types of nozzles could be used to enhance the pollination process.

The dispensing operation is controlled remotely by the control set. In general, a pair of transmitter in the control set and a receiver in the disperser equipped with an adjustable timer are used.

FIGURE 9.4 **(See color insert.)** Sketch of the developed electric pollinator (left) and the pollinator at work (right): (1) extending pole, (2) disperser, (3) pollen pipe, (4) nozzle, and (5) remote controller.

When the timer is set on by the worker through a button on the controller, the blower is set on to a pre-adjusted time and then turns off automatically. Figure 9.4 shows the device at operation in a date palm orchard.

Preliminary evaluation of the tool performance on Barhee cultivar in the Ahwaz region in 2008 showed that although it leads to more fruit set than traditional and mechanical methods, the difference was not significant and thus the tool succeeds in obtaining the functional goal besides the aforementioned advantages. The mean fruit set attained by the developed tool, mechanical pollinator, and traditional method was 68.12%, 62.04%, and 64.94%, respectively. This tool handles 200 cm^3 of pollen mixture by which pollination of about 120 palms (about 1 ha date palm plantation) could be done in each charge by only one worker. The most important features of the new tool include:

1. Light weight (<350 g)
2. Ease of use
3. Low cost
4. Optimized power consumption
5. Independent of pole type
6. No compressed air tank
7. Possibility of pure pollen dispersion
8. Continuous pollen dispersion flow
9. High maneuverability in the palm crown
10. Controllable pollen flow
11. Independent of target (palm crown) height

In overall, the new device was recommended for low-scale date palm farmers, while a precise study of the new device from both time and cost aspects, detailed performance analysis of the device's suitability for other commercial varieties, and evaluation of pure pollen pollinating by this device were also recommended (Mostaan et al., 2010).

A review of research results on mechanization of date palm pollination operation shows that in comparison to traditional methods, mechanized pollinating has the following characteristics:

1. In the traditional method, a worker is able to pollinate only 15–20 palms daily. In contrast, pollination of up to 500 palms or more by only one worker is possible in the mechanized method at the same time (Al-Rawi, 2001).
2. In mechanized method, the operation cost is three times less than the traditional one.
3. Presence of the worker close to clusters is not necessary in pollen application in both traditional and mechanized methods. Less in-site human judgment is required to determine the need and manner of operation and the conditions of its implementation, and most of the required information could be provided from the ground level. Consequently, there is no need to climb the palm in mechanized pollinating and thus the work severity and risk of accidents occurring as with the old method of climbing tall palms are much lower in comparison with the traditional pollination method (Zaid and DeWet, 1999).
4. Time is an important factor ruling fruit set. All cultural operations except pollination for a 25 ha plantation could be completed with ~200 men, whereas pollination requires nearly 700 man-days during the peak period (Perkins and Burkner, 1973). On the other hand, pollen receptivity of female flowers has a time limit. This time limit is changing in different cultivars, but it is 3–5 days after the spathe opening for most of the commercial cultivars. According to the slow pollinating in the traditional method, pollination of all palms is not possible in many cases, while it becomes possible with mechanized pollinating and the pollen receptivity issue will be met (Eeta, 1986).
5. Mechanized pollinating could be done by an unskilled worker with a very little training (Mostaan, 2006).
6. There is no significant difference in fruit quantity and quality between two methods of pollination, and in some cases mechanized pollination leads to more and better yield (Mostaan et al., 2010; El-Mardi et al., 2002; Al-Rawi, 2001; Loghavi, 1993; Eeta, 1988).

Matching pollinating characteristics and operation mechanizing capability criteria shows that there is a high possibility to mechanize date palm pollination and most of the mechanization criteria will be met in the mechanized method.

It should be noted that in some cases, mechanizing an operation faces some limitations that generally originated from the operation condition. These limitations have more technical aspects and usually emerge as the result of the necessity for precise and qualitative judgment in estimation of the need to do the operation, the conditions and manner of its execution, and its interaction with the product. Success in mechanizing an operation is often achieved by providing appropriate and affordable technology for passing these limitations. Major pollination impacts on mechanization and their discussion are as follows (Mostaan, 2006):

1. In traditional pollination, the male flower strands that contain pollen grains are included inside the female clusters as a pollen source. This method ensures the grower from providing the required pollen grain in the time of female flower receptivity and eliminates the need to repeat the process for the same cluster. Whereas in mechanized pollination, only mixed powder of pollen and charge material is used and pollen availability is limited to a small period of time, which is completely weather dependent. This characteristic adds high uncertainty to the pollination process. To tackle this obstacle, the mechanized pollination must be repeated for the inflorescences in the period of receptivity to guarantee an acceptable fruit set.

2. Not all of the female spathes of the individual palm open at the same time. The time interval between the first and last open spathes depends on variety, climate, and nutritional variables and it seems to be more than 1 month. In traditional pollinating method, the worker climbs the palm and thus becomes close to and in contact with the spathes. In this way, it is normal for the farmer to pollinate all cracked and intact spathes at the same time to reduce the cost and need for repeated climbing of the palm. This saves more time for the grower to pollinate as much as possible palms in the pollination period. The immature spathes that were opened by the worker will receive their pollen at the right time of receptivity from the preinserted male flower strand. This guarantees the fruit set and good yield. On the other hand, in mechanized pollinating the pollen is available just for a small period of time after pollen dispersion, which makes it necessary to monitor open spathes for on-time pollination in the 30–45 days of the pollination period. This increases the time and repetition of the operation slightly higher, in contrast with the traditional method. This is the major impact of operation that must be dealt with in mechanized pollination.

Despite significant advantages and in spite of a few restrictions in the way of pollination mechanization that could be solved by current equipments, the mechanized pollinating method was not successful in action. This led to a principal question: Why, in spite of a few reports, this operation is still unmechanized in most date-producing countries? The answer lies in the "reliability" of the method. It can be seen that there are some neglected and unsolved problems that reduce the reliability of the mechanized pollinating method in comparison to the inherent robustness of the traditional method. These problems could be classified into three major problems as follows:

1. Little knowledge among date growers about the benefits and limitations of the mechanized pollinating method in contrast with the traditional method and the manner of dealing with limitations of the new method and lack of instruction regarding mechanized pollinating under grove conditions: A review of research conducted in most date-producing countries shows that there is a big information gap in pollinating the whole plantation with the mechanized method. All research works have focused on pollinating a single cluster and there is still no applicable information regarding the management of the mechanized pollinating operation.
2. Less reliable pollinating equipments: Unfortunately, a proper commercial pollinator has not been developed till now and the operation is mostly done traditionally in date-producing countries. Problems related to both the design and the performance of some developed devices are the major reasons for this problem. Heaviness, existence of an air tank that must be carried by the worker, and in most of them, jamming of the pollinator tubes by pollen mixture are the major problems (the authors' observations). Pollinating devices will be accepted by the growers only if their efficiency and reliability have been proven to be high and the application of pollen by the pollinating devices causes no complexity and delay in operation.
3. Lack of an adequate and reliable source of dry pollen: Mechanized pollination is done by dry pollen. Therefore, enough, suitable, and reliable pollen must be ready at the time of pollination. Currently, each farmer must prepare and maintain his needed pollen by himself. This is a difficult operation and is inapplicable in most cases according to the irregular distribution of male palms inside the plantation, high risk of disease and probability of late maturity of male spathes. Moreover, pollen extraction efficiency in the traditional method is low and in some cases preservation of the collected pollen in appropriate conditions for next year is difficult for the grower. Based on these difficulties, the growers always rely on the fresh pollen and the problem of reliable pollen sources persists.

FACTORS IN MECHANIZED DATE PALM POLLINATION

The success in the development of mechanization in palm pollination is achieved by three essential factors: (a) an adequate and reliable source of dry pollen, (b) an appropriate pollinating device, and (c) clear and accurate mechanized pollinating management guidelines. These three factors form the triangle (Figure 9.5) of mechanized pollination, and without any of them, mechanized pollination loses its reliability (Albozahar, 2003).

Pollen

Extracted pollen is generally used as fresh pollen in its season. It is necessary to store this pollen as the key and determinative material for at least 1 year to have a reliable and compatible system of mechanized pollination. To attain this goal, maximum amount of date palm pollen must be extracted in less time and cost and the pollen must be stored in appropriate conditions to preserve the pollen viability (Eeta, 1993). It seems that it is impractical for all growers to do this operation owing to inherent prevailing problems related to the constitution, culture, and economics of the date sector. This effect causes growers do not rely on mechanized pollinating and makes them use their traditional method in order to avoid production decline. One way to confront this problem is to produce the required pollen in the form of the centralized system which can serve a group of growers. This system can be called a "pollen bank." All activities related to the collection, extraction, storage, and distribution of date palm pollen consistent with regional female cultivars could be done by pollen banks. They may also work beyond the group or region by the establishment of commercial male orchards for the production and sale of certified pollen for other regions.

At least one pollen bank was described by Codekas in 1976. No other pollen bank has been reported in date-producing countries. Thus, the efficiency and performance of such systems are still unclear. Two types of pollen banks have been proposed: (a) cooperative pollen banks (Figure 9.6) and (b) pollen enterprises (Figure 9.7) (Mostaan and Ahmadizadeh, 2008). In overall, the growers must start from small-scale local pollen banks through cooperation in order to supply their pollen for mechanized pollination purpose. The emergence of these pollen banks will lead to the following advantages:

1. Intensive and specialized pollen-gathering system
2. Minimum number of workers for pollen supply
3. Increased efficiency of pollen collection due to an integrated process and minimum handling operations
4. Lower male-to-female palm ratio within commercial orchards and thus more yield and income
5. Possibility of preparing more and high-quality pollen
6. More fruit uniformity due to the usage of more uniform pollen

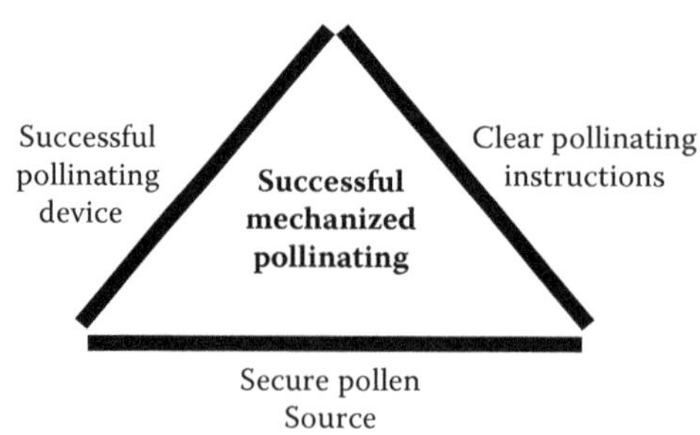

FIGURE 9.5 Triangle of mechanized pollinating. (Adapted from Mostaan, A. 2006. *Zeitun* 170: 55–59. [In Persian].)

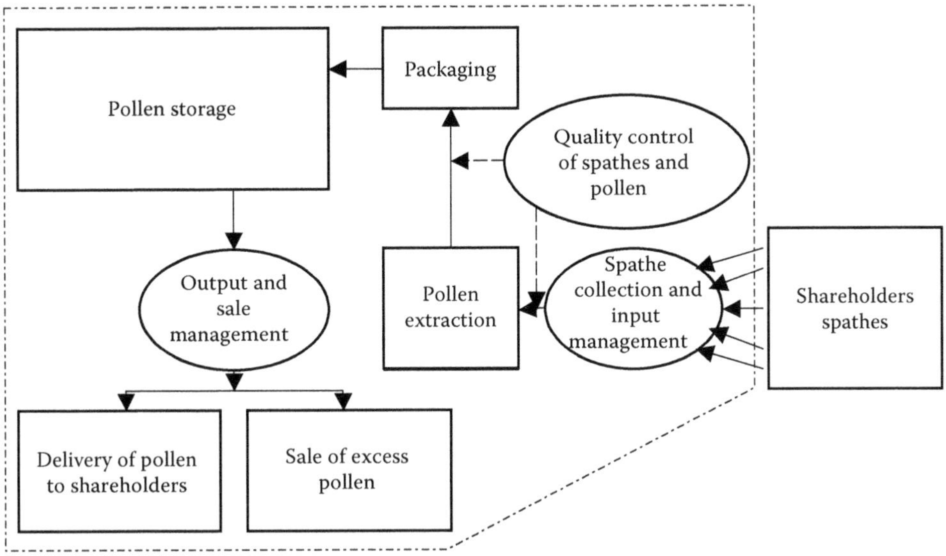

FIGURE 9.6 General proposed model of cooperative pollen banks.

7. Possibility of safe pollen shipment to other regions
8. Greater pollen storage efficiency due to storage of only pollen grains and concentrated pollen storage

The efficiency of pollen-extracting equipments is also of great interest in this topic. The first pollen-extracting machine was developed by Burkner and Perkins in 1975 in the United States. This machine was able to extract 41% more pollen than the traditional method. The basic design of this machine was dominant till now and several pollen-extracting machines were fabricated in different countries including Iran and the UAE. A novel method of pollen extracting has been investigated by the author. This process is based on the extraction of pollen grains from separated male flowers. Test results show that this method leads to 47% more net viable dry pollen than the traditional method.

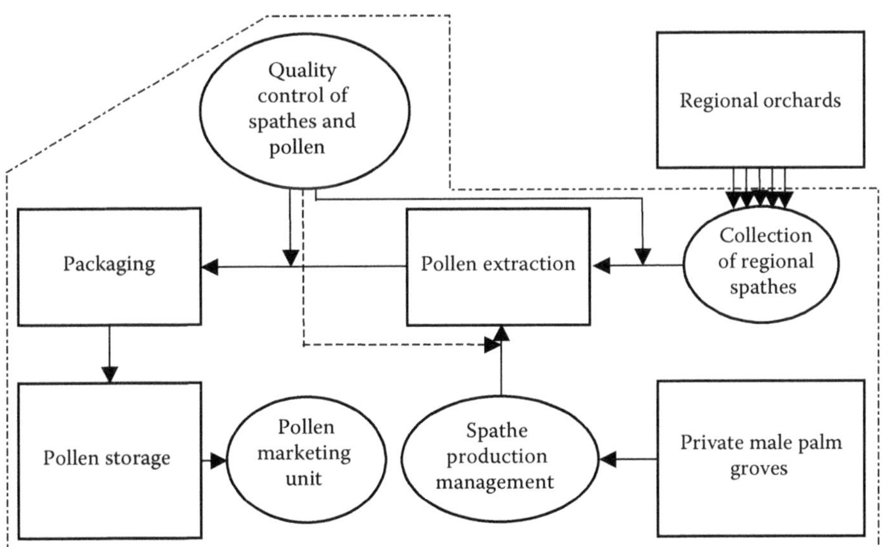

FIGURE 9.7 General proposed model of pollen enterprises.

It also reduces overall extraction time to a maximum of 3 days in comparison with 7–10 days of the pollen-extracting process in previous methods (Mostaan, 2010). Design of more economical pollen-extracting machinery is the next step in this topic to meet small-scale farmers' needs.

Pollinating Equipment

The main disadvantages of current pollinating devices are as follows:

1. High cost: Most date palm orchards are of small-scale type and ownership of some types of pollinating devices is not cost effective.
2. Technical problems such as
   a. Difficulty of use
   b. Jamming of pollinator pipe with pollen mixture, especially in humid weather, which is more likely to occur in pollination season
   c. Low uniformity and consistency in spraying pollen
   d. Instability of the pollinator pole in working with tall palms

Hence, optimization of available pollinating equipments is essential to increase the reliability of mechanized pollinating among growers. However, to better utilize the mechanized pollination system and to increase its compatibility with different regions, palm groves and cultural systems, the development of various types of pollinating equipments with specific capabilities is also an important issue. The following characteristics must be included in order to have a successful pollinator:

1. Relatively cheap (to meet the target farmers)
2. Minimum length of pollen transfer tube
3. Uniformity in pollen spraying
4. High stability of the pollinator pole
5. Higher controllability of the functional parameters

Pollinating Protocols

Pollination of each freshly opened female spathe is the more accurate one, but requires the farmer to have good knowledge of his plantation as well as good record-keeping to ensure the pollination of all spathes. Spraying of the whole female leaf canopy just above the opened spathes is economically feasible and saves time. However, a high rate of aborted fruits could occur when this technique is used (Zaid and DeWet, 1999). Past research on female flowers' receptivity has shown that as a rule of thumb most of the commercial cultivars should be pollinated within 3–5 days after the emergence of spathe cracks. Although this rule is recommended for most cultivars in the traditional method, it leads to maximum pollen consumption in the mechanized method (Mostaan, 2010). During early season pollination, or when the pollination season is characterized by low normal temperatures, it is recommended to alternate the pollination of the sides of the palm at 4–7 day intervals. This overlapping of pollination was shown to yield more reliable results than full palm pollination at one time (Nixon and Carpenter, 1978).

Changing the pollination interval in order to attain good yield while optimizing the pollen application and reducing the operation costs is necessary. There is no optimal mechanized pollinating interval till now and the growers can temporarily tackle this information gap by monitoring the number of cracked spathes at the pollination period. Another way is to mark pollinated spathes by colored powders mixed with applied pollen to mange the spathe pollen intake and hence to reduce the pollen consumption. More phonological studies should be conducted in this regard. When the number of cracked spathes is low, the application of high-diluted pollen in a vast time interval may also save more pollen.

CONCLUSION

The overall status of mechanization in date palm pollination was reviewed. It is shown that dry pollen, a pollinating tool, and pollination protocols could be considered as the three essential and key elements for the successful mechanization of date palm pollination operations. Therefore, planning for the provision of these three elements is essential for date growers. So far, many research works in the field of construction of appropriate tools for pollination have been carried out in different countries, but the technical and information gap still remains in the fields of dry pollen and pollination protocols. These two elements should be considered in the mechanization development plans of date-producing countries. For the supply of dry pollen, local pollen banks are more emphasized. On the other hand, male palm groves of good variety could be established by leading growers in order to provide more and high-quality dry pollen needed in mechanized pollination. This will also lead to a new industry of pollen production in date-growing regions. More intensive research work is also needed to set the mechanized pollination protocols for commercial cultivars.

REFERENCES

Albozahar (Mostaan), A. 2003. Mechanization concepts for mechanization of date palm pollination. *Abstracts of the 9th Date Palm Congress*. Arge-Jadide Bam, Bam, Iran, 27 pp. (In Persian).

Albozahar (Mostaan), A. 2004. Basics of mechanization of date production. *Journal of Agriculture and Technology* 55: 5. (In Persian).

Alexander, D. R. W. 1952. *A Method of Pollinating Dates*. Date Grower's Institute Annual Report No. 29:20.

Almasi, M., S. Kiani and N. Loweimi 2001. *Principles of Agricultural Mechanization*. Hazrat Masomeh Publications. Qom, 248 pp. (In Persian).

Al-Rawi, O. M., 2001. A study on a new US patent date pollinator. *Abstracts of the Second International Conference on Date Palm*. AlAin, United Arabic Emirates, 129 pp.

BenSalah, M. 2000. Major date palm field operations and their effect in improving fruit quality. In *Proceedings of the Regional Workshop on Date Palm Postharvest Processing Technology,* Hamdan, I. Y. and Hegazi, N. A., (ed.), October 11–14 1999. Tehran, Iran. FAO Regional Office for the Near East, Cairo, Egypt.

Brown, G. K. and R. M. Perkins. 1969. *Experiments with Aircraft Methods for Pollinating Dates*. Date Grower's Institute Annual Report No. 46: pp. 35–40.

Brown, R. M. and E. G. V. Perkins. 1972. Experiments with aircraft methods for pollinating dates. *The Punjab Fruit J.* 119: 116–127.

Burkner, P. F. and R. M. Perkins. 1975. *Mechanical Extraction of Date Pollen*. Date Grower's Institute Annual Report No. 52: pp. 3–7.

Codekas, J. P. 1976. *The Pollen Hustlers: Cooperative Ownership Extraction of Date Pollen*. Date Grower's Institute Annual Report No. 53: pp. 23–24.

Dowson, V. H. W. 1982. Date Production and Protection. United Nations FAO Plant Production and Protection Paper No. 35. FAO, Rome, Italy, 294 pp.

Eeta, M. 1985. *Date Palm Pollination Research in Iraq*. Plant & Seed Registration and Certification Institute, Karaj, Iran, 26 pp. (In Persian).

Eeta, M. 1986. *Analysis of Some Tips about Date Palm Pollination*. Plant and Seed Registration and Certification Institute, Karaj, Iran, 51 pp. (In Persian).

Eeta, M. 1988. Final report of determination of best pollen C.V. Ghanami concentration for pollination of date palm spadixes of Estaumran Cultivar. Plant & Seed Registration and Certification Institute, Karaj, Iran, 13 pp. (In Persian).

Eeta, M. 1993. Change in the basic geography of male palms and centralized management in establishment of date palm plantation. Collection of experts and researchers speeches, Agricultural Research Center of Khuzestan, 10 pp. (In Persian).

El-Mardi, M. O., H. Esechie, L. M. Al-Kharousi and K. M. Abdelbasit. 2002. Effect of pollination method on changes in physical and chemical characteristics of date fruit during development. *Agricultural Science* 7(1): 21–27.

Haffar, I. 1999a. *Contribution of Several Duster Parameters to Performance in Mechanical Date Pollen Delivery (Abstract)*. Available on the web: http://asae.frymulti.com/.

Haffar, I. 1999b. *Design and Testing of a Micro-Duster for Date Palm Pollination (Abstract)*. Available on the web: http://asae.frymulti.com/.

Loghavi, M. 1993. Development of a mechanical date pollinator. *AMA* 24(4): 27–32.
Mostaan, A. 2006. Concepts of mechanization development in date palm pollination. *Zeitun* 170: 55–59. (In Persian).
Mostaan, A. 2010. *Study and Performance Evaluation of Process of Pollen Extraction from Date Palm Male Flowers*. PhD dissertation, Modares University, Tehran, Iran. (In Persian).
Mostaan, A. and S. Ahmadizadeh. 2008. Management of sustained production of pollen; the strategic input in date palm cultivation. *Keshavarzi va sanat* 104: 33–37. (In Persian).
Mostaan, A., S. S. Marashi and S. Ahmadizadeh. 2010. Development of a new date palm pollinator. *Acta Hort. (ISHS)* 882: 315–320.
Nixon, R. W. and J. B. Carpenter. 1978. *Growing Dates in the United States*. U.S. Dept. of Agriculture, Agric. Information Bulletin No. 207: USDA. Technical Document, 63 pp.
Obaidi, R. A. 2001. Lightweight date palm pollinator powered by low voltage motors. *Abstracts of the Second International Conference on Date Palm*. AlAin, United Arabic Emirates, p. 127.
Perkins, R. and P. F. Burkner. 1973. *Mechanical Pollination of Date Palms*. Date Grower's Institute Annual Report No. 50: pp. 4–7.
Yost, L. J. 1957. *Pollinator for Fruit Trees*. US patent No. 2,802,302:2 pp.
Zaid, A. and P. F. DeWet, 1999. Pollination and bunch management. In *Date Palm Cultivation*, A. Zaid, (ed.), FAO plant production and protection paper No. 156. FAO, Rome. Italy, pp. 144–205.

10 Design Concepts for Date-Harvesting Process

Hemantha P. W. Jayasuriya

CONTENTS

INTRODUCTION

Mechanization of date harvesting has been of interest among many growers and investigators over the past 40 years. Variable maturity of fruit on a single palm is often a limiting factor in harvest mechanization. As all of the fruits in a bunch do not ripen at the same time, it is necessary to make several pickings during the season. Selective hand-picking of the individual ripe fruits from each bunch is the most expensive cultural operation (Al-Suhaibani et al., 1988).

Many of the cultural operations of a grown date palm tree including harvesting require a human laborer to climb up the tree and work at a considerable height above the ground. This is a dangerous and slow operation further compounded by a severe labor shortage in most date-producing countries. It is a well-understood fact that there is a long due need for a date-harvesting machine/mechanism that can be applied to serve the harvesting operations with the aim of reducing the harvesting losses, enhancing the safety of farmers and harvesters, and minimizing the time spent for harvesting activities. Among the harvesting methods, there have been conventional manual techniques and semi- or fully mechanized methods (Al-Suhaibani et al., 1990; Al-Yahyai, 2009).

The growth habit of date palm (*Phoenix dactylifera* L.) is a cylindrical, nonbranching stem and a relatively tall trunk. The trunk of date palm is composed of vascular bundles held together by connective tissue. Toward the periphery, where the leaf bases are embedded, the tissue tends to become more lignified and tough. A medium-sized tree could be 10–15 m tall, the leaves are pinnate, 3–5 m long, with spines on the petiole and about 150 leaflets, and the leaflets are 30 cm long and 2 cm broad. The full span of the crown ranges from 6 to 10 m. The date palm tree features a slender trunk of more or less constant diameter from the base to the crown. Each year, the old leaves are cut off at the base of the leaf stem. If the bases of the leaves are cut off, the trunk becomes smooth, much smaller in diameter, and more difficult to climb (Zaid, 2002).

Many of the date palm operations such as pollination, thinning, bagging, pruning, harvesting, and spraying for insect control require the farmer to climb the tree and work at a considerable height above the ground. A suitable machine can improve the productivity of the indigenous farmer and hence the quality of the product, give better timeliness and less dirt contamination, and can be considerably safer than climbing trees freehand or with a belt. In this chapter, an innovative machine is proposed that is able to harvest the palm tree in an easy and safe way (Zaid, 2002; Jahromi et al., 2008).

SOME PROPERTIES OF DATE RELATED TO HARVESTING: LITERATURE REVIEW

Facts about Date Palm

Dates are one of the most important fruits in more than 30 countries in the desert regions of the world. The total world date production is ~6800 thousand tons. Over the period of 1999–2001, Egypt, Iran, Saudi Arabia, and Pakistan have been producing 61% and Iraq, Algeria, the United Arab Emirates, Oman, and Sudan 29% of the total world date production (Zaid, 2002; Ampratwum and Al-Siyabi, 2005; Jharomi et al., 2007).

Date Varieties and Specifications

Kader and Hosain (2009) classified the date cultivars based on moisture levels as follows:

1. Soft date cultivars (>30% moisture) include Abada, Amhat, Barhee, Bentaisha, Halawy, Hayany, Honey, Khadrawy, and Medjool (Mejhool)
2. Semidry date cultivars (20%–30% moisture) include Amry, Dayri, Deglet Noor, Khalasa, Sewy, and Zahidi
3. Dry date cultivars (<20% moisture) include Badrayah, Bartamoda, Deglet Beida, Horra, Sakoty, and Thoory

The growth habit of date palm is a cylindrical, nonbranching stem and a relatively tall trunk. The trunk of date palm is composed of vascular bundles held together by connective tissue. Toward the periphery, where the leaf bases are embedded, the tissue tends to become tougher (Barreveld, 1993). The height of the date palm tree is commonly from 10 to 15 m and the trunk has an approximately constant diameter from the base to the crown.

Zaid and de Wet (1999) defined the contents of the date fruit: 60%–65% sugar, about 2.5% fiber, 2% protein, fat, minerals, and pectin substances each <2%, and it is a good source of iron, potassium, and calcium with low sodium. In addition, it contains moderate quantities of chlorine, phosphorus, copper, magnesium, silicon, sulfur, and vitamins.

Maturity Indices

Kader and Hosain (2009) classified five stages in dates leading to the maturity stage for harvesting. The maturity stage of date includes:

1. *Hababouk (aankaziz) stage:* Earliest stage of development: Fruit with a round shape, greenish color, slow growth, takes 4–5 weeks to complete growth.
2. *Kimri (Jamry) stage:* Increased weight, green color, takes 9–14 weeks to complete, not suitable for eating. Moisture content = 85%.
3. *Khalal (balah, biser) stage:* Change from green to yellow or red, takes 3–5 weeks to complete, maximum weight and size, can be boiled and consumed. Moisture content ranges between 45% and 65%.
4. *Rutab stage* (light brown and soft): Change in color to brown or black colors, the tip of the fruit starts ripening, takes 2–4 weeks, decrease in fruit weight, sweet, consumption stage. Moisture content ranges between 30% and 45%.
5. *Tamar stage* (dark brown, soft, and dry): Color is dark brown, soft, lost most water, takes 2–4 weeks to complete, good for storage. Moisture content < 30%.

Date Quality Indices

There are many quality factors in the field of dates, and attention must be given to obtaining high-quality and excellent specifications, which depends mainly on the experience and knowledge of the farms about the needs of all aspects of the palm. And the most important signs of quality are the completeness of the fruit in terms of size, weight, texture, color, flavor, aroma, and appropriate and acceptable for fresh consumption (Zaid and de Wet, 1999). The shelf-life, appearance, and market value depend on the proper harvesting time and method for dates. Some mechanical devices developed (Abounajmi, 2000) have limitations in this and can cause a reduction in quality and market value.

The most important changes in maturity and quality are

1. Change of the color of the crust and core
2. Increase in sweetness
3. Emergence of the distinctive flavor
4. Increase of sugar (glucose and fructose) and total dissolved solids (TDS)
5. Decrease of acidity
6. Decrease of respiratory activity
7. Decrease of humidity
8. Balance of the percentage of acidity to the TDS
9. The size and weight of fruit

Activity of Wind and Rain during the Time of Pollination

Many of the flowering trees can be found in the second half of January, as most varieties bloom in February and the remaining few palm trees flourish in the second half of the month of March. The wind is active during the pollination period, which leads to the failure of the process of fertilization, leading to pollen blown from the female flowers to dry before dispersal on Almiasm in the dry zones. Despite the scarcity of rainfall in Sultanate of Oman, but some rain may fall after the completion of the pollination leading to failure of fertilization by washing away the pollen from the female flowers. And this may also cause explosion of the pollination tube leading to failure in fertilization. So the farms are forced to repeat the process of pollination (Zaid and de Wet, 1999).

High Salinity and Poor Growth and Productivity

High temperatures and scarcity of water in date-growing countries lead to increased evaporation and significantly high concentration of salts in irrigation water, especially because palm areas are irrigated with underground water of high salinity. Usually, the arid zone soils lack organic matter or Aldobal; therefore, water easily penetrates through the soil away from tree roots, disabling sucking by the palm. As a result, the soil lacking in water is not washed on a regular basis. Hence, heavy rains sometimes lead to the erosion of salt to the root region, which increases the bad influence on growth and productivity and the trees may die in the end (Shamsi and Mazloumzadeh, 2009).

Most farmers believe that the palm trees bear with the environmental conditions and land, forgetting that the endurance limit may not produce with it an economical crop. It was decided that palm trees can grow in soils containing 3%–4% salts, and the trees die at 4.8% concentration of salts. We must keep in mind that the greater the concentration of salts in the soil solution, the less productive the palm; whereas the crop reduced to 50% when irrigation water contains 6000 ppm salts (0.6%), it also reduced the crop by 85% when irrigation water contains 12,000 ppm salts (1.2%). The reason for the death of palm at high concentration of salts is the inability to absorb water; whereas an increase in concentration reduces the water potential in the soil solution and so the water cannot move from the soil to the plant, the opposite will happen on the exit of water from the plant into the soil, leading to shrivel and death. The magnitude of the problem increased as the farmers believed that the palm trees bear with the drought; hence they irrigated on separate intervals, which increased the concentration of salts in the soil solution (Zaid and de Wet, 1999; Shamsi and Mazloumzadeh, 2009).

CONVENTIONAL METHODS OF HARVESTING

Manual Harvesting

In the past, the tool that was used for harvesting date was a sickle, which is so simple, dangerous, and uncomfortable to hold and is low in quality. This tool is simple as it is made of rob and dangerous, as it could accidentally cut the operator at any time and so the person who is on the top of the palm could fall.

Dates are harvested by cutting bunches at one time. Pickers climb trees (Figure 10.1) or use ladders (Figures 10.2 and 10.3) to harvest fruits and not all bunches mature at the same time and each palm may be picked up to eight times of 2–3 months.

Mechanical Harvesting

Typical mechanized date-harvesting systems are vehicles equipped with a number of long arms, at the end of which a worker could stand in a basket for picking the fruits (Mazloumzadeha et al., 2008). Depending on the date variety and climate, the dates can be picked by one of the following ways:

1. Bunch cutting is used where all fruits on a bunch ripen simultaneously.
2. Bunch shaking is used when all fruits on a bunch do not ripen at the same time.
3. Selective hand picking is used for varieties with high-quality expensive fruits sensitive to vibration and impact.

The development of date-shaking and -harvesting machines began decades ago (Perkins and Brown, 1967; Levin et al., 1969; Brown et al., 1983; Sarig, 1989). The proposed machinery must also be affordable (Ali and Akyurt, 1998). A suitable machine can improve the productivity of the indigenous farmer and hence the quality of the product, better timeliness, and less dirt contamination, and can be considerably safer than climbing trees freehand or with a belt (Al-Suhaibani et al., 1988).

FIGURE 10.1 **(See color insert.)** Conventional method of harvesting date; the picker climbs the tree to harvest date, with or without a safety belt, a hazardous and dangerous activity.

Al-Suhaibani et al. (1992) designed and evaluated a date service machine that uses the ground as support. Shamsi (1998) developed and tested a climbing machine to harvest and service date palm trees. This machine has three wheels that use the tree trunk as support. Ali and Akyurt (1998) presented several designs for machine systems that enable one to undertake crown-related operations. Sclater and Chironis (2001) presented three gripper designs for gripper robots. They introduced a reciprocating lever mechanism that opens and closes the jaws of a gripper robot and permits it to grasp and release objects.

FIGURE 10.2 **(See color insert.)** Using ladders for selective picking and the use of cloth containers.

FIGURE 10.3 **(See color insert.)** Using a pay-loader bucket.

Perkins and Brown (1964) used a harvesting system in which three men, namely a truck driver, a boom operator, and a bunch cutter or shaker, operate together. Abounajmi (2000) investigated the effect of shaking mode on date separation from a bunch of dates. Sarig (1989) in Israel developed an integrated mechanical system that could harvest the fruits by shaking the tree trunk. A roll-out canvas catching frame was commonly used for collecting the fruit. This method is only suitable for those varieties of dates whose fruits on a bunch do not ripen at the same time.

Shamsi (1990) designed a sprocket-type climbing machine to harvest the dates. The operator sits on a frame and holds on to a handle. He then drives the climber by pushing the pedals with his feet. The study calculations showed that the operator could climb with the machine at a speed of 0.2 m/s.

Al-Suhaibani et al. (1988) reported on the design of a date service machine for Saudi Arabia which was built at Silsoe College, UK. A "U"-shaped platform (Figures 10.3 and 10.4) on the end of a telescopic boom allowed the worker to reach all of the bunches of dates without any additional movement of the platform. Al-Suhaibani et al. (1992) also reported the field test of the machine. It showed that it could harvest a tree in 21 min, which was faster, safer, and easier than hand harvesting.

Research conducted on date harvesting using net bags (Figure 10.5) has shown a significant reduction in losses (Figures 10.6 and 10.7). This emphasizes the need for controlling losses in manual harvesting methods.

Jahromi et al. (2008) reported on the development (computer-based geometrical simulation) of a gripper date palm service machine (Figure 10.8a–c) in order to climb the tree and do special operations at the top. A machine mass perdition system was also developed in order to calculate the total mass of the machine based on the tree height, diameter, and other factors. It was proposed that the machine could be improved to the remote-controlled type considering the safety of the operator.

Development of this type of machines has an advantage as they can be used for multiple operations such as pruning, pollinations, and hand-pick-type harvesting. The height of the tree, canopy

FIGURE 10.4 **(See color insert.)** Using a U-shaped platform.

size, and spread of fruit bunches can be crucial factors when deciding the size of the machine, facilitating the ergonomic design as the operator could easily reach fruits while maintaining comfort and safety.

Objectives toward Date Harvester Designs

Most of the dates are still harvested manually in most parts of the world including the Middle East (Figures 10.1 through 10.3), because a mechanized method that covers all the needs is currently not available. General-purpose lifters of various models are used in some orchards but date growers are not familiar with using many of these machines yet.

The main objective of a date harvesting design project could be to investigate different solutions to the present problems regarding the date harvesting, the selection of the level of mechanization, and other relevant factors. There are a number of factors involved in solving the problem and set goals. Some important factors could be planting and plantation conditions, field terrain

FIGURE 10.5 **(See color insert.)** Net bags used for collecting the ripened fruit.

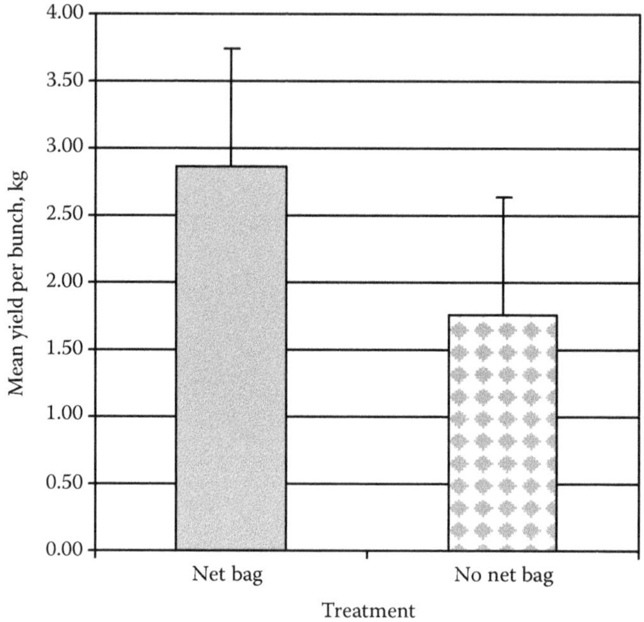

FIGURE 10.6 Effect of fruit collection by using net bags: Bunch yield for Khalas date variety.

conditions, agronomic and soil specifications, type of date cultivars, cultural and socio-economic factors, and safety aspects.

Therefore, the specific objectives for a date harvester design project could be listed as follows:

1. To reduce the time spent for harvest; appropriate field capacity
2. To minimize the labor requirement; different levels of automation can be adopted for the corresponding needs

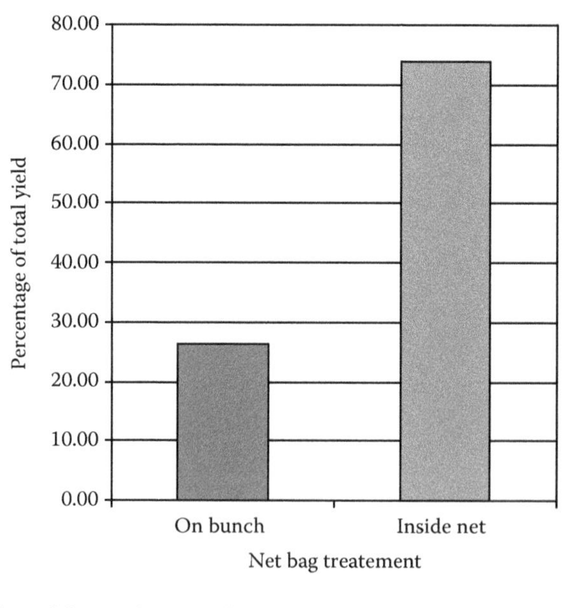

FIGURE 10.7 Distribution of date at the maturity stage under net bag treatment: Khalas date variety.

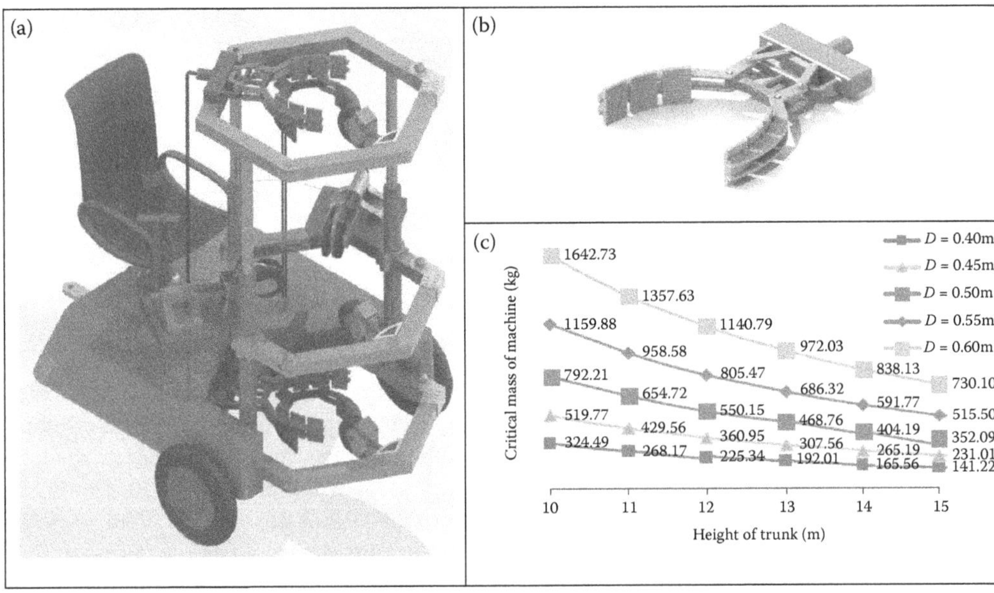

FIGURE 10.8 (a) The designed gripper date palm service machine; (b) the designed gripper; and (c) critical machine mass prediction characteristics for tree height. (From Jahromi, K.M., Mirasheh, R., and Jafari, A. 2008. Proposed lifting model for gripper date palm service machines. *Agricultural Engineering International: the CIGR Ejournal*. Manuscript PM 08 018. Vol. X. October 2008.)

3. To reduce the productivity losses and to improve the product quality
4. To enhance the safety and operating comfort, during work, of the operator as well as others around the field
5. To function as multiobjective, multitasking: such as palm tree preparation, cleaning, pruning, pollinations, and hand-pick type or complete bunch harvesting

PROCEDURAL STEPS FOR A DATE HARVESTER DESIGN WITH EXAMPLES

Design Steps

It is proposed that a conceptual design and the mechanism need to be considered covering the major functional needs. The design approach can consist of several steps and stages, including obtaining the design solution, prototype development, performance testing, and elimination of shortcomings and justification. The design process may have the following general steps:

1. Recognize the problem: Designing a machine which makes the harvesting process safe and easy for the farmer and to improve the quality of the products
2. Define the problem in writing or using sketches or diagrams
3. Develop ideas
4. Gather information: To design the machine, much information needs to be gathered from multiple resources; some of them could be first-hand information from farmers, information of materials, literature reviews, and the theory and equations used in calculations
5. Continue developing and analyzing the ideas on mechanisms
6. Reevaluate the problem and ideas
7. Check standards, codes: The standard guidelines should be followed and the material has to be selected carefully to minimize the costs and ensure the safety
8. Revise or alter the ideas considering limitations and subproblems

9. Decide on a design idea and begin development
10. Bill of quantities, cost estimates
11. Build a test prototype (Prototype 1)
12. Begin testing—Performance evaluation
13. Modifications in prototype (Prototypes 2, 3, 4, etc.)
14. Prepare documents and detail the design: Final documentation, drawings, manuals, and specifications were developed
15. Patent and copyright procedures as necessary
16. Advertise and promote sales programs
17. Collect field information on the product for further improvement

Sample Project

In the sample project, an innovative machine is proposed in which a powered or trailer-mounted hydraulic-lifting system is considered. The specific conditions taken into consideration were: dates in Khalas palm, to utilize this method until the working height of trees is up to 10–15 m (Figure 10.9), properly spaced (Figure 10.10), and to harvest more than 2000 tress in each field.

The new date harvester design should be helpful to farmers, as it avoids the shortcomings of previous designs. Basically, the machine should be more safe, comfortable to operate, and should help in improving the quality of the product. In addition, this should need less effort and land and be labor productive. The base platform with controls, a lifting bucket-platform with auxiliary controls as the operator can control the machine at both locations and a hydraulic system to move the bucket-platform up and down. A single operator should be able to use this machine and the base includes four wheels for mobility from one palm to another, either powered by an engine or pulled by a tractor or truck. The conceptual harvester system is shown in Figures 10.11 through 10.13.

Consideration on Design Concept

Climbing the tree to reach the fruits is the most difficult activity of the date harvesting operation; therefore, the emphasis has been on methods of lifting the operator to the fruit-bunches. The major considerations will be given to

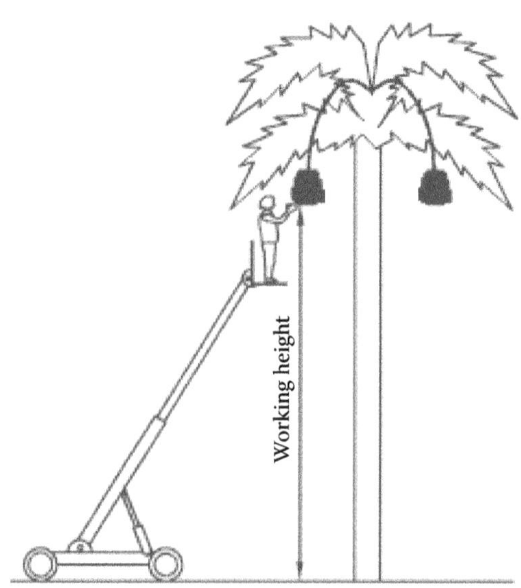

FIGURE 10.9 A simple lifting mechanism for adjustable working height.

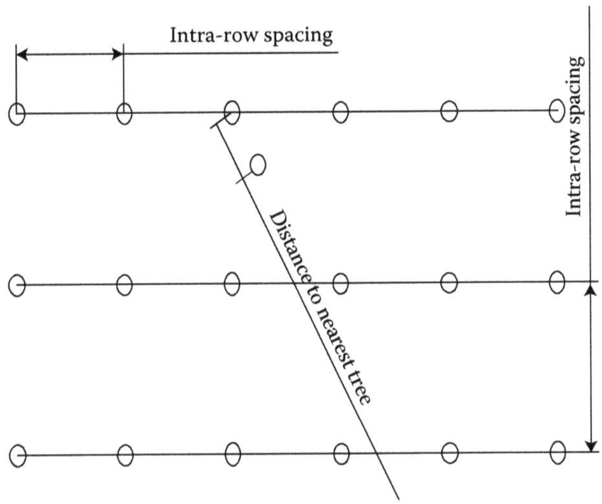

FIGURE 10.10 Proper row and inter-row spacing necessary for mechanization adoption.

1. The operating principle (hydraulic system)
2. Power source and machine type
3. Cost/farmer affordability
4. Field capacity
5. The shape and size
6. Material selection

Design Methodology

Correct design methodology and sequence will be very important to achieve the goals of any project work. Proper reporting and record keeping will help a lot in any revisions or design modification. At the design stage, consideration should be given to different areas:

FIGURE 10.11 Conceptual arrangement for harvesting in a date plantation.

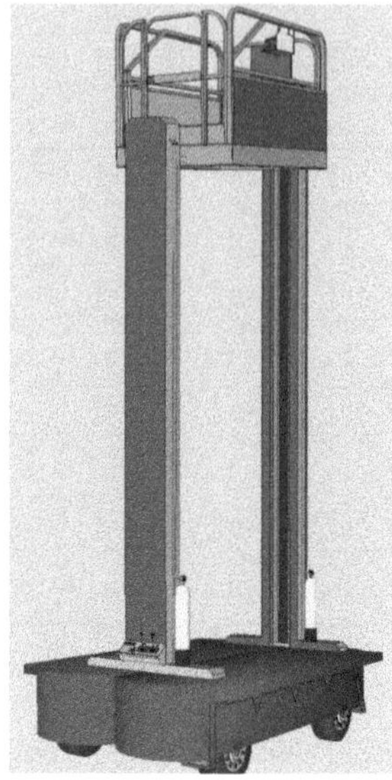

FIGURE 10.12 Harvester with a lifted platform.

Farm considerations:

1. It must be an organized farm for easy adoption of mechanization.
2. The palm trees planted in rows and cross-rows with suitable spacing between each palm tree following agronomic guidelines (i.e., 3 × 5 m).
3. Flat or leveled areas (as possible) between each row for easy tractor movement.
4. Irrigation system should be well organized and located.

Date palm tree considerations:

1. Height range for the design, that is, <15 m—based on the cultivar.
2. The number of date palms trees, that is, 2000 and the date palm cultivar is Khalas.
3. The number of fruit bunches for each palm tree, that is, maybe <16.
4. The mass for each fruit bunch, that is, 10 kg.

Farmer considerations:

1. Operator height range (1.6–2 m).
2. Operator weight range (50–100 kg).
3. Working hours (8:00 a.m. to 5:00 p.m.).

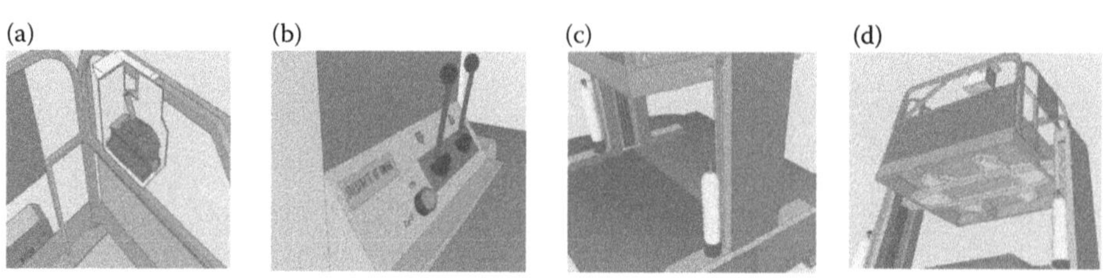

FIGURE 10.13 Date harvester controls; (a) at elevated bucket platform and (b) at the base platform. (c), (d) Arrangement of the hydraulic system and accumulators.

Design Calculations

All design calculations should be performed with care, and each subsystem and elements should be checked for failure by appropriate failure mode following standard design procedures. Standard machine components should be selected as much as possible, minimizing the time and efforts required for the computing steps.

The major results could be the power source and requirement, detailed information on power system, theoretical and actual field capacity, operating conditions, and so on.

Selection of Materials

One of the most important aspects in the design process is selection of the best materials for the need. Systematic selection of materials for the design begins with considering properties and costs of candidate materials. In this stage, each and every component/element should be carefully evaluated for strength, safety, and durability terms.

Drawings and Specifications

Standard engineering drawings and tables of specifications should be made following the coded information. This will help in organizing the manufacturing process of the machine from prototype stage to final model. Suitable drafting software or CAD software can be utilized to achieve this goal.

Bills of Quantities

Each and every component should be accounted for when preparing the bill of quantities. This will enable the estimation of machine cost, operational costs, and subsequently the selling price. This section is an essential part of the design process.

Performance Evaluation

The designed machine cannot be justified without appropriate evaluation of its performances. Performance evaluation is a necessary and beneficial process, which provides annually feedback about the device in order to make new designs, avoiding errors that occurred in the previous stages. Performance evaluations are done at various design stages. First, the performance evaluation of components or subsystems can be done before the assembly of the whole machine. On a whole, the performance will be tested for the first prototype. After revisions or modifications at different prototype stages and of the final model, complete performance tests will be done.

It is important and a challenging task for the design engineer to develop tests for performance evaluations, as standard test procedures may not be available for this specific kind of machinery.

CONCLUSIONS

There is a long due need for development of a date harvesting machine for date growers in many countries of the world. It is clear that the need for mechanical harvesting is one of the important requirements of date growers at this time. Several researchers have attempted to design machines; however, so far there has been no record of date harvesters with perfection.

It is well understood that date harvester design should be done at the regional basis, as certain conditions such as land, cultivar, cultural, and socio-economical factors can influence the features of the machine.

It is also necessary to consider revising the current date cultivation practices, at least for newly planned plantations. The application of the Good Agricultural Practices concept is currently on the move for high-value crops. The date cultivations may need this consideration as mechanization adoption may be impossible for certain date-planning conditions.

REFERENCES

Abounajmi, M. 2000. *Design and Development of an Experimental Shaker for Investigating the Effects of Shaking Mode, Frequency and Amplitude on Date Fruit Detachment*. MS Thesis, Shiraz University, Shiraz, Iran.

Ampratwum, D.B. and Al-Siyabi, B.A.S. 2005. Solar and postharvest equipment for drying and processing dates. *Proceedings of the International Conference on Postharvest Technology and Quality Management*. Arid Tropics, Sultan Qaboos University, Oman, 31 Jan.–2 Feb. 2005, pp. 149–156.

Ali, T. and Akyurt, M. 1998. Design of a Hi-lifter gripper for palm trees. *Applied Engineering in Agriculture* 14(3): 215–221.

Al-Suhaibani, S.A., Babier, A.S., and Kilgour, J. 1990. Design specification of a date palm service machine. *Agricultural Mechanization in Asia, Africa and Latin America* 21(4): 53–60.

Al-Suhaibani, S.A., Babier, A.S., Kilgour, J., and Blackmore, B.S. 1992. Field tests of the KSU date palm machine. *Journal of Agricultural Engineering Research* 51: 179–190.

Al-Suhaibani, S.A., Babaeir, A.S., Kilgour, J., and Flynn, J.C. 1988. The design of a date palm service machine. *Journal of Agricultural Engineering Research* 40(4): 143–157.

Al-Yahyai, R. 2009. Improvement of date palm production in the Sultanate of Oman. *ISHS Acta Horticulturae: III International Date Palm Conference*. Online; http://www.actahort.org/members/showpdf?booknrarnr=736_32.

Barreveld, W.H. 1993. Date palm products. FAO Agricultural Service Bulletin No. 101, Food and Agricultural Organisation of the United Nations, Rome.

Brown, G.K., Sarige, Y., and Perkins, R. M. 1983. Mechanization of date production worldwide. *The 1st International Symposium on Fruit, Nut and Vegetable Harvesting Mechanization*. ASAE Pub. 5-84, ASAE, St. Joseph, Michigan.

Jahromi, K.M., Jafari, A., Rafiee, S., and Mohtasebi, S.S. 2007. A survey on some physical properties of date palm tree. *Journal of Agricultural Technology* 3(2): 317–322.

Jahromi, K.M., Mirasheh, R., and Jafari, A. 2008. Proposed lifting model for gripper date palm service machines. *Agricultural Engineering International: the CIGR Ejournal*. Manuscript PM 08 018. Vol. X. October, 2008.

Kader, A.A. and Hussein, A.M. 2009. *Harvesting and Postharvest Handling of Dates*. Postharvest Technology, Research and Information Center, UC Davis, California.

Levin, J.H., Brown, G.K., Monroe, G.E., and Bilbo, B.C. 1969. Mechanization practice for subtropical fruit-figs, dates, coffee, avocado, etc. In: B.F. Cargill and G.E. Romsiller (eds.). *Fruit and Vegetable Harvesting Mechanization*. Rural Manpower Center, Michigan State University, East Lansing, Michigan, pp. 807–819.

Mazloumzadeha, S.M., Shamsa, M., and Nezamabadi-Pourb, H. 2008. Evaluation of general-purpose lifters for the date harvest industry-based on a fuzzy inference system. *Computers and Electronics in Agriculture* 60 (2008): 60–66.

Perkins, R.M. and Brown, G.K. 1964. Progress in mechanization of date harvesting. Date Growers' Institute Report 41, pp. 19–23.

Perkins, R.M. and Brown, G.K. 1967. Harvesting dates mechanically. *Transactions of the ASAE* 10: 486–488.

Sarig, Y. 1989. An integrated mechanical system for date orchard operation, ASAE, ASAE, St. Joseph, Michigan, pp. 89–1069.

Sclater, N. and Chironis, N.P. 2001. *Mechanisms and Mechanical Devices Sourcebook III*. McGraw-Hill, New York.

Shamsi, M. 1990. *Design and Theoretical Calculation of a Tree Climbing Machine*. Scientific and technologic research organization. Unpublished Project Report. Kerman, Iran.

Shamsi, M. 1998. *Design and Development of a Date Harvesting Machine*. Unpublished PhD Thesis, Silsoe College, Cranfield University, UK.

Shamsi, M. and Mazloumzadeh, S.M. 2009. Studying yield variations using precision farming in a date palm orchard. *Journal of Agricultural Technology* 5(1): 33–39.

Zaid, A. 2002. *Date Palm Cultivation*. FAO Publication No. 156, Rome. Available at: http://www.fao.org/DOCREP/006/Y4360E/y4360e05.htm#bm05.3

Zaid, A. and de Wet, P.F. 1999. Date palm propagation. In: Zaid, A. and Arias, E.J. (eds.). *Date Palm Cultivation*. FAO Plant Production and Protection Paper No. 156, pp. 74–106.

11 Date Marketing

Msafiri Daudi Mbaga

CONTENTS

This chapter explores the marketing of dates from the farm gate to the final consumer both locally and at export markets. This chapter is arranged as follows: The section "Date Fruit Production and Trade" highlights global date production and trade. The section "Date Marketing Channels" looks at date marketing channels. In the section "Marketing Channels," date marketing channels in Oman, Tunisia, Pakistan, and Iraq are presented and briefly discussed. The section "Packaging, Shipment, and Documentation" reviews issues related to dates packaging, shipment, and documentation. The final section is the conclusion.

DATE FRUIT PRODUCTION AND TRADE

Global Date Fruit Production

Date production is an important global agricultural industry with global production reaching 7 million tonnes of fruits in 2007 (FAOSTAT, 2010). Dates are produced mainly in the hot and arid regions of Africa, the Middle East, and Asia, as well as in some parts of Europe and the United States. Processed dates are marketed all over the world for direct consumption and further processing into high-value confectionery products.

The leading global date producers and their respective shares of world production in 2007 (FAOSTAT, 2010) are presented in Table 11.1. Egypt (1313.6 thousand tonnes) is the leading date producer with 19% share of global date fruit production, followed by Iran (1000 thousand tonnes), Saudi Arabia (982.5 thousand tonnes), and the United Arab Emirates (755 thousand tonnes); these

TABLE 11.1
Date Production and Value by the 20 Leading Producers

Rank	Area	Value ($1000)	Production (MT)	Share (%)
1	Egypt	354,221	1,313,696	19.0
2	Iran	313,470	1,000,000	14.5
3	Saudi Arabia	307,998	982,546	14.2
4	UAE	202,188	755,000	10.9
5	Pakistan	156,735	557,524	8.1
6	Algeria	156,735	526,921	7.6
7	Iraq	123,820	440,000	6.4
8	Sudan	104,072	332,000	4.8
9	Oman	77,574	255,871	3.7
10	Libya	39,183	175,000	2.5
11	China	40,751	130,000	1.9
12	Tunisia	34,979	124,000	1.8
13	Morocco	16,562	74,300	1.1
14	Yemen	15,673	53,596	0.8
15	Mauritania	6,896	22,000	0.3
16	Qatar	6,759	21,564	0.3
17	Chad	5,736	18,300	0.3
18	Israel	5,447	17,377	0.3
19	USA	4,858	14,787	0.2
20	Kuwait	4,545	14,500	0.2
21	Others		79,918	1.2
Total global production			6,908,900	100

Source: FAOSTAT. 2010. Available at: http://faostat.fao.org/site/567/DesktopDefault.aspx?PageID=567#ancor

being the "top four" producers with double-digit shares of global date fruit production. The United States and Kuwait hold the 19th and 20th positions, respectively, among the 20 leading global date fruit producers with 0.2% share of the total production. The leading 20 producers produce about 98.8% of global date fruits production (Table 11.1).

Global Date Exports and Imports

The world market for date is expanding. The exporting countries are aggressively working toward increasing their market shares as new markets open up as a result of world market liberalization.

World Date Export

Table 11.2 lists the 20 leading date fruit exporters in 2007 in terms of total value. Tunisia is the leading exporter in terms of value, followed by Iran, Israel, Saudi Arabia, and Pakistan; these being the "top five" export market leaders.

The top 20 export market leaders exported 615.6 million tonnes in 2007, generating 529.2 million dollars in export revenues to the economies of these countries.

Global Date Imports

The global date imports by the 20 leading date importers are as presented in Table 11.3. In 2007, the 20 leading importers imported 493.1 million tonnes of date worth $440.45 million. India is the

TABLE 11.2
World Leading Date Exporters by Value

Rank	Area	Quantity (tonnes)	Value (1000 $)	Unit Value ($/tonne)
1	Tunisia	68,856	164,759	2,393
2	Iran	242,092	83,083	343
3	Israel	9,513	52,592	5,528
4	Saudi Arabia	48,762	40,529	831
5	Pakistan	104,090	38,271	368
6	France	10,470	29,524	2,820
7	Algeria	25,039	23,140	924
8	UAE	38,691	19,988	517
9	USA	3,259	18,863	5,788
10	Germany	2,743	10,246	3,735
11	Iraq	37,063	9,532	257
12	Netherlands	1,775	7,904	4,453
13	Mexico	3,891	7,363	1,892
14	Oman	9,368	5,687	607
15	South Africa	720	4,444	6,172
16	Belgium	789	3,881	4,919
17	Egypt	4,704	3,014	641
18	Jordan	2,806	2,924	1,042
19	Spain	564	2,037	3,612
20	Austria	388	1,469	3,786
Total export quantity and value		615,583	529,250	

Source: FAOSTAT. 2010. Available at: http://faostat.fao.org/site/567/DesktopDefault.aspx?PageID=567#ancor

largest importer of date fruit, importing 253.3 million tonnes in 2007, generating $72.3 million. However, in terms of unit value, India is 20th among the 20 leading importers, implying that India imports low-quality cheap date fruits.

As Table 11.3 shows, all countries ranked 1–9 in terms of unit value are the European Union (EU) countries (Switzerland, Austria, Belgium, Spain, Italy, Germany, the Netherlands, France, and the United Kingdom). EU is a key market for date exports. Although the EU imports of date represent only 10% of world imports in volume, they account for some 30% in value. This reflects the fact that EU import prices for dates are comparatively much higher than the world average.

The leading high-value date varieties in the import markets are *deglet nour* and *medjool.* The rest of the date varieties sold in the international markets are generally referred to as "common dates," especially in the EU market. In the EU, the United Kingdom and Germany are the main importers of common dates. In Germany, traders are interested in common dates because of the low price. In the United Kingdom, the population of Asian origin is thought to have a big influence on the consumption of common dates. These dates are also used by the foodstuff industry in the United Kingdom. While EU imports of common dates are significant, this quantity has not increased substantially over the past years. This seems to indicate that the demand is shifting toward higher-quality dates such as medjool or deglet nour. Deglet nour variety (Figure 11.1) is the most popular variety in EU.

Because of their history and culture, Southern EU countries (France, Spain, and Italy) mainly consume deglet nour date imported from Tunisia and Algeria. Some 90% of the deglet nour produced in the world is imported from Tunisia and Algeria. The remaining 10% is produced in Israel and the United States.

TABLE 11.3
World Leading Date Importers by Value

Rank	Country	Quantity (tonnes)	Value (1000 $)	Unit Value ($/ tonne)	Rank by Unit Value
1	India	253,341	72,359	286	20
2	France	27,439	66,741	2,432	8
3	Morocco	50,473	43,524	862	15
4	United Kingdom	14,357	34,564	2,407	9
5	Germany	10,790	29,768	2,759	6
6	Spain	7,152	21,803	3,049	4
7	Malaysia	15,346	20,974	1,367	12
8	Italy	7,230	20,299	2,808	5
9	Canada	8,605	16,406	1,907	10
10	UAE	12,266	14,654	1,195	14
11	Russian Federation	22,375	13,628	609	18
12	Indonesia	15,549	12,055	775	17
13	USA	8,646	11,308	1,308	13
14	Australia	6,657	10,401	1,562	11
15	Netherlands	3,787	10,254	2,708	7
16	Switzerland	2,022	9,957	4,924	1
17	Belgium	3,029	9,699	3,202	3
18	Austria	2,440	7,940	3,254	2
19	Jordan	8,534	7,242	849	16
20	Turkey	13,092	6,874	525	19
Total import quantity and value		493,130	440,450		

Source: FAOSTAT. 2010. Available at: http://faostat.fao.org/site/567/DesktopDefault.aspx?PageID=567#ancor

Medjool date variety (Figure 11.2) has been known in the EU since the early 1990s. Imports of medjool are very low in comparison to the imports of deglet nour, but they have been rapidly increasing over the years. The United States and Israel today share the European market, whereby the United States is the foremost producer. Some other countries (e.g., Namibia) have also started producing medjool.

FIGURE 11.1 Deglet nour date variety. (http://sweetenergy.com/php/famDGL.php)

FIGURE 11.2 Medjool date variety. (UNIDO. 2004. *Date's Market*. Available at: http://exchange.unido.org/upload/1726_Date.pdf)

DATE MARKETING

Food marketing, according to Kohls and Uhl (2001), is "the performance of all business activities involved in the flow of food products and services from the point of production until they are in the hands of consumers." For dates, these activities include harvesting, processing, packaging, and transportation/shipment to local or export markets.

HARVESTING

Harvesting is one of the important marketing functions that has direct impact on date quality and hence the ultimate price. Date fruit harvesting is done either manually or with some degree of mechanization, especially for big farms. Harvesting needs to be faultless and clean since it significantly affects the rest of the downstream processes such as processing, packing, and marketing.

Whole dates are harvested and marketed at the three stages of their development. The three stages are the sweet *Khalaal*, the *Rutab*, and the *Tamr*, as indicated in Figure 11.3.

The choice for harvesting at one or another stage depends on varietal characteristics, climatological conditions, and market demand. The degree of perishability decreases from the sweet Khalaal, to Rutab and Tamr. Therefore, the level of processing necessary to prolong the storage ability is higher for Khalaal, followed by Rutab, and then Tamr.

When dates reach the Khalaal stage, they are regarded to be ready for trading as "fresh" fruit (Zaid and Arias-Jiménez, 2002). The Khalaal being the first date fruits arriving at the market at the beginning of the harvesting season usually find a ready market, especially in the traditional date-producing countries. Because of their perishability, the markets for sweet Khalaal dates are normally of a local nature (Figure 11.4). The marketing channel is therefore mostly direct from the farm to the local consumer, especially in places where transport and logistics are inefficient. However, in countries such as the United States, where transport logistics are efficient (e.g., in California), a postal market service of sweet Khalaal exists through which Khalaal dates are distributed all over the United States (Figure 11.5). The customers are mainly foreigners who are acquainted with this type of fruit. Date varieties suitable for marketing at Khalaal stage are Barhee, Zaghlool, Hayany, and Khalas. Of these varieties, only Barhee is sold in England, France, and Australia, while the other two are mostly consumed locally where they are produced.

FIGURE 11.3 Three major stages of maturity in which dates are consumed: Khalaal, Rutab, and Tamr. (Adapted from Barreveld W.H. 1993. *Date Palm Products* FAO Agricultural Services Bulletin no. 101, Food and Agriculture Organization of the United Nations, Rome. M-17 ISBN 92-5-103251-3. Available at http://www.fao.org/docrep/t0681E/t0681e00.htm#con)

At the Rutab stage, date have passed one stage further in its maturation process on the palm tree and hence are less risky than the sweet Khalaal. However, the handling and transportation of Rutab is both difficult and expensive because it is produced in comparatively short periods and it is delicate and perishable. With local markets likely to become saturated and market expansion over longer distances hampered by costs of special treatment and freight, a local surplus situation can very easily arise. Once this happens, Rutab dates can be either preserved by cold storage or turned into Tamr.

At the Tamr stage, dates have dried down on the palm to below 24% moisture content, and its flesh has firmed through moisture loss but is pliable. The fruit at this stage is nonperishable because microorganisms will not be able to grow on it. However, it is still susceptible to insect infestation, moisture uptake, and its consequences during subsequent storage, unless special precautions are taken.

FIGURE 11.4 Sweet Khalaal on sale in a local market in Bahrain. (Adapted from Barreveld W.H. 1993. *Date Palm Products*. FAO Agricultural Services Bulletin no. 101, Food and Agriculture Organization of the United Nations, Rome. M-17 ISBN 92-5-103251-3. Available at http://www.fao.org/docrep/t0681E/t0681e00.htm#con)

FIGURE 11.5 Packing of sweet Khalaal for postal market sales in California. (Adapted from Barreveld W.H. 1993. *Date Palm Products*. FAO Agricultural Services Bulletin no. 101, Food and Agriculture Organization of the United Nations, Rome. M-17 ISBN 92-5-103251-3. Available at http://www.fao.org/docrep/t0681E/t0681e00.htm#con)

After harvest, dates have to be sorted at the farm level before they are shipped to a local market or to a processing/packing plant. Dates already sorted can be put in containers for transportation to the packing plants or to the market.

MARKETING CHANNELS

A marketing channel describes the movement of a product or commodity from the site of production to the place of consumption. It may include transportation, handling and storage, ownership transfers, processing, and distribution. The marketing channels for dates include initial processing at farm level after harvest, transport to the local market directly (especially for the highly perishable sweet Khalaal) or to the packing plant, processing and packaging at factory level and transport to the port for export. There is no one universal set of marketing channels because each country is unique and institutions involved operate differently under different sets of infrastructure. Therefore, marketing channels differ from one country to another. The differences are sometimes very minor. In order to have a clear understanding of this, date marketing channels in four countries (Oman, Tunisia, Iraq, and Pakistan) are presented in this chapter.

Date Marketing Channels in Oman

The marketing channels for Omani dates include on-farm selling, retailers, local markets, date factories, and export as shown in Figure 11.6a. There are therefore mainly three channels through which date flow from the farm to local and foreign consumers/export market. Dates can be sold at the farm gate and from there dates are either marketed directly to the final consumer at the local market or to the local wholesale market (especially sweet Khalaal).

Alternatively, dates can be marketed directly to wholesalers and from there to either retailers before reaching the local market or directly to the local market. Next is the export market channel. In this channel, dates are sold directly to the processing factories for processing and packaging before being shipped to export markets. Each of these channels or outlets attracts a different share of the whole date market.

Generally, it is how dates are produced, harvested, sorted, graded, processed, packaged, and transported that determines their final market price and hence value (Figure 11.6b). In other words, it is the efficiency, safety, and quality management of date marketing channels that ultimately determine the final market value of the dates.

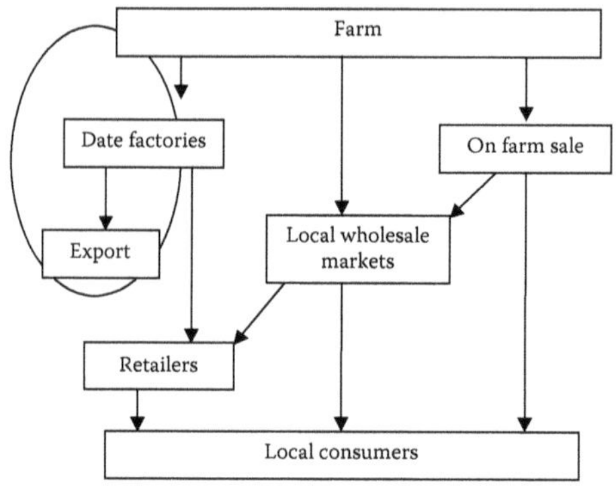

FIGURE 11.6a Omani date marketing channels with date export supply chain indicated by circle. (Data from Oman Ministry of Agriculture: A Country Report on the Development of Date Production and Marketing, 2003.)

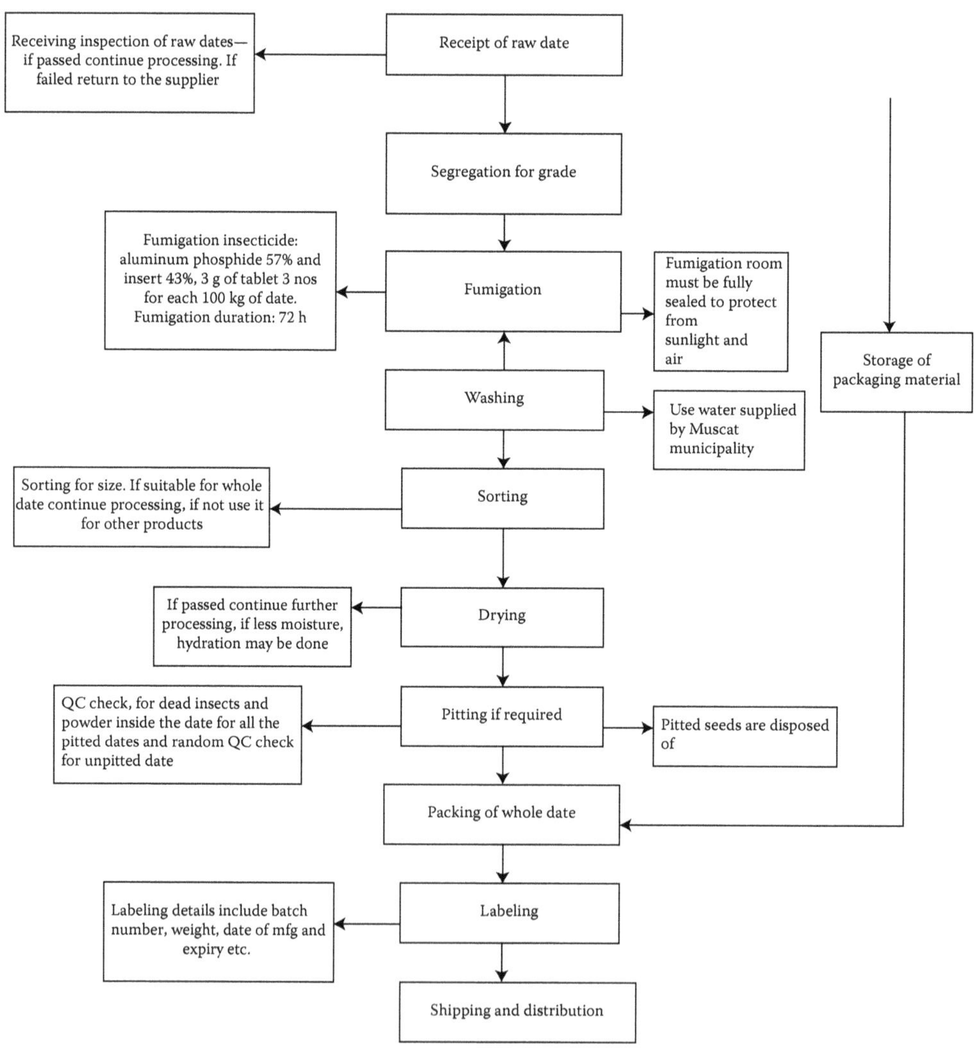

FIGURE 11.6b Processing and handling of dates in Oman. (Adapted from United dates factory, 2007, http://goldendateoman.com)

Tunisia Date Marketing Channels

Tunisia is the world-leading producer of deglet nour and possesses ~50% of the world's deglet nour palm trees. The marketing channels of dates are somewhat complex. Many agents, including producers, collectors, conditioners, exporters, wholesalers, etc., are involved (Figure 11.7a).

Date marketing channels in Tunisia (Figure 11.7a) look much more organized and involve many agents performing different functions. All these functions are intended to deliver a high-quality final product to the market at a reasonable price to the final consumer.

Processing and handling of dates in Tunisia (Figure 11.7b) are very well organized, and dates are handled very carefully and efficiently at each stage of the marketing channel. This may partly explain why Tunisia is very successful in date exports to the EU market. The EU market customers demand high-quality products and the competition for a share of that market is very stiff.

Date Marketing Channels in Pakistan

The marketing channels for dates in Pakistan are presented in Figure 11.8. The marketing channels include both processing and handling of dates. The marketing channels in Pakistan look somewhat different from the channels in Oman and Tunisia. In Pakistan, there are preharvest contractors.

The preharvest contractors are the main assembling/consolidating agents. Usually, these traders conduct visits to individual date farmers before harvest and enter into contract by fixing lump-sum amount for purchase of all or some of the dates, and pay some advancing money. The preharvest contractors usually have their fixed trading partners, the wholesale commission traders. The wholesale

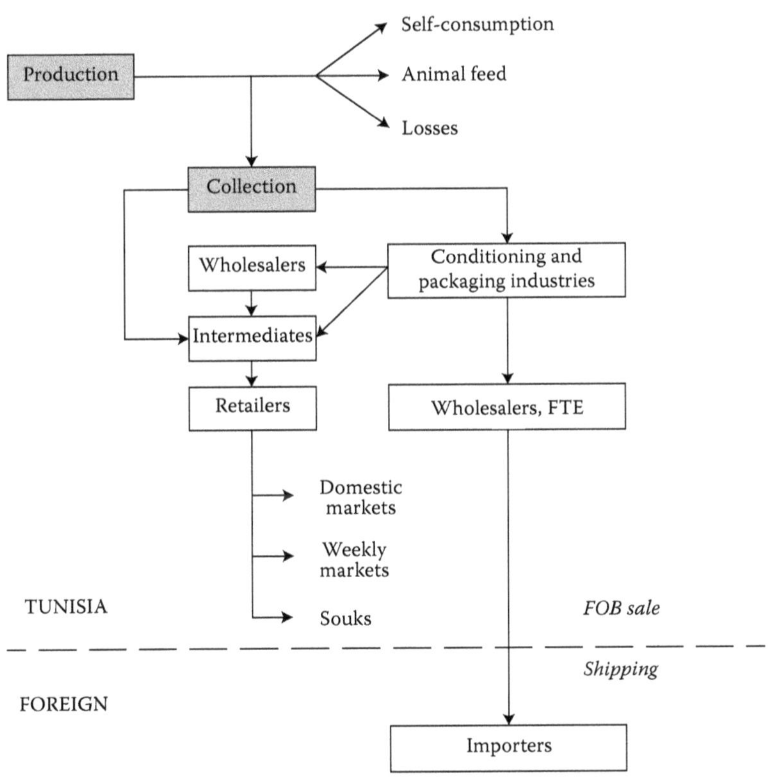

FIGURE 11.7a Date marketing channels in Tunisia. (Adapted from Martinez, M. Garcia and Gharbi, M., and Laajimi, A. 2004. Food safety and quality systems in the Tunisian olive oil and date sectors: An exporters view. In: *Quality Assurance, Risk Management and Environmental Control in Agricultural and Food Supply Networks*. Universitat Bonn, Germany, pp. 325–333. ISBN 3-932887-65-4.)

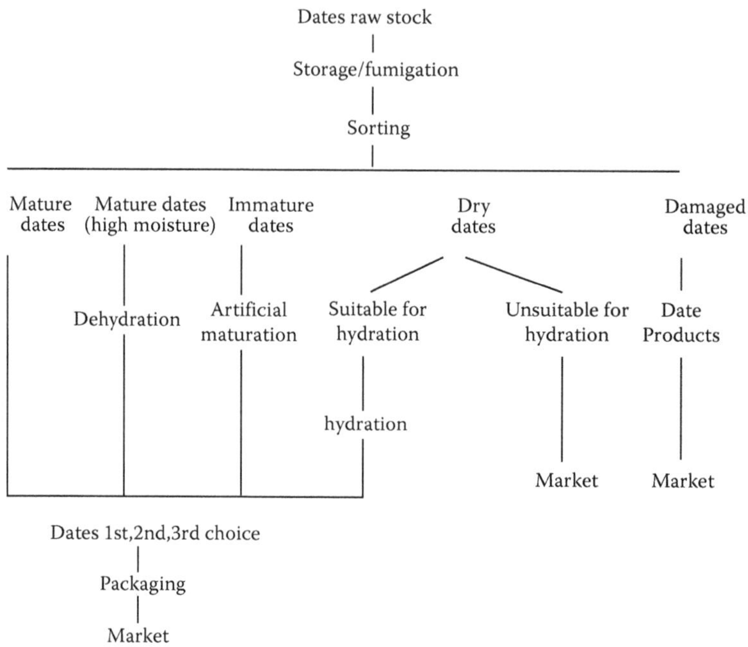

FIGURE 11.7b Processing and handling of dates in Tunisia. (Adapted from Barreveld W.H. 1993. *Date Palm Products*. FAO Agricultural Services Bulletin no. 101, Food and Agriculture Organization of the United Nations, Rome. M-17 ISBN 92-5-103251-3. Available at http://www.fao.org/docrep/t0681E/t0681e00.htm#con)

commission traders finance the preharvest contractors for expenses such as payment of advance money to the date farmers.

These preharvest contractors therefore play a very important role of assembling and quality control.

Date Marketing Channels in Iraq

The Iraq dates (Khalaal and Tamr) marketing channels are presented in Figure 11.9a and b. The bulk of the dates (80%) flowing through the marketing channels in Iraq are dried dates (Tamr). The

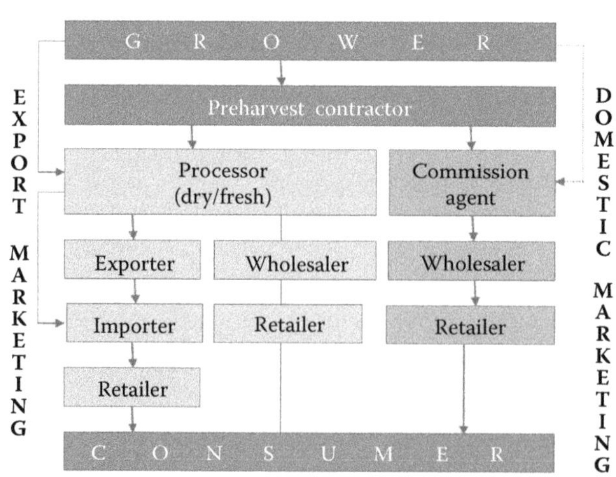

FIGURE 11.8 Pakistan date marketing channels. (Adapted from Pakistan Horticulture Development & Export Board. 2008. *Date Marketing Strategies*. Available at http://www.phdeb.org.pk/MktStrategies/Date.pdf)

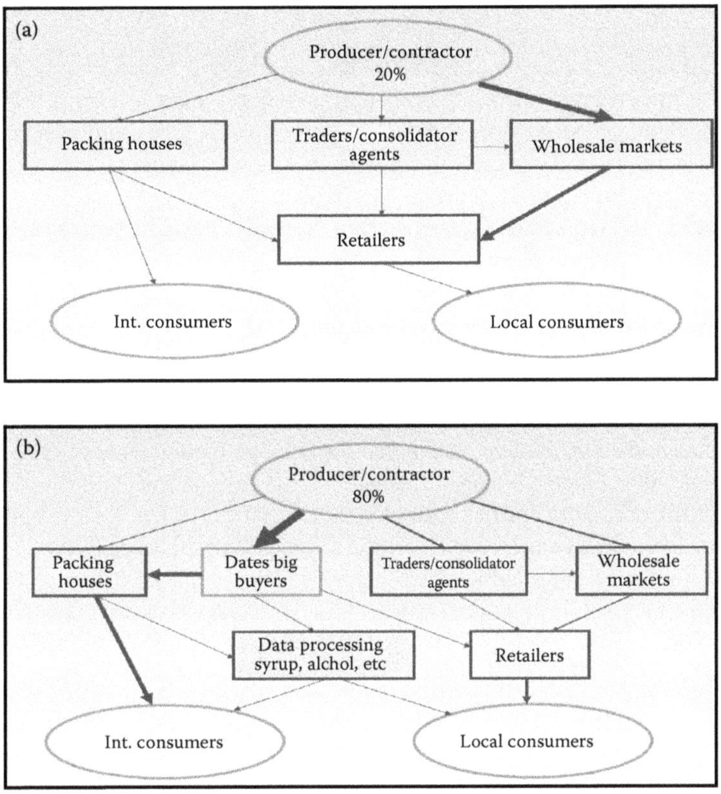

FIGURE 11.9 (a) Iraq marketing channels of fresh dates (Khalaal). (b) Iraq marketing channels of dried dates (Tamr). (Adapted from Jabarin A. and F.A. Hussain. 2004. *Iraqi Date Industry Marketing and Post-Harvest Issues* Unpublished document, September. Agricultural Restoration & Development Program for Iraq and National Program for Propagation and Improvement of Date Palm. Available at http://www.usaid.gov/iraq/contracts/pdf/ARDI14-IraqiDateIndustry.pdf)

Khalaal dates which are naturally perishable comprise only 20% of the total quantity of dates going through the marketing channels. The agents and institutions handling both dried (Tamr) and fresh (Khalaal) dates are more or less the same.

PACKAGING, SHIPMENT, AND DOCUMENTATION

PACKAGING

Packing is an important stage in both the traditional and modern methods of marketing. There is packing for long-distance transportation (local and export) and packing for the final consumers at supermarkets. Packing for long-distance transportation is intended at making it possible to transport the fruit by various means. It could range from packing in baskets made of palm leaflets for local transportation, to the use of modern packing containers for transportation by air or by sea for export. Aside from meeting international standards, packing materials should protect the fruit when packed so that it remains in good condition under various circumstances and for various periods of time.

Packing can also be done with the aim of promoting marketing (Liu, 2003). For example, supermarkets tend to carry small packages ranging from 120 to 500 g. Small-sized packages are considered convenient to consumers compared to a 2 or 5 kg package.

Shipment

All shipments are usually carried out according to the planning and direction of both the local and the export markets. It is important to choose the cheapest transport which will bring the fruit to the client at lower cost and without compromising the quality of the fruit. This is important so as to stay competitive against other suppliers. The types of shipment include overland, sea, overland and sea, and air.

Sometimes, overland transport to markets is possible, especially to local or regional markets. Where this is possible the fruit must be transported in a way that will protect it from the environment and, if necessary, it should be in refrigerated trucks.

Shipment by sea in containers is an efficient and relatively cheap means of transportation for export markets. By this means of shipment, the fruit is protected from the environment from the moment it leaves the producer to the moment it reaches the customer's door. The containers are usually refrigerated by cold air flowing horizontally over the layers of fruit. This air is distributed uniformly throughout the container.

With the overland and sea shipment, refrigerated trucks go from the supplier to a port where it is loaded to a ship that will transport the product further to its destination. This method is more expensive than shipment in containers but it is usually faster.

Shipment by air is the most expensive of all the shipment alternatives but it is sometimes inevitable when the fruit must be supplied at short intervals. This is one of the means of transportation for Khalaal fresh dates from California (Figure 11.5). Transport to and from the airport must also be taken into account.

Documentation

All the shipments must be documented in detail to ensure speedy transfer to the client, especially during export. Besides documents for the customs, payment, and transport, it is also important to add a phytosanitary certificate stating that the fruit is healthy. This document is issued in every country by the relevant authorities and it certifies that the fruit is not infected by pests or diseases and meets the standards of the importing country.

Dates for export must be certified by an internationally recognized certifying body and must be recognized by the buyer. Quality standards for dates have been set by different bodies. The Codex Alimentarius, for example (Appendix I), sets permitted levels of residues of pesticides (MRL = minimum residue limit) according to types of material and species of fruits, vegetables, and other foodstuffs. The Codex Alimentarius is published by FAO and WHO and has been ratified by most of the 146 member countries. Date exporters also must comply with

1. ISO 9000 and 140000. The ISO 9000 family is primarily concerned with quality management and ISO 14000 is concerned with environmental management.
2. HACCP (hazardous analytical critical control point) quality systems.
3. EUREP-GAP (Euro-retailer produce—good agricultural practices policies and surveillance systems.

There are other requirements such as CE-MARK and traceability that countries have to comply with nowadays in order to export. All these have to be studied very carefully and must be complied with in order to export, especially to EU and North American markets.

CONCLUSION

Date production is an important global agricultural industry with global production of over 7 million tonnes. In 2007, for example, the 20 leading date producers exported 615.6 million tonnes, generating 529.2 million dollars in export revenues. This chapter looks at the marketing channels for dates.

Marketing channels are a system of marketing institutions that promotes the physical flow of goods and services, along with ownership title, from producers (farmers) all the way to the consumers at the lowest cost possible.

Date marketing channels for Oman, Tunisia, Pakistan, and Iraq are reviewed in this chapter. Furthermore, issues related to date packaging, shipment, and documentation are also discussed briefly.

APPENDIX: CODEX ALIMENTARIUS FOR DATES*

CODEX STANDARD FOR DATES

CODEX STAN 143-1985

1. SCOPE

This standard applies to commercially prepared whole dates in pitted or un-pitted styles packed ready for direct consumption. It does not apply to other forms such as pieces or mashed dates or dates intended for industrial purposes.

2. DESCRIPTION

2.1 Product Definition

Dates are the product prepared from sound fruit of the date tree (*Phoenix dactylifera* L.), which fruit:

(a) is harvested at the appropriate stage of maturity;
(b) is sorted and cleaned to remove defective fruit and extraneous material;
(c) may be pitted and capped;
(d) may be dried or hydrated to adjust moisture content;
(e) may be washed and/or pasteurized; and
(f) is packaged in suitable containers to assure preservation and protection of the product.

2.2 Varietal Types

Varietal types are classified as:

(a) **Cane sugar varieties** (containing mainly sucrose) such as Daglat Nuur (Deglet Noor) and Daglat Beidha (Deglet Beidha).
(b) **Invert Sugar varieties** (containing mainly invert sugar—glucose and fructose) such as Barhi (Barhee), Saiidi (Saidy), Khadhraawi (Khadrawy), Hallaawi (halawyj, Zahdi (Zahidi), and Sayir (Sayer).

2.3 Styles

Styles may be classified as:

(a) unpitted; and
(b) pitted.

* http://www.codexalimentarius.net/download/standards/256/CXS_143e.pdf

2.4 Sub-Styles

Sub-stytes are as follows:

(a) **Pressed**—dates which are compressed into layers using mechanical force.
(b) **Unpressed or Loose**—dates which are free-flowing or packaged without mechanical force or compression.
(c) **Clusters**—dates with the main bunch stem attached.

2.5 Size Classification (Optional)

Dates may be designated as to size names in accordance with the following charts:

(a) **Unpitted dates**

Size	No. of dates in 500 g
Small	More than 100
Medium	80 to 100
Large	less than 80

(b) **Pitted dates**

Size	No. of dates in 500 g
Small	More than 100
Medium	80 to 100
Large	less than 80

3. ESSENTIAL COMPOSITION ANO QUALITY FACTORS

3.1 Composition

3.1.1 Optional Ingredients

Glucose syrup, sugars, flour, vegetable oils.

3.2 Quality factors

3.2.1 General Requirements

Dates shall be prepared from such fruit and under such practices that the finished product shall possess a characteristic colour and flavour for the variety and type, be of proper stage of ripeness, be free of live insects and insect eggs and mites and meet the following additional requirements:

(a) **Moisture content**	**Maximum**
Cane Sugar varieties	26%
Daglat Nuur	30% (not processed in accordance with 2.1[d][e])
Invert Sugar varieties	30%
(b) **Size (minimum)**	
Unpitted Dates	4.75 grammes
Pitted Date	4.0 grammes
(c) Pits (Stones) (in Pitted Style)	Not more than two pits or 4 pieces of pit per 100 per
(d) **Mineral impurities**	Not more than 1 g/kg

3.2.2 Definition of Defects

(a) **Blemishes**	Scars, discoloration, sunburn, dark spots, blacknose or similar abnormalities in surface appearance affecting an aggregate area greater than that of a circle 7 mm in diameter.
(b) **Damaged**	(Unpitted dates only)—dates affected by mashing and/or tearing of the flesh exposing the pit or to such an extent that it significantly detracts from the visual appearance of the dates.
(c) **Unripe Dates**	Dates which may be light in weight, light in colour, have shrivelled or little flesh or a decidedly rubbery texture.
(d) **Unpollinated Dates**	Dates not pollinated as evidenced by thin flesh, immature characteristics and no pit in unpitted dates.
(e) **Dirt**	Dates having embedded organic or inorganic material similar to dirt or sand in character and affecting an aggregate area greater than that of a circle 3 mm in diameter.
(f) **Insects and mites**	Dates damaged by insects or mites or contaminated by damage and contamination the presence of dead insects or mites, fragments of insects or mites or their excreta.
(g) **Scouring**	Breakdown of the sugars into alcohol and acetic acid by yeasts and bacteria.
(h) **Mould**	Presence of mould filaments visible to the naked eye.
(i) **Decay**	Dates that are in a state of decomposition and very objectionable in appearance.

3.2.3 Allowance for Defects

The maximum allowances for the defects defined in 3.2.2 shall be:
A total of 7% by count of dates with defect (a)
A total of 6% by count of dates with defects (b), (c) and (d)
A total of 6% by count of dates with defects (e) and (f)
A total of 1% by count of dates with defects (g), (h) and (i)

3.3 Lot Acceptance

A lot will be considered as meeting the quality criteria requirements of the standard when:

(a) there is no evidence of live infestation; and
(b) the sub-sample, as taken in conformity with Sub-samples for Examination and
(c) Testing in relevant Codex texts on methods of analysis and sampling, meets the general requirements of sub-section 3.2.1 and does not exceed the allowances for the respective defects in sub sections 3.2.2 and 3.2.3, except that, with respect to size requirements, 5% by count (5 dates out of 100) may weigh less than the specified minimum.

4. FOOD ADDITIVES

		Maximum Level
4.1 Glycerol	)	In accordance with GMP (see also Section 3.1.1)
4.2 Sorbitol	)	

5. HYGIENE

5.1 It is recommended that the product covered by the provisions of this standard be prepared and handled in accordance with the appropriate sections of the Recommended International Code of Practice - General Principles of Food Hygiene (CAC/RCP 1-1969), and other Codes of Practice recommended by the Codex Alimentarius Commission which are relevant to this product.
5.2 To the extent possible in Good Manufacturing Practice, the product shall be free from objectionable matter.
5.3 When tested by appropriate methods of sampling and examination, the product:

- shall be free from microorganisms in amounts which may represent a hazard to health;
- shall be free from parasites which may represent a hazard to health; and
- shall not contain any substance originating from microorganisms in amounts which may represent a hazard to health.

6. WEIGHTS AND MEASURES

Containers shall be as full as practicable without impairment of quality and shall be consistent with a proper declaration of contents for the product.

7. LABELLING

In addition to the requirements of the General Standard for the Labelling of Prepackaged Foods (CODEX STAN 1-1985), the following specific provisions apply:

7.1 The Name of the Food

7.1.1 The name of the product shall be "Dates" or "Dates coated with Glucose Syrup".
7.1.2 The style shall be indicated as "pitted'" or "unpitted", as is applicable.
7.1.3 The name of the product may include the name of the varietal type, such as "Hallawi", "Saher", "khadhrawi", "Daglat", "Noor", "Barhee", or others, the sub-style as "pressed" or "unpressed", and the size designation as "small", "medium" or "large".

8. METHODS OF ANALYSIS AND SAMPLING

See relevant Codex texts on methods of analysis and sampling.

8.1 Special Provisions for Sampling of Dates

8.1.1 Gross Sample

Select at random not less than 2 individual packages per each 1,000 kg portion of the lot. From each individual package draw a sample of 300g and in any case sufficient to obtain a gross sample of not less than 3,000g. Use the gross sample for checking carefully for live infestation and general cleanliness of the product prior to its examination for compliance with other provisions of the standard.

8.1.2 Sub-Samples for Examination and Testing

Mix the gross sample well and take small quantities at random from many different places as follows:

For moisture test	500 grammes
For pits (in pitted style)	100 dates
For specified defects and size requirements	100 dates

REFERENCES

Barreveld W.H. 1993. *Date Palm Products*. FAO Agricultural Services Bulletin no. 101, Food and Agriculture Organization of the United Nations, Rome. M-17 ISBN 92-5-103251-3. Available at http://www.fao.org/docrep/t0681E/t0681e00.htm#con

FAO/WHO. 1985. *Food Standards*, Codex Alimentarius. *CODEX STAN 143–1985*. Available at http://www.codexalimentarius.net/download/standards/256/CXS_143e.pdf

FAOSTAT. 2010. Available at: http://faostat.fao.org/site/567/DesktopDefault.aspx?PageID=567#ancor

Jabarin A. and F.A. Hussain. 2004. *Iraqi Date Industry Marketing and Post-Harvest Issues*. Unpublished document, September. Agricultural Restoration & Development Program for Iraq and National Program for Propagation and Improvement of Date Palm. Available at http://www.usaid.gov/iraq/contracts/pdf/ARDI14-IraqiDateIndustry.pdf

Kohls R.K and J.N. Uhl. 2001. *Marketing of Agricultural Products*. 9 edition, Prentice Hall (July 29, 2001). ISBN-10:0130105848, ISBN-13:978 0130105844.

Pakistan Horticulture Development & Export Board. 2008. *Date Marketing Strategies*. Available at http://www.phdeb.org.pk/MktStrategies/Date.pdf

Liu P. 2003. *The Marketing Potential of Date Palm Fruits in the European Market*. FAO Commodity and Trade Policy Research Working Paper No. 6. Raw Materials, Tropical and Horticultural Products Service Commodities and Trade Division 2003.

UNIDO. 2004. *Date's Market*. Available at http://exchange.unido.org/upload/1726_Date.pdf

Zaid A. and E.J. Arias-Jiménez. 2002. *Date Palm Cultivation*. FAO Plant Production and Protection paper 156 rev. 1. Food and Agricultural Organization of the United Nations Rome, 2002. Available at: http://www.fao.org/DOCREP/006/Y4360E/Y4360E00.HTM

Part II

Processing

FIGURE 1.3 World map of date producing countries according to area distribution (shaded gray) and production quantity (columns).

FIGURE 7.1 Damaged date palm orchards due to salinity in coastal area of Batinah, the Sultanate of Oman.

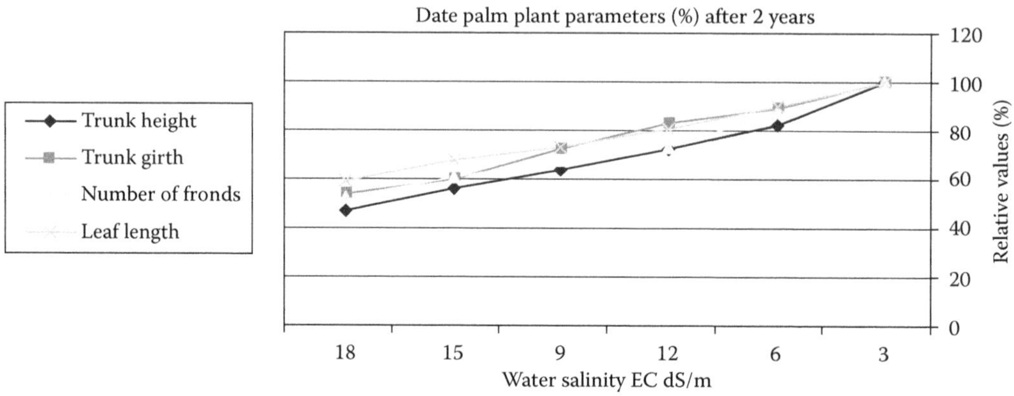

FIGURE 7.2 Effect of water salinity on plant characters of date palm. (Adapted from Al-Rasbi, SAR. 2010. *Evaluation of the Growth of Date Palm Seedlings Irrigated with Saline Water in Sultanate of Oman*. PhD Thesis, Department of Landscape Ecology and Nature Conservation, University of Kassel, Germany.)

FIGURE 9.1 Traditional pollination in Iran; left: cv. Barhee, Ahwaz; right: cv. Mozafti, Jiroft.

FIGURE 9.2 Manual dry pollen dispersion method: A fine net bag filled with dry pollen was fastened on a stick and shaken remotely on open female inflorescences by the worker.

FIGURE 9.3 Mechanical pollinator at work in date palm pollination, Ahwaz, Iran.

FIGURE 9.4 Sketch of the developed electric pollinator (left) and the pollinator in work (right): (1) extending pole, (2) disperser, (3) pollen pipe, (4) nozzle, and (5) remote controller.

FIGURE 10.1 Conventional method of harvesting date; the picker climbs the tree to harvest date, with or without a safety belt, a hazardous and dangerous activity.

FIGURE 10.2 Using ladders for selective picking and the use of cloth containers.

FIGURE 10.3 Using a pay-loader bucket.

FIGURE 10.4 Using a U-shaped platform.

FIGURE 10.5 Net bags used for collecting the ripened fruit.

FIGURE 13.11 Dried dates (18.5% MCWB) ready for unloading (July 27–28, 2009).

FIGURE 13.12 Dried products ready for unloading from the dryer.

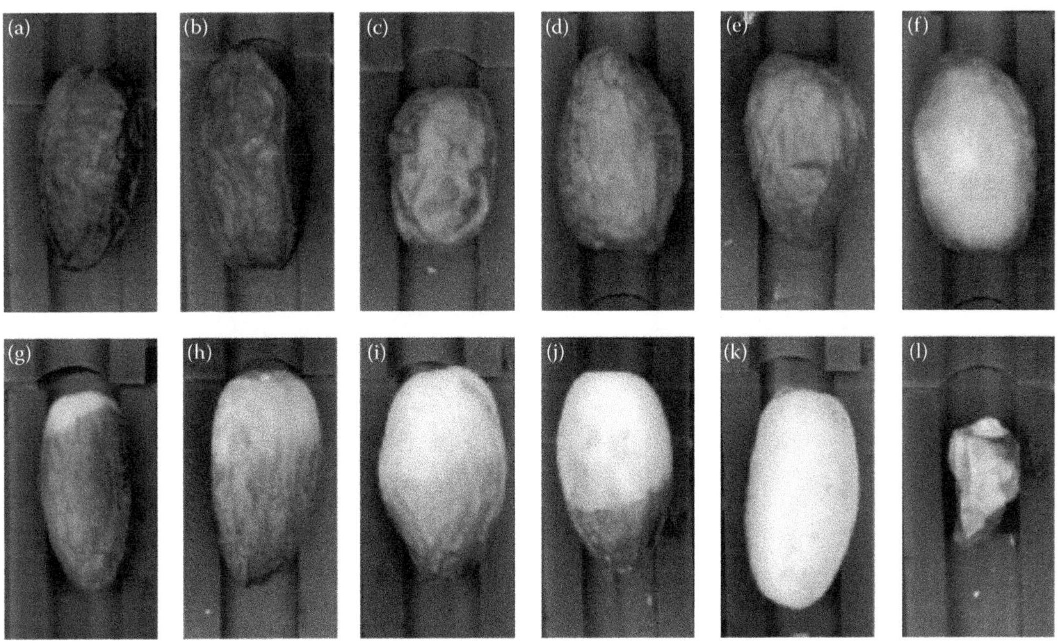

FIGURE 15.3 Classification of date samples according to seven maturity levels used by the Bard Valley Medjool Date Grower Association. Grades: Ready for packing (a and b); slight skin separation (c); soft with light red or orange color (d and e); partial red (f); yellow head (g); yellow spot (h); parial yellow (i and j); yellow (k) and trash (l). (Adapted from Lee, D. et al. 2008a. *Journal of Food Engineering* 88: 364–372.)

FIGURE 24.3 Vespasian coin celebrating the victory over the rebels. The legend says: IVDÆA CAPTA (Judea captured).

FIGURE 24.4 Antonius Felix issued a coin showing date palm leaf.

12 Drying Characteristics, Rehydration, and Color Changes of Palm Fruits during Storage

Kolawole O. Falade and Emmanuel S. Abbo

CONTENTS

INTRODUCTION

Fresh dates contain high moisture (about 60%), thus limiting the shelf life; therefore, drying is necessary. Matured date fruits are allowed to partially dry on the trees before they are harvested and further sun dried to enhance their keeping quality and storage. However, problems associated with sun drying are well documented (Guine and Castro 2002; Doymaz 2004, 2005). The quality of sundried products under dusty conditions becomes very poor and nonuniform with a low yield. Due to persistent rain and stormy conditions, a large amount of the harvested dates becomes moldy, fermented, dusty, and damaged by birds and insects. The growers become disheartened and consequently the rate of further cultivation and propagation of the variety is impaired badly. To minimize the postharvest damages, streamline the production, and improve the quality of the produce, the adaptation of appropriate scientific measures becomes imperative (Saleem et al. 2005). Dehydration in low-cost oven driers at controlled temperatures and flow of air will ensure an appropriate level of moisture in dates and better preservation of the product quality.

Drying of dates is necessary because the fresh date contains high moisture (60%) and therefore a short shelf life. Date fruits at *tamr* stage have low moisture content (about 10%–25%), which makes them safe for long-term storage (Zaid and de Wet 2001). Dehydration in oven driers at controlled temperatures and air velocity will ensure an appropriate level of moisture in dates and better preservation of the product quality. However, reduction of moisture to a very low level will produce fruits too hard to eat and inappropriate for some consumers, and may necessitate rehydration. Rehydration is a complex process aimed at the restoration of previously dried materials in contact with water. It is generally accepted that the degree of rehydration is dependent on the degree of cellular and structural disruption. During dehydration, irreversible rupture and dislocation occur, resulting in loss of integrity and hence a dense structure of collapsed, greatly shrunken capillaries with reduced hydrophilic properties as reflected by the inability to rehydrate fully (Lewicki 1998a). Predrying treatments and drying reduce the changes in the structure and composition of plant tissues (Lewicki 1998b), which results in impaired rehydration properties.

One of the problems associated with poor-quality dried date palm fruit is browning. Dates because of their high sugar content and high moisture content at the fresh stage (*khalaal*) darken during drying and storage due to nonenzymatic browning. Browning of dates occurs during drying and storage, and this lowers its acceptability. Sulfiting has been used to retard enzymatic and nonenzymatic browning in fruits during drying and storage. Although sulfiting of fresh fruits has been found to be very effective in preserving the color of products by retarding enzymatic and nonenzymatic browning reactions, its extensive use in foods has lately been challenged for human health reasons (Krokida et al. 2001). FDA had proposed that maximum residual sulfur dioxide levels of 2000 ppm be permitted in dried fruits (FDA 1988). The use of sulfites at lower permissible levels and air-drying at controlled conditions can help preserve the color of dates during drying and storage.

Color stability of a dried fruit can be measured quantitatively in terms of CIE L^*, a^*, b^* values, color difference value, hue angle, and chroma values (Joubert et al. 2001). The L^* value is a measure of lightness from completely opaque (0) to completely transparent (100), a^* is a measure of redness and b^* of yellowness. Changes in chroma reflect a bias toward the dominant color component (a^* or b^*). The chroma expresses the color intensity with a heavy bias toward the dominant color and hence depicts the brownness of the products (Bakker et al. 1994). Thus, the objectives of this work were to investigate the rehydration characteristics of three varieties of dried date fruits and the effect of sulfite pretreatment on color stability of dehydrated date fruits during storage at $30 \pm 2°C$.

MATERIALS AND METHODS

Three date palm fruits (*Phoenix dactylifera* L.) cultivars including Red soft type, Tempo 2, and Tempo 3 (*Zabia*) were obtained from the Nigerian Institute for Oil Palm Research (NIFOR) Date Palm Research Substation in Dutse, Nigeria. Fresh mature ripe date palm fruits were collected, sealed in 100 μm low-density polyethylene and kept at low temperature during transportation ($8 \pm 2°C$) and prior to use. Average moisture contents of Red soft type, Tempo 2, and Tempo 3 were 59%, 60%, and 65% wet basis, respectively.

DETERMINATION OF THE EQUIVALENT RADIUS AND THE SHAPE FACTOR OF DATE PALM FRUIT

The equivalent radius and the shape factor (ψ) were determined according to the methods of Mohsenin (1980) and Hebbar and Rastogi (2001). The major and minor axes of 10 randomly chosen date palm fruits from each variety were measured using Vernier calipers. The values of the major and minor axes were used to calculate the equivalent radius and the shape factor (ψ) as shown by Falade and Abbo (2007).

DETERMINATION OF DRYING CHARACTERISTICS OF DATE PALM FRUIT

Fresh mature ripe date palm fruits were weighed and dried in a Gallenkamp (model OV-160) cross flow air oven drier at different temperatures of 50°C, 60°C, 70°C, and 80°C and at an air velocity of 1.5 m^2/s and the same orientation of product. Drying commenced when constant temperature was achieved in the drier. During drying, preweighed date samples were weighed at 10 min intervals for the first hour, followed by 30 min for the next hour and subsequently at hourly intervals until three consecutive constant weights (equilibrium) were achieved. On attaining equilibrium, samples were further dried in the oven at 100°C for 24 h (Koyuncu et al. 2004). Equilibrium moisture content was calculated as the difference between equilibrium weight and bone dry weight divided by the bone dry weight. Drying data were expressed as the drying rate versus moisture content (dry basis), and as the moisture ratio versus drying time. The drying rate was calculated as the amount of water removed per unit time (h). The effective moisture diffusion coefficient (D_{eff}) was determined by plotting ln MR as a function of time from the integrated form of Fick's equation (Equation 12.3), and the activation energy as shown by Babalis and Belessiotis (2004) and Senadeera et al. (2003) was adopted (Equation 12.4).

The experimental drying data for the determination of diffusivity coefficients were interpreted by using Fick's second diffusion (Equation 12.1) model (Falade and Abbo 2007)

$$\frac{\delta M}{\delta t} = D_{\mathrm{eff}}\left(\frac{\delta^2 M}{\delta x^2}\right) \tag{12.1}$$

$$\mathrm{MR} = \frac{M - M_{\mathrm{e}}}{M_0 - M_{\mathrm{e}}} = \exp(-kt) \tag{12.2}$$

For a thin layer, the solution of Equation 12.1, with the assumption of moisture migrating only by diffusion, negligible shrinkage, constant temperature and diffusivity coefficient, and long drying (or rehydration) times, are given below (Equation 12.3) (Crank 1975; Senadeera et al. 2003):

$$\mathrm{MR} = \frac{M - M_{\mathrm{e}}}{M_0 - M_{\mathrm{e}}} = \frac{6}{\pi^2}\sum_{n=1}^{\infty}\frac{1}{n^2}\exp\left(-n^2\pi^2\frac{D_{\mathrm{eff}}t}{\psi R_{\mathrm{eq}}^2}\right) \tag{12.3}$$

where M, M_0, and M_e are the instantaneous, initial, and equilibrium moisture content, respectively. MR (moisture ratio) is the unaccomplished moisture change defined as the ratio of the free water still to be removed at any time to the total free water available initially.

This solution is the most widely investigated theoretical model in the thin-layer drying of different foods.

The effective diffusivity (D_{eff}) thus obtained can be expressed by using an Arrhenius-type equation (Maskan et al. 2002; Babalis and Belessiotis 2004) in the form

$$D_{\mathrm{eff}} = D_0\exp\left(\frac{-E_{\mathrm{a}}}{R_g T_{\mathrm{abs}}}\right) \tag{12.4}$$

The E_a and D_0 coefficients can subsequently be related to the air drying and rehydration temperature by applying nonlinear regression analysis.

According to Falade and Abbo (2007), the shape of the date palm fruit was likened to spherical geometry. For arbitrarily shaped bodies, the equivalent radius was calculated as

$$R_{\mathrm{s}} = \frac{3V}{S_{\mathrm{p}}} = \left(\frac{3V}{S_{\mathrm{s}}}\right)\left(\frac{S_{\mathrm{s}}}{S_{\mathrm{p}}}\right) = \psi R_{\mathrm{eq}} \tag{12.5}$$

where V and S_p are the volume and the surface area of the fruit, respectively, and S_s is the surface area of the sphere having volume equal to that of the sample. ψ is the shape factor, R_s is the radius of the sphere, and R_{eq} is the equivalent radius of the date palm fruit (characteristic length in the case of slabs).

DETERMINATION OF REHYDRATION CHARACTERISTICS AND CAPACITY OF DEHYDRATED DATE PALM FRUIT

Rehydration characteristics of dates were determined as described by Falade and Abbo (2007). Fresh batches of date palm fruits were dehydrated until they attained equilibrium weight at 60°C for about 48–60 h. Dehydrated samples were rehydrated in distilled water at 15°C, 30°C, and 45°C according to the method of Rastogi et al. (2000, 2004) maintaining a date palm fruit:water ratio of 1:25 w/w. High date palm fruit:water ratio was adopted to minimize the influence of leaching during rehydration. Samples were weighed initially, immersed in water, removed from water blotted, and weighed after 0.25, 0.50, 0.75, 1.00, 1.50, 2.00, 3.00, 4.00, 5.00, and 7.00 h and subsequently until equilibrium (17–25 h) depending on rehydration temperature. After rehydration, samples were dried at 100°C for 24 h (Koyuncu et al. 2004). From the data, moisture contents (kg water/kg dry matter) of date palm fruits during rehydration were calculated. All the experiments were performed in triplicate and average values were reported. From the rehydration curves, the effective moisture diffusion coefficient and activation energy were estimated as described in Equations 12.3 and 12.4.

Also, the rehydration capacity was determined as percentage water gained (Doymaz 2004), using Equation 12.6

$$\mathrm{RC} = \frac{W_{\mathrm{absorbed}}}{W_{\mathrm{removed}}} \times 100 \tag{12.6}$$

where

RC = rehydration capacity (%)

W_{absorbed} = weight of water absorbed during rehydration

W_{removed} = weight of water removed during drying.

EVALUATION OF COLOR CHANGES IN UNTREATED AND SULFITED DRIED DATE PALM FRUIT VARIETIES DURING STORAGE

Batches of fresh date palm fruits were dipped in varying concentration of sodium metabisulfite (200, 400, and 600 ppm) for 10 min. Both untreated and sulfited date palm fruits were air dried at 60 ± 2°C for 24 h. Dried samples were sealed in 50 μm low-density polyethylene and stored at 30 ± 2°C for 5 weeks. L*, a*, and b* values were directly read from a ColorTec-PCM™ colorimeter (Accuracy Microsensors Inc., Pittsford, New York, USA) which had been standardized. CIE Lab tristimulus values, L* (lightness), a* (redness), and b* (yellowness) were measured, for both fresh and dehydrated samples, at three positions randomly selected for each sample. The color difference value (ΔE), color intensity (chroma) (ΔC), and hue angle were calculated from the following equations (de Man 1990; Hunt 1991):

$$\Delta E = \sqrt{\Delta L^{*2} + \Delta a^{*2} + \Delta b^{*2}} \tag{12.7}$$

$$\Delta C = \sqrt{\Delta a^{*2} + \Delta b^{*2}} \tag{12.8}$$

$$\mathrm{Hue} = \tan^{-1}\left(\frac{b}{a}\right) \tag{12.9}$$

RESULTS AND DISCUSSION

Effect of Variety on Moisture Content, Moisture Ratio, and Drying Rates of Date Palm Fruits

Initial moisture contents of Red soft, Tempo 2, and Tempo 3 varieties were 59.2%, 59.9%, and 64.5% wet basis, respectively. Moisture content of date palm fruits decreased steadily throughout the drying period. Nondimensional moisture ratio (MR) was calculated from the drying data. A plot of the moisture ratio versus time during air drying of date palm fruits at 60°C is shown in Figure 12.1a. Tempo 2 and Red soft varieties recorded steeper moisture ratio curves than Tempo 3, showing that Red soft and Tempo 3 varieties took a longer time to dry than Tempo 2. The moisture ratio reduced in a similar fashion to that observed for the moisture content versus time (Falade and Abbo 2007).

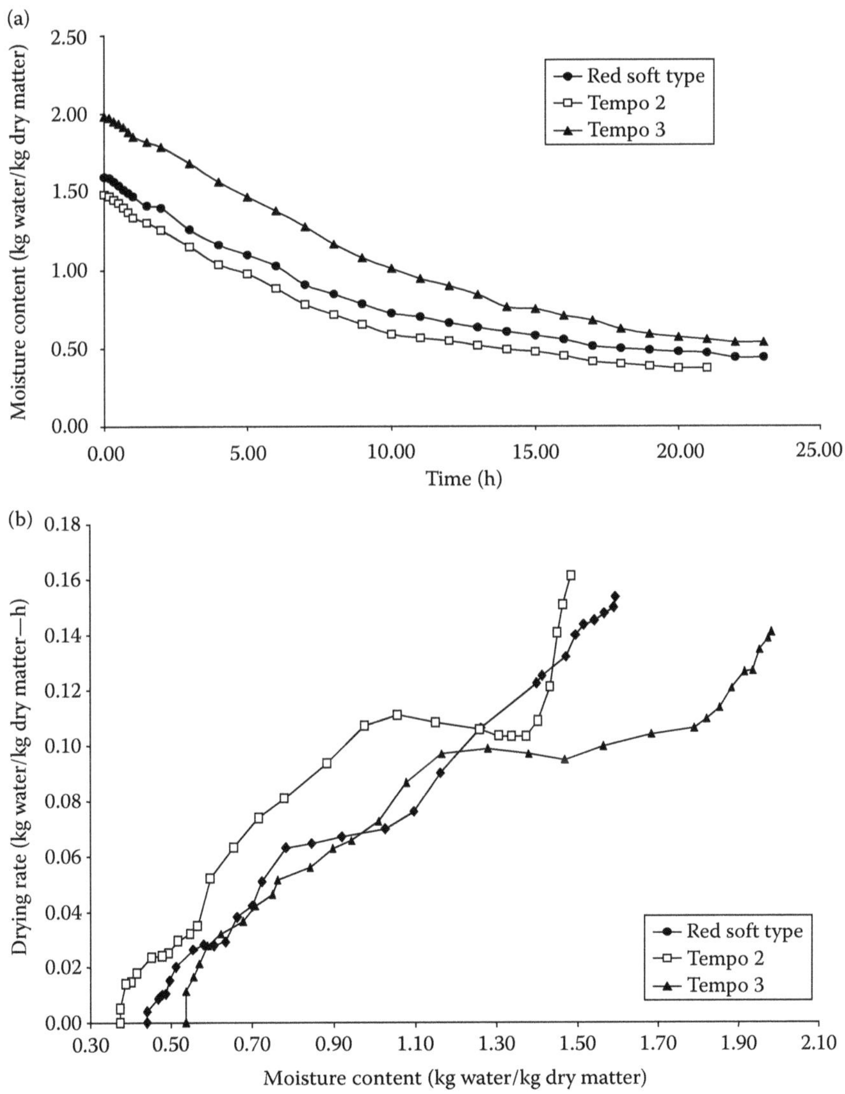

FIGURE 12.1 Effect of varietal differences on drying rates of date palm fruit dried at 60°C: moisture content vs. time (a); and drying rate vs. moisture content (b).

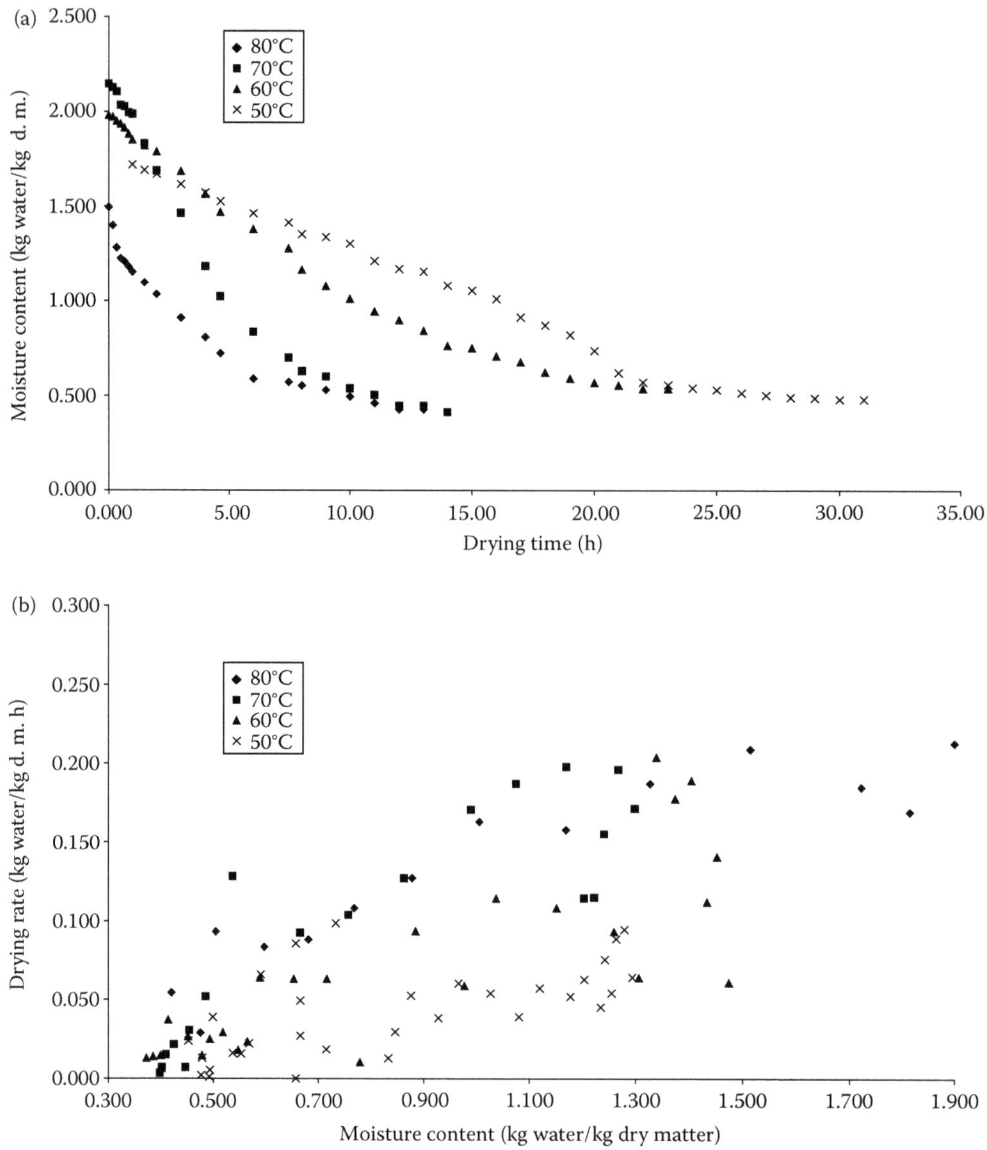

FIGURE 12.2 (a) Effect of drying temperature on moisture contents of Tempo 3 date cultivar. (b) Effect of temperature on drying rate of Tempo 2 date fruit cultivar.

Drying rates for the three varieties were calculated and plotted against moisture content (dry basis). Tempo 2 showed higher drying rates than Red soft type and Tempo 3 varieties of date palm fruit (Figure 12.1b) (Falade and Abbo 2007). Generally, the drying rate of date palm fruit declined rapidly until equilibrium was reached. It was noted that there was no constant rate-drying period during the drying of date palm fruits. Drying took place predominantly in the falling rate period. This indicated that diffusion was the main physical mechanism governing moisture movement in date palm fruits. This trend is typical of most fruits (Sankat and Castaigne 2004). The results were in agreement with the existing literature on drying of various food products (Doymaz 2005; Wand and Xi 2005; Senadeera et al. 2003). Differences in the moisture ratio and drying rates could be due variation in composition (dry matter contents) and surface area of the date palm fruits.

Effect of Drying Temperature on Moisture Ratio and Drying Rates of Date Palm Fruits

Drying of date palm fruits at higher temperatures resulted in shorter drying times. Falade and Abbo (2007) reported that the drying curve at 70°C was steeper than that at other temperatures (50°C, 60°C, and 80°C), showing that the drying time decreases with increasing air drying temperature. This indicated that temperature has a direct effect on the moisture contents and drying time of date palm fruits. However, the drying curve at 70°C was steeper than 80°C, showing probably a change in drying mechanism due to case hardening. The influence of temperature on Red soft and Tempo 2 date palm fruits was similar to the trend in Tempo 3. Higher drying temperatures also showed steeper drying rate curves, indicating that a higher drying temperature promoted faster drying (Figure 12.2b). Although 80°C was the highest temperature in this study, the steepest drying rate curve was obtained at 70°C (Falade and Abbo 2007). The results obtained are consistent with those reported in the literature (Njintang and Mbofung 2003; Babalis and Belessiotis 2004) in which air temperature is considered the most important factor affecting the drying rate. Drying rate was considerably increased when higher temperatures were used (Babalis and Belessiotis 2004). The change in drying rate trend of date palm fruits air dried at 80°C could be due to the case hardening effect, which was obvious at this temperature. The observed case hardening could be due to the high solid sugar content of dates, which hindered moisture movement at the higher drying temperature.

Effect of Variety on Moisture Diffusivity and Activation Energy of Date Palm Fruit during Drying

In calculating the effective moisture diffusion coefficient (D_{eff}) of date palm fruits at different air-drying temperatures, ln MR was plotted against drying time (Equation 12.3) and D_{eff} was extrapolated. It was reported that plots satisfactorily describe the drying behavior over the moisture ratio range ($0 \leq MR \leq 1.0$) (Falade and Abbo 2007). The effect of cultivar on moisture diffusivity of date palm fruits during drying is shown in Figure 12.3a. The results showed that D_{eff} varied from 5.89×10^{-10} to 1.78×10^{-9} m²/s, 3.22×10^{-9} to 8.16×10^{-9} m²/s, and 7.53×10^{-9} to 2.98×10^{-8} m²/s for Red soft type, Tempo 2, and Tempo 3, respectively (Falade and Abbo 2007). Tempo 3 consistently showed higher D_{eff} than Red soft type and Tempo 2 at all drying temperatures. D_{eff} values mostly lie within the general range of 10^{-12}–10^{-8} m²/s reported for most food materials (Zogzas et al. 1994, 1996; Babalis and Belessiotis 2004). Effective moisture diffusivity increased with increasing air-drying temperature. Generally, lower D_{eff} values were obtained at 50°C rather than at higher temperatures for the three types of dates. During the first phase of drying, water diffusion was mainly due to the mass transport mechanism. The surface was saturated with moisture and the core moisture had to replace the surface, as it evaporated. As the drying progresses, the surface moisture was completely evaporated; vapor diffusion of water could be the predominant mechanism resulting in higher values of D_{eff} (Ramesh et al. 2001).

Activation energy is the energy needed to initiate mass diffusion in food (Mittal 1999). The temperature dependence of diffusivity is described by the Arrhenius-type equation (Equation 12.4). Activation energies for mass diffusion during air drying were 35.17, 29.50, and 44.02 kJ/mol for Red soft, Tempo 2, and Tempo 3, respectively (Falade and Abbo 2007). Generally, values for the activation energy lie within the general range of 12.7–110 kJ/mol for most food materials (Zogzas et al. 1996). Differences in E_a for the drying of dates could be due to the variation in geometry (surface

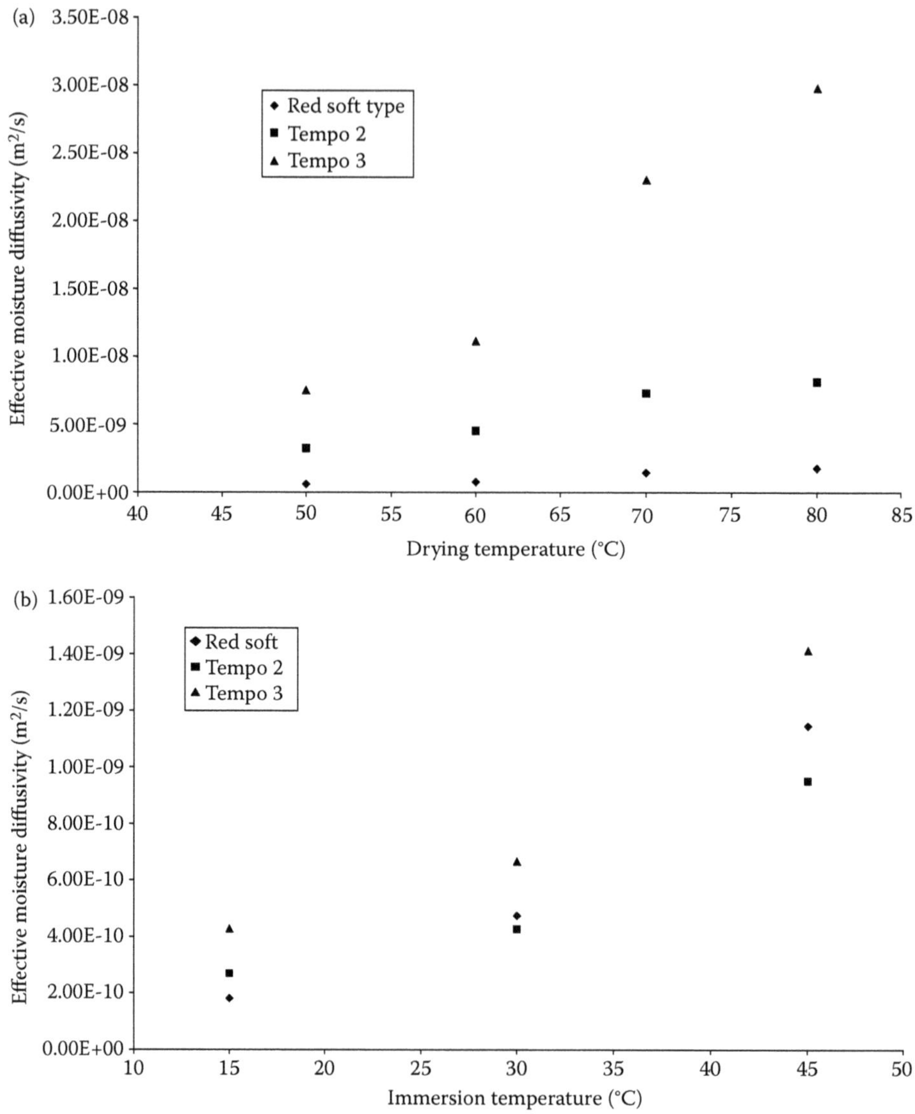

FIGURE 12.3 Effective moisture diffusivities of date palm fruit during air drying (a) and rehydration (b) at different temperatures.

area, shape factor) and composition. Tempo 3, which had a surfaced area of 3.704×10^{-3} m^2 and a shape factor of 0.857 ± 0.04, had the highest activation (44.02 kJ/mol). The surface area and shape factor of Tempo 2 and Red soft type were 3.547×10^{-3} m^2 and 0.61 ± 0.02, and 3.095×10^{-3} m^2 and 0.74 ± 0.04, respectively.

Effect of Variety on Moisture Diffusivity and Activation Energy during Rehydration of Date Palm Fruits

In modeling the rehydration process, it is considered that water transport from the surface to the center of the product takes place mainly by diffusion (Sanjuan et al. 2004). Hence Fick's second law of diffusion (Equation 12.1) was used to model the rehydration characteristics of date palm fruit. Equation 12.3 was also used to estimate the effective moisture diffusivity during rehydration of dried date palm fruits. Plots of ln MR against rehydration time showed a linear relationship, which is influenced by the temperature of the rehydrating medium. The effect of cultivar on effective moisture diffusivity during rehydration of dried date palm fruits is shown in Figure 12.3b. Generally, D_{eff} increased with increasing rehydration temperature. D_{eff} varied from 1.80×10^{-10} to 1.15×10^{-9} m^2/s, 2.69×10^{-10} to 9.51×10^{-10}, and 4.27×10^{-10} to 1.41×10^{-9} m^2/s for Red soft type, Tempo 2, and Tempo 3, respectively, for the rehydration temperature range of 15–45°C (Falade and Abbo 2007).

Relationships between D_{eff} and temperature during rehydration were also represented by the Arrhenius-type equation (Equation 12.4). Calculated activation energies required to initiate moisture diffusion during rehydration of dried dates were 50.85, 46.29, and 30.29 kJ/mol for Tempo 2, Red soft, and Tempo 3, respectively (Falade and Abbo 2007). Thus, the activation energy required for mass diffusion during rehydration was in the order Tempo 2 > Red soft type > Tempo 3 (Falade and Abbo 2007). Activation energies for air drying of date were lower than that for rehydration for the date palm fruits. This could be attributed to the difficulty of restoring the dried date to their fresh state due to cell disruption which took place during drying (Falade and Abbo 2007). Moreover, mass transfer during rehydration was lower than that in air drying; hence, more energy would be required during rehydration. Irreversible rupture and dislocation that occurred during dehydration may have resulted in loss of integrity and hence a dense structure of collapsed, greatly shrunken capillaries with reduced hydrophilic properties as reflected by the inability to rehydrate fully (Lewicki, 1998).

REHYDRATION CAPACITY OF DATE PALM FRUITS

During the rehydration process, sample weight increased. Generally, Tempo 3 showed a higher rehydration capacity compared to Red soft type and Tempo 2. Table 12.1 shows the rehydration capacity (expressed as percentage) of date palm fruits. Results showed that the rehydration capacity increased with rehydration time and temperature. Absorption of water by dry material can be considered a complex procedure in which temperature plays a major role in water absorption capacity (Abdel Kader 1995). Soaking at elevated temperatures decreased the required time to attain maximum water sorption capacity and moisture content at which it was achieved. According to Krokida et al. (2003), during rehydration, water temperature influences

TABLE 12.1
Effect of Cultivar on Rehydration Capacity of Date Palm Fruits

	Rehydration Capacity (%)								
	15°C			30°C			45°C		
Time (h)	Red Soft	Tempo 2	Tempo 3	Red Soft	Tempo 2	Tempo 3	Red Soft	Tempo 2	Tempo 3
0.00	0.0	0.0	0.0	0.0	0.0	0.0	0.0	0.0	0.0
0.25	9.4	4.6	5.0	8.0	8.2	7.8	9.2	15.2	11.0
0.50	12.2	8.7	9.4	11.2	12.0	13.9	15.6	25.1	18.8
0.75	15.8	13.5	16.5	13.0	15.0	17.0	23.6	37.6	27.9
1.00	19.2	16.0	21.6	16.8	19.5	22.0	28.7	40.1	30.3
1.50	22.0	19.5	27.1	30.0	25.6	27.6	33.7	47.7	38.8
2.00	22.9	22.6	31.3	33.3	30.3	34.7	39.2	54.9	42.9
3.00	32.4	30.2	41.0	39.3	39.6	44.8	56.2	63.4	51.8
4.00	39.1	35.1	47.9	45.3	47.9	54.1	59.8	68.1	58.5
5.00	45.8	40.4	54.4	50.6	53.4	60.1	63.1	70.9	63.4
7.00	55.5	48.7	62.7	58.1	60.6	65.6	65.8	73.7	67.2
17–25	76.2	66.8	77.8	69.2	67.2	74.5	69.2	76.6	69.8

TABLE 12.2
Effect of Sulfite Concentration on Color Parameters of Red Soft Cultivar of Date Palm Fruits

Sulfite conc. (ppm)	Storage Time (Weeks)	L*	a*	b*	ΔE	ΔC	Hue Angle
0	Fresh	52.52	13.67	31.75	–	–	66.71
	Dried	31.65	2.01	5.41	35.58	28.81	69.70
	1	34.27	1.97	4.14	35.11	29.99	64.54
	2	35.03	2.08	2.19	36.25	31.75	46.67
	3	34.91	2.35	1.76	36.57	32.06	37.23
	4	35.33	2.59	1.90	36.18	31.84	36.13
	5	37.09	2.66	3.98	33.62	29.87	53.31
200 ppm	Dried	31.77	3.46	6.24	34.44	27.48	60.95
	1	33.54	3.50	5.93	33.62	27.75	59.53
	2	34.32	3.32	5.22	33.80	28.48	57.51
	3	34.42	3.21	3.28	35.32	30.33	45.57
	4	34.69	3.02	3.22	35.29	30.45	46.94
	5	35.48	2.77	4.19	34.19	29.64	56.48
400 ppm	Dried	33.31	4.60	0.63	37.68	32.41	77.70
	1	34.06	3.93	0.90	37.25	32.36	12.95
	2	34.49	2.64	2.19	36.34	31.55	39.69
	3	35.04	2.26	2.98	35.55	30.95	52.85
	4	36.27	2.06	4.38	33.88	29.73	64.85
	5	37.40	1.93	6.06	32.04	28.25	72.33
600 ppm	Dried	37.26	4.09	7.57	30.16	26.02	61.61
	1	37.07	4.24	3.97	33.16	29.34	40.03
	2	36.70	3.90	3.70	33.65	29.70	43.53
	3	36.06	2.41	3.86	34.29	30.08	57.99
	4	36.04	1.91	4.16	34.22	30.00	65.36
	5	36.00	1.57	4.43	34.14	29.88	76.64

the rehydration rates and equilibrium moisture content. However, the maximum rehydration capacity observed, after 17–25 h of immersion, decreased with decreasing rehydration temperature. The maximum rehydration capacity attainable decreased with increasing temperature. Prolonged rehydration in water may have resulted in leaching of fruits solids into water at higher temperatures. This agrees with the results of Femenia et al. (2000) that broccoli showed maximum water sorption capacity when rehydrated at lower temperatures. Visual observation showed swelling of the date palm fruits during rehydration. Absorption of water by polymers and mobilization of other material components cause swelling of the matrix to the extent that the size increased and a rebuilding of structure could be observed (Lewicki et al. 1997). Water added during rehydration plasticizes the glassy matrix and a collapse of structure manifests itself. Rehydration of dried tissues is a complex procedure composed of three simultaneous processes: the process of water getting imbibed into dried material, the swelling, and the leaching of solutes (Lewicki 1998a; McMinn and Magee 1997).

Even after 17–25 h of immersion in water to facilitate rehydration, the maximum rehydration capacities were between 66.8% and 77.8%. It is generally accepted that the degree of hydration is dependent on cellular and structural dislocation during drying. During drying, irreversible cellular rupture and dislocation occur, resulting in loss of integrity and hence a dense structure of collapsed, greatly shrunken capillaries with reduced hydrophilic properties, as reflected by the inability to imbibe sufficient water to rehydrate fully (Jayaraman et al. 1990). In practice, most changes caused

TABLE 12.3
Effect of Sulfite Concentration on Color Parameters of Tempo 2 Cultivar of Date Palm Fruits

Sulfite conc. (ppm)	Storage Time (Weeks)	L*	a*	b*	ΔE	ΔC	Hue Angle
0	Fresh	38.13	11.03	15.23	–	–	54.09
	Dried	37.26	5.52	6.03	10.78	10.72	47.47
	1	37.07	5.17	6.86	10.25	10.22	53.06
	2	36.70	5.02	9.02	8.65	8.65	60.95
	3	36.06	4.67	10.76	7.78	7.78	66.50
	4	36.04	5.04	9.21	8.50	8.50	61.35
	5	36.00	5.48	6.96	9.96	9.96	51.78
200 ppm	Dried	37.07	5.52	12.45	6.26	6.17	66.13
	1	37.42	5.17	10.10	7.82	7.79	62.85
	2	37.80	5.02	8.50	9.03	9.02	59.39
	3	37.85	4.67	10.57	7.89	7.88	66.13
	4	37.79	5.04	12.71	6.51	6.50	68.36
	5	38.16	5.48	15.76	5.58	5.58	70.85
400 ppm	Dried	37.22	8.27	4.58	11.05	11.01	28.81
	1	37.71	6.68	4.83	11.28	11.27	35.75
	2	37.80	5.95	8.00	8.84	8.84	53.27
	3	39.11	5.52	9.62	7.93	7.87	60.11
	4	40.20	4.34	10.65	8.37	8.11	67.88
	5	41.00	3.67	12.49	8.36	7.85	73.61
600 ppm	Dried	36.48	6.35	8.42	8.43	8.27	53.06
	1	37.62	7.17	9.83	6.66	6.64	53.87
	2	37.77	7.29	11.25	5.47	5.46	57.00
	3	37.57	7.01	7.82	8.45	8.43	48.24
	4	37.38	5.71	5.76	10.89	10.86	45.28
	5	37.26	4.71	4.92	12.12	12.09	46.12

by predrying and drying treatments are irreversible, and rehydration cannot be considered simply as a process reversible to dehydration (Lewicki 1998b).

During reconstitution of dehydrated products, the amount and rate of water absorption determine to a considerable extent the sensorial properties and preparation time (Maskan 2001). Rehydration characteristics of a dried product are used as a quality index and they indicate the physical and chemical changes during drying as influenced by processing conditions, sample pretreatment, and composition (Feng and Tang 1998).

Effect of Sulfite Pretreatment on Color Stability of Dried Date Palm Fruits

The color (CIE L*, a*, and b* values) of date palm fruits, which had been immersed in a sodium metabisulfite solution of various concentrations (200, 400, and 600 ppm) prior to air drying and packaging in 50 μm low-density polyethylene, was monitored during storage. L*, a*, and b* values of the dried fruits during storage were affected by the sulfite solution concentration in which date palm fruits were immersed prior to air drying. Generally, the L* value increased with increasing sulfite pretreatment concentration and storage time of dried dates (Tables 12.2 through 12.4). Generally, the a* value of dried dates decreased during storage (Tables 12.2 through 12.4) and showed a better parameter that could be used to measure the color changes in the dates. Yellowness, indicated by the b* value, showed a more complex relationship. Generally, b* values decrease with

TABLE 12.4
Effect of Sulfite Concentration on Color Parameters of Tempo 3 Cultivar of Date Palm Fruits

Sulfite conc. (ppm)	Storage Time (Weeks)	L*	a*	b*	ΔE	ΔC	Hue Angle
0	Fresh	36.59	12.29	11.95	–	–	44.2
	Dried	34.5	4.82	6.71	9.36	9.12	54.27
	1	34.1	4.15	4.51	11.3	11.03	47.46
	2	34.1	3.11	4.26	12.23	11.98	53.87
	3	33.89	2.72	9.46	10.25	9.89	73.96
	4	33.58	2.43	11.07	10.35	9.9	77.63
	5	35.84	3.04	12.15	9.28	9.25	75.96
200 ppm	Dried	36.56	7.31	12.18	4.99	4.99	59.09
	1	38.59	6.76	7.6	7.32	7.04	48.24
	2	38.87	6.22	7.14	8.07	7.74	48.99
	3	35.52	5.49	8.33	7.78	7.7	56.66
	4	35.46	5.34	10.02	7.3	7.21	61.99
	5	38.78	6	10.65	6.42	6.42	60.67
400 ppm	Dried	36.22	7.95	10.86	4.49	4.48	53.87
	1	39.21	6.79	9.37	6.62	6.08	54.07
	2	37.1	4.32	7.39	9.2	9.18	59.68
	3	35.85	3.82	7.25	9.71	9.69	62.24
	4	35.64	3.79	7.72	9.54	9.49	63.77
	5	37.86	4.89	8.4	8.31	8.21	59.83
600 ppm	Dried	36.85	5.56	10.18	6.97	6.96	61.35
	1	37.13	3.7	10.06	8.81	8.8	69.81
	2	37.58	3.05	9.53	9.6	9.55	72.28
	3	40.4	3.3	10.8	9.83	9.06	73
	4	41.6	4.51	11.4	9.27	7.8	68.43
	5	41.82	5.58	12.19	8.51	6.72	65.46

increasing sulfite pretreatment concentration in Red soft and Tempo 3 dried dates. However, the b* value increased with increasing concentration of sulfite pretreatment (Tables 12.2 through 12.4). According to Joubert et al. (2001), an investigation of the effect of temperature and fruit moisture content on sulfured Bon Chresien pears showed that lightness (L*), yellowness (+b*), and redness (+a*) decreased during storage. The L* parameter is used in several studies to indicate browning of foods including peaches (Kluter et al. 1994), pears (Saper and Douglas 1987), apples (Bolin and Steele 1987), and sultanas (Canellas et al. 1993).

The calculated color difference for date palm fruits generally increased during storage. However, dates pretreated with 600 ppm metabisulfite showed a decrease in color difference during storage. The calculated color difference (ΔE) for date palm fruits generally increased during storage. However, dates pretreated with 600 ppm metabisulfite showed a decrease in color difference during storage (Tables 12.2 through 12.4). Calculated chroma (ΔC) and hue angle in Tempo 2 and Tempo 3 varieties decreased with increasing concentration of metabisulfite pretreatment prior to air drying. However, chroma values (Tables 12.3 and 12.4) and hue values (Tables 12.2 through 12.4) of untreated date palm fruits increased during storage, indicating browning of the date palm fruits. Baysaal et al. (2003) have stated that hue angle values indicate the degree of browning. Joubert et al. (2001) showed that hue angle and chroma decreased during storage.

The effect of cultivars on hue angle of the dried sulfited date palm fruits packaged in 100 μm low-density polyethylene can be observed in Tables 12.2 through 12.4. Red soft type showed a higher hue angle than Tempo 2 and Tempo 3. However, Tempo 2 indicated a lower hue angle than Tempo 3. Red soft dates showed higher Chroma and color difference values than Tempo 2 and Tempo 3 cultivars. Generally, dried date palm fruits showed increased browning during storage as shown by the hue angle, chroma (ΔC) and color difference (ΔE) values, particularly by the dates pretreated in lower sulfite concentration (200 ppm). Browning of Red soft date cultivar was confirmed by the darkening visual appearance during storage.

CONCLUSION

Air drying and rehydration of date palm fruits could be modeled using Fick's diffusion model. It would be possible to attain faster drying by increasing the drying temperature up to 70°C, after which the drying mechanism changes, probably due to case hardening. Rehydration of dried date palm fruits recorded lower diffusivity and higher activation energy than air drying. Activation energies for drying were 35.17, 29.50, and 44.02 kJ/mol, while the E_a for rehydration were 46.29, 50.85, and 30.29 kJ/mol for Red soft type, Tempo 2, and Tempo 3, respectively.

The rehydration capacity of dehydrated date palm fruit was investigated by rehydrating date palm fruits in water at 15°C, 30°C, and 45°C. The rehydration capacity increased with increasing rehydration temperature. Tempo 3 had the highest rehydration capacity (expressed as percentage), while Red soft and Tempo 2 varieties were similar. The concentration of sulfite dip of date palm fruits prior to air drying influenced the CIE color (L*, a*, and b*), color difference (ΔE), chroma (ΔC), and hue angle during storage. Untreated date palm fruits showed increased chroma and hue angles during storage, indicating increased browning of date palm fruits during storage. Generally, the calculated chroma, color intensity, and hue angle in date palm fruits decreased with increasing concentration of metabisulfite pretreatment, indicating decreased browning of the dates during storage.

REFERENCES

Abdel Kader, Z. M. 1995. Study of some factors affecting water absorption by faba beans during soaking. *Food Chemistry* 53:235–238.

Babalis, S. J. and Belessiotis, V. G. 2004. Influence of drying conditions on the drying constants and moisture diffusivity during the thin-layer drying of figs. *Journal of Food Engineering* 65:449–458.

Bakker, J., Bridle, P., and Bellworthy, S. J. 1994. Strawberry juice: A study of the quantitative and qualitative pigment composition of juices from 39 genotypes. *Journal of Science of Food and Agriculture* 64:31–37.

Baysaal, T., Icier, F., Ersus, S., and Yildiz, H. 2003. Effects of microwave and infrared drying on the quality of carrot and garlic. *European Food Research Technology* 218:68–73.

Bolin, H. R. and Steele, R. J. 1987. Non-enzymatic browning in dried apples during storage. *Journal of Food Science* 52:1654–1657.

Canellas, J., Rosello, C., Simal, S., Soler, L., and Mulet, A. 1993. Storage conditions affect quality of raisins. *Journal of Food Chemistry* 58:805–809.

Crank, J. 1975. *The Mathematics of Diffusion* (2nd ed.). Oxford, UK: Clarendon Press.

de Man, J. M. 1990. *Principles of Food Chemistry*. New York: AVI Publishing Co.

Doymaz, I. 2004. Convective air-drying characteristics of thin layer carrots. *Journal of Food Engineering* 61:359–364.

Doymaz, I. 2005. Drying characteristics and kinetics of okra. *Journal of Food Engineering* 69:275–279.

Falade, K. O. and Abbo, E. O. (2007). Air-drying and rehydration of date palm (*Phoenix dactylifera*) fruits. *Journal of Food Engineering* 79, 724–730.

FAO. 1982. Plant production and protection paper No. 35. *Date Production and Protection* (p. 294). Rome: FAO.

FDA. 1988. Sulfiting agents: Affirmation of GRAS status. *Food and Drug Administration Fed Reg.* 53:51065–51084.

Femenia, A., Bestard, M. J., Sanjuan, N., Rossello, C., and Mulet, A. 2000. Effect of rehydration temperature on the cell wall components of broccoli (*Brassica oleracea* L. Var. italica) plant tissues. *Journal of Food Engineering* 46:157–163.

Feng, H. and Tang, J. 1998. Microwave finish drying of diced apples in a spouted bed. *Journal of Food Science* 63:679–683.

Guine, R. P. F. and Castro, J. A. A. M. 2002. Pear drying process analysis: Drying rates and evaluation of water and sugar concentrations in space. *Drying Technology* 20:1515–1526.

Hebbar, H. U. and Rastogi, N. K. 2001. Mass transfer during infrared drying of cashew kernel. *Journal of Food Engineering* 47:1–5.

Hunt, R. W. G. 1991. *Measuring Colour*, 2nd Edition (pp. 75–76). New York: Ellis Horwood.

Jayaraman, K. S., Das Gupta, D. K., and Baru Rao, N. 1990. Effect of pretreatment with salt and sucrose on the quality and stability of dehydrated cauliflower. *International Journal of Food Science and Technology* 25:105–119.

Joubert, E., Wium, G. L., and Sadie, A. 2001. Effect of temperature and fruit-moisture content on discoloration of dried, sulphured Bon Chretien pears during storage. *International Journal of Food Science and Technology* 36:99–105.

Kluter, R. A., Nattress, D. T., Dunne, C. P., and Popper, R. D. 1994. Shelf life evaluation of cling peaches in retort pouches. *Journal of Food Science* 59:849–854, 865.

Koyuncu, T., Serdar, U., and Tosun, I. 2004. Drying characteristics and energy requirement for dehydration of chestnut (*Castanea sativa* Mill). *Journal of Food Engineering* 62:165–168.

Krokida, M. K. and Marinos-Kouris, D. 2003. Rehydration kinetics of dehydrated products. *Journal of Food Engineering* 57(1):1–7.

Krokida, M. K., Maroulis, Z. B., and Saravacos, G. D. 2001. The effect of the method of drying on the colour of dehydrated products. *International Journal of Food Science and Technology* 36:53–59.

Lewicki, L. 1998b. Effect of pre-drying treatment,drying and rehydration on plant tissue properties: A review. *International Journal of Food Properties* 1(1):1–22.

Lewicki, P. P. 1998a. Some remarks on rehydration of dried foods. *Journal of Food Engineering* 36:81–87.

Lewicki, P. P., Witrowa-Rajchert, D., and Mariak, J. 1997. Changes of structure during rehydration of dried Apples. *Journal of Food Engineering* 32:347–350.

Maskan, A., Kaya, S., and Maskan, M. 2002. Hot air and sun drying of grape leather (pestil). *Journal of Food Engineering* 54:81–88.

Maskan, M. 2001. Drying, shrinkage and rehydration characteristics of kiwifruits during hot air and microwave drying. *Journal of Food Engineering* 48:177–182.

McMinn, W. A. M. and Magee, T. R. A. 1997. Physical characteristics of dehydrated potatoes. *Journal of Food Engineering* 33:49–55.

Mittal, G. S. 1999. Mass diffusivity of food products. *Food Research International* 15:19–66.

Mohsenin, N. N. 1980. *Physical Properties of Plant and Animal Materials: Structure, Physical Characteristics and Mechanical Properties* (pp. 54–67). New York: Gordon Breach Science Publishers.

Njintang, Y. N. and Mbofung, C. M. F. 2003. Development of taro (*Colocasia esculenta* (L.) Schott) flour as an ingredient for food processing: Effect of gelatinization and drying temperature on the dehydration kinetics and colour of flour. *Journal of Food Engineering* 58:259–265.

Omoti, U. and Okolo, E. C. 2000. Date palm research in Nigeria: Progress and priorities. In: *Proceedings of the Date Palm International Symposium*, Windhoek, Namibia, pp. 147–148.

Ramesh, M. N., Wolf, W., Tevini, D., and Jung, G. 2001. Influence of processing parameters on drying of spice paprika. *Journal of Food Engineering* 49:63–72.

Rastogi, N. K., Angersbach, A., Niranian, K., and Knorr, D. 2000. Rehydration kinetics of high pressure treated and osmotically dehydrated pineapple. *Journal of Food Science* 65(5):838–81.

Rastogi, N. K., Nayak, C. A., and Raghavarao, K. S. M. S. 2004. Influence of osmotic pre-treatments on rehydration characteristics of carrots. *Journal of Food Engineering* 65:287–292.

Saleem, S. A., Baloch, A. K., Baloch, M. K. Baloch, W. A., and Ghaffoor, A. 2009. Accelerated ripening of Dhakki dates by artificial means: Ripening by acetic acid and sodium chloride. *Journal of Food Engineering* 70:61–66.

Sanjuan, N., Bon, J., Clemente, G., and Mulet, A. 2004. Changes in quality of dehydrated broccoli florets during storage. *Journal of Food Engineering* 62:15–21.

Sankat, C. K. and Castaigne, F. 2004. Foaming and drying behaviour of ripe bananas. *Lebensm.-Wiss. U.-Technologie* 37:517–525.

Sapers, G. M. and Douglsa, F. W. 1987. Measurement of enzymatic browning at cut surfaces and in juices of raw apple and pear fruit. *Journal of Food Science* 52:1262–1285.

Senadeera, W., Bhandari, B. R., Young, G., and Wijesinghe, B. 2003. Influence of shapes of selected vegetable materials on drying kinetics during fluidized bed drying. *Journal of Food Engineering* 58:277–283.

Wang, J. and Xi, Y. S. 2005. Drying characteristics and drying quality of carrot using a two-stage microwave process. *Journal of Food Engineering* 68:505–511.

Zaid, A. and de Wet Origin, P. F. 2001. Geographical, distribution and nutritional values of date. In: *Date Palm Cultivation* (Edited by A. Zaid). FAO Plant Production and Protection Paper, 156 Rev. 1. Food and Agricultural Organisation of the United Nations, Rome.

Zogzas, N. P., Maroulis, Z. B., and Marinos-Kouris, D. 1994. Moisture diffusivity methods of experimental determination: A review. *Drying Technology* 12:483–515.

Zogzas, N. P., Maroulis, Z. B., and Marinos-Kouris, D. 1996. Moisture diffusivity data compilation in foodstuffs. *Drying Technology* 14:2225–2253.

13 Solar Tunnel Dryer
A Novel Dryer for Dates

M. A. Basunia and H. H. Handali

CONTENTS

INTRODUCTION

Drying is an important operation in agriculture because it is the cheapest, easiest, and most common method of preserving and storing perishable agricultural products. There are various methods and techniques available to dry fruits, grains, and other agricultural products. Each method has its own advantages and limitations. Dried products are becoming popular as these are a good alternative to freshly harvested products.

The world total dates production is about 5 million tons (Ahmed and Mirani, 2008). Freshly harvested dates with high moisture content must be dried to preserve them unless they are to be consumed within a few hours of harvesting, or placed in low-temperature storage. Most of the dates are still dried by the traditional method of open air natural sun drying in date-growing countries. In this method, dates are spread on date leaf mats. This method of drying normally takes 14–21 days (Ampratwum, 2003). The traditional open air natural sun drying methods often yield poor quality. In most cases the drying yard is not properly fenced. So, the product is not protected against dust, rain, and wind, or even against insects, birds, rodents, and domestic animals while drying. Soiling, contamination with microorganisms, and infection with disease-causing germs take place during drying by these methods. Dew may accumulate on the surface of the dates and this causes mold growth if the date is not covered properly at night. Color change occurs and drying becomes slow. The quality of dried dates is likely to be varied by partial overdrying and possible contamination by dust, sand, and insect infestation. The dates dried in this way of open air natural sun drying have short shelf-life and may not be free from contamination. A study showed that about 22% of the total dates produced in Pakistan are wasted every year due to unavailability of proper drying, processing, and storage facilities (Ahmed and Mirani, 2008). The solar drying system provides higher temperature compared to open air natural sun drying, and the temperature in the solar drying system is controllable. In general, date-producing countries receive a high amount of solar radiation. Solar drying facilities combine the advantages of traditional and industrial methods, namely low investment costs and high product quality. Ampratwum (2003) reported the construction procedures and test results of a natural convection solar date dryer in Oman, but due to lower drying rates and capacity, it has not gained popularity in Oman. A successful new solar tunnel dryer was designed and developed at the University of Hohenheim, Germany, to meet the drying requirements of small farmer cooperatives (Lutz et al., 1987; Esper et al., 1994, 1996). This dryer is classified as a solar dryer that has been successfully tested under field conditions in about 30 countries under different climatic conditions in drying various agricultural products (Schirmer et al., 1996; Mastekbayeva et al., 1998; Bala and Mondal, 2001; Basunia and Abe, 2001a,b; El-Sebaii et al., 2002; Bala et al., 2003). This new dryer design eliminated the dependence on grid electricity since the power consumption of the fan could be supplied by batteries or solar PV panels. Unfortunately, a solar tunnel dryer has not yet been established for drying of dates in date-producing countries, where solar energy is abundant and can be used for drying of dates and other agricultural products. This chapter deals with the design and construction procedure of a solar tunnel dryer and its performance evaluation in drying of dates.

The original design of the dryer was 20 × 2 m, but with a capacity that is suitable for use in a large farm. In order to adapt its design for small- and medium-scale rural farmers of date-growing countries, a scaled down 12 × 2 m prototype of the tunnel dryer was designed and fabricated for a detailed experiment at the Department of Soils, Water and Agricultural Engineering, Sultan Qaboos University, Oman.

DESIGN AND CONSTRUCTION OF THE DRYER

A 12-m-long and 2-m-wide tunnel was designed and constructed to dry about 200 kg of freshly harvested dates per batch (Figure 13.1). The prototype solar tunnel dryer consists of a flat plate air heating solar collector and drying tunnel, fabricated as a single unit. The base of the tunnel was made of a wooden frame. The light-weight aluminum frames were used as the upper structure for the entire tunnel to support the transparent plastic cover.

The tunnel was placed on concrete block substructures 700 mm above the ground surface (Figures 13.2 through 13.6). One half of the nearly air-tight tunnel base was used as the flat plate air heating solar collector (6 × 2 m = 12 m^2) and the remaining half as a dryer. The metallic net (Figure 13.2) was used as the supporting structure at the base of the entire tunnel for the plywood planks in addition to the wooden base frames. Over the wooden base, black-painted metallic sheets (0.9 × 2.0 m) of thickness 0.25 mm (Figure 13.4) were used as the absorber plates in the collector

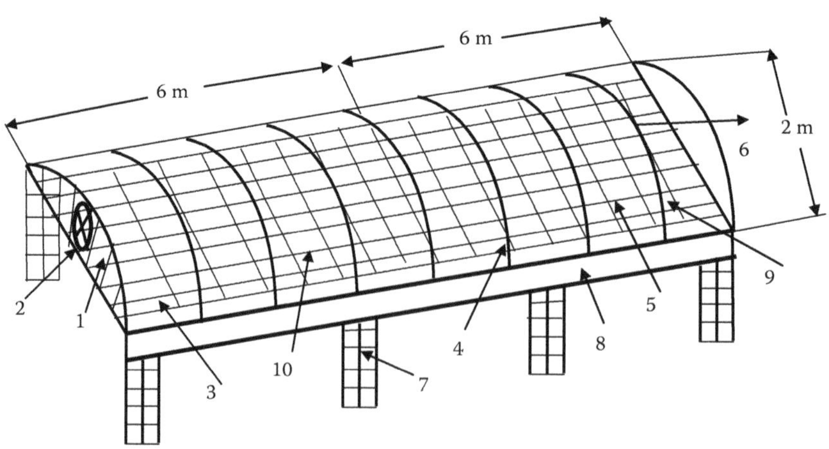

FIGURE 13.1 A schematic diagram of a solar tunnel dryer (1—Air inlet to the collector, fan; 2—South side wooden cover; 3—Collector part [12 m^2]; 4—Light-weight aluminum frame; 5—Dryer part [12 m^2]; 6—Air outlet from the dryer; 7—Concrete block substructures; 8—Wooden frame to support the bends and base of the tunnel; 9—Metallic wire mesh net over the wooden base in the dryer part; 10—Absorber plate [black-painted metallic sheet over the wooden base], not shown in figure, 12 m^2).

FIGURE 13.2 Part of the tunnel (under construction).

part of the tunnel. The light-weight steel wire mesh net was spread over the wooden base of the dryer part of the tunnel to dry the desired product.

A 0.2-mm-thick UV-stabilized colorless polyethylene sheet was used as the transparent cover over the entire tunnel. Thus, the entire tunnel became almost air-tight except the inlet opening (south side) for fixing a fan and the exit side (north side) for the moist air when it was covered with a thin polyethylene sheet.

The solar radiation passes through the upper transparent plastic cover. The black-painted metallic sheets act as solar heat collectors and absorb heat. As a result, the air inside the tunnel becomes hot particularly in the collector part of the tunnel. Then, this hot air was forced by a fan from the collector region (south side) to pass over the product in the drying region (north side) of the tunnel.

FIGURE 13.3 Tunnel placed on concrete block substructures showing the dryer part without the wooden base over the metallic net.

FIGURE 13.4 Tunnel placed on concrete block substructures showing the collector part (black-painted metallic sheets).

FIGURE 13.5 A solar tunnel dryer with transparent plastic cover.

FIGURE 13.6 A complete solar tunnel dryer (the solar panel is used to energize a 40 Wt axial flow fan).

The product also directly receives solar radiation in addition to the heat supply from the collector part of the tunnel. Thus, this dryer is considered as a mixed mode type solar dryer.

A solar-powered fan (40 W) was installed at the holes made on the wooden cover plank, 150 mm above the absorber metallic sheet at the air-input side of the tunnel. The fan can also be operated by an electric power source, instead of using solar panels. Due to extremely low resistance when forcing the air over the products to be dried, only 20–60 W fan or fans are sufficient to maintain the required air flow and the drying air temperatures within 65°C.

Prior to the operation of the dryer, nine thermocouples were installed to record the temperatures at different locations within the tunnel. Two thermocouples were also used to record the ambient dry- and wet-bulb temperatures. A pyranometer of Moll–Goregynstic type was used to measure the solar radiation incident on a horizontal surface. The thermocouple probes, anemometer, and pyranometer were connected through the interface of an AD (analog to digital) converter and then to a personal computer for data collection.

DRYING EFFICIENCY OF THE DRYER

The drying efficiency of the dryer was calculated from the measured solar energy input to the collector and the electrical energy supplied by the single fan during operations. The solar radiation incident directly on the drying portion of the tunnel was not considered for the calculation of the drying efficiency as this energy had little effect on drying.

The drying efficiency was calculated from the following equation:

$$\eta_d = \frac{100(W_r \cdot L_v)}{(P_e + S_t \cdot A_c \cdot R_b \cdot \eta_c \cdot t)}$$

$$\eta_d = \frac{100(W_r \cdot L_v)}{P_e + S_t \cdot A_c \cdot R_b \cdot \eta_c \cdot t}$$

where η_d is the drying efficiency in percentage, W_r is the mass of the moisture removed from the product in kg, L_v is the latent heat of vaporization of water in MJ/kg, P_e is the total electrical energy required to run the AC fan/fans during operation in MJ, S_t is the measured total solar irradiation on a horizontal surface in MJ/(h m^2), A_c is the area of the collector in m^2, R_b is the geometric factor at the test site, η_c is the collector efficiency in decimal at the test site, and t is the total drying time in hours.

NO-LOAD TESTS (TEST WITHOUT PRODUCT)

The dryer was placed in the open farmyard of the Agricultural Experimental Station (AES) of Sultan Qaboos University. The no-load tests with and without fan were conducted to know the temperature and air flow characteristics in different weather conditions and also the temperature gradient in the both collector and dryer regions of the tunnel. Figure 13.7 shows the variation of the ambient, collector, and dryer temperatures with the time of the day. The drying air temperature could be easily raised by some 5–30°C above the ambient temperature at an air flow rate of 0.5–1.0 m/s. The difference between the drying air temperature and ambient temperature gradually increased from morning till mid-day and then gradually reduced in the afternoon (Figure 13.7). The highest temperature of 69°C was observed at around 1 p.m. This indicates that the solar tunnel dryer can be easily used to dry dates. In no-load tests, the maximum difference between the average temperatures of the dryer and collector parts was about 2°C. There was almost no temperature gradient in the vertical direction of the whole tunnel, both in the collector (Figure 13.8) and the dryer

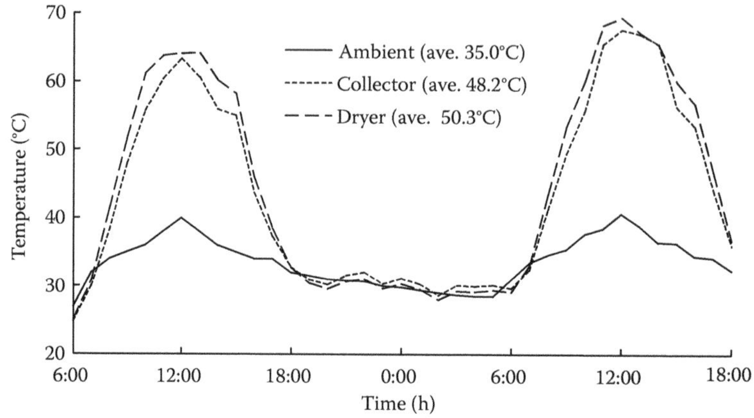

FIGURE 13.7 Variation of ambient, dryer, and collector air temperatures with time of the day (August 8–9, 2008) under no-load conditions (with fan).

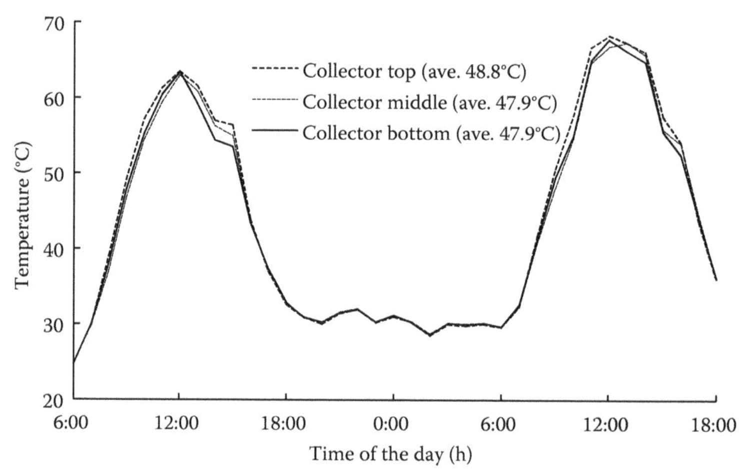

FIGURE 13.8 Variation of temperatures at the bottom, middle, and top (vertical temperature profile) of the collector center with time of the day (August 8–9, 2008) under no-load conditions (without product).

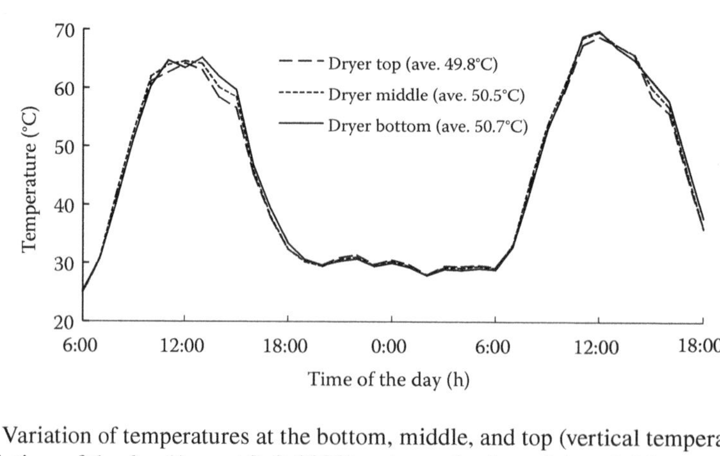

FIGURE 13.9 Variation of temperatures at the bottom, middle, and top (vertical temperature profile) of the dryer center with time of the day (August 8–9, 2008) under no-load conditions (without product).

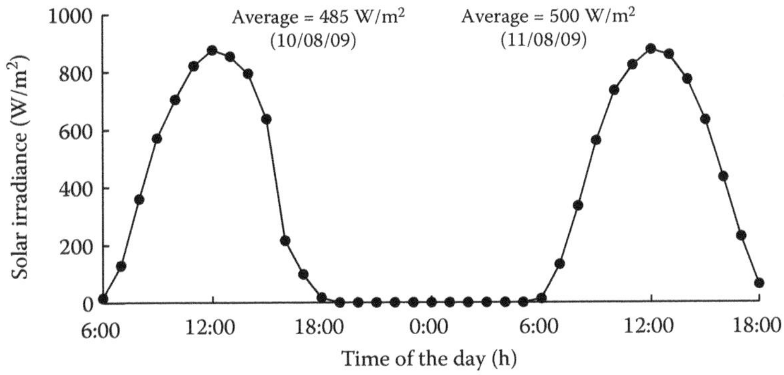

FIGURE 13.10 Variation of solar radiation with time of the day (August 8–9, 2008) under no-load conditions (without product).

(Figure 13.9) parts of the tunnel. Negligible temperature gradients were also observed in the horizontal directions through the bottom, middle, and top of the tunnel. This indicated the uniformity of temperatures inside the entire tunnel.

A no-load test was also conducted without using a fan. It was also observed that the highest temperature inside the tunnel became 75°C when the fan was not in operation. This indicated the operation of the dryer without a fan would cause overdrying of the product, particularly during 11:00 a.m. to 3:00 p.m. Figure 13.10 shows the variation of the measured total (beam and diffuse) solar radiation on the horizontal surface with the time of the day. It clearly indicated that an average 500 W/m^2 solar energy was incident on the horizontal surface.

TEST WITH FRESHLY HARVESTED DATES

The full-load test with 190.0 kg freshly harvested dates (Khalas variety) was conducted to study the dryer performance on July 27–28, 2009 (Figures 13.11 and 13.12). The average initial moisture content of the freshly harvested dates collected from the Agricultural Farm of the University was 32.8% (wet basis). The dates were spread on a wire mesh net in a single layer thickness placed over the plywood bottom of the dryer section of the tunnel. The drying was started at 6:30 a.m. and continued till 6:30 p.m. The approximate sun rising and setting times were 6:00 a.m. and 7:00 p.m., respectively. After the termination of the first day drying, the product was kept undisturbed overnight in the dryer closing both the sides of the tunnel by polyethylene sheets so that air could not pass through the tunnel. The moisture content was reduced to 21.5% from the initial 32.8% wet basis in 12 h drying on the first day. The next day, drying was started at 6:30 a.m. and continued till 5:00 p.m. when the products reached their final moisture content of 18.5% (w.b.) (Figure 13.12). Samples were collected for moisture content determination at the end of drying and the moisture content was found to be about 18.5% (wet basis). The moisture contents were determined by keeping the date samples for 24 h in an oven at 100°C (AOAC, 1984). The variation of the dryer and collector air temperature and the ambient air temperatures with drying time (6:00–19:00 h/day) in 2 days is shown in Figure 13.13. The variation of solar radiation with drying time is shown in Figure 13.14. The drying air temperature inside the entire tunnel and the ambient temperature were 52.0°C and 36.2°C, respectively. The average total radiation on a horizontal surface was 490.9 W/m^2. The average temperature in the collector part, particularly at the inlet of the collector (44.6°C), was lower than the average temperature in the dryer part (55.2°C) because of the presence of a fan at the air entrance side of the collector. The average temperature at the exit of the collector (i.e., the entrance of the dryer) was 56.2°C (Figure 13.13).

FIGURE 13.11 **(See color insert.)** Dried dates (18.5% MCWB) ready for unloading (July 27–28, 2009).

To compare the drying rates of a solar tunnel dryer and natural convection open air drying, two separate tests were conducted with about 100 g samples under the same weather conditions. The samples from the tunnel dryer and open air drying were taken at 1 h intervals for the first 4 h of drying starting from 6:30 a.m. and then at 2 h intervals till 6:30 p.m. to measure the moisture content against the drying time. The variation of moisture contents in solar tunnel drying and natural open air sun drying with time is shown in Figure 13.15.

The moisture removal rate in the solar tunnel dryer was much higher than that in open air natural sun drying (Figure 13.15). The amount of water removed in 2-day drying from the product on reducing its moisture content from 32.8% to 18.5% wet basis was 32.8 kg. Latent heat of vaporization of water is 2.5 MJ/kg, and it was considered that the energy requirement in low-temperature drying would be very close to the latent heat of vaporization of water. The total energy input to the fan in 20 h was 2.88 MJ. The average measured total (beams and diffuses) solar radiation incident on a horizontal surface in 2 days was 1.77 MJ/(m^2h).

The geometric factor and efficiency of the collector for Muscat, Oman was considered as 1.50% and 35%, respectively (Duffie, 1991; Sodha, 1994). The area of the collector was 12.0 m^2 and the time of drying was 20 h. The average drying efficiency of the solar tunnel was found to be 36.4% in drying dates. The total capacity of the dryer is estimated to be about 200–220 kg of freshly harvested dates, but 190.0 kg was loaded. Therefore, the drying area of the dryer was not fully utilized, which explains the low efficiency of the drying system. The average collector,

FIGURE 13.12 **(See color insert.)** Dried products ready for unloading from the dryer.

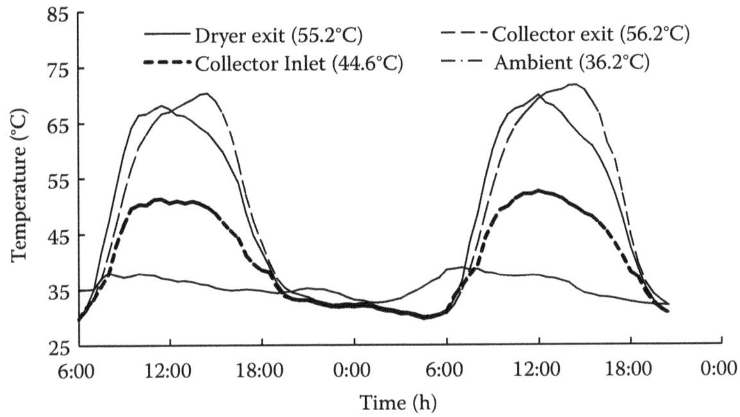

FIGURE 13.13 Variation of ambient, dryer, and collector temperatures with time of the day (July 27–28, 2009) in drying freshly harvested dates.

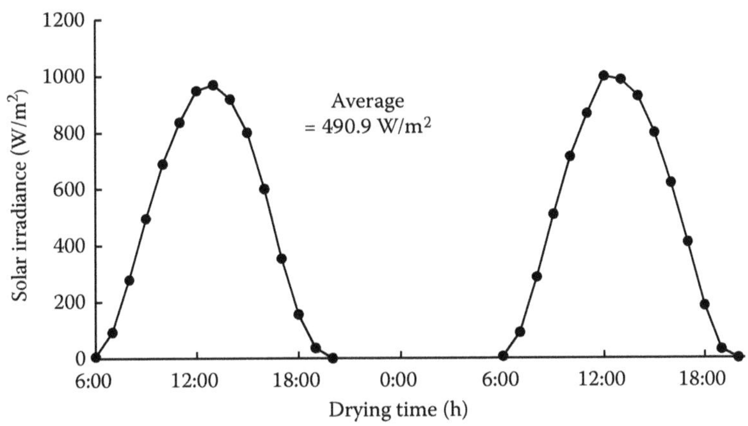

FIGURE 13.14 Variation of solar radiation with time of the day (July 27–28, 2009) in drying freshly harvested dates.

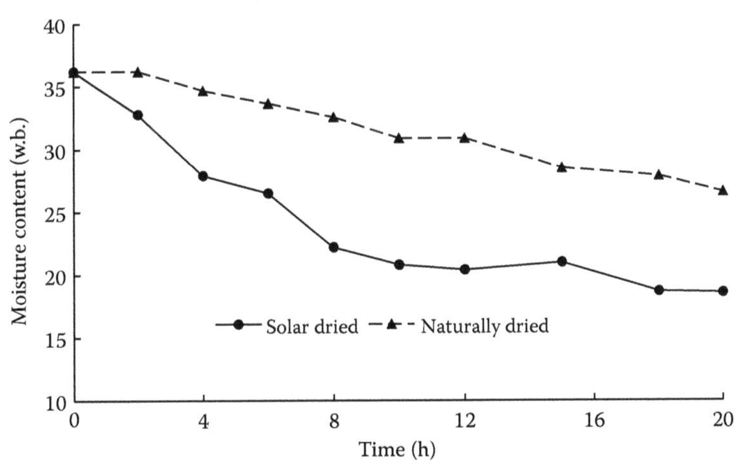

FIGURE 13.15 Variation of moisture content with drying time using a solar tunnel dryer and natural open air sun drying.

dryer, and ambient air temperatures in 2-day drying were 56.2°C, 55.2°C, and 36.2°C, respectively, and the average relative humidities were 34.0%, 46.5%, and 55.0%, respectively. It was also observed that the average relative humidity of the air leaving the dryer was fairly low (46.5%), while the ambient air relative humidity was 55.0%. This indicated that the drying potential of the drying air leaving at the exit of the dryer was not fully utilized. The relative humidity of the air leaving the dryer could be raised to the relative humidity of the ambient air if the length of the dryer was more than 12 m. In the overflow drying system (solar tunnel dryer), it is not possible to fully utilize the harvested energy as the drying air is passed over the product to be dried.

The much softer skin of the dates allows easy diffusion of moisture through it to the drying air at the initial stage of drying. The thickness of the date's muscles also contributed to faster drying, as the average thickness was only about 2 mm. In general, the drying efficiency was reduced considerably during the final stage of drying. This is due to much less mass transfer from the inner part of the dates to the surface after it is dried to a certain extent. This indicates that the diffusion of moisture from inside the product to the surface of the product becomes much more difficult.

CONCLUSION

This chapter has described the design, construction, and experimental investigation of a solar tunnel dryer for dates. The no-load tests indicated that the drying temperature can be easily raised to 5–30°C above the ambient temperature while the average air flow velocity inside the tunnel was 0.5–1.0 m/s. The average drying air temperature could be easily attained and maintained within 40–65°C, while the average ambient temperature is around 35.0°C.

The experiments were carried out with 190.0 kg freshly harvested dates and the performance of the dryer was compared to open air natural sun drying. It was possible to reach the moisture content level for safe storage within <2 days (20 h) with the solar tunnel dryer against 5–7 days in open air natural sun drying. The improvement in the quality of dates in terms of color, brightness, flavor and taste, and food value was distinctly recognized.

The thermal efficiency of the dryer was found to be approximately 36.4%. The economic analysis showed that a medium-scale farmer having 150–200 date trees could easily earn an additional $3000 per season if the farmer dried the product instead of selling it as freshly harvested products.

These investigations show that the solar tunnel dryer can be used for low-temperature drying of dates and other agricultural products in rural areas of date-producing countries.

ACKNOWLEDGMENT

The financial support provided by IG/AG/SWAE/08/01 for this study conducted at the College of Agricultural and Marine Sciences, Sultan Qaboos University is gratefully acknowledged.

REFERENCES

Ahmed, M. and A.A. Mirani. 2008. Development and performance of a solar-cum-gas fired dates dryer. *Agricultural Mechanization in Asia, Africa and Latin America*, 39(4), 63–68.

Ampratwum, D.B. 2003. Development of solar technology for drying dates. *Final Research Report*, Sultan Qaboos University, Muscat, Oman.

AOAC. 1984. *Official Methods of Analysis* (14th edition). Association of the Official Analytical Chemists, Washington, DC.

Bala, B.K. and M.R.A. Mondal. 2001. Experimental investigation of solar tunnel dryer. *Drying Technology*, 19(2), 1–10.

Bala, B.K., M.R.A. Mondal, B.K. Biswas, B.L. Das Choudhury, and S. Janjai. 2003. Solar drying of pineapple using solar tunnel dryer. *Renewable Energy*, 28, 183–190.

Basunia, M.A. and T. Abe. 2001a. Thin-layer solar drying characteristics of rough rice under natural convection. *Journal of Food Engineering*, 47, 295–301.

Basunia, M.A. and T. Abe. 2001b. Design and construction of simple three-shelf solar rough rice dryer. *Agricultural Mechanization in Asia, Africa and Latin America*, 32(3), 54–59, 66.

Duffie, J.A. and W.A. Beckman. 1991. *Solar Engineering of Thermal Process*. A Wiley-International Publication. John Wiley and Sons, New York.

El-Sebaii, A.A., S. Aboul-Enein, M.R.I. Ramadan, and H.G. El-Gohary. 2002. Empirical correlations for drying kinetics of some fruits and vegetables. *Energy*, 27, 845–859.

Esper, A., O. Hansel, and W. Muhlbauer. 1994. *PV-driven Solar Tunnel Dryer, Agricultural Engineering Conference*, Bangkok, December 6–9.

Esper, A., W. Muhlbauer, W. Rakwchian, and S. Janjai. 1996. Introduction of solar tunnel dryer for drying tropical fruits in Thailand. Paper presented at *The International Seminar on Financing and Commercialization of Solar Energy Activities in South and East Asia*, Kunming, China, August 24–31.

Lutz, K., W. Muhlbauer, J. Muller, and G. Reisinger. 1987. Development of multi purpose solar crop dryer for arid zones. *Solar and Wind technology*, 4, 417–428.

Mastekbayeva, G.A, M.A. Leon, and S. Kumar. 1998. Performance evaluation of a solar tunnel dryer for chili drying. Paper presented at *The ASEAN Seminar and Workshop on Drying Technology*, Phitsanulok, June 3–5.

Schirmer, P., S. Janjai, A. Esper, R. Samitabhindu, and W. Muhlbauer. 1996. Experimental investigation of the performance of the solar tunnel dryer for drying bananas, *Renewable Energy*, 2, 119–129.

Sodha, M.S. and R.I. Chandra. 1994. Solar drying systems and their testing procedures: A review, energy conservation and management, *Journal of Energy Conversion and Management*, 35(3), 219–267.

14 Processing of Dates by Dehydration and Microwave Drying

D. G. Rao, T. Hariharan, and S. Feroz

CONTENTS

DRYING OF DATES

Date palm (*Phoenix dactylifera*) is an important fruit crop of the Arabian peninsula and is closely associated with the cultural and religious practices of the region. Its production has increased considerably in the last three decades, making it imperative to develop various postharvest technologies for its preservation and handling.

Drying is one of the age-old methods of preservation of foods by removing the moisture from them without affecting the desirable qualities of foods to a great extent. Many fruits including date berries are dried by using mechanical dryers after a certain amount of preprocessing such as cleaning, washing, and so on. The moisture content in the dates varies considerably. Fresh dates contain about 79% moisture depending on the variety and extent of maturity, while the dried ones contain

7%. Dehydration becomes necessary as they (dates) contain excess moisture and are not likely to be consumed immediately or stored under refrigeration. The normally acceptable and desirable moisture content for Deglet Noor variety of dates acceptable to the consumer with soft texture is 23%–25% (Rygg, 1975). The quality aspects of dates are assessed (Boubekri et al., 2009) on the basis of:

- Mouth feel of structure
- Hand-tested elasticity of berries
- Visual color

Mechanical dryers operate at 40–70°C except in some cases. Hot air is generally used to dry in convective dryers. The heat source could be electrical heating or by steam or by burning agro wastes or furnace oils. Hot air of a certain temperature and humidity is passed over the dates. Moisture from the dates absorbed by the air will be disposed off through air vents along with the exhaust air. Various types of industrial dryers are generally used in food processing and are detailed in the literature (Rao, 2010). Of them, the cabinet tray dryer (shown in Figure 14.1) is very popular and is frequently used for processing at small-scale levels. For large-scale processing, continuous tunnel dryers are used.

Dehydration is carried out in maturation rooms or in specialized cabinets (shown in Figure 14.1 for small quantities) and tunnel dryers for large-scale operations. For the drying of soft dates, a temperature of 65°C is recommended to ensure a reasonable drying rate, since the date berries by nature are rather resistant to moisture loss. Relative humidity is maintained between 40% and 60%, mainly to avoid case-hardening and also to achieve some amount of fuel economy. After the date berries are subjected to preprocessing steps (viz., washing, removing the pits, and cutting into half pieces lengthwise), they are dried at about 40°C for 20 h. However, in the final stage of drying in a batch dryer, the temperature is raised to obtain pasteurizing and glazing effects, which would also help kill insects.

MOISTURE CONTENT AND EMC–RH DATA

Moisture is an important parameter in dates, and it varies considerably depending upon the variety of dates. It may vary anywhere between 85% in the early *kimri* stage to <10% in dry desert dates. Most of the dates are marketed in the fresh form. Hence, equilibrium moisture content

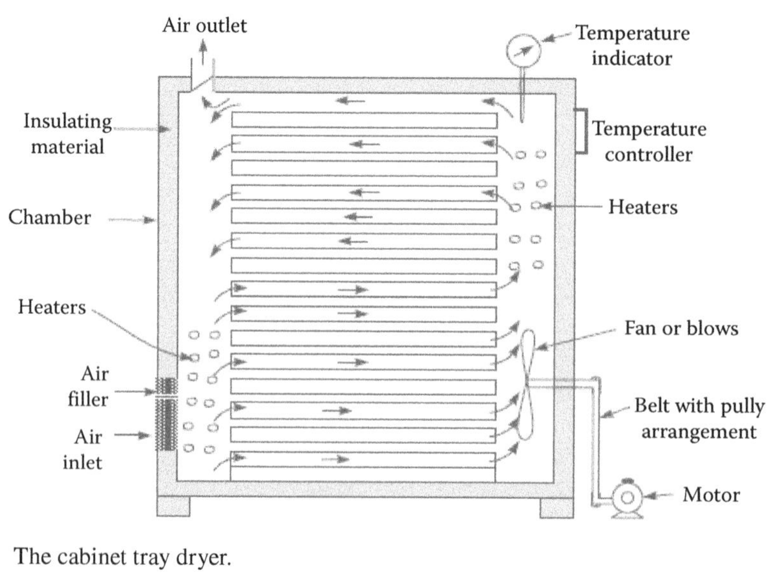

FIGURE 14.1 The cabinet tray dryer.

(EMC)–relative humidity (RH) data play a vital role. Normally, molds cannot grow in an atmosphere of 70% RH. From knowledge of EMC–RH data, it is possible to make out what should be the safe moisture content of dates for storage.

Figure 14.2 shows such data (FAO Corporate Document Repository, 1993). From the figure, it can be observed that about 24% moisture content corresponds to 70% RH. However, the moisture content data were found to be on the lower side for immature Indian dates for which the initial moisture content (IMC) was 15.9%, corresponding to an EMC of 75.9% (Kulkarni et al., 2008). Belarbi et al. (2000) reported the water desorption isotherms for 11 varieties of dates at a temperature of 25°C collected from different parts of the world in the RH range (water activity) of 11%–90% and analyzed the isotherm data on the basis of standard models, namely, those of Guggenheim–Anderson–Brunauer (GAB) and Brunauer–Emmett–Teller (BET). The results are satisfactory. Myhara et al. (1998) reported the moisture sorption isotherms for two cultivars of Omani dates (viz., *Fard* and *Khalas*) at three different temperatures, which displayed a crossing effect due to the dissolution of crystalline sugars at higher temperatures and moisture contents.

VARIOUS METHODS OF DRYING DATES

The moisture content can be reduced and brought to safe limits by dehydration or *drying*. There are various methods of drying dates to safer moisture levels which will be discussed in subsequent sections. There are also various other methods of preservation by preparing a number of shelf-stable products. Essentially the variety and maturity status of dates determine the type of pretreatments and type of drying methods to be adopted. There are various types of dryers available for industrial purposes, which obviously indicate various types of drying methods as well. Dates are dried essentially by

i. Solar drying/yard drying
ii. Hot air drying/cabinet tray drying
iii. Vacuum drying
iv. Freeze drying
v. Microwave drying

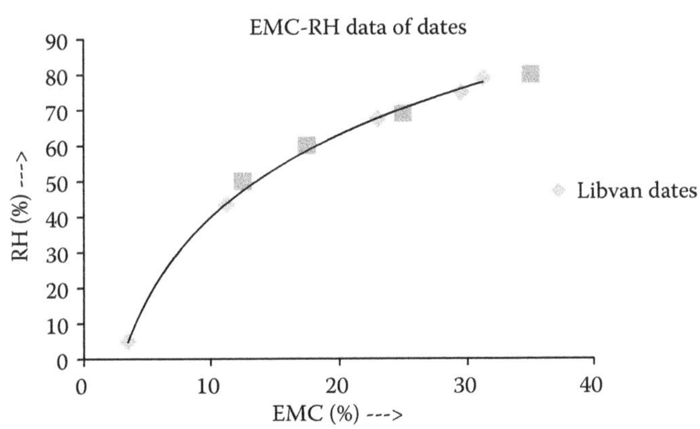

FIGURE 14.2 EMC–RH data for dates. (Adapted from FAO Corporate Document Repository. 1993. Whole dates. Ch. 1. http://www.fao.org/docrep/t0681e/t0681e03.htm. Accessed on July 12, 2010.)

Solar Drying

Solar heat is used extensively for drying various food materials including dates, which are mostly grown in the Arabian peninsula, where the day temperatures are usually high. Boubekri et al. (2009) used a solar dryer that was made in the form of a cabinet tray dryer. The heating source was only solar radiation with which they could run the experiments at controlled temperatures of 37°C, 50°C, and 60°C. Such studies were also reported by Hussain (2005) from Pakistan. Figure 14.3 shows a typical sun drying setup for dates in Bahrain at field level. We also notice dates being dried at domestic level in small quantities (Figure 14.3c).

Hot Air Drying/Cabinet Tray Drying

It is a frequently adopted industrial method of drying dates. Dates with 23% moisture content can be stored for long periods. Generally, cabinet tray dryers are used (shown in Figure 14.1) in which the trays are loaded with dates for drying. The dryer contains a number of trays (usually multiples of 12). The trays are loaded evenly with the material to be dried, put in a cabinet, and pushed into the hot air chamber.

Falade and Abbo (2007) studied air drying and rehydration characteristics of Nigerian dates in the temperature range of 50–80°C, and reported that most of the moisture migration took place in the falling rate period. The moisture migration was modeled using Fick's law and the results of effective diffusivities were reported. Kulkarni et al. (2008) reported on the drying of Indian dates, which were not fully mature at the time of harvesting. Blanching at 96°C before drying helped in preserving dates with good sensory qualities. The drying kinetics could be modeled using *thin layer drying models* for *Khalas* date variety (Abdalla and Al-Amri, 1999; Boubekri et al., 2009). Hassan and Hobani (2000) have also reported on the application of the thinlayer drying model for drying of dates (*Sukkar*i and *Sakie* cultivars) at three drying temperatures, 70°C, 80°C, and 90°C. They applied three drying models, namely, the exponential model, the Page equation, and the approximation of the diffusion model. The Page (1949) equation was found to be better at predicting the moisture ratio as a function of drying time and drying temperature for two date cultivars.

Thin layer drying models are usually applied for small grains, where actually the material should be spread in the form of a thin layer. Such studies conducted earlier by Rangroo and Rao (1992) for drying of *thoria* seeds (*Brassica campestris* L.) were successful. But the above three works (Abdalla and Al-Amri, 1999; Boubekri et al., 2009; Hassan and Hobani, 2000) have also conclusively proved the application of thin layer drying models for drying of dates.

Vacuum Drying

It is indeed a very useful technique for drying of food materials (both for berries and for pulps), which are heat sensitive in nature. Amellal and Benamara (2008) reported on the vacuum drying of date pulp in the form of cubes at 60°C, 80°C, and 100°C. The drying kinetics were described by standard models such as the Newton model and the Henderson and Pabis model.

Freeze Drying

Few freeze drying studies on dates have been reported in the literature even though the pulp is dried in a freeze dryer for long storage. Some studies have been reported on freeze drying of date pollens for future pollination. The pollens stay for even 2 years (Boughediri et al., 1995).

Microwave Drying

Microwave drying is also an important approach for drying of dates in view of its (microwave drying) obvious advantages, but all the studies reported in the literature are restricted to laboratory

FIGURE 14.3 (a) Sun drying (yard drying) of dates in garden itself after harvest in Bahrain. (b) Sun drying of dates on elevated platforms after harvest in Bahrain. (c) Sun drying of dates at the field and domestic levels.

scale only. The authors are not aware of any industrial applications of microwaves in processing of dates even though microwaves are used in various other industrial operations in food processing. This will be discussed in subsequent sections.

MICROWAVE DRYING/HEATING

Microwaves have been used in the last half century for heating or drying purposes in food processing in view of their distinct advantages. In fact, compared to any other industrial area of usage, microwaves found excessive applications in food processing for various purposes, namely, pasteurization, sterilization, blanching, baking, curing, sintering, thawing, and so on, in addition to heating and drying. An extensive review of the works on applications of microwaves in food processing is given by Decareau (1985).

We restrict ourselves here to giving a brief introduction to microwave heating and subsequently its application in drying of dates.

MICROWAVE ENERGY

Microwaves are electromagnetic waves of definite frequencies that were allocated by the Federal Communications Commission (USA) so that they do not interfere with radio frequencies. Thus, they differ from radio frequencies. The purpose for which these microwave frequencies are used has been categorized as Industrial, Scientific, and Medical (ISM) frequencies. These are shown in Table 14.1. The units used for measuring microwave frequencies are mega-Hertz, abbreviated as MHz.

Of the above microwave frequencies, 915 and 2450 MHz are assigned to microwave ovens. But it is an interesting question how these frequencies of electromagnetic waves can be utilized for heating purpose. Water consists of dipoles having a configuration of two charges, which are equal in magnitude but opposite in sign, and are held at a fixed distance from each other. In water, oxygen constitutes the negatively charged atom, while hydrogen constitutes the positively charged atoms. This kind of configuration is known as a dipole. Thus, when the water molecule is subjected to an electric field that is highly fluctuating in nature, naturally the dipoles are forced to change their orientation. Since the electric field is of very high frequencies which are of the order of a few mega-Hertz (which is approximately of the order of 21.45 billion times per second), the water molecules subjected to such a highly fluctuating electric field would also be forced to reorient themselves (or rotate) some millions of times every second. This is shown conceptually in Figure 14.4.

TABLE 14.1
Microwave Frequency (in MHz) Assigned for ISM Use

	Frequency	Countries Permitted
Radio	13.56 ± 6.68 kHz	
	27.12 ± 16.0 kHz	
	40.68 ± 20 kHz	
Microwave	896.0 MHz	UK
	915.0 ± 25 MHz	North and South America
	2450.0 ± 50 MHz	All over the world
	5800.0 ± 75 MHz	
	24,225.0 ± 125 MHz	

Source: Adapted from Decareau, R.V. 1985. *Microwaves in the Food Processing Industry.* Orlando Academic Press, Inc.

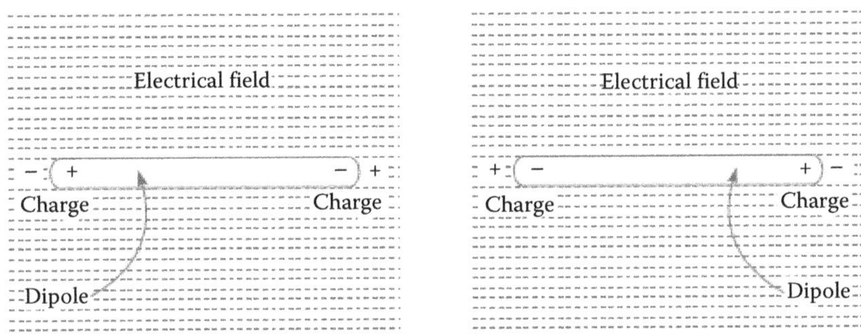

FIGURE 14.4 Change of dipole orientation with changing electrical charge.

These high-frequency rotations would generate heat in the body due to *molecular friction*. Thus, the heating of the material has nothing to do with the heating sources or container of the food material. The heat is generated within the food materials because of the moisture content in the foods unlike in the other modes of heat transfer. So, if the material is totally dry (i.e., free of water molecule dipoles), microwaves cannot and do not generate heat in such materials, no matter how long the material is exposed to microwaves.

Theory of Microwave Heating

The food material is heated by absorption of some amount of microwave energy by the dipole molecules of the food. The amount of energy absorbed by the foods to generate heat is considered to have been *lost* by the microwaves. Such energy lost by microwaves is determined by the *loss factor* of the food material. Thus, the loss factor is a desirable property of the material. Dielectric properties of various food materials have been summarized by Ayappa et al. (1991).

Microwave heating takes place by generation of heat from inside the food material, which results in migration of moisture. Thus the overall heating process is a classical case of *simultaneous heat and mass transfer*. The dielectric properties of the food materials are dependent on the moisture content, and they also determine the absorption of microwave energy in the electric field (Young, 1969; Tong et al., 1992).

Advantages and Limitations of Microwave Heating

When a comparison is made between conventional heating and microwave heating, we find a number of *benefits* that would come from the microwave heating (Rao, 2010; Sanga et al., 2000):

i. The heating time is considerably reduced because heating takes place from inside the food materials. The size of the equipment also reduces considerably because of a faster heating rate.
ii. The running costs of microwave heating are less and the energy efficiency is considerably high. There are no comprehensive reports available on energy saving; however, a conservative estimate states that the energy conversion efficiency in microwaves is of the order of 70%–85%.
iii. The heating control is instantaneous because of which the moment the power supply is switched off, heating also stops unlike in conventional heating systems.

However, the microwave heating system is not free from limitations. The following are a few of the limitations:

i. The microwave heating systems involve high initial costs for the equipment. This is one of the major reasons for it not being popular at the industrial scale. The lack of dielectric properties of foods also adds further complication to the design of robust industrial heaters.
ii. The sample size and shape affect the absorption of microwaves and the heating characteristics. This sometimes develops hot spots in the food materials.
iii. Microwave bumping and explosion phenomena are often found in reheating of self-stable foods.

MICROWAVE DRYING OF DATES

From the above discussion on microwave heating/drying, it is evident that microwave drying could obviously be used for drying fruits such as dates. Most of the technological advantages of microwave heating may not be realizable for date drying, but the quickness in drying is an obvious advantage, provided the local heating/excessive heating is properly taken care of. Hardly any systematic studies have been reported on the microwave drying of dates except for the works of Benamara from Algeria. Benamara et al. (2009) have reported on the microwave drying of date pulp with the main purpose of making date powder and comparing it with the hot air-dried results in terms of browning (due to caramelization of sugars) of the pulp/powder. The studies were conducted with date pieces of $15 \times 15 \times 3$ mm^3 size and weighing approximately 50 g. The term "moisture content" was defined as the ratio of the moisture removed to the removable moisture content, and was evaluated by assuming the date piece to be an infinitely long slab.

$$\mathrm{MR} = \frac{W - W_e}{W_0 - W_e} = \frac{8}{\pi^2}\left[\exp\left(-\pi^2 \frac{Dt}{4X^2}\right)\right] \tag{14.1}$$

However, the processing of microwave drying data is questionable in terms of the application of Equation 14.1 for the moisture ratio involving the diffusivity coefficient, which in turn is dependent on temperature of drying. Obviously, in microwave drying, we cannot achieve temperature control during heating. It varies considerably as heating continues (Tong et al., 1992).

There are no published works available on the microwave drying of date pieces as they are (i.e., after deseeding). Hence, work was undertaken by the authors to supplement the data.

MATERIALS AND METHODS

Date Sample

Date berries were harvested during the season July–August 2010 from the campus of Caledonian College of Engineering. The berries are of wild growth and hence are purported to represent a typical wild growing variety. The initial moisture content of reasonably matured berries was determined to be 63.93 on the dry weight basis by the oven-drying method of AOAC (1997).

Microwave Oven

The microwave oven is a domestic model (LG make, model MS 1947C, 2450 MHz frequency) of 19 L capacity with the provision for step change of input wattage.

Experimental Method

Microwave drying: Reasonably matured dates with yellowish brown colour were taken for microwave drying studies. The dates were deseeded, cut into two halves, and approximately 20 g of samples were loaded into the microwave oven. The wattage was set at 160 W. Moisture content versus time data were noted for further processing. The studies were repeated with the next higher range of wattages (320 and 480 W). Further studies with an increase in wattage were not possible as the material was getting caramelized.

Some studies were also conducted using fully matured dry dates, approximately 165 g (purchased from a local supermarket [The Sultan Centre]) and having a moisture content of 25.3% (wet basis) for drying in the microwave oven keeping the wattage at 320 and 480 W for a period of 9 min to allow a small amount of caramelization to take place, which in turn developed a taste similar to that of a toffee.

Cabinet tray drying: Drying studies were also conducted in a laboratory-scale tray dryer with a single tray (0.09 m^2 tray area) fitted with a thermostatic temperature control system with heating arrangement. Air was blown by an air blower. Moisture data were collected with drying time at a temperature of 70°C.

Results and Discussion

Microwave Drying of Dates

Data on moisture content (X) versus time t (in seconds) of drying of dates (cut pieces) were collected and the data are shown in Figure 14.5. The moisture content (X) was defined on the dry basis as follows:

$$X = (\text{kg of moisture present/kg of dry weight of solids}) \tag{14.2}$$

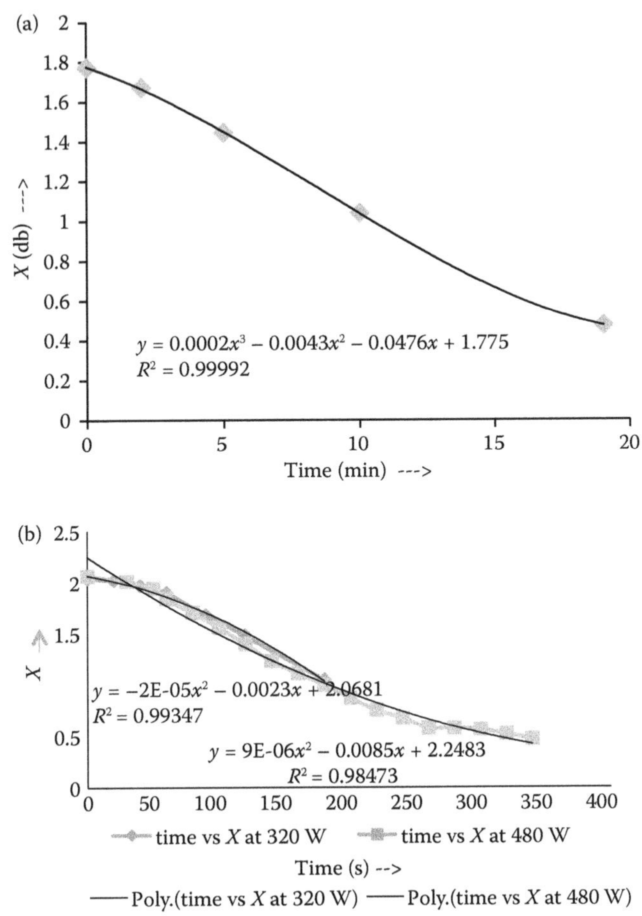

FIGURE 14.5 (a) Microwave drying data at 160 W. (b) Microwave drying data at 320 and 480 W.

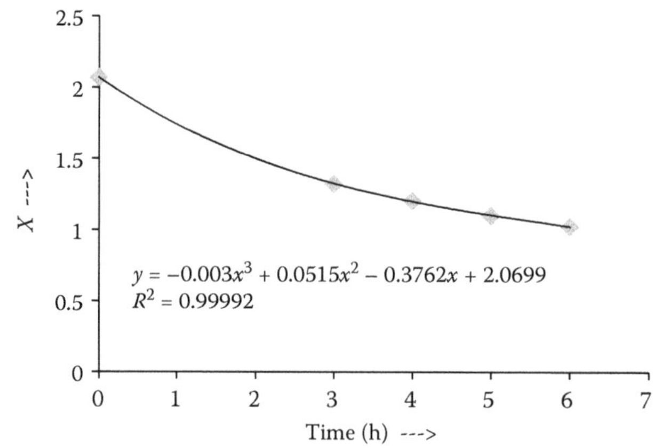

FIGURE 14.6 Cabinet tray-drying data at 70°C.

The moisture content steadily fell with time as expected. With an increase in microwave power, the fall in moisture content was steep.

Data were also collected in a cabinet tray dryer on *X* versus *t* at 70°C, and are plotted in Figure 14.6. The fall in *X* was typical, but the time taken was longer. For example, for moisture content (db) to fall down from 2.0 to 1.0, it took 6 h in the cabinet tray dryer, whereas for a similar moisture reduction, it was almost 540, 175, and 150 s for microwave powers of 160, 320, and 480 W, respectively.

The fall in moisture in the microwave oven did not show any typical pattern as was noticed for the cabinet tray-drying data shown in Figure 14.6. A clear difference in the drying data could be noticed in microwave drying and cabinet tray drying. In the former, the curves were convex upwards (Figure 14.5), whereas in the latter, it was concave (convex downwards in Figure 14.6). This showed that the decrease in moisture content was slow during the later part of drying in a cabinet tray dryer, whereas it was reverse in the case of microwave drying. However, the microwave drying time was much less as compared to the cabinet tray drying. Obviously, solar (or sun drying) drying would take much longer times.

The drying data at 160 W were processed to find the rate data ($r = \Delta X/\Delta t$), and are plotted in Figure 14.7a; they did not show any typical trend of the drying curve (as shown in Figure 14.8). Subsequently, we tried to evaluate the rate data by using the expression

$$r = -\mathrm{d}X/\mathrm{d}t \tag{14.3}$$

For this purpose, the drying data at 320 and 480 W were fitted to polynomial equations, and were differentiated to find the rate of drying, which was given by ($-\mathrm{d}X/\mathrm{d}t$). Even when the data were plotted (Figure 14.7b and c), they did not show any typical trend of the drying curve as shown in Figure 14.8. This could be due to nonuniform drying conditions in the oven during microwave drying when the temperature was not constant; in contrast, temperature increased as time progressed. Hence, the peculiar trend of rate curves. But it is interesting to note that even the cabinet tray dryer data did not yield the expected curve even though the drying data (moisture content vs. time) were of a typical nature. The drying rates were evaluated by Equation 14.3. It could probably be due to the fact that the unbound moisture content in dates could be negligible.* Thus, what we had was only bound moisture, which oozed out during the falling rate period. Hence, the plot in Figure 14.7d

* It is the unbound moisture which is responsible for the constant rate period, which is shown as a typical horizontal line BC in Figure 14.8.

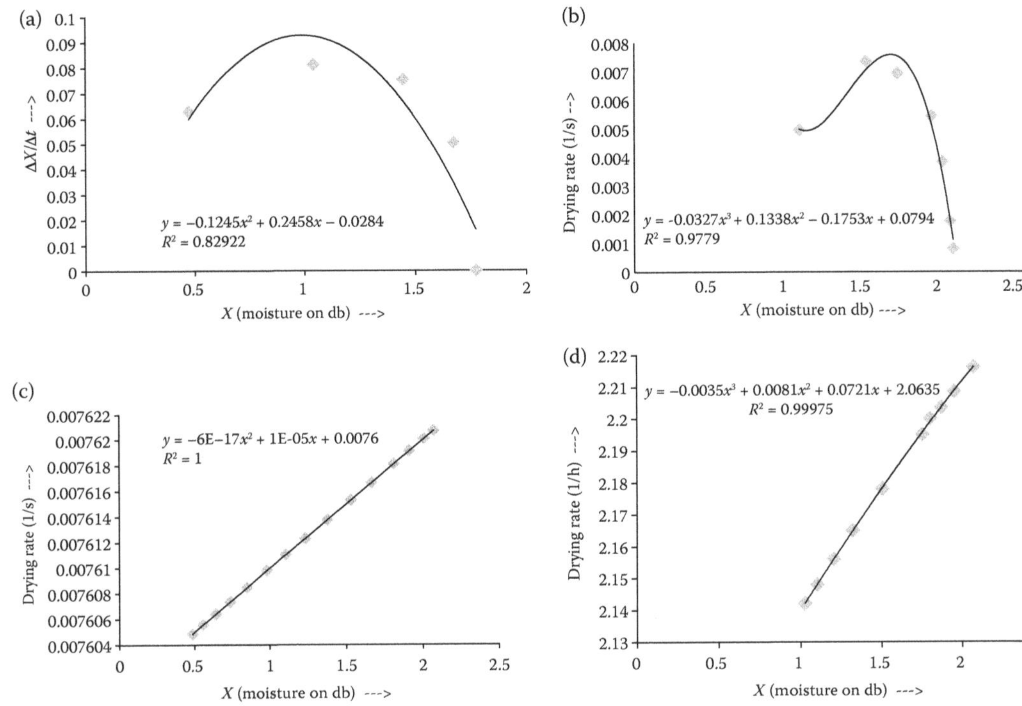

FIGURE 14.7 (a) Rate curve for microwave drying at 160 W. (b) Rate curve for microwave drying at 320 W. (c) Rate curve for microwave drying at 480 W. (d) Rate curve for cabinet tray drying at 70°C.

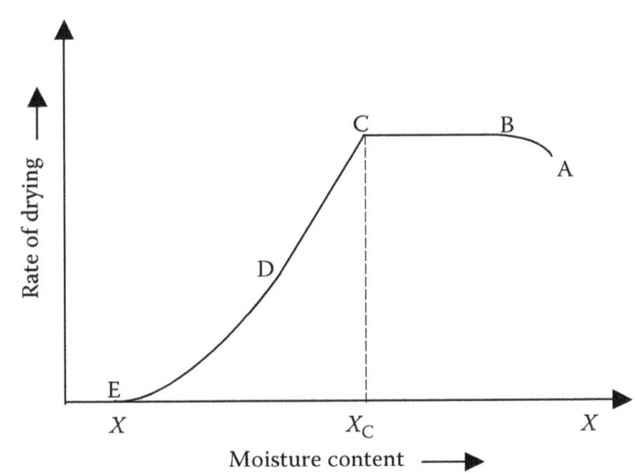

FIGURE 14.8 Typical drying rate curve.

showed the falling rate period. Falade and Abbo (2007) also noticed that drying of dates had taken place in the falling rate period.

Sensory Studies on Microwave-Dried Dates

The microwave-dried dates developed a typical toffee-like flavor that was much appreciated by the panelists when sensory studies were conducted. However, in these studies, raw dates could not be used. Dates purchased from a local supermarket with an IMC of 25.3% were used along with the seeds inside for microwave drying and sensory studies. Sensory studies were conducted for the appearance, color, flavor, taste, chewiness, and overall acceptability. Even though the panelists

TABLE 14.2
Sensory Score for the Evaluation of Microwave Dried Dates

Parameter	Sensory Score[a]
Appearance	5.66
Color	5.66
Flavor	7.5
Texture	6.25
Sweetness	7.33
Sorness	—
Astringency	—
Bitterness	—
Taste	7.33
Overall acceptability	7.66
Defects, comments, and suggestions if any	—

Signature & Name Date:

Score: 9: Like extremely; 8: Like very good; 7. Like good; 6: Like moderately; 5: Neither like nor dislike; 4: Dislike moderately; 3: Dislike fairly; 2: Dislike very much; 1: Dislike very much

Source: Adapted from Amerine, M.A., Pongborn, R.M., and Roessler, E.B. 1965. *Principles of Sensory Evaluation of Food*. Academic Press, New York.

[a] Score on 9-point Hedonic scale.

were not trained people, their reports showed very encouraging results with an overall acceptability of 7.66 in the 9-point Hedonic scale (Amerine et al., 1965). However, most of the panelists preferred the toffee flavor that has developed due to microwave heating. The results (shown in Table 14.2) indicated the possibility of making fruit toffees from dates, which has not been attempted/reported so far. In fact, it is pertinent to note that dates are considered as a candy that grows on the tree.

CONCLUSION

Dates production is an important agricultural activity in the Arabian peninsula. The production is excessive during the harvesting season, and hence needs processing for preservation and utilization during the rest of the year. Dehydration is an important processing method for preservation. Of the various drying methods, microwave drying seems to be a potential method even though it is not yet industrially practiced for drying of dates. The present work reported some studies on the moisture loss during microwave drying of dates in a domestic microwave oven. Advantages and limitations of microwave drying were indicated.

Microwave-dried dates developed a flavor similar to that of a fruit toffee with soft texture and good mouth feel, which are desirable attributes of a fruit toffee. Such studies have not been reported so far, and further studies are needed to standardize the process parameters.

REFERENCES

Abdalla, K.N. and Al-Amri, A.M.S. 1999. Thin-layer drying of *Khalas* date variety. *Agric. Mech. Asia Afr. Lat. Am.* 30(1), 47–50.

Amellal, H. and Benamara, S. 2008. Vacuum drying of common date pulp cubes. *Drying Technol.* 26(3), 378–382.

Amerine, M.A., Pongborn, R.M., and Roessler, E.B. 1965. *Principles of Sensory Evaluation of Food.* New York: Academic Press.

AOAC. 1997. *Official Methods of Analysis.* 16th Ed., Washington, DC: Association of Official Analytical Chemists.

Ayappa, K.G., Davis, H.T., Crapiste, G., Davis, E.A., and Gordon, J. 1991. Microwave heating: An evaluation of power formulations. *Chem. Eng. Sci.* 46(4), 1005–1016.

Belarbi, A., Aymard, Ch., Meot, J.M., Themilin, A., and Reynes, M. 2000. Water desorption isotherms for eleven varieties of dates. *J. Food Eng.* 43(2), 103–107.

Benamara, S., Khireddine, H., Amellal, H., and Djouab, A. 2009. Drying and browning of date pulp during hot air and microwave drying. *Afr. J. Food Agric. Nutr. Dev. (AJFAND)* 9(5), 1161–1173.

Boubekri, A., Benmoussa, H., and Mennouche, D. 2009. Solar drying kinetics of date palm fruits assuming a step-wise temperature change. *J. Eng. Sci. Technol.* 4(3), 292–304.

Boughediri, L., Larrival, M.T.C., and Dore, J.C. 1995. Significance of freeze drying in long term storage of date palm pollen. *Grana* 34, 408–412.

Decareau, R.V. 1985. *Microwaves in the Food Processing Industry.* Orlando, FL: Academic Press, Inc.

Falade, K.O. and Abbo, E.S. 2007. Air drying and rehydration characteristics of date palm (*Phoenix dactylifera* L.) fruits. *J. Food Eng.* 79(2), 724–730.

FAO Corporate Document Repository. 1993. Whole dates. Ch. 1. http://www.fao.org/docrep/t0681e/t0681e03.htm. (Accessed on July 12, 2010).

Hassan, H. and Hobani, A.I. 2007. Thin layer drying of dates. *J. Food Process Eng.* 23(3), 177–189.

Hussain, S. 2005. Solar dryer for ripening and drying of dates. http://www.solarfood.org/solarfood/pages/displayProject.xsp?id=37. (Accessed on July 10, 2010).

Kulkarni, S.G., Vijayanand, P., Aksha, M., Reena, P., and Ramana, K.V.R. 2008. Effect of dehydration on the quality and storage stability of immature dates (*Phoenix dactylifera*). *LWT—Food Science and Technology* 41(2), 278–283.

Myhara, R.M., Taylor, M.S., Slominski, B.A., and Al-Bulushi, I. 1998. Moisture sorption isotherms and chemical composition of Omani dates. *J. Food Eng.* 37(4), 471–479.

Page, G.E. 1949. Factors influencing the maximum rates of air drying shelled corn in thin layers. MS Thesis, Purdue University, Lafayette, IN.

Rangroo, S. and Rao, D.G. 1992. Drying of Toria (*Brassica campestris* var. *toria*): Part 2—Drying conditions. *J. Food Eng.* 17, 59–69.

Rao, D.G. 2010. *Fundamentals of Food Engineering.* New Delhi: Prentice-Hall of India, pp. 213–224.

Rygg, G.L. 1975. Date development, handling and packaging in United States, in *Agricultural Handbook No. 482*, Agr. Res. Service, US Department of Agriculture. (http://www.fao.org/docrep/t0681e/t0681e03.htm).

Sanga, E., Mujumdar, A.S., and Raghavan, G.S.V. 2000. Principles and applications of microwave drying, in *Drying Technology in Agriculture and Food Science*, Ed. Mujumdar, A. S., New Hampshire: Science Publishers, Inc., pp. 253–289.

Tong, C.W., Sheen, S.A., Fu, Y.F., Goedeken, D.L., and Lund, D.B. 1992. Microwave heat transfer in Food, in *Advances in Food Engineering Operations*, Eds. Singh, R.P. and Wirakartakusumah, MA, FL: Taylor & Francis, Inc., pp. 149–164.

Young, J.H. 1969. Simultaneous heat and mass transfer in porous hygroscopic solids. *Trans. ASAE* 12, 720–725.

15 Quality Assessment of Dates by Computer Vision Technology

A. Manickavasagan and R. Al-Yahyai

CONTENTS

INTRODUCTION

In computer vision technology, the image of an object is captured, enhanced, processed, and analyzed to obtain useful information about the object. Different types of cameras such as monochrome, color, near-infrared (NIR), mid-infrared, thermal infrared, infrared hyper spectral, x-ray, and others are used to acquire images remotely and characterize the internal and external qualities of the object instantaneously. The basic principle and components of a computer vision system are explained in Figures 15.1 and 15.2. Computer vision is a reliable, quick, nondestructive, noncontact (which means that samples are not in physical contact while testing) and low-cost method of quality evaluation. In many applications, this technology has replaced manual grading and inspection by providing objective decision-making criteria. It also has the potential to detect the invisible surface quality and internal characteristics of the products while using an appropriate imaging system.

Even though computer vision is a relatively new technology, it has become very popular in various fields of science including medical diagnosis, automatic manufacturing, remote sensing, and robot guidance. Recent developments in electronics have improved the potential applications of this technique in quality inspection, classification, and evaluation of a wide range of food products. Computer vision technology provides a high level of flexibility and repeatability in quality analysis of food at a relatively lower cost (Gunasekaran 1996).

COMPUTER VISION IN FOOD INDUSTRIES

Computer vision is a proven technology for the evaluation of various qualities of food and agricultural materials and is used in various food-related industries. This technology has the potential to evaluate internal and external qualities of food systems from micro- to macrolevels. In external quality evaluation, visual attributes such as the shape, size, color, and surface texture are analyzed from the captured images. Chemical components such as starch, protein, oil, sugar, and moisture

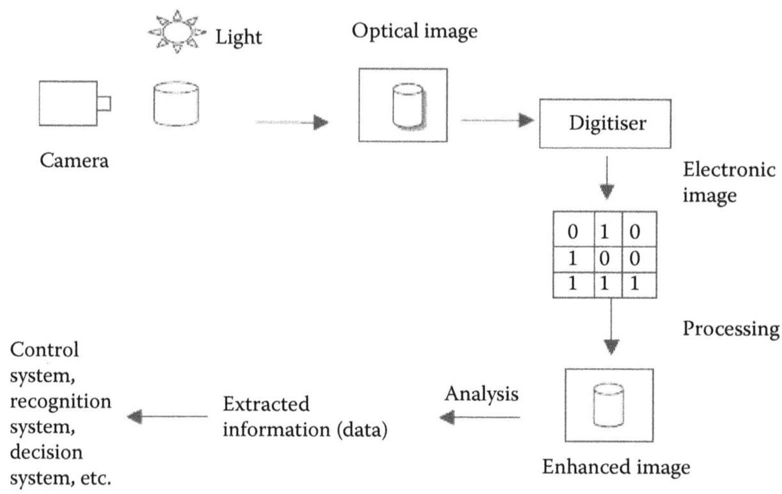

FIGURE 15.1 Principle of a computer vision system. (Adapted from Brosnan, T. and D. Sun. 2002. Inspection and grading of agricultural and food products by computer vision systems—A review. *Computers and Electronics in Agriculture* 36: 193–213.)

are analyzed with the help of unique imaging systems. This approach is highly beneficial in automation of food-handling and food-processing operations. In North America, the food sector is one of the top 10 industries using computer vision technology (Gunasekaran 1996).

Monochrome and color cameras have been widely used in computer vision systems to classify the products (mainly external qualities) based on the surface phenomena such as size, shape, color, texture, and other visible characteristics. These cameras are relatively cheaper compared to infrared and other cameras. Monochrome cameras are used in applications where color features are not required for classification. An additional advantage of this system is rapidity in data transfer and processing due to the size of the image. Manickavasagan et al. (2008) developed a monochrome imaging system to identify eight wheat classes and obtained 92%–96% classification accuracy.

Computer vision systems using color cameras have wide application in grading and quality evaluation of food materials based on visual color characteristics. In the handling and processing of fruits, evaluation of bicolor fruit in mango (Kang et al. 2008), grading of strawberry (Liming and Yanchao 2010), grading of apple (Leemans et al. 2002), sorting of Satsuma mandarin (Blasco et al. 2009), quality evaluation of orange (Kondo et al. 2000), and peel defects in citrus (Blasco et al.

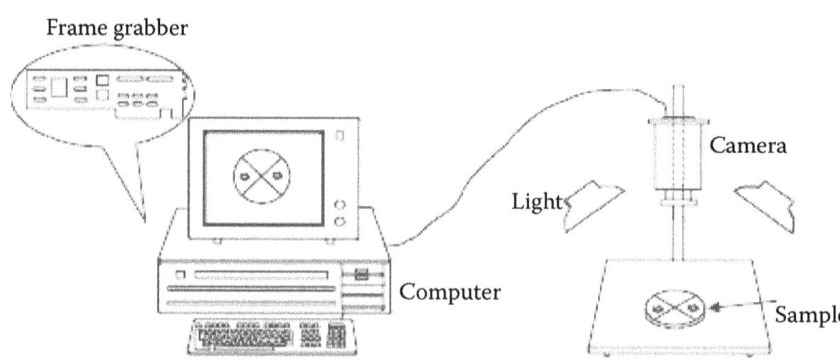

FIGURE 15.2 Components of a computer vision system. (Adapted from Brosnan, T. and D. Sun. 2004. *Journal of Food Engineering* 61: 3–16; Wang, H.H. and Sun D.W. 2002. *Journal of Food Engineering* 51: 305–310.)

2007) are some applications that have used color cameras. Similarly, the use of color cameras has been successful in prepared foods such as quality evaluation of pizza (pizza base, sauce spread, and topping analysis) (Sun and Brosnan 2003a, b), melting characteristics of cheese (Wang and Sun 2002), browning of cheese (Wang and Sun 2003), color of potato chips (Pedreschi et al. 2006), and shrinkage of potato slice during drying (Yadollahinia et al. 2009). Moisture content estimation in dehydrated shrimp (Mohebbi et al. 2009), sorting of salted cod fillets (Kohler et al. 2002), sorting of fish (Zion et al. 1999), and classification of guppies (fish variety) gender (Zion et al. 2008) are some of the potential applications of the color vision system in the seafood industry.

In the electromagnetic spectrum, the NIR radiation covers the wavelength range of 780–2500 nm. When the NIR region of the light hits a sample, it may be reflected, absorbed, and transmitted. The relative contribution of each phenomenon depends on the chemical constitution and physical characteristics of the sample. The reflectance, absorbance, and transmittance can be correlated with the chemical and physical properties of the samples. NIR spectroscopy has been successfully used for the quality evaluation of many food products. NIR imaging techniques have the added advantage of spatial data collection in addition to the spectral information, which is beneficial for online and nondestructive quality evaluations. Nicolai et al. (2007) have reviewed the quality evaluation of fruits and vegetables by means of NIR spectroscopy. Some of the quality characteristics of fruits and vegetables measured successfully with NIR were sugar in banana, essential oil in coriander, carotenoid in carrot, maturity in citrus fruit, hardness in guava, pectin in pear, internal flesh color in kiwifruit, moisture content in mushroom, sweetness index in peach, and oil constituents in pepper (Nicolai et al. 2007). This technology also has the potential to determine two or more quality attributes simultaneously by proper calibration. Pedro and Ferreira (2007) developed an NIR spectroscopy system to determine solids, sugars, and acidity of tomato simultaneously. With the advancement of NIR technology, data analysis, and classification models, it is possible to determine the quality parameters with higher accuracy. Rodriguez-Saona et al. (2001) developed rapid analyzing techniques to determine individual sugars in fruit juices with 99.9% accuracy using Fourier transform NIR and multivariate techniques. NIR methods with different calibration and classification approaches have been widely used to determine the sugar levels in foods. For example, sugar content of apples, bayberry juice, citrus fruits, grapes, and fruit juices have been measured by NIR spectroscopy (Yan-de et al. 2007; Xie et al. 2009; Tewari et al. 2008; Fernandez-Novales et al. 2009; Rambla et al. 1997).

Brosnan and Sun (2004) reviewed the quality inspection of food products by computer vision technology and listed some of the advantages and drawbacks of this technology. Generation of precise descriptive data, quickness, objective measurement, reduction of tedious human involvement, consistency, efficiency, cost effectiveness, automation of many labor-intensive processes, simplicity, robustness, nondestructiveness, and recording of data for future analysis are the major advantages of this system. However, the identification of objects in unstructured scenes and the requirement for artificial lighting in dark conditions are the major challenges of this system in quality assessment of food.

COMPUTER VISION FOR QUALITY ASSESSMENT OF DATES

Although computer vision technology has been developed for several commodities to determine various quality aspects, it requires algorithm development, calibration, and validation for individual products. Only a limited number of works on computer vision technology for dates are available.

Physical Properties of Dates

Jahromi et al. (2008) used image processing technology to measure the linear dimensions such as the length, width, and thickness of Mazafati cultivar dates in Iran. They used a Sony CCD (charge-coupled device) color camera (model TRV 225E) for image acquisition and Visual Basic (version 6) to analyze the images. From the basic measured units, they also calculated the projected area and sphericity to develop processing machinery for that variety. The image analysis approach has potential to investigate not only the macroexternal dimensions, but also the microcellular quality of dates. Shomer et al. (1998) studied the cellular damage in Madjhoul date variety after frozen storage for 10 months. They used an electron microscope for taking the images of the specially prepared date samples. It was determined that freezing at the higher temperature range resulted in rupture of various cellular compartments, whereas in fruits frozen at the lower temperature range (between −35°C and −50°C) both intracellular membranes and cell walls were intact even after 10 months of storage.

Color Attributes of Dates

The color of date fruits is an important property to determine the maturity index (readiness to harvest), cultivars, some extent sugar content, and other surface qualities. Automatic classification of dates requires detailed information including color intensity and distribution of each cultivar (Fadel et al. 2006). In general, dates are harvested at different maturity levels and subjected to grading immediately after harvesting. For grading based on maturity, color is the major factor as it is highly correlated with maturity. The color of dates ranges from yellow to dark red with many shades in between based on the maturity level (Lee et al. 2008a). The grades of Medjool dates based on maturity level (Bard Valley Medjool Date Grower Association standards) are given in Figure 15.3.

Fadel et al. (2006) measured the color properties of five date varieties (Lolo, Khalas, Berhi, Fard, and Bomaan) from UAE to develop the classification protocols. The images of the date samples were captured using a color camera (model: Mavica, Sony; resolution: 1912 × 916 pixels). The RGB color values were analyzed for the tested varieties. It was determined that the R component was suitable for discrimination of Berhi from Bomann and Berhi from Fard, the G component was good for the classification of Fard and Berhi, and Fard and Bomann. Similarly, the B component was suitable for differentiating Fard and Berhi; Fard and Bomaan; Lolo and Berhi; and Khalas and Berhi.

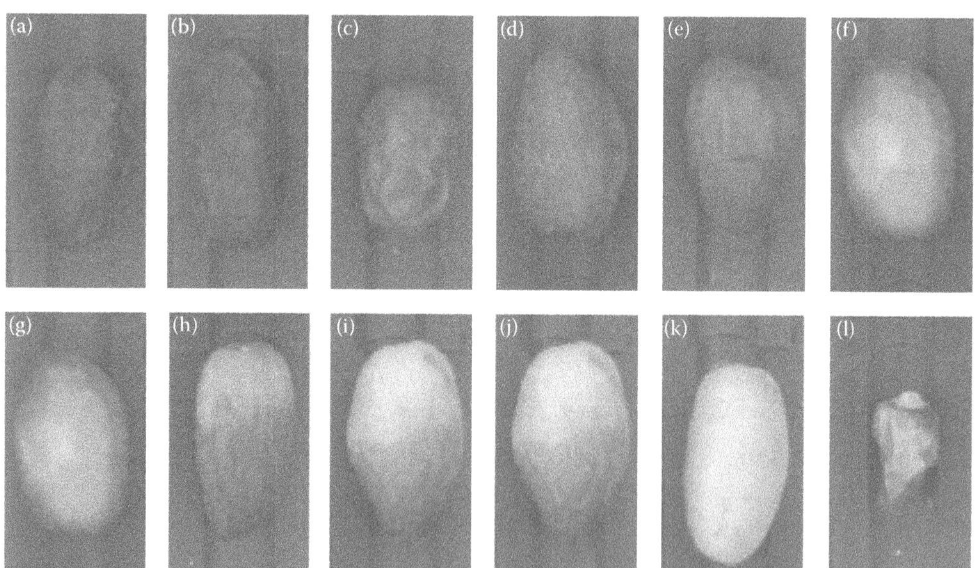

FIGURE 15.3 (See color insert.) Classification of date samples according to seven maturity levels used by the Bard Valley Medjool Date Grower Association. Grades: Ready for packing (a and b); slight skin separation (c); soft with light red or orange color (d and e); partial red (f); yellow head (g); yellow spot (h); parial yellow (i and j); yellow (k) and trash (l). (Adapted from Lee, D. et al. 2008a. *Journal of Food Engineering* 88: 364–372.)

Lee et al. (2008a) developed a color space conversion method and color index distribution analysis technique for the automated evaluation of date maturity. The system was calibrated by an experienced grader in which a set of colors of interest was specified. An overall accuracy of about 90% was achieved while identifying the seven grades based on the maturity level. It was stated that the developed system was easy for a human operator to specify and adjust the color preference settings for different color groups.

The sugar content is considered as the most important commercial characteristics of dates (Fadel 2002). Preliminary research results depict a good correlation between the surface color of dates and the sugar content. Fadel (2008) developed algorithms to determine the sugar content of dates (Lolo and Bomoaan varieties) based on the color analysis. RGB components of the images of the date samples acquired with a CCD color camera were correlated with actual values of fructose and glucose measured by HPLC. It was determined that a minimum accuracy of 86% was achieved while determining the sugar content of dates using RGB components.

Grading and Inspection of Dates

Throughout date-producing and date-processing countries, date grading and inspection are, in general, conducted manually (Lee et al. 2008b), which involve a large number of graders and inspectors. The obtained results are therefore subjective depending mostly on the inspector's experience and the work environment. Some prototype models have been developed and tested successfully even in commercial factories for the automated grading and inspection of dates based on various attributes.

A computer-vision-based grading system was designed and developed by Ohali (2011). The developed prototype model and proposed layout are shown in Figure 15.4. RGB images were taken by a CCD color camera, and the external quality features were extracted from the acquired images. The features were used to classify the dates into three grades (grades 1, 2, and 3) determined by experienced graders. A backpropagation neural network was used to classify the dates based on grade, and about 80% classification accuracy was achieved. In another study, a computer-vision-based prototype mechatronic system was developed for grading Saudi Arabian date varieties (Sukkari and Maneefi) (Al-Janobi, 2010) (Figure 15.5). The developed unit had four sections: the feeder, lighting system, imaging system, and grading mechanism. The grading or sorting mechanism, installed at the end of the belt conveyer, was operated by a transistor–transistor logic (TTL) signal from the computer to push the graded date into the corresponding grade boxes. The grading capacity of the developed prototype was ~108 kg/h. The average grading accuracy (for grades A, B, and C) for Sukkari variety was 88% and that for Maneefi variety was 93% while using a feedforward multilayer perceptron neural network trained with a backpropagation algorithm.

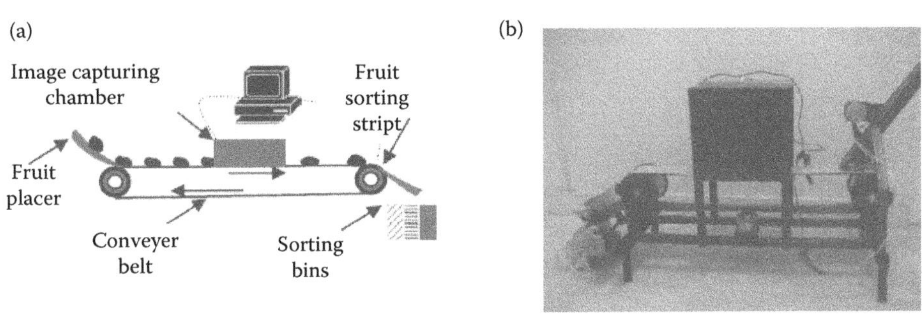

FIGURE 15.4 Computer vision system for grading of dates: (a) mediated layout; (b) developed prototype. (Adapted from Ohali, Y.I. 2011. Computer vision-based date fruit grading system: Design and implementation. *Journal of King Saud University—Computer and Information Sciences* 23: 29–36.)

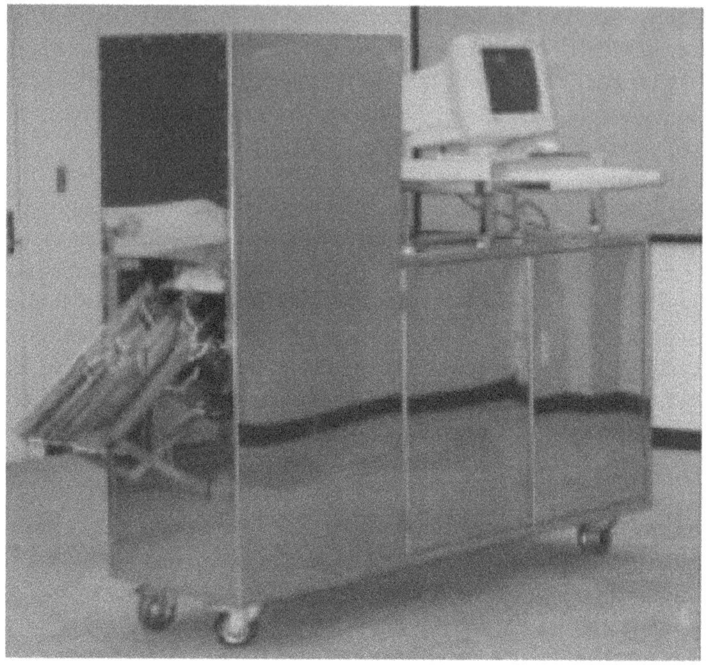

FIGURE 15.5 Prototype mechatronic system for the inspection of dates. (Adapted from Al-Janobi, A.A. 2010. *A Prototype Mechatronic System for Inspection of Date Fruits.* Unpublished Report. Department of Agricultural Engineering, King Saud University, Saudi Arabia. Available at: http://colleges.ksu.edu.sa/Papers/Papers/003%20Mvip%20cd.pdf. Accessed on May 31, 2011.)

Another major quality attribute in the grading of dates is skin delamination. Percentage of skin delamination and size are the two major criteria in the grade assignment of high-quality Medjools dates. Lee et al. (2008b) developed a machine vision system for automatic date grading using reflective NIR imaging. They used a Hitachi NIR-extended CCD camera (750–1200 nm) to obtain reflectance images of the top surface of dates. The block diagram for the image processing algorithm is given in Figure 15.6. The surface reflectance of the date fruit was depending on the skin delamination level. Figure 15.7 shows the images of date samples with and without skin delamination during various stages of image analysis. The percentage of delamination was calculated and

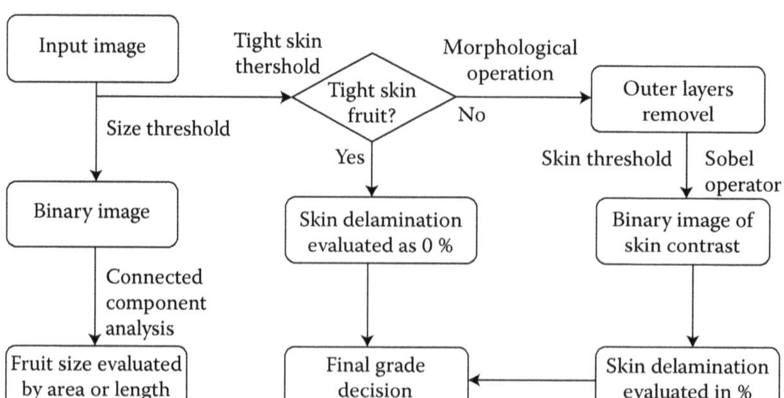

FIGURE 15.6 Block diagram of image processing algorithm for the determination of skin delamination in dates. (Adapted from Lee, D. et al. 2008b. *Journal of Food Engineering* 88: 388–398.)

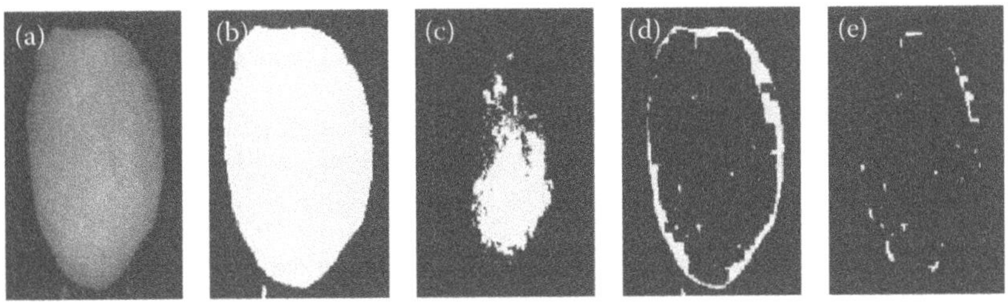

i. Date with no skin delamination

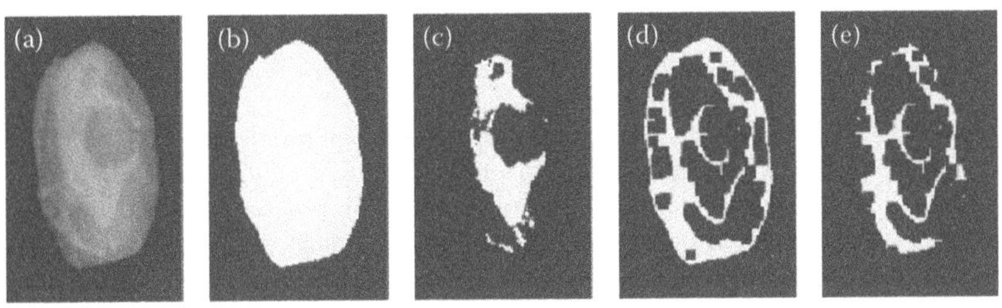

ii. Date with severe skin delamination

FIGURE 15.7 Processing of images of dates with and without skin delamination. (a) Original NIR image; (b) fruit and background separation; (c) binary image after "tight skin" threshold to the original image; (d) edge detection with sobel operator; (e) final skin delamination image. (Adapted from Lee, D., R., Schoenberger, J. Archibald, and S. McCollum. 2008b. *Journal of Food Engineering* 88: 388–398.)

used in grade determination. The overall accuracy of the system was in the range of 79%–95% for various grades. Also, the developed system showed an improvement in grading accuracy of ~10% over human grading and reduction in labor cost by almost 75% and grading time.

CONCLUSIONS

There are ample opportunities to develop computer-vision-based quality evaluation procedures at various stages of the date supply chain from the field to packaging houses. Intensified research is required to determine the capability and accuracy of various imaging systems and date quality attributes. Yield estimation in fields, size grading, color grading, surface defect detection, internal defect detection, sugar content and moisture content determination, and variety identification are some of the potential applications of computer vision technology in the date industry. Before replacing the existing manual inspection methods by the automated computer vision technique, both methods should be used in parallel at least for two seasons in order to test the performance and consistency of the system for a wide range of products.

ACKNOWLEDGMENT

This work was supported by The Research Council Grant—No. RC/AGR/SWAE/11/01 (Development of computer vision (CV) technology for quality assessment of dates in Oman).

REFERENCES

Al-Janobi, A.A. 2010. *A Prototype Mechatronic System for Inspection of Date Fruits*. Unpublished Report. Department of Agricultural Engineering, King Saud University, Saudi Arabia. Available at: http://colleges.ksu.edu.sa/Papers/Papers/003%20Mvip%20cd.pdf. Accessed on May 31, 2011.

Blasco, J., N. Aleixos, S. Cubero, J. Comez-Sanchis, and E. Molto. 2009. Automatic sorting of (*Citrus unshiu*) segments using computer vision and morphological features. *Computers and Electronics in Agriculture* 66: 1–8.

Blasco, J., N. Aleixos, and E. Molto. 2007. Computer vision detection of peel defects in citrus by means of a region oriented segmentation algorithm. *Journal of Food Engineering* 81: 535–543.

Brosnan, T. and D. Sun. 2002. Inspection and grading of agricultural and food products by computer vision systems—A review. *Computers and Electronics in Agriculture* 36: 193–213.

Brosnan, T. and D. Sun. 2004. Improving quality inspection of food products by computer vision—A review. *Journal of Food Engineering* 61: 3–16.

Fadel, M.A. 2002. Validation of computerized date sugar content estimation. In *Poceedings of the World Congress of Computers in Agriculture and Natural Resources* (March 13–15, 2002, Iguauacu Falls, Brazil), F.S. Zazueta and J. Xin (eds). ASAE Publication No. 701P0301. pp. 127–135.

Fadel, M.A. 2008. Sugar cintent estimation of date (*Phoenix dactylifera*, L.) fruits in tamr stage. *Agricultural Engineering International—CIGR Journal* X: 1–9.

Fadel, M., L. Kurmestegy, M. Rashed, and Z. Rashed. 2006. Fruit color properties of different cultivars of dates (*Phoenix dactylifera*, L.). *Agricultural Engineering International—CIGR Journal* VIII: 1–9.

Fernandez-Novales, J., M. Lopez, M. Sanchez, J. Morales, and V. Gonzalez-Caballero. 2009. Shortwave-near infrared spectroscopy for determination of reducing sugar content during grape ripening, winemaking and aging of white and red wines. *Food Research International* 42: 285–291.

Gunasekaran, S. 1996. Computer vision technology for food quality assurance. *Trends in Food Science and Technology* 7: 245–256.

Jahromi, M.K., S.S. Mohtasebi, A. Jafari, R. Mirasheh, and S. Rafiee. 2008. Determination of some physical properties of date fruit (cv. Mazafati). *Journal of Agricultural Technology* 4(2): 1–9.

Kang, S.P., A.R. East, and F.J. Trujillo. 2008. Colour vision system evaluation of bicolour fruit: A case study with 'B74' mango. *Postharvest Biology and Technology* 49: 77–85.

Kohler, A., A. Skaga, G. Hjelme, and H.J. Skarpeid. 2002. Sorting salted cod fillets by computer vision: A pilot study. *Computers and Electronics in Agriculture* 36: 3–16.

Kondo, N., U. Ahmad, M. Monta, and H. Murase. 2000. Machine vision based quality evaluation of Iyokan orange fruit using neural networks. *Computers and Electronics in Agriculture* 29: 135–147.

Lee, D., J.K. Archibald, Y. Chang, and R. Greco. 2008a. Robust color space conversion and color distribution analysis techniques for date maturity evaluation. *Journal of Food Engineering* 88: 364–372.

Lee, D., R., Schoenberger, J. Archibald, and S. McCollum. 2008b. Development of machine vision system for automatic date grading using digital reflective near-infrared imaging. *Journal of Food Engineering* 88: 388–398.

Leemans, V., H. Magein, and M.F. Destain. 2002. On-line fruit grading according to their external quality using machine vision. *Biosystems Engineering* 84: 397–404.

Liming, X. and Z. Yanchao. 2010. Automated strawberry grading system based on image processing. *Computer and Electronics in Agriculture* 71: S32–S39.

Manickavasagan, A., G. Sathya, D.S. Jayas, and N.D.G. White. 2008. Wheat class identification using monochrome images. *Journal of Cereal Science* 47(3): 518–527

Mohebbi, M., M. Akbarzadeh-T, F. Shahidi, M. Moussavi, and H. Ghoddusi. 2009. Computer vision system (CVS) for moisture content estimation in dehydrated shrimp. *Computers and Electronics in Agriculture* 69: 128–134.

Nicolai, B.M., K. Beullens, E. Bobelyn, A. Peirs, W. Saeys, K.I. Theron, and J. Lammertyn. 2007. Nondestructive measurement of fruit and vegetable quality by means of NIR spectroscopy: A review. *Postharvest Biology and Technology* 46: 99–118.

Ohali, Y.I. 2011. Computer vision based date fruit grading system: Design and implementation. *Journal of King Saud University—Computer and Information Sciences* 23: 29–36.

Pedreschi, F., J. Leon, D. Mery, and P. Moyano. 2006. Development of a computer vision system to measure the color of potato chips. *Food Research International* 39: 1092–1098.

Pedro, A.M.K and M.M.C. Ferreira. 2007. Simultaneously calibrating solids, sugar, and acidity of tomato products using PLS2 and NIR spectroscopy. *Analytical Chimica Acta* 595: 221–227.

Rambla, F.J., S. Garrigues, and M.D.L. Guardia. 1997. PLS-NIR determination of total sugar, glucose, fructose and sucrose in aqueous solutions of fruit juices. *Analytica Chimica Acta* 344: 41–53.

Rodriguez-Saona, L.E., F.S. Fry, M.A. McLaughlin, and E.M. Calvey. 2001. Rapid analysis of sugars in fruit juices by FT-NIR spectroscopy. *Carbohydrate Research* 336: 63–74.

Shomer, I., H. Borochov-Neori, B. Luzki, and U. Merin. 1998. Morphological, structural and membrane changes in frozen tissues of Madjhoul date (*Phoenix dactylifera* L.) fruits. *Postharvest Biology and Technology* 14: 207–215.

Sun, D. and T. Brosnan. 2003a. Pizza quality evaluation using computer vision—Part 2 Pizza topping analysis. *Journal of Food Engineering* 57: 91–95.

Sun, D. and T. Brosnan. 2003b. Pizza quality evaluation using computer vision—Part 1. Pizza base and sauce spread. *Journal of Food Engineering* 57: 81–89.

Tewari, J.C., V. Dixit, B. Cho, and K.A. Malik. 2008. Determination of origin and sugars of citrus fruits using genetic algorithm, correspondence analysis and partial least square combined with fiber optic NIR spectroscopy. *Spectrochimica Acta. Part A: Molecular and Biomolecular Spectroscopy* 71: 1119–1127.

Wang, H.H. and Sun D.W. 2002. Melting characteristics of cheese: Analysis of effects of cooking conditions using computer vision technology. *Journal of Food Engineering* 51: 305–310.

Wang, H.H. and Sun D.W. 2003. Assessment of cheese browning affected by baking conditions using computer vision. *Journal of Food Engineering* 56: 339–345.

Xie, L., X. Ye, D. Liu, and Y. Ying. 2009. Quantification of glucose, fructose and sucrose in bayberry juice by NIR and PLS. *Food Chemistry* 114: 1135–1140.

Yadollahinia, A., A. Latifi, and R. Mahdavi. 2009. New method for determination of potato slice shrinkage during drying. *Computers and Electronics in Agriculture* 65: 268–274.

Yan-de, L., Y. Yi-bin, F. Xiaping, and L. Huishan. 2007. Experiments on predicting sugar content in apples by FT-NIR technique. *Journal of Food Engineering* 80: 986–989.

Zion, B., A. Shklyar, and I. Karplus, 1999. Sorting fish by computer vision. *Computers and Electronics in Agriculture* 23: 175–187.

Zion, B, V. Alchanatis, V. Ostrovsky, A. Barki, and I. Karplus, 2008. Classification of guppies' (*Poecilia reticulata*) gender by computer vision. *Aquacultural Engineering* 38: 97–104.

16 Nonchemical Methyl Bromide Alternatives in Dates' Processing Sector

Mohsen Ahmed El Mohandes

CONTENTS

INTRODUCTION

Before reviewing methyl bromide (MB) alternatives in general and nonchemical alternatives in particular in dates' production and processing sector, it is of importance to shed light on current status of world dates production, the Montreal Protocol (MP) on the substances that deplete the ozone layer (a protocol to the Vienna Convention for the Protection of the Ozone Layer) and the impact of related decisions on the use of MB and its alternatives for disinfestation of dates. This is to link the national, regional, and international compliance requirements of this treaty and the requirements for developing and improving the dates industry.

Dates (*Phoenix dactylifera* L.) are a staple food in the diet of many countries, especially in the Near East and North Africa, since they are a major source of carbohydrates (Pennington, 1989; Reynes et al., 1994; Al-Hooti et al., 2002). Date palm cultivation and production are concentrated mainly in South West Asia and North Africa where arid regions and high temperature climate prevail in these regions. Estimated global date production is about 7,109,974 tons (FAOSTAT, 2008). Two-third of the total world date production (about 4,739,982.7 tons, 66.7%) come from Asia whereas one-third (2,369,991.3 tons, 33.3%) comes from Africa. It can be concluded that about 96.35% of the worldwide date production is concentrated in 13 out of a total of 37 date-producing countries. About 86.06% of the world production is produced in eight major countries, namely Egypt (1,326,133 tons), Iran (1,006,406 tons), Saudi Arabia (986,000 tons), the United Arab Emirate (755,000 tons), Algeria (552,765 tons), Pakistan (680,107 tons), Iraq (476,318 tons), and Sudan (336,000 tons) whereas 10.4% is produced in another five countries namely Oman (255,871 tons), Libya (150,000 tons), China (135,000 tons), Tunisia (127,000 tons), and Morocco (72,700 tons). The remaining 3.54% of date production comes from other 24 producing countries. However, these figures do not reflect the crop value or quality in each country where a country of small-producing capacity may produce fruits of the best quality and highest value in the world trade market (FAOSTAT, 2008).

MP is an international treaty entered into force in January 1989, signed by 196 countries so far, and it is one of the most successful treaties in terms of implementation and follow-up mechanisms. Some amendments to this treaty have been considered at Meetings of the Parties (MOP) in 1990 (London), 1991 (Nairobi), 1992 (Copenhagen), 1993 (Bangkok), 1995 (USA), 1997 (Montreal), and 1999 (Beijing). The treaty aims to protect the ozone layer; the layer that protects life on earth from harmful ultraviolet radiation that causes many human diseases and deformities such as certain skin cancers, lack of human's natural immunity and vulnerability to certain eye diseases such as eye opacity. This is in addition to other diseases that affect both animals and plants. MP works to protect the ozone layer by phasing-out the ozone-depleting substances (ODS) involving, but not limited to, chlorofluorocarbons, hydrochlorofluorocarbons, halons, carbon tetrachloride, methyl chloroform, bromochloromethane and MB. On the sideline of the United Nation's International Day for Preservation of the Ozone Layer, scientists indicated that the layer could potentially be fully restored over the next four decades, culminating in a 2050 recovery (Enviro News, 2010) if the parties to the MP complied with the Convention's obligations.

Bromine exists exclusively as bromide salts in diffuse amounts in crustal rock. Bromine may be economically recovered from bromide-rich brine wells and from the Dead Sea waters where the concentration up to 50,000 ppm (Oumeish, 1996; Radwan, 2008). The most important properties of this fumigant are that it is heavier than air (molecular weight of 94.94 g/mol), colorless, odorless with water solubility degree of 13,400 mg/L at 25°C. The gas boils at 3.6°C (Albemarle Corporation, 1999). The gas is marketed in containers ranged from cylinders of 50 kg weight to cans of 1.5 lbs. MB for soil fumigation purposes contains 2% of chloropicrin as a warning agent whereas the gas for commodities treatment is a pure gas.

MB is used to eliminate insects, microbes, rodents, weeds, nematodes, viruses, mites, and all other insect pests responsible for causing much damage to crops and stored food products/commodities. The major problem with MB is its ozone-depleting potential and bromine residues in the soil. They are absorbed by plants and end up in the edible plant parts. Moreover, the high toxicity of MB to mammals limits its application to nondomesticated areas and only by licensed persons. Other problems resulting from MB applications are (i) the severe boomerang effect, (ii) the toxicity of bromine residues to several plant species such as carnations, (iii) the elimination of beneficial microorganisms such as mycorrhizal fungi, plant growth-promoting bacteria, bacteria-transforming ammonium nitrogen to nitrate nitrogen, and so on (http://www.minagric.gr/greek/data/malathr1.doc). Additionally, MB has substantial harmful effects to the environment which is not limited to those associated with depletion of the ozone layer such as the residual effects in and on treated commodities, effects on beneficial soil flora that play prominent role in facilitating the key nutrients for plant nutrition and growth, influence on ground water pollution with MB residues as a result of soil washing. In 2005, Methyl Bromide Technical Option Committee (MBTOC) indicated that MB consumed for soil fumigation represented 87% of the total amount used while commodities and structures sectors consumed 13%. Quarantine and preshipment (QPS) applications are not included (UNEP, 2006).

According to the United Nations Environment Programme's Scientific Panel, it is estimated that MB is responsible for ~5%–10% of worldwide ozone depletion with ozone-depleting potential of 0.6 (UNEP Home Page). Having recognized the environmental challenges and threats, the international community reached a consensus in 1997 on regulatory measures aimed at global MB phase-out as an ODS under the MP. The MP's schedule requires that developed countries (known as non-Article 5 countries) phase-out MB by 2005 and developing countries (known as Article 5 countries) by 2015. The parties agreed to exempt MB applications in QPS sector since there are no suitable alternatives available so far that could replace this fumigant. Other exempted MB applications include use as a chemical feedstock and critical uses (Glassey, 2008). MBTOC recognized the problem of MB phase-out in nonarticle 5 countries, and, consequently, has published a handbook on Critical Use Nominations

(CUNs) for MB. Accordingly, countries may apply for CUNs after complete cessation of MB (i.e., beyond 2005 in developed countries and 2015 in developing countries) (TEAP and MBTOC, 2003).

MP AND MB ALTERNATIVES IN DATES' SECTOR

In general, alternatives have been identified by MBTOC of the United Nations Environment Programme as "any practices or treatments that can be used in place of MB." These alternatives are categorized to (1) existing alternatives that include those alternatives in present or past use in some regions and (2) potential alternatives that include those in the process of investigation or development (UNEP, 2006). The Committee assumed that an alternative demonstrated its efficiency in one region of the world is applicable in any other region unless there were obvious constraints to the contrary. In other words, an effective alternative can be applied anywhere unless there is no difficulty in terms of environmental conditions and atmosphere or in terms of target lesions to be eliminated that preclude the use of this alternative. In dates' industry, there are some factors that may cause a potential alternative to MB unsuitable for dates' disinfestation such as (1) technical and economical infeasibility of the alternative, (2) geographical and climatic variations between different areas where date-processing facilities are located, (3) limitations of some alternatives due to its effect on characteristics of the final products, and (4) negative economic impact to the producers and processors as alternative fumigants often require more time compared to MB.

According to MBTOC Assessment Reports (UNEP, 2002, 2006), alternatives to MB have been identified for about 95% of controlled uses in 2005. Yet, MBTOC has not identified available, technically and economically effective alternatives for high-moisture fresh dates. Following the difficulty in finding viable alternatives for stabilizing high-moisture dates, the 15th MOP adopted XV/12. Decision on the use of MB for treatment of high-moisture dates (UNEP, 2003). This decision reads the following:

Recognizing that in its 2002 report, the MBTOC has explicitly acknowledged that there is currently no alternative to the use of MB for high-moisture dates that is in use in any country in the world.

Recognizing also that parties which consume over 80% of their MB for high-moisture dates cannot meet the Protocol's MB control schedule without production losses for that important cash crop for their countries.

Recognizing further the need for further work to be undertaken to demonstrate alternatives to MB for high-moisture dates,

1. That the Implementation Committee and MOP should defer the consideration of the compliance status of countries that use over 80% of their consumption of MB on high-moisture dates until 2 years after the Technology and Economic Assessment Panel formally finds that there are alternatives to MB that are available for high-moisture dates.
2. That the above provision shall apply so long as the relevant Party does not increase consumption of MB on products other than high-moisture dates beyond 2002 levels, and the Party has noted its commitment to minimizing the use of MB for dates to the extent necessary to ensure effective control of pests.
3. To request the Executive Committee to consider appropriate demonstration projects for alternatives on high-moisture dates, and to ensure that the results of those projects are shared with the Technology and Economic Assessment Panel.

Based on the aforementioned resolution, it is clear that the resolution deals with the exception of MB use for treatment of high-moisture dates only, that is, Deglet Noor (Figure 16.4), and not any other dates varieties. The Regional Office for West Asia recognized this critical issue and consequently, via its Compliance Assistance Programme (CAP), discussed through different activities applications of MB and its alternatives in this vital sector at national and regional levels. The main objective was to help date-producing countries define their needs for gradual MB phasing-out in dates' sector before 2015. The outcomes of conducted activities indicated clearly that (1) world dates' production increases dramatically as a result of growing new varieties of high productivity, application of developed agricultural practices and increasing interest of many countries to include date palm cultivation in their national agricultural policies (2) consumption of MB is increased steadily in the last few years (2007–2010) and is expected to exacerbate in the next 20 years as all dates' producers will have to store the majority of dates' production for almost a year, particularly in the Arab region, as Holy Ramadan month is coming before harvesting season. It is noteworthy that 60%–65% of dates production in Arab countries is consumed in the Holy month whereas the rest 35%–40% is consumed around the year (3) long periods of storage under high temperatures and humidity that characterizes these countries make stored dates more vulnerable to infestation by various pests and insects, specially Coleoptera and Lepidoptera (Ahmed, 1993), causing the necessity for more fumigation treatments, and (4) most of date-producing countries consume large and unjustifiable quantities of MB for disinfestation of semidry and dry varieties (where commercially feasible and technically effective alternatives are already available) which play an important socioeconomic role in these communities. Consequently, MB use is expected to increase dramatically in the future (UNEP/ROWA, 2010, unpublished report). These applications are considered as anticommitment practices as the countries committed to reduce MB use gradually until complete phase-out by 2015. Hence, the problem is that despite proximity of the deadline for phasing-out MB use in all controlled applications including dates, except for the exempted uses, we cannot address clear and potential policies, programs, or strategies in most dates-producing countries for testing different alternative(s) as a step to adopt some of these alternatives at the national level commensurate with the circumstances of each country in terms of climatic conditions, available dates' cultivars, type of associated pests, and economic feasibility related to those alternative(s). These circumstances, in addition to absence of clear information and effective alternatives, would undoubtedly expose dates industry to real threats where MB would not be available in the markets after in 2015, except for uncontrolled uses. Based on information available, few ongoing researches and projects are taking place in some date-producing countries to reach viable, efficient, and cost-effective alternatives for dates' disinfestation, particularly for high-moisture dates. However, no recommendations are available to apply certain alternative(s) on commercial scale, so far (El-Mohandes, 2010a).

OBSTACLES IN ADOPTING ALTERNATIVES FOR DATE TREATMENT

Date production is in real danger due to (1) spread of insects, pathogens, and diseases that infect this important crop in every development stage including postharvest pests; (2) the lack of projects and researches that aim to develop date palm cultivation and production, preserve the current and develop new cultivars through new agricultural technology and develop new postharvest preservation methods. This is mainly due to the lack of fund allocated for research in these areas. When converting from fumigating dates with MB to adopting the use of any other chemical or nonchemical alternative, it is of importance for dates' producers to study carefully some factors that must be taken into consideration. These factors include locally enacted regulatory laws and legislation, technical aspects, and effectiveness of the adopted alternative(s) as pests disinfectant for dates preservation, economic considerations, and feasibility of adopting alternatives as a substitute to MB, health effects of substitutes on workers and finally safety aspects which must be taken into account to ensure safety of workers, consumers, and the environment. Evaluating the level of market acceptance and consumer demand for dates treated with a given alternative(s) is also of great importance that needs to be considered. Once the above considerations are determined and evaluated, dates producer may address and identify the following to specify the proper alternative(s) to be adopted:

1. Pests need to be controlled.
2. Successful applications of MB alternatives, especially those applied in countries of similar environmental and climatic conditions and type of pests. Studying the use of a mixture of combined alternatives to control the targeted pests is another option.
3. Advantages and disadvantages of each type of candidate alternatives/combinations available for use.
4. Experts and specialists at national and regional levels experienced in applications of MB alternatives to ensure availability of technical and practical expertise when needed.
5. Once addressed, dates producers can make their decisions on the type of alternative(s) to be adopted and transition modality from MB- to non-MB-based dates' industry. However, there are some factors that play an inhibiting role in adopting MB alternatives in the dates-processing sector.
6. The national organizational regulations and procedures followed for registering chemicals/alternative(s) are generally the most important factor limiting the availability of MB alternatives. Registration of new chemicals/alternatives usually requires a long time until these alternatives are available in the markets. The situation becomes more difficult when registration of these alternatives is required for the treatment of food commodities and products due to possible direct and indirect impacts on human health resulting from chemical residues associated with treated products.
7. Lack of adequate studies, publications that cover the technical aspects related to application of alternatives, which contribute effectively to good practice in application of these alternatives to harvest the best possible results comparable to those obtained through the use of MB.
8. Uncertainty regarding future applications of alternatives after MB phasing-out as a result of lack of feasibility studies for each type of available alternatives. Availability of such studies may form a base for dates' producers to rely upon and help learning more about the most suitable alternative(s) that can be adopted for disinfestation of dates that at least help achieving economic return comparable to that obtained through MB application.
9. Lack of training and awareness-raising programs and policies aimed at informing dates producers on types, availability, and technical competence of different alternatives.
10. Unfamiliarity with the side effects of MB alternatives in general on quality and marketing characteristics of treated dates' products. Application of certain alternative(s) might affect the taste, smell, color, degree of consistency total solids, sugar content, and other properties that may negatively affect the marketing value.
11. The importance of achieving and keeping chemical safety of food products within limits of international and/or national standards that does not endanger the health and lives of consumers. Dates are consumed primarily by human and contribute also in animal fodder through wastes and by-products of dates-processing operations, where animal products end up into the human food chain.

MB ALTERNATIVES

MB alternatives, for dates' treatment, can be divided into two main categories: the first category is nonchemical alternatives whereas the second category is chemical alternatives. Nonchemical alternatives are characterized as environmentally friendly alternatives and have no negative impacts on environment or to human health as products treated with those alternatives are free of chemicals' and pesticides' residues. However, the use of these alternatives need a technical experience in the application in order to prevent injury and efficiently eliminate pests that could infest the dates during storage period or manufacturing without any impact on quality characteristics and features of the products or on the marketing value of treated dates. A logical protocol to select, evaluate, and adopt an alternative is developed in Figure 16.1. Low temperatures (cold treatment), heat treatment (Figure 16.2) using relatively high temperatures, modified atmospheres (MAs), and irradiation are the most promising and potential nonchemical alternatives. Chemical alternatives include the use of certain chemical fumigants such as phosphine, sulfuryl fluoride, ethyl formate, and ethyl iodide. These alternatives may be used individually or in combination of two or more substances to ensure comparable efficacy and effectiveness to that of MB in preventing pests' and microbial infestation. However, reinfestation of fumigated dates has been observed while doubt arose about the efficiency of chemical alternatives as the results achieved through conducted experiments were obtained via maintaining precise experimental conditions that might not be applicable for common producers (Al-Abid et al., 2010). It is well

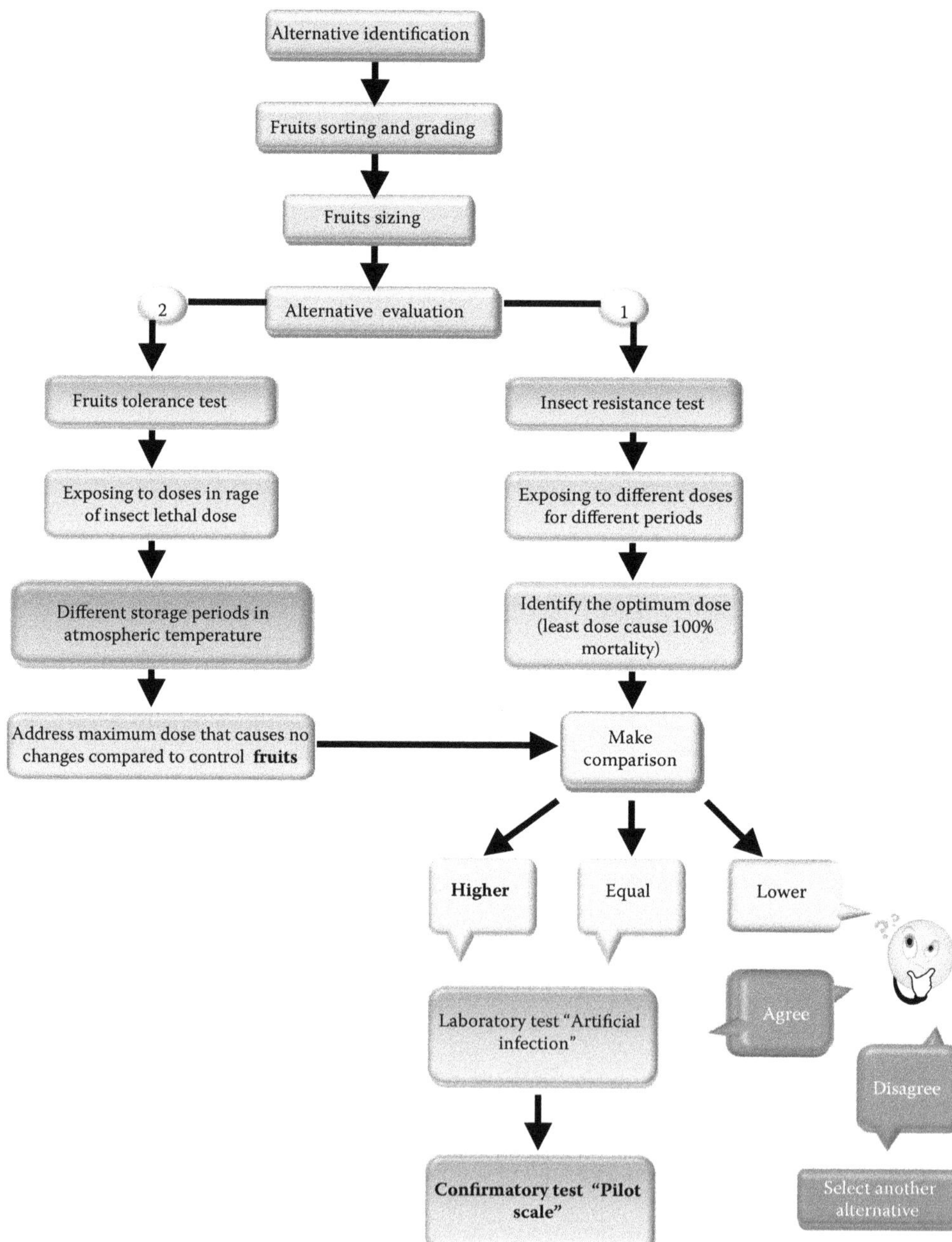

FIGURE 16.1 Steps to select and evaluate an alternative to MB.

FIGURE 16.2 Small unit for heat treatment of dates.

known that dates are vulnerable to spoilage and damage during the storage period as a result of undesirable changes in the physical and chemical characteristics that affect the virtual shape, taste, smell, and nutritional value of the fruits. Browning reactions are also considered of the common changes in dates during that period which, with other properties, affect validity of the products for consumption and marketing value of dates on domestic, regional, and international levels. In the present chapter, emphasis will be on nonchemical alternatives' category.

NONCHEMICAL ALTERNATIVES

The traditional practice of preserving dates, at "rutab" stage, for prolonged storage is based on pressing into containers. This operation represents preventive actions against insects, microbes, and hence deterioration as air will be removed and biological infestations will be restricted to a large extent. However, the resulting sticky mass is less suitable for introducing dates in international markets,

FIGURE 16.3 Dates samples irradiated in burlap sacks and foam plates.

FIGURE 16.4 Deglet Noor dates in branches.

where the consumer expects an individual date fruits. Hence, during the course of combating stored products pest, there have been at least a dozen of materials used as fumigants (Rajendran, 2001; Navarro, 2006) that keep that property for the fumigated dates. However, most of the available fumigants have limitations in use due to various reasons. For example, according to MP, MB is scheduled to be phased out worldwide by 2015 (UNEP, 2002). Sulfuryl fluoride is known to have limited efficiency on egg stages of insects and already declared as a potent greenhouse gas (Andersen, 2009). Phosphine is the most widely used fumigant due to its low cost and ease of use is threatened by the development of pests of high resistance (Schlipalius et al., 2006; Aurelio et al., 2007; Lilford et al., 2009). Methyl iodide is known as neurotoxic, developmentally toxic, and an endocrine disrupter (UNEP, Ozonews, 2010). Also, research for new fumigants becomes restricted due to concerns related to chemical residues in treated products and the environment. Accordingly, under the given limitations and constraints of fumigants, it is anticipated and recommended to give more attention to develop and adopt new nonchemical alternatives that are economically feasible, sustainable, safe for applicators, and environmentally friendly.

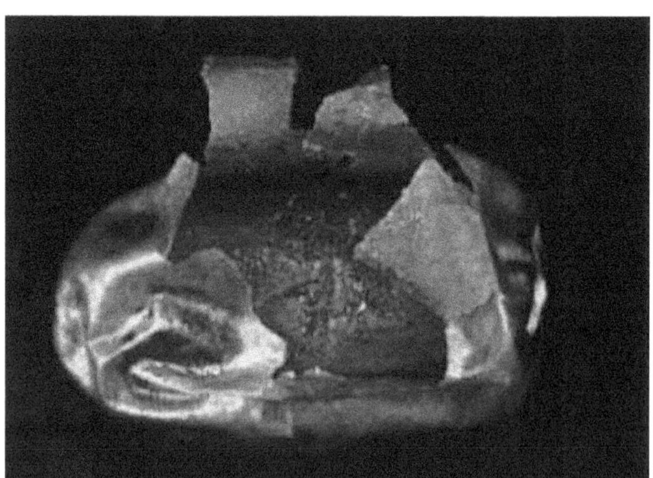

FIGURE 16.5 Blistering effect phenomenon.

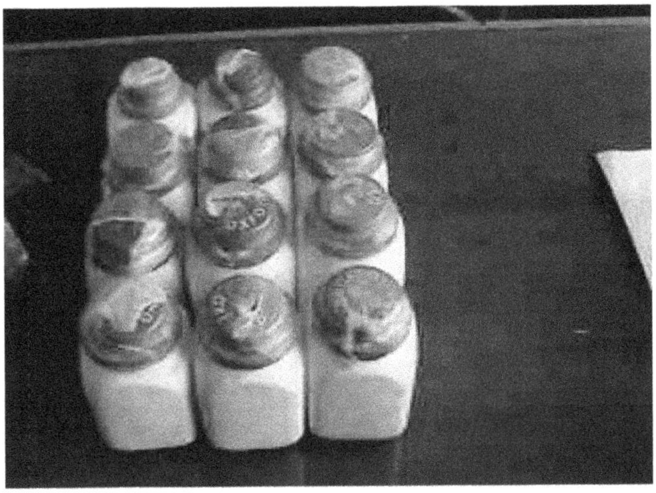

FIGURE 16.6 Dates samples irradiated in plastic bottles (El-Mohandes, unpublished data).

COLD TREATMENT

Prior to the use of MB and other fumigants, cold treatment or cool storage was used to a large extent in controlling wide range of insect pests that infest stored products including dates. Nowadays, pest control using cold treatment is receiving growing interest as a nonchemical approach to replace MB and to prolong shelf life of dates via stopping or retarding biochemical and biological processes through slowing down enzymatic reactions and biological activities of microorganisms and insect pests with an advantage of leaving no residues on treated products. The extent to which these activities go further or slow down depends mainly on dates characteristics, temperature, and moisture content. For example, for the latter factor, it has been indicated that dates of 20% moisture content

FIGURE 16.7 Accelerator.

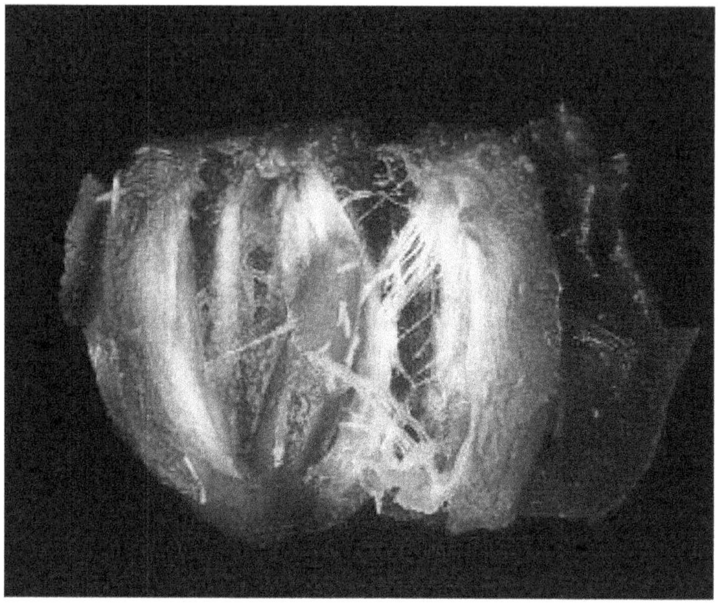

FIGURE 16.8 Nitidulid beetle larvae-infested date fruit.

can be kept for a year at 4.5°C while dates of 30% moisture content would not remain longer than 4 months at the same temperature (Barreveld, 1993).

This technique for dates' preservation is used in transit ports for export/import activities using cold storages (Figures 16.12 and 16.13); also a lot of dates' processing facilities in date-producing countries contain refrigerated warehouses to maintain dates' fruits after fumigation with other fumigants such as phosphine gas until processed and packed. This technology relies on introducing cold air into the storage through special cooling units placed inside the storage room and designed to adjust the degree of cooling and relative humidity to the required levels to suit the nature and quality characteristics of the stored variety. Some important factors should be considered when using cold treatment including dates variety, physical and chemical properties, nutritional characteristics, the marketing value, storage period or in other words the cooling period. Also, modality of

FIGURE 16.9 Confused flour beetle *Tribolium confusum*.

FIGURE 16.10 Suriname beetle *Oryzaephilus surinamensis.*

storage, that is, whether the fruits will be stored as loose bulk or packed products and the type and size of packaging used should be taken into account (UNEP, 2002).

Cold treatment is used efficiently in controlling tropical pests that infest the dates during storage as these candidates often characterized by its low resistance to cold treatment compared with those of temperate and cool regions which are highly resistant to low temperatures. Pests' resistance to low temperatures depends on the type of pest, growth stage, and its adaptation ability to temperature changes. Looking at the relationship between temperature patterns and behavior of warehouse pests in general, these pests cannot reproduce or increase numerically in temperatures below 18°C (except for Granary weevil, *Sitophilus granarius*, that can continue reproduction at temperatures as low as 15°C), at temperatures below 5°C pests are dormant and immobile whereas all warehouse pests dying at temperatures from −10°C to −18°C (Paul and Noel, 2002).

In practice, the common followed practice is storing dates in temperatures ranging from 2°C to 5°C or from zero to −5°C. However, it has been stated that in the range from 2°C to 5°C larvae of certain pests can stay for up to 85 days whereas the range from zero to −5°C proved not enough to kill all pests that infest dates such as *Carpophilus* spp. The larvae of this pest can resist and remain dormant under such temperature and then return active again when the temperature reaches higher levels that allow reactivity. Merchant grain beetles, such as saw-toothed grain beetle (*Oryzaephilus surinamensis*, Figure 16.10) and confused flour beetle (*Tribolium confusum*, Figure 16.9) that infest

FIGURE 16.11 Drying dates under plastic sheets.

dates during storage causing much damage over short time, are among the most sensitive pests to low temperatures. Currently, lower temperatures are used for storing dates as MB alternative by freezing at −10°C to −18°C which had greater impact on preventing pests' incidence (Donahaye et al., 1991). Freezing dates under such degrees (−10°C or −18°C) are more efficient on pests' mortality if the treatment is carried out in presence of low-oxygen level of 2.8% (Donahaye et al., 1992). Freezing at −10°C under low concentration of oxygen for 10.5 h was enough to kill all lesions and associated growth stages while same results were obtained after only 2.25 h when treated at −18°C with low-oxygen level. Also, freezing at low-oxygen level not only killed all pests, but led all pests to migrate from inside the fruits (Donahaye et al., 1992) to the external environment contributing in obtaining insect-free dates, hence increasing product acceptability and marketing value. Other effective regime for application of freezing technology for dates' preservation was tested in Saudi Arabia where stored dates subjected to temperature ranges from −25°C to −30°C for 7 days in a refrigerated warehouse followed by storing at −10°C for 3 days and then storing at 25°C for 3 days. Upon completion of this regime, treated dates were completely free of pests, fruits' temperature was similar to the ambient temperature and dates' fruits were then ready for subjecting to different manufacturing processes (El-Mohandes, 2010b).

It is recommended when using this technique is to expose the dates to relatively low temperature before storing under cooling to reduce the potential impacts of chilling injury on quality characteristics and properties of treated dates (Hatton, 1990; Wang, 1990). These effects can also be reduced or avoided by lowering oxygen and raising carbon dioxide levels before or during exposure to cooling temperatures (Forney and Lipton, 1990). Additionally, it is also recommended to maintain the proper relative humidity in the cold storage to prevent either drying out or moisture uptake of the date. Also, an important factor that would increase application efficiency of cold treatment as an alternative technology to MB for disinfestation of dates is the uniform distribution of cool air within the warehouse, especially in warehouses with large areas and storing capacities. This uniform air distribution can be achieved by using fans. However, some concerns are accounted on cold treatment that restricting the widespread of this technology, (1) it takes long time for dates' flesh to reach the required degree of cooling as a result of high proportion of total

FIGURE 16.12 Cold storage for dates disinfestation.

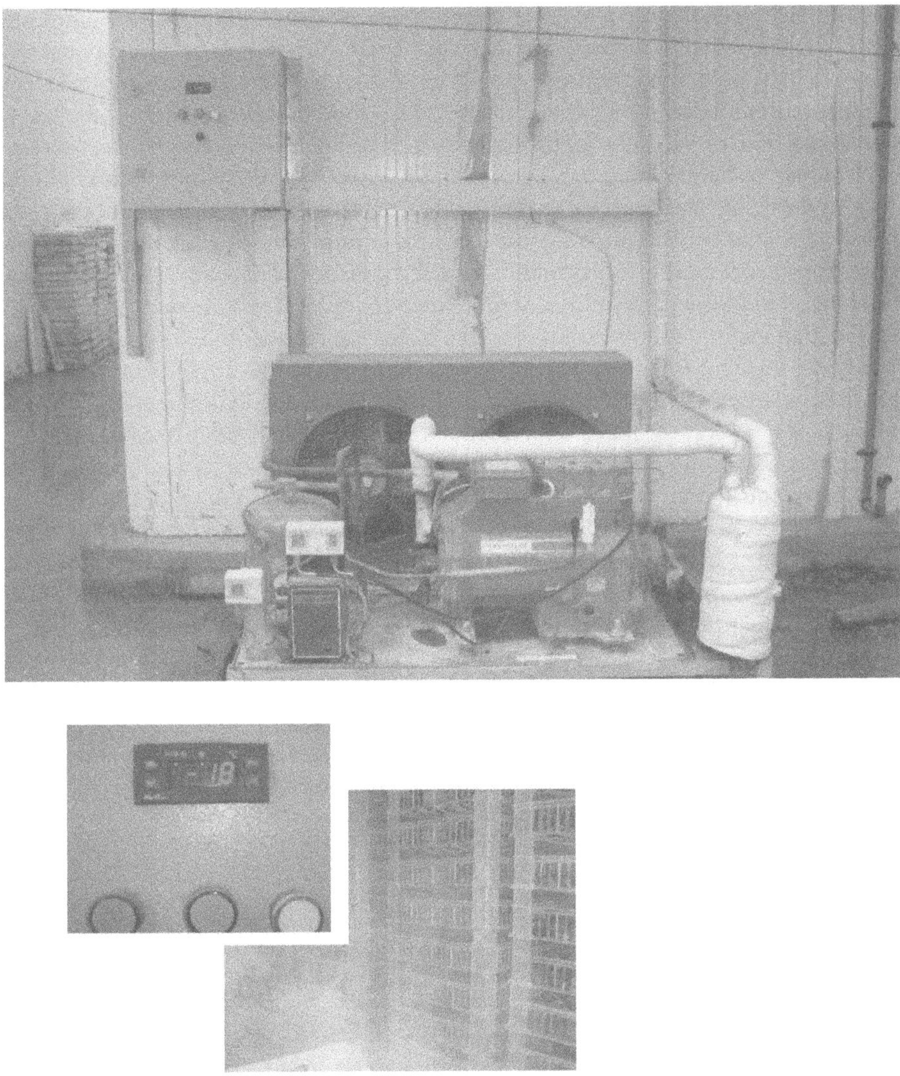

FIGURE 16.13 Cold storage for dates preservation at –18°C.

solids (sugars) in the fruits. Additionally, it also needs long time after treatment for the fruits to reach the atmospheric temperatures to be ready for manufacturing processes, and (2) the high consumption of energy that leads to raise the production's cost which in turn is reflected on the price of the final product.

HEAT TREATMENT

The use of heat in sterilization and killing of insect pests is also not a new technology as it was used in the past in sterilization of mills and other structures. It is also used as disinfestation treatment of dry and durable food products against pests, insects, and microorganisms. It has been accepted by the U.S. Department of Agriculture as disinfestation treatment against pests for perishables and is used as disinfectant against Mediterranean fruit fly (Couey, 1989) and stored products (Roesli et al., 2003).

It is well known that each type and development stage of insect pests has minimum, optimum, and maximum temperature for growth and activity. Table 16.1 indicates the response of stored-product insects to different temperature levels. These different levels of temperature sensitivity or resistance are important information that must be understood in order to address the level of thermal treatment and time of exposure to be applied to eliminate certain type(s) of pests need to be eradicated. The use of heat, which can be combined with CO_2, as a nonchemical alternative to MB is a highly efficient treatment in killing insect pests and microorganisms of stored dates (AlterBromide, 2008). It contributes to further the conservation and storage period without affecting the characteristics of quality, nutritional and market values of dates' products. Several advantages are attributed to the use of heat treatment of dates disinfestation of which is that the workers and technicians can get into the storage during the treatment process to fix any potential malfunctions ensuring the thermal treatment is proceeding properly, or to sample from the treated dates during and after heat treatment to evaluate the treatment efficiency. Also, heat is environmentally safe treatment and leaves no residues on treated dates, the problem often associated with chemicals. However, there are some factors that limit the spread of heat use as an alternative technique including (1) time required for the treatment, where it needs several hours for the fruits' temperature to reach the targeted level that eliminates pests, plus exposure time which is calculated when fruits' temperature reaches the targeted degree. This is in addition to posttreatment time required to lower the fruits' temperature to the ambient environment. In this respect, Navarro et al. (2004a,b) indicated that the target temperature of 50°C could be achieved within 1 h after introduction of the dates into the drier with an additional 2 h to achieve emigration and mortality of the pests, (2) the belief that heat treatment, or any other alternative, is not as efficient as MB and that it is not possible to achieve the same results obtained using this fumigant, (3) the unavailability of reliable economic feasibility studies that indicate the actual cost of heat treatment compared with MB, which help the producers and manufacturers of dates adopt appropriate and

TABLE 16.1
Response of Stored-Product Insects to Temperature[a]

Zone	Temperature (°C)	Effects
Lethal	62	Death in <1 min
	50–62	Death in <1 h
	45–50	Death in <1 day
	35–42	Population die out
Suboptimum	35	Development stops
	33–35	Slow development
Optimum	25–33	Maximum rate of development
Suboptimum	20–25	Slow development
	13–20	Development slow or stops
Lethal	3–13	Death in days (unacclimatized) and movement stops
	−10 to −5	Death in weeks to months if acclimatized
	−25 to −15	Death in minutes, insects freeze

Source: Banks, H.J. and Fields, P.G. 1995. Physical methods for insect control in stored grain ecosystems. In: D.S. Jayas, N.D.G. White, and W.E. Muir (Eds.), *Stored Grain Ecosystems*. Marcel Dekker, Inc., New York, NY, pp. 353–409; Fields, P.G. and Muir, W.E. 1996. Physical control. In: B. Subramanyam and D.W. Hagstrum (Eds.), *Integrated Management of Insects in Stored Products*, Marcel Dekker, Inc, New York, NY, pp. 195–221.

Note: Species, stages of development and moisture content of food will influence the response to temperature.

correct decisions, (4) the belief in negative impact of heat on both equipment and plant or facility where heat is applied.

One of the common ways for dates' disinfestation using heat treatment, in addition to the traditional method that depends on the use of direct solar radiation or drying under plastic sheets (Figure 16.11), is the use of hot air whether dry (with a relative humidity below 60%), wet (with about 90% relative humidity), or air saturated with water vapor (Krista and Mangan, 2000, http://aciar.gov.au/files/node/2248/PR050%20part%2014.pdf). The required heat levels can be generated by different means such as electrical, steam, or propane heaters (AlterBromide, 2008). Electric or steam heaters are placed inside the warehouse whereas propane gas heaters are placed outside the structure while hot air stream is introduced into the warehouse through a piping system. As the technology is developed to introduce the hot air faster (2 m or more per second), heat treatment now takes less time than before (Sharp et al., 1991), until the temperature of treated dates reaches the degree required to kill the pest without affecting the quality and physiological properties of the treated products. Some countries are currently developing heat-treatment technologies for different applications, including timber, dates and other treatments to reduce the economic cost aiming at spreading the commercial use of this technology in various sectors.

Some hypothesis illustrated the mode of action of heat treatment on pests that include phospholipids' membranes become more fluid, the structure of proteins and hence enzymes are adversely affected by high temperature, pH is temperature dependent, and water stress may be a crucial factor between 35°C and 43°C (Paul and Noel, 2002). The tests indicated that the use of temperature at 65°C for 1 min is enough to kill all life stages of different insects and that the exposure time required increases with temperature decline. To achieve 90% mortality of Nitidulids (Figure 16.8), the insect must be subjected to 49°C for 4–20 min depending on dates' moisture content, while *Carpuphilus hemipterus* that infests the dried fruits is more resistant to heat and therefore requires exposure time of 25–60 min at 50°C to reach 100% mortality. *Codra cautella* has medium tolerance to heat where 33 min of exposure to 60°C is required to achieve 100% mortality. From dates-processing technology's point of view, the appropriate temperature for dates' treatment ranges from 35°C to 55°C as temperature below 35°C increases the time needed to kill all growth stages of insects whereas temperature above 55°C causes chemical and physical changes (Navarro et al., 2004a,b; Finkelman, et al. 2006). Examples of these possible changes are discoloration and blistering effect (Figure 16.5), the phenomena that separates the skin from the flesh of the fruit. Dates' fruits usually require from 1 to 2 h to reach the required degree (50–55°C) depending on dates variety and related specific characteristics.

Some studies were carried out to assess the impact of heat as MB alternative on dates' storage and quality characteristics, migration of insects from fruits to external environment, preventing insect infestation and on larvae of *Carpuphilus hemipterus*. Heat treatment of dates at 55°C for about 2.5 h led to 100% migration of insects from the fruits and to 100% insect mortality of sap beetle (Coleoptera: Nitidulidae) thereby achieving similar results obtained using MB fumigation (Rafaeli et al., 2006). However, treatment of dates (Deglet Noor var.) at 50°C for 2 h caused 100% mortality of targeted pest but also caused a noticeable dry skin (Ducom and Ciesla, 2008). These findings clearly indicate that the marketing and quality properties of each variety should be considered before choosing heat treatment as an alternative application to MB. The immediate disinfestation of dates upon arrival to the processing facility is important step to break the chain of insect infestation and choose the treatment that drives the insect exit from fruits and migrate to the external environment as well as killing all populations and development.

There are some circumstances that should be considered to increase the efficiency of heat treatment, and hence achieve the targeted objectives, (1) preventing the entry of pests to the warehouses through proper hygiene procedures, (2) treatment of warehouse's wall cracks by a spray of insecticides before heat treatment to ensure elimination of hiding pests where the heat does not reach the required level in these places, (3) raising the temperature gradually in order not to affect quality properties of the dates variety, (4) calculating the exposure time when the fruits' flesh reaches the required level (5) using fans in the warehouses to ensure uniform distribution of the heat across the structure especially at ground level, (6) leaving dates' fruits after completion of heat treatment for several hours to reach the ambient temperature. The efficiency of heat treatment may also be checked using pheromone or food traps before and after the heat treatment.

Finally, when a decision is taken to apply heat or cold treatment as MB alternatives for dates preservation and storage, it is recommended to go through the following steps to assess the appropriate temperature and exposure time required to achieve dates' disinfestation, preserving quality characteristics and marketing values of treated cultivars:

1. Identify the type of pest(s) need to be eliminated.
2. Test the pest(s) resistance level to estimate the range of resistance to high or low temperatures to be used.
3. Test dates' resistance to temperature by exposing samples of dates to different temperatures for different exposure periods to determine both maximum temperature and length of time that do not give rise to a significant difference between characteristics of treated and untreated samples. To test the resistance of dates' fruits correctly you must:
   a. Classify the fruits according to the size (small, medium, and large) where fruit size affects the temperature level and exposure time required to kill pests without affecting characteristics of the product.
   b. After heat or cold treatment, keep the treated samples under normal temperatures to simulate what happens on ground during marketing, transporting, storing, and sale. After storing under normal temperature, assess the storage period of dates, incidence of infestation, quality characteristics, and marketing properties including color, hardness concentration of sugars, and total solids and weight change.
4. Perform confirmatory test, using predetermined temperature and time, on infected fruits after which mortality of different phases of the insect(s) and dates' quality properties should be evaluated.

MODIFIED ATMOSPHERES

The use of MAs has dated back to ancient Egyptians in the form of hermetic storage of grains by which grain itself through respiration creates an atmosphere rich in carbon dioxide and low in oxygen (White and Leesch, 1996). More recently, the concept has been used in different parts of the world for the fumigation of a variety of commodities including grains, pulses, tree nuts, dried fruits, coffee and cocoa beans, spices, medicinal herbs, geophytic bulbs and historic artifacts (Adler et al., 2000), vegetables, and fresh meats. More to the point, it has moved fresh products at both the wholesale and retail level from bulk storage to controlled atmosphere packaging (Testin, 1988). Applications of MAs/CAs (controlled atmosphere storage) are preferred for the treatment of organic food products and are performed mostly in transportable polyvinylchloride (PVC) structures, Cocoons or Volcanic Cubes. In museums, high MAs application is used to eradicate the structural pests (Emekci, 2010).

Controlled atmosphere is storage method in which concentration of the atmosphere major components (oxygen, carbon dioxide, and nitrogen) as well as temperature and humidity are regulated and is selectively controlled to increase shelf life. MA is the practice of modifying the composition of the internal atmosphere of a package in order to improve the shelf life (WikiAnswers, http://wiki.answers.com/Q/What_is_the_difference_between_ ontrolled_atmosphere_packaging_CAP_and_Modified_atmosphere_packaging_MAP). In the case of produce, package atmosphere is affected by the transmission rates of the packaging material and changes in storage temperatures. Higher temperatures lead to higher respiration rates, creating lower O_2 levels in the package atmosphere and higher concentrations of CO_2. Hence, the atmosphere inside the package is modified but not controlled (http://www.ppcind.com/pdfFiles/MAP%20vs.%20CAP.pdf). The technique relies on changing or modifying the composition of the atmospheric gases in the facility creating a desired environment to control storage pests, increasing the shelf life of the products through slowing down the respiration, and hence, keeping the quality of the stored products. This can be achieved by increasing the concentration of CO_2 and/or N_2 and lowering O_2 at normal or altered atmospheric pressure for a certain exposure time enough to kill the targeted insect pests in the storage. Also, the natural effects of respiration (commodity, molds or insects) to reduce O_2 and increase CO_2 which is known as Hermetic storage can be used. The use of MAs as a treatment for eradicating insect pests and prolonging shelf life of stored produce is influenced by insect response and commodity tolerance. The insect response to MAs depends on the species, development stage and age, atmospheric temperature, relative humidity, the partial pressure, time of exposure, and atmospheric composition used as treatment (Navarro and Donahaye, 2005). Commodity tolerance to MAs is influenced by storage temperature, levels of O_2 and CO_2, commodity respiration rate, commodity resistance to gas diffusion, and commodity-soluble solids content (Ke and Kader, 1991). Some beneficial effects on taste and texture of treated products, as a result of the dissolution of CO_2 into moisture at the tissue surface, can be achieved. Also, the absorption of CO_2 will lower pH, reducing the growth opportunity of harmful organisms (Testin, 1988).

To have an insecticidal toxic effect of MAs on storage pests, a high CO_2 concentration must be contained in the atmosphere of the treatment enclosure, often >60% (Navarro and Donahaye, 2005) with low O_2 concentration below 3% and preferably <1% if rapid kill is required (Banks and Annis, 1990; Fleurat, 1990). This statement was confirmed by a preliminary laboratory experiment conducted to evaluate the use of CO_2 in air as MB alternative for disinfestation of stored dates (var. Sewi, semidry, 22%–24% moisture content) at Tamr stage and its effect on controlling adults of *Oryzaephilus surinamensis* and *Tribolium confusum* (Hashem et al., unpublished data). The results indicated that 100% mortality was achieved after 36 h at CO_2 concentrations of 75% whereas at CO_2 concentrations of 65% and 85% mortality of 100% was achieved after 48 h exposure. In Israel, dates served as a model to demonstrate the feasibility of storing dried fruit for 4.5 months under atmosphere of 60%–80% CO_2 in the chamber in bulk or packed on pallets containing dates of vars. Hallawi, Hadrawi, Zahidi, Derei, and Ameri. The results indicated that insect population was effectively controlled and the quality of the treated dates insignificantly changed compared to controls stored at −18°C (Navarro et al., 2001). However, other research on disinfestation of figs showed that MAs applications using high CO_2 in flexible PVC units (Volcani cubes) were effective but exposure time of 5 days was needed to attain a complete kill (Emekci et al., 2007).

Combined applications of MAs with other treatments had significant effect on reducing disinfestation exposure time to 4 h such as combination with cold and low-pressure treatments (Navarro et al., 1998) as well as elevated heat treatment (Donahaye et al., 1994; Whiting et al., 1991). Combined MAs with cold storage and heat treatments offer little advantage over insecticidal effect to offset their additional costs. Also, combined application with high positive pressure ranging between 10 and 37 bars enhanced the performance of MAs where significant reduction in exposure time to a few hours was obtained (Emekci, 2010). The cost of high-pressure chambers limits the use of this method only to the treatment of valuable commodities such as dried fruits, nuts, spices, herbs, and cocoa beans (Adler et al., 2000; Navarro, 2006). Moreover, chemical fumigants may be used in combination with MAs to reduce the lethal exposure time and maintain the dates quality. Phosphine tablets in the form of aluminum phosphide were used in combination with MAs (CO_2 in air) to reduce the exposure time and achieve the 100% mortality of postharvest pests, that is, *Oryzaephilus surinamensis* and *Oryzaephilus surinamensis* and preserve quality of treated dates. Application of CO_2 at concentrations of 55%, 65%, 75%, and 85% with half-dose of PH_3 caused 100% mortality, of all stages of tested insects, after only 6 h of exposure. The use of phosphine tablets alone at the recommended dose did not cause complete mortality to either stage of any insect when exposed to the same period. The treatments slightly reduced content of free phenols, total sugars, and reducing sugars content were increased whereas total amino acids remained unchanged (Hashem et al., unpublished data).

VACUUM TREATMENTS

Vacuum, or low pressure, is one of the potential alternate treatments for disinfestation of durable and perishable commodities. Several investigators studied the use of vacuum as disinfestation treatment (Bare, 1948; Calderon et al., 1966; Calderon and Navarro, 1968; Navarro and Calderon, 1979; Friedlander and Navarro, 1983; Mbata and Phillips, 2001; Mbata et al., 2004; Al-Abid et al., 2010). Low pressure kills insects through achieving low-oxygen concentration in treatment enclosure where the primary mechanisms of action of vacuum treatment on insect survival are shown to be lack of oxygen and dehydration due to removal of water vapor (Navarro, 2006) or structural breach of cell membranes (Navarro and Calderon, 1979). Previous work indicated that vacuum of 50 mm Hg, equivalent to 1%–2% oxygen, can kill insect stages (eggs and pupae) between 12 and 200 h at temperature from 37°C down to 5°C. Treatment time can be reduced to <4 h at 25°C when vacuum at 50 mm Hg combined with fumigation. Experiments with eggs of the Indian meal moth, *Plodia interpunctella*, lesser grain borer, *Rhyzopertha dominica*, found that application of ethyl formate with vacuum confirmed the later fact (Thomas et al., 2007). Other disinfestation treatment of post-harvest tested vacuum combined with propylene oxide (Navarro et al., 2004a,b). The need for massive, rigid, and expensive structures withstanding the low pressures was the main barrier in using low pressures at the large-scale commercial level. With the development of flexible plastic units, GrainPro Cocoons®, sufficiently low pressures (25–50 mm Hg absolute pressure) to kill the insects can be obtained using a commercial vacuum pump and maintained for indefinite periods of time (Finkelman et al., 2003). This technology, known as vacuum-hermetic fumigation (V-HF), is currently in use at commercial level for pest treatment of organic soybeans and flours in Israel (Navarro, 2006), and, accordingly, is recommended to be tested with dates for the purpose of disinfestation and quality preservation.

IRRADIATION

Irradiation is a physical treatment through which energy sources of various types are used to eliminate the different development stages of pests using low doses that affect neither consumer's health nor the properties of treated product and commodities. Sterilization using irradiation is called "cold sterilization" where such treatment does not heat up the treated products. Irradiation includes the use of ionizing or nonionizing radiation. Ionizing radiation is electromagnetic energy that generates electrically charged particles in treated products. Examples of these rays are γ rays, x-rays, and accelerated electrons. Nonionizing radiation, such as microwave radiation, is electromagnetic waves or photons with frequencies between 300 MHz (0.3 GHz) and 300 GHz and of energy <12 eV that do not ionize the exposed materials or products (Wikipedia).

γ- and x-rays are also highly efficient for use in quarantine applications because of the high penetration capacity through treated goods and products, taking into account that Gamma has more penetration capacity compared to x-rays. Gamma rays generate from Cobalt-60 or Cesium-137 and have frequency of $>10^{19}$ H (Wikipedia) and energies above 100 keV and wavelength <10 p.m. X-rays are of wavelength in the range of 0.01–10 nm and energies in the range of 120 eV to 120 keV. Therefore, γ-rays are shorter and therefore have more applications in agriculture sector. Microwave radiations are of long wavelengths range from as long as 1 m to as short as millimeter.

When adopting irradiation technology as disinfestation treatment (e.g., Figure 16.7), it is essential to determine the radiation dose required. Radiation dose is the amount of energy absorbed by irradiated products. It is important that the minimum dose absorbed in the fruits is enough to kill all growth stages of all pests and microorganisms whereas the upper limit of the absorbed dose in fruits is less than the dose that could cause unwanted effects on physical, chemical, and virtual characteristics of treated products/fruits. Radiation dose is measured by Gray (Gy) or KiloGray (kGy). Food and Agriculture Organization/International Atomic Energy Agency (FAO/IAEA) promotes the use of radiation, recommending minimum doses of <4000 Gy for nematodes, <300 Gy for most insects,

100 Gy for *Diptera*, and 150–320 Gy for mites (Paul and Noel, 2002) as at such levels most fruits are not damaged (TEAP, 2000). Commodity tolerance to irradiation varies according to growing location and season, cultivar, fruit maturity, irradiation dose, and irradiation dose rate (Burdit, 1982).

Irradiation technology for food in general and dates in particular gained growing interest due to its close relevance to the food security file for countries and populations, the hot file which drew attention of the international community to reduce the poverty and malnutrition, especially in developing countries. This file is closely linked to agricultural production, the sector in which date production represents one of its pillars in many countries, particularly the Arab region. Dates, like any other agricultural product, are characterized by seasonality and short-term storage as it is perishable and undergo both quantitative and qualitative losses (i.e., amount produced, the nutritional value, and quality characteristics) which are clearer in soft and semidry varieties. Accordingly, irradiation is considered of the promising technologies that are increasingly adopted worldwide as highly effective preservation technique because of its efficacy in prolonging storage period and conservation of irradiated products, including dates. The lethal effect of irradiation on pests and microbes is due to changes in the nature of the DNA, cell membranes, proteins, and enzymes and thus disruption in cell functions that lead to death. Physiological and biochemical studies on the red flour beetle, *Tribolium castaneum* (Coleoptera: Tenebrionidae), indicated that peroxide values, total free amino acid content, protein subunits molecular weights, and the amino acid composition were changed when exposed to radiation leading to insect's DNA damage (Lu et al., 2010).

However, despite FAO/IAEA/WHO Study Group on High Dose Irradiation concluded that foods treated with doses >10 kGy can be considered safe and nutritionally adequate when produced under good manufacturing practice (FAO/IAEA/WHO, 1999), this technology is still not accepted in majority of developing countries as disinfestation treatment that could replace MB in dates' industry. Some doubts of adopting the commercial application of this technology for disinfestation of dates rely on different factors including:

1. Public reluctance to consume irradiated dates due to possible negative health impacts and implications of treated products.
2. Presence of insect development stages inside the irradiated products as irradiation does not lead the pest to emigrate the product to the external environment.
3. Adopting irradiation technology requires Ministerial decree and concomitant decisions and legislation to ensure safety application and security.
4. The need for high expertise in applications, practices, and maintenance procedures.
5. Relatively high costs of irradiation treatment compared with other chemical treatments.

The MP regime and MB phasing-out schedule have encouraged research on food irradiation (Esteves et al., 1999; Johnson and Marcotte, 1999; Hallman, 2000, 2001; Kwon et al., 2004; UNEP, 2006), which has been identified as an alternative (IPPC, 2003) due to its important role in agricultural development in general and in dates' industry in particular. Application of irradiation in dates-processing aims mainly at:

1. Contributions to food security of communities, particularly in developing countries, by reducing postharvest losses of dates which in turn contribute in maximizing the benefits through reducing both economic losses and price rates, and hence, reduce the gap between food supply and demand.
2. Securing food safety through providing insects/microbes-free dates. Insect injury of dates undoubtedly negatively affects acceptability of such dates and thereby increases losses. On the other hand, microbial infection where fungi are the most dangerous makes it more likely these dates contain aflatoxins that cause human diseases when consumed.

3. Prolonging dates' preservation periods with maintaining nutritional value, quality characteristics, and marketing values of different varieties.
4. Compliance with the requirements of regional and international trade treaties that require pesticide-free products as well as quarantine requirements in accordance with laws and regulations of each State. This is to ensure that imported dates are free of all forms of insect injuries and harmful microorganisms to prevent entry of such pests to the importing countries.
5. Provide an alternative to the soon phased-out fumigant, and hence, assisting States and Governments to fulfill their commitments to phase out the use of this ODS in accordance with the terms of the MP to protect the ozone layer.
6. The use of irradiation as safe replacement to different chemicals used for treatment and fumigation of dates and hence ensures chemical-free products.
7. The need to adopt new technologies to control postharvest pests and its growing resistance to chemical pesticides which requires increased doses that have harmful impacts on consumers.

Determination of suitable irradiation dose is of particular importance for disinfestation of stored dates, based on specific characteristics of each variety, where low doses ineffectively disinfest treated dates whereas high doses may cause not only a rapid death of pests but also unfavorable quality changes and health impacts and increase the costs. Accordingly, irradiation experiments focus particularly on determination of the least effective dose against insects' development stages, doses effects on dates' quality and its impact on consumer's health. The Joint FAO/IAEA/WHO study group (1999) concluded that food irradiated to any dose appropriate to achieve the intended technological objective is both safe to consume and nutritionally adequate, but we would say it must not exceed 10 kGy. However, it has been stated that maximum permitted irradiation dose for preservation of dates is 2.0 kGy (IAEA, 1991). Depending on the level of security required in commercial operations, fruit could receive up to three times the minimum absorbed dose for disinfestation (Hallman and Martinez, 2001).

The effect of irradiation on chemical properties of several dried foods has been studied (Duarte and Rupnow, 1994; Byun et al., 1995; Ahmed, 2001; Farkas, 2001; Al-Bachir, 2004); however, not much information is available on the effect of irradiation dates' preservation (Al-Kahtani et al., 1998). Few conducted experiments aimed at determining the effective dose for dates' disinfestation and/or its effects on chemical composition and quality properties. With regard to the effective dose, tests on preventing the development of *Oryzaephilus surinamensis, Ephestia cautella*, and *Batrachedra amydraula* in dates at all stages of their development were conducted showing 100% mortality obtained with irradiation at 20 krad (1 krad = 10 Gy) at the most damaging developmental stages of the species, except for late-instar larvae, which show a high resistance to γ-radiation (El-Sayed and Baeshin, 1983; Ahmed, 1981). The latter conclusion has given rise to proposals for combined treatments such as fumigation/irradiation, heat/irradiation (Ahmed et al., 1982, 1986), cool storage/irradiation (Wahid et al., 1987), all aimed at keeping radiation to the minimum and exploiting the complementary effectiveness of the control methods. Irradiation dose of 3.0 kGy was efficient in preventing growth of insects and fungi as well as production of aflatoxins to obtain dates-free poisons (Emam et al., 1994). Recent study has shown that a minimum dose of 0.45 kGy was required to cause the death of *P. interpunctella* in dates, "Boufeggous" variety (Azelmat et al., 2005).

Compositional and quality changes as a result of irradiation indicated that doses up to 150 krad retained the initial aroma of acid and nonacid fractions and that no major qualitative or quantitative changes in other components were detected (Jaddou and Al-Hakim, 1977, 1978). The same conclusion was recorded with Iraqi dates irradiated with dose of 500 krad as no significant changes in carbohydrates, protein, amino acids, carotenoids, or pectin content were detected (Auda, 1980). In Morocco, irradiating dates' samples (Medjoul var.) using doses of 0.6, 0.9, 1.8 kGy of γ-rays caused insignificant differences between irradiated and nonirradiated samples in dry matter, fats, and protein contents. However, amino acids and dry matter content were reduced in irradiated samples after 8 months of storage compared to control samples. Irradiation using dose of 0.9 kGy increased the proportion of total sugars and glucose while fructose remained unaffected in samples stored for 8 months (Azelmat et al., 2006).

The impact of irradiation doses used for dates disinfestation on human health was studied, indirectly on insects. Interaction between γ-radiation (100 krad) and phosphine was tested for possible toxicological effects by feeding treated dates to *Ephestia cautella*. Results showed a comparable development, fecundicity, and fertility of the insects fed on the control, fumigated, irradiated, and fumigated/irradiated dates (Al-Hakkak et al., 1985). Long-term feeding of irradiated dates (200 krad) on five generations of *Ephestia cautella* did not show significant differences in development from egg to adult, female fecundicity, mating frequency, and egg hatch leading to the conclusion that the irradiated diet had no genetic effects (Al-Hakkak et al., 1984).

Other experiments focused on studying dates disinfestation as well as chemical and quality changes. γ-Irradiated dates (Sewi var.) packed in different packaging materials (kenaf bags, plastic bottles [Figure 16.6], and foam [Figure 16.3]) artificially infected by both adults and larvae of flour beetle (*Tribolium confusum*) were tested for disinfestation efficiency and quality changes. The results showed that irradiation dose of 1.0 kGy did not achieve 100% mortality of insect immediately after irradiation but this was achieved 8 days postirradiation. Irradiation dose of 2.5 kGy led to immediate 100% mortality after treatment. Irradiation reduced the content of free phenols and total sugars compared to control untreated samples while content of reducing sugars was increased in irradiated samples (El-Mohandes, 2010c).

In West Asian countries, where table dates are a major diet, limited experiments had been conducted on disinfestation of dates using Gamma rays and microwave energy to eliminate pests and prolong conservation. An experiment using gamma rays was conducted in Saudi Arabia on dates (Khalas var.) artificially infected by saw-toothed grain beetle (*Oryzaephilus surinamensis*), flour beetle (*Tribolium confusum*), or combination of the two insects. Exposing fruits to doses of 0.3 kGy was enough to kill populations in samples infected with one type of insects while exposing fruits to doses of 0.9 kGy were sufficient to disinfest samples infected with combination of both. Irradiation contributed in lowering the microbial populations on irradiated fruits for 6 months posttreatment. Regarding the effect of irradiation on sensory and quality characteristics, the results showed that irradiation had no effect on treated fruits; however, proportion of sugars have relatively declined after treatment and increased gradually with increasing storage period by Al-Kahtani et al. (1998). In other experiment on Sefri and Sokkari var., treatment of dates with dose of 0.25 kGy prevented the development of eggs, larvae and the pupae of saw-toothed grain beetle, lesser and great date moth to the next stage and was lethal for full insect stage. This dose did not cause any changes in the nutritional value, taste, or flavor of treated varieties. In third experiment, exposing dates' samples (Al Shebi var.) to doses of 4–6 kGy reduced numbers of microbes from 4.9×10^4 to <10 cell/g without affecting the taste of treated samples. In Jordan, irradiating dates (Medjoul var.) using 0.25 kGy stopped development of eggs and all life stages of *Oryzaephilus surinamensis* without affecting the nutritional value, taste, and smell properties of dates samples stored for 12 months after treatment (UNEP/ROWA, 2010, unpublished report).

On the other hand, dates undergo physiological changes during storage, that is, darkening. It is also known as browning reactions, a phenomenon that should be prevented due to its adverse effect on acceptability and marketing value. It is caused by three mechanisms, enzymatic oxidative browning (polyphenolase enzyme), nonenzymatic oxidative browning by involving the more complex tannins or by nonoxidative browning in which sugars and proteins play a role (Maillard reaction). There is a positive correlation between the degree of darkening and increased moisture content, storage temperature, and time as clearly demonstrated from experimental work (Rygg, 1957). Preventive measures to reduce darkening are focused on enzyme inactivation (heat treatment), exclusion of air (pressing, or storage in an inert gas atmosphere), and low-temperature storage. In this respect, microwave energy was tested for dates' disinfestation and discoloration. Microwaves can be used to disinfest pests at 915 MHz, but great care must be taken not to overheat the fruits (Ikediala et al., 1999). A portable pallet irradiator has been developed to treat up to 4160 kg of product/h and tested in the United States (TEAP, 2000). In Tunisia, a study has been conducted to assess the use of microwave as an alternative to MB in controlling moth (*Ectomyelois ceratoniae*) as a requirement for dates (Deglet Nour var.) export to Western countries. Thermal properties, quality characteristics, and effect on larvae in treated samples were examined. The results showed that microwave radiation was absorbed in soft dates (25% moisture content) more than dry dates leading to higher temperature, indicating that fruit content of moisture has significant impact on heating efficiency and pests' killing. When dates of homogenous moisture content exposed to microwave for 55 s, the internal and external temperature of treated dates reached >52°C and killed 100% of pest's larvae and eggs. Exposing a mixture of soft fruits (25%) and dry fruits (13.5%) to microwave energy for 90 s heated soft dates to >100°C, while dry fruits temperature was less than the degree required to kill the insect larvae. The chemical analyses also showed that samples exposed to microwave energy did not differ in their characteristics from that of nonexposed samples (Zouba et al., 2009).

Electron beams were also tested for its efficacy on dates' disinfestations as several studies reported its mortal effect on common pests (Salimov et al., 2000; Cleghorn et al., 2001; Todoriki et al., 2006; Al-Abid et al., 2010). Previous studies found electron beam irradiation doses of 0.5–1.0 kGy maintained the quality of treated fruits and had no adverse effect on its physico-chemical and nutritional properties (Moreno et al., 2006; Egea et al., 2005). Comparison study to evaluate the effect of electron beams versus microwave, steaming, and fumigation by phosphine on dates' disinfestation, microbial and mold counts and quality properties of treated samples concluded that electronic beams reduced the microbial loads, caused 100% mortality of pests, maintained quality of treated samples and increased concentration of antioxidants compared with control treatment. Electron beams doses of 1.5 MeV/1.0 kGy and 1.5 MeV/2.0 kGy had lowest total microbial and mold populations in comparison with other treatments (Al-Abid et al., 2010).

CONCLUSION

Worldwide ban on MB is no doubt threatening the future of dates industry particularly in Arab region where dates represent a major diet in these communities. Currently, alternative chemical fumigants such as phosphine and sulforyl fluoride are the fumigants of choice due to its low cost, availability, ease of use, and global acceptance as a residue-free treatment. However, in the case of phosphine, some postharvest insect species acquired resistance against phosphine which increased chances for other insects to develop resistance. The use of nonchemical alternatives is very promising approach for controlling insect pests as it introduces harmless, technically effective, economically feasible, and environmentally friendly technologies that produce chemical-free dates products.

REFERENCES

Adler, C., Corinth, H.G., and Reichmuth, Ch., 2000. Modified atmospheres. In: Bh. Subramanyam and D. Hagstrum (Eds), *Alternatives to Pesticides in Stored Product IPM*, Kluwer Academic, New York, USA, pp. 105–146.

Ahmed, M., 1993. Up-to-date status of food irradiation. *Radiation Physics and Chemistry*, 42, 245–251.

Ahmed, M., 2001. Disinfestation of stored grains, pulses, dried fruits and nuts, and other dried foods. In: R.A. Molins (Ed), *Food Irradiation: Principles and Applications*, Wiley, New York, pp. 77–112.

Ahmed, M.S., Hameed, A.A., and Kadhum, A.A., 1986. Disinfestation of commercially packed dates by a combination treatment. *Acta Alimentaria*, 15(3), 221–226.

Ahmed, M.S.H., 1981. Investigations on insect disinfestation of dried dates by using gamma radiation—A review. *Date Palm Journal*, 1(1), 107–116.

Ahmed, M.S.H., AL-Hakkak, Z.S., AL-Maliky, S.K., Kadhum, A.A., and Lamooza, S.B., 1982. Irradiation processes in food irradiation. *Food Science and Technology Abstracts*, 14, 8C 331 J.

Al-Abid, M., Gohs, U., Al-Farsi, M., Al-Amri, M., and Al-Rawahi, F., 2010. Disinfestation of dates using electron beams in comparison with other treatments. *Proceedings of the Fourth International Date Palm Conference*, Acta Horticulture (ISHS), No. 882, pp. 569–576.

Al-Bachir, M., 2004. Effect of gamma irradiation on fungal load, chemical and sensory characteristics of walnuts (*Juglans regia* L.) *Journal of Stored Products Research*, 40, 355–362.

Albemarle Corporation, 1999. Methyl Bromide—Technical. http://www.albemarle.com/TDS/Fine_chemistry-services/BC0060F_Methyl_Bromide.pdf

Al-Hakkak, Z.S., Ali, S.R., Ahmed, M.S.H., and AL-Maliky, S.K., 1984. Wholesomeness studies with a full diet of irradiated dates using the insect *Ephestia cautella* (Walker). III: Effects of long-term feeding. *Date Palm Journal*, 3(1), 337–347.

Al-Hakkak, Z.S., Hussain, A.F., and Murad, A.M.B., 1985. Effect of gamma irradiation of phosphine-fumigated dates on the development of *Ephestia cautella* (Walker) (Lepidoptera: Pyralidae). *Journal of Stored Products Research*, 21(1), 21–23.

Al-Hooti, S.N., Sidhu, J.S., Al-Saqer, J.M., and Al-Othman, A., 2002. Chemical composition and quality of date syrup as affected by pectinase/cellulase enzyme treatment. *Food Chemistry*, 79, 215–220.

Al-Kahtani, H., Abu-Tarboush, H.M., Al-Dryhim, Y.N., Ahmed, M.A., Bajaber, A.S., El-Shami, Adam, E., and El-Mojaddidi, M.A., 1998. Irradiation of dates: Insect disinfestation, microbial and chemical assessments, and use of thermoluminescence technique. *Radiation Physics and Chemistry*, 53(2), 181–187.

AlterBromide, 2008. Alternatives: Heat treatment. http://www.alterbromide.org/albro/rw/document/item1309.do.

Andersen, M.P.S., 2009. Sulfuryl fluoride: Atmospheric chemistry and global warming potential. In: *2009 Annual International Research Conference on Methyl Bromide Alternatives and Emissions Reductions*, pp. 9–13, November 2009, San Diego, California, USA, 1–1.

Auda, H., 1980. Date irradiation in Iraq. *Food Irradiation Information*, No. 10, Baghdad.

Aurelio, M., Pimentel, G., Faroni, Leda R. D'A, Totola, M.R., and Guedes, N.R.C., 2007. Phosphine resistance, respiration rate and fitness consequences in stored-product insects. *Pest Management Science*, 63, 876–881.

Azelmat, K., ElGarrouj, D., Mouhib, M., and Sayah, F., 2006. Irradiation of 'Boufeggous' dates: Effects on chemical composition during storage. *Postharvest Biology and Technology*, 39(2), 217–222.

Azelmat, K., Sayah, F., Mouhib, M., Ghailani, N., and ElGarrouj, D., 2005. Effects of gamma irradiation on fourth-instar *Plodia interpunctella* (Hübner) (Lepidoptera: Pyralidae). *Journal of Stored Products Research*, 41, 423–431.

Banks, H.J. and Annis, P.C., 1990. Comparative advantages of high carbon dioxide and low oxygen types of controlled atmospheres for grain storage. In: M. Calderon and R. Barkai-Golan (Eds), *Food Preservation by Modified Atmospheres*, CRC Press, Boca Raton, FL, pp. 93–122.

Banks, H.J. and Fields, P.G., 1995. Physical methods for insect control in stored grain ecosystems. In: D.S. Jayas, N.D.G. White and W.E. Muir (Eds), *Stored Grain Ecosystems*, Marcel Dekker, Inc., New York, pp. 353–409.

Bare, C.O., 1948. The effect of prolonged exposure to high vacuum on stored tobacco insects. *Journal of Economic Entomology*, 41, 109–110.

Barreveld, W.H., 1993. *Date Palm Products*, Food and Agriculture Organization of the United Nations Agricultural Services Bulletin No. 101.

Burdit, A.K., 1982. Food irradiation as a quarantine treatment of fruits. *Food Technology*, 36, 51–62.

Byun, M.W., Kang, I.J., Kwon, J.H., Hayashi, Y., and Mori, T., 1995. Physicochemical properties of soybean oil extracted from gamma irradiated soybeans. *Radiation Physics and Chemistry*, 46, 659–662.

Calderon, M. and Navarro, S., 1968. Sensitivity of three stored-product species exposed to different low pressures. *Nature*, 218, 190.

Calderon, M., Navarro, S., and Donahaye, E., 1966. The effect of low pressures on the mortality of six stored-product species. *Journal of Stored Products Research*, 2, 135–140.

Cleghorn, D.A., Nablo, S.V., Ferro, D.N., and Hagstrum, D.W., 2001. Electron beam treatment parameters for control of stored products insects. *Radiation Physics and Chemistry*, 63(3), 575–579.

Couey, H.M., 1989. Heat treatment for control of postharvest diseases and insect pests of fruits. *HortScience*, 24, 198–202.

Donahaye, E., Navarro, S., and Rindner, M., 1994. The influence of temperature on the sensitivity of two nitidulid beetles to low oxygen concentrations. In: E. Highley, E.J. Wright, H.J. Banks, and B.R. Champ (Eds), Stored products protection. *Proceedings of the Sixth International Working Conference on Stored-product Protection*, pp. 17–23, April 1994, Canberra, Australia, CAB International, Wallingford, UK, pp. 88–90.

Donahaye, E., Navarro, S., and Rindner, M., 1991. The influence of low temperatures on two species of *Carpophilus* (Col., Nitidulidae). *Journal of Applied Entomology*, 111, 297–302.

Donahaye, E., Navarro, S., Rindner, M., and Dias, R., 1992. An artificial feeding site to investigate emigration of nitidulid beetles from dried fruits. *Journal of Economic Entomology*, 85(4), 1990–1993.

Duarte, P.R. and Rupnow, J.H., 1994. Gamma irradiated dry bean (*Phaseolus vulgaris*) starch: Physicochemical properties. *Journal of Food Science*, 59, 839–843.

Ducom, P. and Ciesla, Y., 2008. Preliminary tests on quality of high moisture dates with five alternatives to methyl bromide. http://mbao.org/2009/proceedings/071DucomPCieslaMBAO2009AlternativesMeBrdates.pdf.

Egea, M., Murcia, M., Sanchez-Bel, P., Romojaro, F., and Martinez-Madrid, M., 2005. Effect of electron beam ionization on shelf life of apricot. *Acta Horticulturae*, 682, 1211–1218.

El-Mohandes, M.A., 2010a. Dates' processing between degradation and challenge in light of the montreal protocol. *The Blessed Tree*, 2(2), 36–39.

El-Mohandes, M.A., 2010b. Methyl bromide alternatives for dates' production and processing in the Arab world. *The Blessed Tree*, 2(3), 44–49.

El-Mohandes, M.A., 2010c. Methyl bromide alternatives for dates' disinfestation. *Proceedings of the Fourth International Date Palm Conference*. Acta Horticulture (ISHS), No. 882, pp. 555–562.

El-Sayed, S.A. and Baeshin, N.A., 1983. Feasibility of disinfestation of date fruits produced in Saudi Arabia by gamma irradiation. *Proceedings of the First Symposium on the Date Palm*, Saudi Arabia.

Emam, O.A., Farag, S.E.A., and Hammad, A.I., 1994. Comparative studies between fumigation and irradiation of semi-dry date fruits. *Food/Nahrung*, 38(6), 612–620.

Emekci, M., 2010. Section: Fumigation, modified atmospheres and hermetic storage. *10th International Working Conference on Stored Product Protection*. Julius-Kühn-Archiv, 425, pp. 303–313.

Emekci, M., Ferizli, A.G., Tütüncü, S., and Navarro, S., 2007. The applicability of controlled atmospheres as an alternative to methyl bromide fumigation of dried fruits in Turkey. In: E.J. Donahaye, S. Navarro, C. Bell, D. Jayas, R. Noyes, and T.W. Phillips (Eds), *Proceeding of the International Conference on Controlled Atmosphere and Fumigation in Stored Products*, Gold-Coast Australia, August 8–13, 2004, FTIC Ltd. Publishing, Israel, pp. 159–166.

EnviroNews, 2010. *Ozone Layer Study Predicts 2050 Recovery*. http://www.enviro-news.com/news/ozone-layer-study-predicts-2050-recovery.html.

Esteves, M.P., Andrade, M.E., and Empis, J., 1999. Detection of prior irradiation of dried fruits by electron spin resonance (ESR). *Radiation Physics and Chemistry*, 55, 737–742.

European Environment Agency. *Ozone Depleting Substances*. http://www.eea.europa.eu/publications/92-9157-202-0/page302.html.

FAO/IAEA/WHO, 1999. International conference on ensuring the safety and quality of food through radiation processing. *Food and Environmental Protection Newsletter*, 2(2), p. 11.

FAOSTAT, 2008. http://faostat.fao.org/default.aspx.

Farkas, J., 2001. Radiation decontamination of spices, herbs, condiments, and other dried food ingredients. In: R.A. Molins (Ed), *Food Irradiation: Principles and Applications*, Wiley, New York, pp. 291–312.

Fields, P.G. and Muir, W.E., 1996. Physical control. In: B. Subramanyam and D.W. Hagstrum (Eds), *Integrated Management of Insects in Stored Products*, Marcel Dekker, Inc, New York, pp. 195–221.

Finkelman, S., Navarro, S., Isikber, A., Dias, R., Azrieli, A., Rindner, M., Lotan, Y., and Debruin, T., 2003. Insects control of cocoa pests using a novel vacuum approach. In: P.F.A. Credland, D.M. Armitage, C.H. Bell, P.M. Cogan and E. Highley (Eds), *Proceedings of the Eighth, International Working Conference on Stored product Protection*, July 22–26, 2002, York, UK, CAB International, Wallingford, UK, pp. 579–582.

Finkelman, S., Navarro, S., Miriam R., and Dias, S., 2006. Use of heat for disinfestation and control of insects in dates: laboratory and field trials. *Phytoparasitica*, 34(1): 37–48.

Fleurat Lessard, F., 1990. Effect of modified atmospheres on insects and mites infesting stored products. In: M. Calderon and R. Barkai-Golan (Eds), *Food Preservation by Modified Atmospheres*, CRC Press, Boca Raton, FL, pp. 21–38.

Forney, C.F. and Lipton W.J., 1990. Influence of controlled atmospheres and packaging on chilling sensitivity. In: C.Y. Wang (Ed), *Chilling Injury of Horticultural Crops*, CRC Press, Boca Raton, FL, pp. 257–267.

Friedlander, A. and Navarro, S., 1983. Effect of controlled atmospheres on the sorbitol pathway in *Ephystia cautella* pupae. *Experientia*, 39, 744–746.

Glassey, K., 2008. Official uses of methyl bromide. Biosecurity New Zealand. http://www.biosecurity.govt.nz/commercial-transport-and-border-management/facilities/official-use-of-methyl-bromide.

Hallman, G. and Martinez, P., 2001. Ionizing irradiation quarantine treatment against Mexican fruit fly (Diptera: Tephritidae) in citrus fruit. *Postharvest Biology and Technology*, 23, 71–77.

Hallman, G.J., 2000. Expanding radiation quarantine treatments beyond fruit flies. *Journal of Agricultural and Forest Entomology*, 2, 85–95.

Hallman, G.J., 2001. Irradiation as a quarantine treatment. In: R.A. Molins (Ed), *Food Irradiation: Principles and Applications*, Wiley, New York, pp. 113–130.

Hatton, T.T., 1990. Reduction of chilling injury with temperature manipulation. In: C.Y. Wang (Ed), *Chilling Injury of Horticultural Crops*, CRC Press, Boca Raton, FL, pp. 269–280.

http://www.minagric.gr/greek/data/malathr1.doc. Soil fumigation with methyl bromide: Advantages and disadvantages. Malathrakis, N.E.

http://www.ppcind.com/pdfFiles/MAP%20vs.%20CAP.pdf. Modified atmosphere packaging vs. controlled atmosphere packaging.

IAEA, 1991. Analytical Detection Methods for Irradiated Foods. IAEA, TECDOC-578, Vienna.

Ikediala, J.N., Tang, J., Neven, L.G., and Drake, S.R., 1999. Quarantine treatment of cherries using 915 MHz microwaves: Temperature mapping, codling moth mortality and fruit quality. *Postharvest Biology and Technology*, 16, 127–37.

IPPC, 2003. International Plant Protection Convention (IPPC), Guidelines for the Use of Irradiation as a Phytosanitary Measure, FAO, Rome, publication no. 18.

Jaddou, H. and AL-Hakim, M., 1977. Effect of irradiation on the flavour of Iraqi dates. *International Symposium on Food Preservation by Irradiation*, Wageningen, the Netherlands. Report IAEA- STI/PUB/470.

Jaddou, H. and AL-Hakim, M., 1978. Chromatographic analysis of sugars from irradiated Iraqi dates. *International Congress of Food Science and Technology*, Kyoto, Japan, 17–22 September 1978.

Johnson, J.A. and Marcotte, M., 1999. Irradiation control of insect pests of dried fruits and walnuts. *Food Technology*, 53, 46–51.

Ke, D. and Kader, A.A., 1991. Potential of controlled atmospheres for postharvest insect disinfestation of fruits and vegetables. *Postharvest News and Information*, 3, 31–37.

Krista, C.S. and Mangan, R., 2000. Disinfestation: Effect of non-chemical treatments on market quality of fruit. USDA-ARS Crop Quality and Fruit Insects Research Unit, USA.

Kwon, J.H., Kwon, Y.J., Byun, M.W., and Kim, K.S., 2004. Competitiveness of gamma irradiation with fumigation for chestnuts associated with quarantine and quality security. *Radiation Physics and Chemistry*, 71, 41–44.

Lilford, K., Fulford, G.R., Schlipalius, D., and Ridley, A., 2009. Fumigation of storedgrain insects—A two locus model of phosphine resistance. In: *18th World IMACS/MODSIM Congress*, July 13–17, 2009, Cairns, Australia, pp. 540–546.

Lu, H., Zhou, J., Xiong, S., and Zhao, S., 2010. Effects of low-intensity microwave radiation on *Tribolium castaneum* physiological and biochemical characteristics and survival. *Journal of Insect Physiology*, 56(9), 1356–1361.

Mbata, G.N. and Phillips, T.W., 2001. Effects of temperature and exposure time on mortality of stored-product insects exposed to low pressure. *Journal of Economic Entomology*, 94, 1302–1307.

Mbata, G.N., Phillips, T.W., and Payton, M., 2004. Mortality of eggs of stored product insects held under vacuum: Effects of pressure, temperature and exposure time. *Journal of Economic Entomology*, 97, 695–702.

Moreno, M., Castell-Perez, M., Gomes, C., Da Silva, P., and Moreira, R., 2006. Effect of electron beam irradiation on physical, textural and microstructural properties of Tommy Atkins mangoes (*Mangifera indica* L). *Journal of Food Science*, 71(2), E80.

Navarro, S., 2006. New global challenges to the use of gaseous treatments in storedproducts. In: I. Lorini, B. Bacaltchuk, H. Beckel, D. Deckers, E. Sundfeld, J. P. dos Santos, J. D. Biagi, J. C. Celaro, L. R. D'A. Faroni, L.de O. F. Bortolini, M. R. Sartori, M. C. Elias, R. N. C. Guedes, R. G. da Fonseca, and V. M. Scussel (Eds), *Proceedings of the Ninth International Working Conference on Stored Product Protection*, October 15–18, 2006, Campinas, Brazil, Brazilian Post-harvest Association, Campinas, Brazil, pp. 495–509.

Navarro, S. and Calderon, M., 1979. Mode of action of low atmospheric pressures on *Ephestia cautella* (Wlk.) pupae. *Experientia*, 35, 620–621.

Navarro, S. and Donahaye, E., 2005. Innovative environmentally friendly technologies to maintain quality of durable agricultural produce. In: S. Ben-Yehoshua (Ed), *Environmentally Friendly Technologies for Agricultural Produce*. CRC Press, Taylor & Francis Group, Boca Raton, FL, pp. 205–262.

Navarro, S., Donahaye, E., Rindner, M., and Azrieli, A., 1998. Disinfestation of nitidulid beetles from dried fruits by modified atmospheres. *Proceedings of the Annual International Research Conference on Methyl Bromide Alternatives and Emission Reductions*, December 7–9, 1998, Orlando, FL, pp. 68_1–68_3.

Navarro, S., Donahaye, J.E., Rindner, M., and Azrieli, A., 2001. Storage of dates storage of dates under carbon dioxide atmosphere for quality preservation. In: E.J., Donahaye, S. Navarro and J.G. Leesch (Eds), *Proceedings of the International Conference on Controlled Atmosphere and Fumigation in Stored Products*, Fresno, CA. 29 October–3 November 2000, Executive Printing Services, Clovis, CA, pp. 231–239.

Navarro, S., Finkelman, S., Rindner, M., and Dias, R., 2004a. Emigration and control of nitidulid beetles from dates using heat. *Integrated Protection of Stored Products IOBC Bulletin/wprs*, 27(9), 219–225.

Navarro, S., Isikber, A., Finkelman, S., Rindner, M., Azreali, A., and Diaz, R., 2004b. Effectiveness of short exposures of propylene oxide alone and in combination with low pressure of carbon dioxide against *Tribolium castaneum* (Herbst) (Coleoptera: Tenebrionidae). *Journal of Stored Products Research*, 40, 197–205.

Oumeish, Y., 1996. Climatotherapy at the Dead Sea in Jordan. *Clinics in Dermatology*, 14, 659.

Paul, G.F. and Noel, D.G.W., 2002. Alternatives to methyl bromide treatments for stored-product and quarantine insects. *Annual Review of Entomology*, 47, 331–59.

Pennington, J.A.P., 1989. *Bows and Church's Food Values of Portions Commonly Used.* J. 8., Lippincot Co., Philadelphia, PA, p. 97.

Radwan A., 2008. The water balance of the Dead Sea: An integrated approach. *Hydrological Processes*, 14, 145.

Rafaeli, A., Kostukovsky, M., and Carmeli, D., 2006. Successful disinfestation of sap-beetle contaminations from organically grown dates using heat treatment: A case study. *Phytoparasitica*, 34(2), 204–212.

Rajendran, S., 2001. Alternatives to methyl bromide as fumigant for stored food commodities. *Pesticide Outlook*, 12, 249–253.

Reynes, M., Bouabidi, H., Piombo, G., and Risterucci, A.M., 1994. Caractérisation des principales variétés de dattes cultivées dans la région du Djerid en Tunisie, *Fruits*, 49, 289–298.

Rodriguez, J.C., Bartosik, R.E., Cardoso, M.L., and Crocce, D., 2008. Factors affecting carbon dioxide concentration in interstitial air of wheat stored in hermetic plastic bags (silo-bag). In: D.L. Guo, S. Navarro, Y. Jian, T. Cheng, J. Zuxun, L. Yue, L. Yang and W. Haipeng (Eds), *Proceedings of the 8th International Conference on Controlled Atmospheres and Fumigation in Stored Products*, September 21–26, 2008, Chengdu, China, Sichuan Publishing House of Science and Technology, Chengdu, China, pp. 589–592.

Roesli, R., Subramanyam, B., Fairchild, F.J., and Behnke K.C., 2003. Trap catches of stored product insects before and after heat treatment in a pilot feed mill. *Journal of Stored Products Research*, 39, 521–540.

Rygg, G.L., 1957. The relation of moisture content to rate of darkening in Deglet Noor dates. *Date Growers' Institute*, 34, 12–13.

Salimov, R., Cherepkov, N., Kuksanov, N., and Koznetzov, S., 2000. The use of electron accelerators for radiation disinfestations of grain. *Radiation Physics and Chemistry*, 57, 625–627.

Schlipalius, D.I., Collins, P.J., Mau, Y., and Ebert, P.R., 2006. New tools for management of phosphine resistance. *Outlooks on Pest Management*, 17, 52–56.

Sharp, J.L., Gaffney J.J., Moss, J.I., and Gould, W.P., 1991. Hot-air treatment device for quarantine research. *Journal of Economic Entomology*, 84, 520–527.

TEAP, 2000. Montreal Protocol on Substances That Deplete the Ozone Layer: UNEP Technology and Economic Assessment Panel, April 2000 Report United Nations.

TEAP and MBTOC, 2003. Handbook on Critical Use Nominations for Methyl Bromide. Technological and Economical Assessment Panel and Methyl Bromide Technical Options Committee. http://hq.unep.org/ozone/teap/reports/mbtoc/.

Testin, R.F., 1988. The use of controlled atmosphere packaging for extending shelf life of fruits, vegetables and other fresh products. International Trade Center, Packdata Factsheet No.1. Modified from "Extended Shelf Life Conference, ESL 188, held on 9 and 10 February 1988 in Orlando, Florida, USA.

Thomas, W.P., Pajshekhar, H., Siwei, L., and Guy, J.H., 2007. Vacuum as a methyl bromide alternative for disinfestation of durable and fresh commodities. In: E.J. Donahaye, S. Navarro, C. Bell, D. Jayas, R. Noyes, and T.W. Phillips (Eds), *Proceedings of the International Conference on Controlled Atmosphere and Fumigation in Stored Products*, Gold-Coast, Australia, August 8–13, 2004, FTIC Ltd. Publishing, Israel, pp. 557–566.

Todoriki, S., Hasan, M., Miyanoshita, A., Imamura, T., and Hayashi, T., 2006. Assessment of electron beam-induced DNA damage in larvae of chestnut weevil, *Curculio sikkimensis* (Heller) (Coleoptera: Curculionidae) using comet assay. *Radiation Physics and Chemistry*, 75, 292–296.

UNEP, 2002. Report of the Methyl Bromide Technical Options Committee (MBTOC). 2002 Assessment Report. UNON/Publishing Section Services/Nairobi.

UNEP, 2003. Report of the Fifteenth Montreal Protocol Meeting of the Parties to the Montreal Protocol on the Substances that Deplete Ozone Layer. Nairobi, 10–14 November. http://ozone.unep.org/Meeting_Documents/mop/15mop/15mop-9.e.doc.

UNEP, 2006. Montreal Protocol on Substances That Deplete the Ozone Layer. Assessment Report on the Methyl Bromide Technical Options Committee (MBTOC).

UNEP Home Page. Methyl Bromide Phase-Out in Africa. http://www.unep.org/roa/Projects_Programmes/ozone/MB/About/Intro.asp.

UNEP OZONEWS, 2010. A fortnightly electronic news service on ozone protection and implementation of methyl bromide.

Wahid, M., Sattar, A., Neelofar Atta, S., Khan, I., and Ehlermann, D.A.E., 1987. Radiation disinfestation and quality of dried fruits. *Acta Alimentaria*, 16(2), 159–166.

Wang, C.Y., 1990. Alleviation of chilling injury of horticuitural crops. In: C.Y. Wang (Ed), *Chilling Injury of Horticultural Crops*, CRC Press, Boca Raton, FL, pp. 281–302.

WikiAnswers. http://wiki.answers.com/Q/What_is_the_difference_between_Controlled_atmosphere_packaging_CAP_and_Modified_atmosphere_packaging_MAP.

Wikipedia. http://en.wikipedia.org/wiki/radiation.

White, N.G. and Leesch, J.G., 1996. Chemical control in integrated management of insects in stored-products. In: B. Subramanyam and D.W. Hagstrum (Eds), *Integrated Management of Insects in Stored-Products*, Marcel Dekker, Inc., New York, pp. 41–70.

Whiting, D.C., Foster, S.P., and Maindonald, J.H., 1991. The effects of oxygen, carbon dioxide and temperature on the mortality responses of *Epiphyas postvittana* (Lepidoptera: Tortricidae). *Journal of Economic Entomology*, 84, 1544–1549.

Zouba, A., Khoualdia, O., Diaferia, A., Rosito, V., Bouabidi, H., and Chermiti, B., 2009. Microwave treatment for postharvest control of the date moth *Ectomyelois ceratoniae. Tunisian Journal of Plant Protection*, 2(4), 173–184.

17 Processing of Dates into Value-Added Products

P. Vijayanand and S. G. Kulkarni

CONTENTS

INTRODUCTION

Date is an important commercial crop grown in various parts of the world. The date fruits are relished for their sweet, succulent, and exotic flavor. The fruits are processed into a wide variety of products. Processed products produced from dates include dry dates, date bars, date syrup, date juice concentrate, date jam, date butter, date candy, date chutney, date relish, date pickles, date wine, and so on. Date oil and date coffee are some of the by-products produced from date seeds. This chapter describes the processing, composition, and quality characteristics of various products produced from dates.

DRY DATES

Dry dates are also known as "Chhoharas" or "Khalal matbuuk." Yellow-colored dates are boiled and dried till they become very hard and wrinkled. Dried dates are sweet in taste and have a long storage life (Kalra et al., 1978). "Chibchaab" in Southern Iraq is a boiled and dried date variety found in all date-growing places along the north-eastern shore of the Persian Gulf and also in lower Iraq (Dowson and Aten, 1962). Boiling of dates of Hillawi, Khadrawi, Medjool, and Thoory varieties for 4–10 min in water is reported to yield dry dates of superior quality (Chohan et al., 1972). Decrease in reducing sugars and increase in nonreducing sugars were observed when the boiling period of dates increased. Blistering of skin, that is, separation of skin from the pulp was noticed in case of dates treated with hot water (Kalra et al., 1976).

A process for preparation of dehydrated dates from immature dates was reported. Dates grown in the Kutch region of Gujarat are harvested before maturation, that is, at Khalal stage to prevent spoilage caused due to the onset of monsoon. Blanching of immature dates in water at 96 ± 1°C and subsequent dehydration at 60 ± 2°C for 18–20 h resulted in good-quality dehydrated dates as compared to the dates dried without heat treatment. Dehydrated dates were acceptable with respect to color, flavor, taste, and overall quality. Dehydrated dates contained total sugars of 52%, reducing sugars of 41.5%, tannins 1.35%, and ascorbic acid 3.37 mg/100 g. Equilibrium relative humidity of the dehydrated dates is found to be 75.9% with an initial moisture content of 15.9%. Dehydrated dates packed in 75 μm low-density polyethylene packaging material are shelf stable for 6 months at room temperature (Kulkarni et al., 2008).

Dried Date Product

"Tamar Eddin" is a dried date product in the form of sheets, prepared from "Reazaiz" variety of dates. "Tamar" dates are pitted manually and the edible portion homogenized with water in the ratio of 1:3 for 2 min in a blender. The homogenate is filtered through cloth and dried in a hot air drier at 65°C until sheets of "Tamar Eddin" are formed with a final moisture content of 18%. Addition of 1% citric acid and a moderate amount of 167–355 ppm sulfur dioxide to the date pulp resulted in a product with acceptable color, texture, and taste. "Tamar Eddin" can be consumed either in dry form or after it is rehydrated by soaking or simmering in water.

DATE BARS

Date bars fortified with almonds, sesame seeds, oat flakes, and skim milk powder can be prepared. Date fruits at the tamar stage are washed, pitted, and macerated in a food processor to obtain a paste. Date paste, powdered almonds, and roasted oat flakes are blended into a homogeneous mass using a Hobart planetary mixer. The mixture is sheeted to 1 cm thickness using a rolling pin and cut into 4 × 1 cm bar using a rolling pizza cutter. The bars are wrapped individually in cellophane and stored in airtight plastic containers. Similarly, fortified date bars can be made using skim milk powder and sesame seeds (Al-Hooti et al., 1997b).

DATE SYRUP

Date syrup, also known as "dibis," is the extract of mature ripe dates. The extract is concentrated to a total solids content of 70% by weight and contains no additives or preservatives. Date syrup has a long shelf life at room temperature due to the presence of high sugar content which acts as a natural preservative.

"Siwi" dates of tamar stage yielded the best-quality date syrup with extraction at 80°C for 3 h (Table 17.1). Date extract is filtered and concentrated under vacuum at 55°C to a total solids of 63%–70%. Citric acid (0.5%) was added during the concentration of date extract (Abd El-Mohsen and Nezam El-Din, 1995).

Date syrup contains glucose and fructose in almost equal quantities. It has 70% total solids and an energy value of 245 kcal/100 g (Table 17.2). Date syrup is dark brown in color, which is a result of the interaction of the protein with the reducing sugars of the dates. Development of color, flavor, and aroma of date syrup is enhanced by high temperature. The concentration of glucose in

TABLE 17.1
Composition of Sugars of Siwi Dates at Different Stages

	Khalal	Rutab	Tamar
Reducing sugar (%)	57.50	62.20	61.50
Nonreducing sugar (%)	18.35	12.72	11.10
Total sugars (%)	75.85	74.92	72.60

Source: Abd El-Mohsen, M. and Nezam El-Din, M.M. 1995. *Egyptian Journal of Food Science* 23(3): 229–239.

date syrup is 51.6%. When temperature falls below 25°C, glucose begins to crystallize, imparting a granular texture to the syrup. These crystals can be easily dissolved by warming the syrup to 50–60°C. This principle is used for the preparation of set date syrup. Viscosity of date syrups at different temperatures varies depending on the origin of dates and the concentration of the syrup.

A process for the preparation of date syrup from "tamar" stage dates was reported. Washed dates are steamed for extracting the pulp. The extracted pulp is treated with pectolytic enzymes under optimized conditions. The extracted date juice is concentrated in a thin-film vacuum evaporator to a total soluble solids (TSS) of more than 75 °Brix. The composition of date syrup indicated that it is a good source of carbohydrates with highly acceptable color and flavor (Table 17.3). Date syrup packed in glass bottles was found to be stable during the storage of 1 year at room temperature (Vijayanand et al., 2006a).

A process for preparation of date juice concentrate from immature dates was reported. Immature date fruits with less commercial value were used for processing into date juice concentrate. Immature dates were crushed in a fruit mill and treated with 0.1% pectinase enzyme for 120 min to obtain maximum juice. Date juice was found to be rich in reducing and total sugars. The juice was pasteurized at 85°C to inactivate the enzyme, cooled, and centrifuged at 3000 rpm to obtain a clear juice. Clarified date juice is concentrated in a thin-film evaporator to a TSS of 76 °Brix in a thin-film vacuum evaporator (Table 17.4). Date juice concentrate stored at −20°C for 6 months indicated no significant changes in TSS, acidity, ascorbic acid, total sugars, and pH. Date juice concentrate can be reconstituted for preparing ready-to-serve date juice beverages with acceptable sensory quality (Kulkarni et al., 2010).

TABLE 17.2
Composition of Date Syrup

Component	Per 100 g
Water	30
Carbohydrate (total)	67
Fructose	32
Glucose	32
Pectin	1
Protein	1
Mineral matter	1.3
Energy (kcal)	245
Density	1.35

TABLE 17.3
Composition of "Tamar" Stage Dates and the Date Syrup

Composition	Dates	Date Syrup
TSS (°Brix)	73.20	79.56
Total acidity (%)	0.47	0.8
Reducing sugars (%)	57.93	57.32
Total sugars (%)	63.07	61.42
Ascorbic acid (mg/100 g)	15.97	46.47
Tannins (%)	0.39	0.18

Source: Adapted from Vijayanand, P. et al. 2006a. *Process for Preparation of Date Syrup*, (unpublished data). Central Food Technological Research Institute, Mysore, India.

TABLE 17.4
Composition of Date Juice Concentrate from Immature Dates

Total Soluble Solids (°Brix)	76.0 ± 0.32
pH	4.6 ± 0.05
Acidity (% citric acid)	0.33 ± 0.01
Reducing sugars (%)	63.0 ± 0.20
Total sugars (%)	67.0 ± 0.30
Nonreducing sugars (%)	3.8 ± 0.02
Ascorbic acid (mg/100 g)	13.0 ± 0.14
Tannins (%)	1.16 ± 0.03

Source: Kulkarni, S.G., Vijayanand, P., and Shubha, L. 2010. *Journal of Food Science and Technology* 47: 2.

Uses of Date Syrup

Date syrup is used as a sweetening agent with a characteristic flavor (Table 17.5). It can be used as a substitute for honey, maple syrup, malt syrup, molasses, glucose syrup, invert sugar, high fructose syrup, and all forms of crystalline sugar, when a sweet taste accompanied by the delicate aroma of the mature date fruit is desirable. Date syrup is used in various processed foods. Date syrup has been used to replace sucrose in yellow and chocolate-flavored layer cakes at 0%–100% replacement levels. Incorporation of date syrup in cake caused deterioration in cake volume, cell structure, grain, texture, color, and flavor. Date syrup was also found to delay gelatinization, but to a lesser extent than did sucrose (Almana and Mahmoud, 1991).

DATE JELLY

Date jelly is prepared by boiling date juice and sugar to TSS of 65 °Brix. Citric acid, pectin, ascorbic acid, and benzoic acid are added during the preparation of date jelly. The "Ruzeis" variety is used for the preparation of date jelly. Dates are cleaned, pitted, and soaked in sufficient water with 1:2 ratio at 25°C for 5 min, and drained and ground to produce date paste. Date juice is prepared by adding water to date paste in the ratio of 3:1 (w/w). The mix is boiled gently with continuous stirring

TABLE 17.5
Utilization of Date Syrup in Different Food Products

Dairy products	Yoghurt, milk shakes, ice cream, 15% date syrup solids are used to sweeten and flavor the products
Desserts	Crème caramel, mousse, sundae, etc. can be made using date syrup as a sweetener
Jam and marmalade	Substitution of sugar by date syrup in jam and marmalade
Soft drinks	10%–15% date syrup solids used for ready-to-drink beverages, e.g., lemon, lime, and apricot
Sauces and relishes	A variety of sauces can be prepared by blending date syrup with different ingredients
Confectionary	Date syrup used as binder and as a sweetener for cereal and fruit bars. It can also be used as an ingredient instead of invert sugar for filled chocolate bars. It could also replace part of the sugar in toffee and fudge
Famine-relief instant drink	Date syrup formulated with skim milk powder for instant mixing with water and distribution to famine-struck areas
Halwa	Partial substitution of sugar and glucose syrup with date syrup resulting in natural flavor
Baked goods	As a sweetener and flavoring agent for biscuits, fruit loaves, cakes, and bakery filling

for 5 min and filtered through to remove the fiber and other impurities. Date juice is added with sugar at a ratio of 1:1. Citric acid and pectin are dissolved separately in and added to the boiling mix. Boiling is continued till TSS of 65 °Brix is reached. Ascorbic acid (0.1%) and benzoic acid (0.01%) are dissolved in small quantities of hot syrup and added to the jelly. Date jelly is filled hot in glass bottles and stored (Yousif et al., 1990).

DATE JAM

Date fruits at the "rutab" stage were used for the preparation of date jam. Dates are pitted, mixed with equal amount of water, and cooked for 45 min. The pulp is extracted and a sugar-to-date pulp ratio of 55:45 was used. Citric acid is used to adjust the pH of pulp and sugar mixture to 3.2. The pulp extract and sugar mixture are cooked with constant stirring to a TSS of 68 °Brix. Date jam is poured hot into clean dry glass jars, and sealed, labeled, and stored at room temperature.

DATE BUTTER

Date fruits at the "tamar" stage were used for the preparation of date butter (Al-Hooti et al., 1997d). The fruits are pitted, mixed with equal amount of water, and cooked for 45 min for extracting date pulp. A sugar-to-date pulp ratio of 40:60 is used. Citric acid is added to adjust the pH to 4.5–4.7. Date pulp extract and sugar mixture are cooked with constant stirring to obtain a TSS of 74–75 °Brix. Date butter is poured hot into clean dry glass jars, and sealed and stored at room temperature. Date butter is an intermediate moisture product with a water activity ranging from 0.61 to 0.66. Saudi date cultivars are evaluated for production of date butter. It is suggested that the use of orange, banana, or almond flavorings resulted in date butter of acceptable quality (Khatchadourian et al., 1986).

DATE CANDY

A process for date candy from immature dates was developed. Immature dates, which are highly astringent, were used for the preparation of date candy. Dates are washed and blanched in boiling water. Blanched date fruits are boiled in sugar syrup of 60 °Brix in the ratio of 1:3 for 10 min. The fruits are equilibrated for 6 h in the syrup. The syrup is filtered and concentrated to 65 °Brix by boiling. Citric acid is added at the rate of 0.2 g/kg of syrup and the dates are added into the syrup. After the syrup is equilibrated to >70 °Brix, the date candy is drained and dried in a hot air drier at 60°C for 3 h. The date candy is packed in flexible pouches and sealed (Vijayanand et al., 2006b).

DATE CHUTNEY

Dates are sorted, washed in tap water, and peeled using an abrasive peeler. Peeled date fruits are pitted manually and each fruit is cut into eight longitudinal slices. Slices are mixed with spices and fried lightly in mustard oil for 25–30 min with constant stirring till the fruit is softened. After adding vinegar, heating is continued on a medium flame for another 10 min with constant stirring. Sugar is added and the chutney cooked further to a temperature of 105–106°C. Potassium sorbate is dissolved in a small amount of water and added to the mixture. Date chutney is poured hot at 85–90°C into clean sterilized dry glass jars. Date chutney prepared from "kimri" and "khalal" fruits had a water activity of 0.62 and 0.63, respectively. Chutney prepared from "khalal" stage fruit is found to be much softer in texture than the chutney made from "kimri" stage fruit (Al-Hooti et al., 1997a).

DATE RELISH

Date fruits at the "tamar" stage are pitted, mixed with an equal amount of water, and cooked for 45 min. "Rutab" dates are washed, pitted, and chopped into circular rings. Carrots, onions, and ginger are peeled, chopped, and cooked for 15–20 min on low heat. Salt, sugar, and seasonings are then added and cooking was continued for another 20 min. Pectin, skim milk powder, and acetic acid are added to the mixture and cooked to a temperature of 105–106°C. Potassium sorbate is dissolved in a small amount of water and stirred into the cooked mass. Date relish is poured hot into clean dry glass jars that are capped and labeled (Al-Hooti et al., 1997a).

DATE PICKLES

Date fruits are washed, dried, and cut into four uniform longitudinal slices. The slices are blanched in boiling water for 2 min and excess water is drained off. The dried slices are used for the preparation of pickle in oil. The spices, fenugreek, fennel, nigella, and cumin, are fried in mustard oil to a light-brown color. Salt, turmeric, red chilli powder, and date fruit slices are added to the cooled fried spice mixture. Potassium sorbate is dissolved in a small quantity of water and added. Date pickle is filled into clean dry glass jars and sealed. The sealed jars are kept in the sun daily for 6 h during daytime for a week, after which the samples are stored at room temperature. Sensory analysis of the pickle-in-oil samples made from "kimri" and "khalal" stage fruits showed that "kimri" pickles were acceptable for up to 1 month of storage, but the overall acceptability dropped after 5 months of storage. The pickles made from "khalal" stage fruit were acceptable only initially (Al-Hooti et al., 1997c).

Date Pickle in Vinegar

The pitted, peeled date fruit rings are packed into glass jars which are then hot filled at 85–90°C with filling medium containing 3% salt, 20% sugar, 1% acetic acid, and 0.05% potassium sorbate. The jars are sealed and pasteurized at 71°C for 15 min in a water bath. The jars are then cooled, labeled, and stored. Acidity and salt contents decreased significantly in the "kimri" pickles after 4 months of storage (Al-Hooti, 1997c). The water activities for pickle-in-vinegar samples made from "kimri" and "khalal" fruits were 0.83 and 0.75, respectively. Unpollinated Siwi dates pickled containing 10% lemon fruit and 10% green pepper were found to be highly acceptable (Abd El-Mohsen and Nezam El-Din, 1996).

DATE WINES

A number of alcoholic fermented date products have been developed in Egypt. The most important among these are "dakkai," "sherbote," and "nabit" wines. Dakkai is prepared by mixing clean whole date fruits with lukewarm water in an earthenware jar which is closed tightly and left to ferment for 3–4 days. The contents are then strained to give "dakkai," a pale-yellowish wine.

Date fruits to be used in the preparation of "sherbote" and "nabit" wines are first boiled in water with dates-to-water ratio of 1:3 for 3–4 h, until the flesh of the fruit practically dissolves. The boiled dates are strained through a cheesecloth to obtain brown syrup. The residue is boiled again for an hour, strained, and the extract is combined with the first extract. The combined syrup is cooled to room temperature and then used for fermentation. Ground cinnamon and ginger tied in cloth bag are added to the syrup to be fermented in the preparation of "sherbote." The jar containing the syrup is closed tightly and fermentation is left to proceed for 1–3 days. In case of "nabit," sorghum malt tied in a cloth bag is added to the syrup. The fermentation jar is tightly covered, sealed with mud, and buried in a pit in the ground. Fermentation of the syrup takes between 2 and 3 days. Both "sherbote" and "nabit" are served without further straining. A microbiological study of the wines indicated the presence of "Torulopsis" in "sherbote," *Saccharomyces* in nabit, and *Candida* and *Saccharomyces* in dakkai. Acetic acid bacteria, namely, *Acetobacter* and *Gluconobacter* were found in all the three wines. *Leuconostoc*, a lactic acid bacteria, was isolated from all three wines. *Lactobacillus* was isolated from "sherbote" and "dakkai" and *Streptococci* from "nabit." In addition, *Bacillus* and *Klebsiella* were isolated from "dakkai." Analyses of liquid samples and headspace vapor of date wines using gas chromatography and mass spectrometry indicated that in addition to ethanol and lactic and acetic acids, the samples showed the presence of acetaldehydes, ethyl acetate, methyl acetate, chloroform, diacetyl, furfural, and methanol (Mohamed and Hamid, 1984). Fermentation of date extracts to ethanol and vinegar is also carried out in batch and continuous membrane reactors. Sugars in extracts of dates are converted into ethanol by *Saccharomyces cerevisiae*. Ethanol is subsequently fermented to acetic acid by *Acetobacter aceti* (Meharia and Cheryan, 1991).

DATES-IN-SYRUP

Peeled, pitted date fruits can be packed in glass jars with 50 °Brix sugar syrup. Citric acid is added to adjust the pH of the syrup to 2.8. Hot syrup is poured into the jars containing the fruits, and then sealed and stored. The samples were found to be free from aerobes, molds, coliforms, and Enterobacteriaceae after 3 months of storage with acceptable shape, texture, and flavor of the fruit (Al-Hooti et al., 1997d). An acceptable dates-in-syrup from three Saudi date cultivars can be prepared with or without the addition of various synthetic flavors (Khatchadourian et al., 1986).

DATE SEEDS

The presence of appreciable quantities of proteins, minerals, and especially fat makes the date seeds a valuable raw material for the production of animal feeds (Rygg, 1975). Concentration of "potassium" was highest in date seeds followed by magnesium and calcium. Among the microelements, iron content was highest followed by zinc, manganese, copper, and sodium. Fatty acids, lauric, myristic, palmitic, stearic, oleic, linoleic, and linolenic acids were present at <1% level. Date seeds are also a good source of essential amino acids which can be included in animal feed formulations (Al-Hooti et al., 1998).

DATE COFFEE

A coffee-like beverage was prepared from date stones, which were washed, dried, roasted, and ground. The ground material was infused to produce the beverage. Optionally, ground cardamom may be added. The beverage had a pleasant flavor and was found to be free from caffeine (Al-Dossary, 1991).

DATE OIL

Oil from date seeds is extracted by pulverizing in a mill and cooking in a steam kettle. The flour, which contains 8% oil, is then processed with a solvent. Date seed oil is yellow in color and can be refined and bleached to a lighter shade. Date seed oil is suitable for making soaps. Deoiled cake contains a high percentage of proteins and can be used as animal feed (Joshi, 1996).

CONCLUSION

Date palm (*Phoenix dactylifera* L.) has been an important staple food crop in various parts of the world, especially the middle eastern countries for centuries. Dates at the "rutab" and "tamar" stages are consumed directly by the human population with little or no processing. The production of processed date products is growing rapidly due to the nutritional quality and convenience. Development of new processed date products would enhance the economic importance of this crop in the international trade. The introduction of new processed date products is important due to the increase in the production and need for value addition to the dates. A number of processed date products were developed. Date paste has been used to replace wheat flour at various levels in the manufacture of bread and cookies. Dates-in-syrup, a product that resembles fruit preserves, is another feasible product that can be made from dates. Cane sugar in cakes can be replaced with date syrup at varying levels of production. Date syrup finds application in a number of processed food products. A few attempts have been made in this direction, including the production of date wine, milk-date drinks, soft drinks, ice cream and frozen desserts, and date juice. Date toffee, date bars, and date instant beverages are some of the novel products which can find acceptability and variety in the processed foods market. Utilization of dates which are rich in natural nutritional compounds for producing processed products can result in value addition and availability of a new range of products. It will have a considerable commercial value in the nutraceutical product segment.

REFERENCES

Abd El-Mohsen, M. and Nezam El-Din, M.M. 1995. Technological study on Dibis production from the Siwi date. *Egyptian Journal of Food Science* 23(3): 229–239.

Abd El-Mohsen, M. and Nezam El-Din, M.M. 1996. Study on the utilization of unpollinated Siwi dates. *Egyptian Journal of Food Science* 24(2): 147–165.

Al-Dossary, S.A.A. 1991. Process for manufacture of a coffee like beverage from dates date coffee. *German Federal Republic Patent Application* DE 3931504 A1.

Al-Hooti, S., Sidhu, J.S., Al-Otaibi, J., and Al-Ameeri, H. 1997a. Processing of some important date cultivars grown in United Arab Emirates into chutney and date relish. *Journal of Food Processing and Preservation* 21: 55–68.

Al-Hooti, S., Sidhu, J.S., Al-Otaibi, J., Al-Ameeri, H., and Qabazard, H. 1997b. Date bars fortified with almonds, sesame seeds, oat flakes and skim milk powder. *Plant Foods for Human Nutrition* 51(2): 125–135.

Al-Hooti, S., Sidhu, J.S., Al-Otaibi, J., Al-Amiri, H., and Qabazard, H. 1997c. Utilization of date fruits at different maturity stages for variety pickles. *Advances in Food Science* 19(1/2): 1–7.

Al-Hooti, S., Sidhu, J.S., Al-Saqer, H., Al-Amiri, H., and Qabazard, H. 1997d. Processing qualities of important date cultivars grown in the United Arab Emirates for jam, butter and date-in-syrup. *Advances in Food Science* 19(1/2): 35–40.

Al-Hooti, S., Sidhu, J.S., and Qabazard, H. 1998. Chemical composition of seeds of date fruit cultivars of United Arab Emirates. *Journal of Food Science and Technology* 35(1): 44–46.

Almana, H.A. and Mahmoud, R.M. 1991. Effect of date syrup on starch gelatinization and quality of layer cakes. *Cereals Food World Research* 36(12): 1010–1012.

Chohan, G.S., Jawanda, J.S., and Bakhshi, J.C. 1972. Studies on the curing of dates at Abohar. *The Punjab Horticultural Journal*, 12(1): 46–53.

Dowson, V.H.W. and Aten, A. 1962. *Dates Handling, Processing and Packing*. FAO, Rome.

Joshi, G.V. 1996. Date. The tree of paradise. *Plants*. pp. 122–123.

Kalra, S.K., Jawanda, J.S., and Munshi, S.K. 1976. Studies on the curing of dates (*Phoenix dactylifera*) by hot water treatment. *Indian Journal of Horticulture* 33(2): 146–151.

Kalra, S.K., Munshi, S.K., and Jawanda, J.S. 1978. Studies on preparation of dry dates (chhoharas). *Journal of Food Science and Technology* 15(2): 43–46.

Khatchadourian, H.A., Sawaya, W.N., Safi, W.J., and Al-Shalhat, A.F. 1986. Date products. In *Dates of Saudi Arabia*, Sawaya. W.N. (ed.). Ministry of Agriculture and Water, Riyadh, Saudi Arabia, pp. 95–142.

Kulkarni, S.G., Vijayanand, P., Aksha, M., Reena, P., and Ramana, K.V.R. 2008. Effect of dehydration on the quality and storage stability of immature dates. *LWT Food Science and Technology* 41: 278–283.

Kulkarni, S.G., Vijayanand, P., and Shubha, L. 2010. Effect of processing of dates into date juice concentrate and appraisal of its quality characteristics. *Journal of Food Science and Technology* 47: 2.

Meharia, M.A. and Cheryan, M. 1991. Fermentation of date extracts to ethanol and vinegar in batch and continuous membrane reactors. *Enzyme and Microbial Technology* 13(3): 257–261.

Mohamed Z.A. and Hamid, A.D. 1984. A microbiological study of Sudanese date wines. *Journal of Food Science* 49: 459–467.

Rygg, G.I. 1975. Date development, handling and packing in the United States. *Agriculture Hand Book No. 482*, US Development of Agriculture (USDA), Washington, DC, 56 pp.

Vijayanand, P., Kulkarni, S.G., Nambiar, P.C.S., and Ramana, K.V.R. 2006a. *Process for Preparation of Date Syrup*, (unpublished data) Central Food Technological Research Institute, Mysore, India.

Vijayanand, P., Kulkarni, S.G., and Ramana, K.V.R. 2006b. *Process for Preparation of Date Candy*. (unpublished data) Central Food Technological Research Institute, Mysore, India.

Yousif, A.K., Abou, Ali, M., and Bou Idreese, A. 1990. Processing, evaluation and storability of date jelly. *Journal of Food Science and Technology* 27(5): 264–267.

18 Exploring the Possibility of Biofuel Production from By-Products and Wastes Generated from Date Industry

D. Ramesh, S. Karthikeyan, and G. Chinnanchetty

CONTENTS

The date palm (*Phoenix dactylifera* L.) belongs to the Arecaceae or Palmae family and *Phoenix* is the genus name of 17 palm species (Burnie et al., 2006), and more than 600 varieties are available in the world (Ahmed et al., 1995; Zaid, 1999). It is native to the Canary Islands, Atlantic Ocean, off the coast of Northeast Africa. The world production of dates has increased from about 4.6 million tonnes in 1994 to 6.9 million tonnes in 2004 and expectations are that their cultivation will continue to increase (Al-Farsi and Lee, 2008). This indicates that the consumption and demand for dates are increasing every year, and the production of dates is also increased to meet the increasing demands. The date palm starts to produce fruits at an average age of 5 years, and continues production with an average yield of 400–600 kg/tree/year for up to 60 years (Shinwari, 1993; Al-Shahib and Marshall, 2003). It yields date fruit, which is composed of an edible fleshy pericarp and the seed. The food products produced from date fruits are pitted dates, date powder, date syrup, chocolate-coated dates, date confectionery, and so on.

The wastes generated from the date fruit-processing industry are date seeds, rejected date fruits, immature date fruits, and waste date pulp depending upon the unit operations of date fruit processing. The wastes produced in the date-processing industry and their present usage are furnished in Figure 18.1.

DATE PITS

The date pits are one of the by-products obtained from date fruit after removing the edible fleshy pericarp. The date pits are also called date seeds, pips, kernels, or stones. The weight of date pits

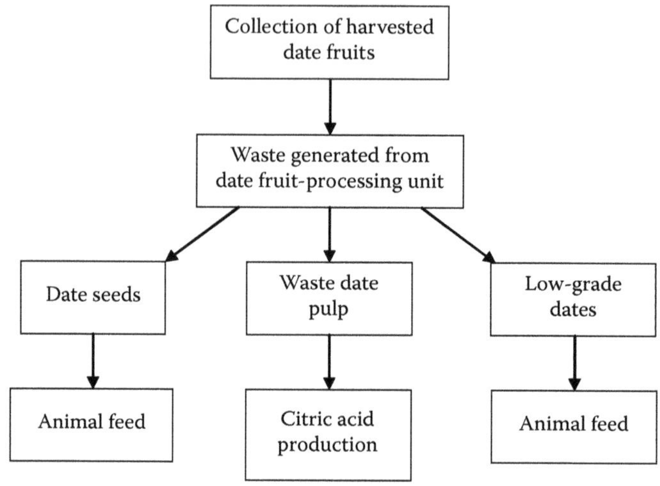

FIGURE 18.1 Process flowchart for waste generation in a date fruit-processing industry and its current utilization.

ranging from 4.6% to 15% of whole date fruit weight depends upon varieties (Mossa et al., 1986; Al-Showiman, 1990; Almana and Mahmoud, 1994; Hussein et al., 1998). Date pits are odorless and have a light-to-dark-brown color and a bland taste with slight bitterness depending upon varieties (Hamada et al., 2002).

Date pits are rich in protein (5.1%), fat (9.0%), dietary fiber (73.1%), phenolics (4.0%), and antioxidants (Al-Farsi and Lee, 2008). After extracting oil from seed, date seed meal is used in animal feeds (Rygg, 1977) for cattle, sheep, camel, and poultry. The date powder can be used as substitute for coffee (caffeine-free drinks). It can be used as dirt road construction, activated carbon, wood-preserving agents, and so on. It is also used in medicines for treating cloudiness of the cornea and to relieve headaches and hermicronia (Fayadh and Al-Showiman, 1990). Although date pits are largely disposed to the environment as waste, possibilities exist to utilize such wastes to increase the income of the growers (Al-Farsi and Lee, 2008; Al-Qaessi, 2010).

WASTE DATE FRUIT PULP

The waste fruit pulp is one of the by-products generated from the date fruit-processing industry next to date pits. Date pulp is produced from the processing of dates into date syrup, which represents nearly 50% of the weight of the date syrup (Mazaheri Assadi and Nikkhah, 2002). Waste date fruit pulp and other wastes coming out of date fruit-processing plants can be used for the production of many pharmaceuticals, citric acid, and ethanol (Hui et al., 2006).

LOW-GRADE DATES

The low-grade dates are composed of low- and second-grade dates, which are rejected from the date fruit-processing industry due to their inferior qualities. They are not consumed by human because of several factors, viz nonappreciated texture (too soft or too hard), contamination by fungus, and infestation by insects, or simply because they are disregarded in comparison with more attractive dates (Besbes et al., 2010). At present, very little use is made of these by-products and they are either discarded or used in limited cases for animal feed (Besbes et al., 2006). Scientific studies on "second- and low-grade dates" were especially focused on their biological transformation aiming the production of biomass and various other compounds such as citric acid, oxytetracycline, and ethanol (Besbes et al., 2006).

TECHNOLOGIES FOR BIOFUEL PRODUCTION FROM DATE WASTES

BIOFUEL

Biomass provides a clean, renewable energy source that could dramatically improve our environment, economy, and energy security (Demirbas, 2009). Biofuel can be derived from renewable biological materials and it may be in the form of gas or liquid or solid fuels. Biofuels include bioethanol, biobutanol, biodiesel, and bio-oil. Liquid biofuels offer alternative fuel option in terms of environmental benefits since it is free from nitrogen and sulfur compounds (Katikaneni et al., 1995). Liquid biofuels are primarily used in vehicles but can also be used in engines or fuel cells for electricity generation. The added advantages of biofuel include energy security, environmental concerns, foreign exchange savings, and socioeconomic issues related to the rural sector. Due to their environmental merits, the share of biofuel in the automotive fuel market will grow fast in the next decade (Kim and Dale, 2005).

Biomass energy technologies use organic matter or plant matter to produce energy with a lower level of greenhouse gas emissions than fossil fuel sources (Sheehan et al., 1998). Present conversion technologies that are being applied to high-moisture biomass mainly include anaerobic digestion, ethanol fermentation, and thermal pyrolysis technologies. New technologies are being developed, and they include high-rate anaerobic digesters for solid organic materials (Zhang and Zhang, 2002). The low-moisture biomass, such as rice hulls, nut shells, and fruit pits, are generally better suited for thermal conversion technologies, such as combustion and gasification. Many of the low-moisture food residues have already found their way to thermal biomass-to-energy conversion units (Gary Matteson and Jenkins, 2007). In the case of date fruit-processing industry, date seed and lost dates are come under low-moisture food residues and waste date pulp is coming under high-moisture biomass category.

BIODIESEL FROM DATE SEED OIL

Date seeds contain 7%–10% oil (Al-Hooti et al., 1998) and about 20% crude fiber. The oil extracted from date pits is pale yellowish-green with a pleasant odor and it has cosmetic, pharmaceutical, food products, and industrial applications (Ghazanfari et al., 2005). The major fatty acid found in date seed oil is oleic acid (Al-Showiman, 1990; Devshony et al., 1992). Fair amounts of lauric acid, myristic acid, and palmitic acid ranging between 15.4% and 23.8%, 7.42% and 11.8%, and 6.96% and 10.2%, respectively, were also observed (Al-Showiman, 1990). The oxidative stability of date seed oils was higher than that of most vegetable oils and comparable to that of olive oil (Besbes et al., 2004). The density and viscosity of raw date seed oil is higher than that of diesel fuels.

Raw vegetable oils from different oil crops were tried for running diesel engines by many researchers (Goering et al., 1982; Korus et al., 1982; Peterson et al., 1983; Bagby, 1987; Juneja and Singh, 1995; Kalam and Masjuki, 2002). However, they encountered problems like higher viscosity, poor volatility, and poor flow characteristics of vegetable oils, which caused severe engine deposits, injector choking, and piston sticking in diesel engines (Peterson et al., 1983). Acrolein is a highly toxic substance emitted from the engine during thermal decomposition of glycerol present in vegetable oils (Schwab et al., 1987). These problems can be eliminated by transesterification of vegetable oils to form biodiesel (Perkins et al., 1991) and also the fuel properties of biodiesel can be made closer to the diesel fuel. As its name indicates, the use of this fuel in diesel engine alternates to diesel fuel.

Transesterification technology provides biodiesel as a renewable fuel and glycerol as a by-product, which can be used as a raw material for different chemical products. Biodiesel is produced by transesterification of large branched triglycerides of seed oil into smaller straight chain molecules of methyl esters (biodiesel), in the presence of the catalyst (Figure 18.2). There are three stepwise reactions with intermediate formation of diglycerides and monoglycerides resulting in the production of 3 mol of biodiesel and 1 mol of glycerol. The biodiesel is biodegradable, nontoxic, has

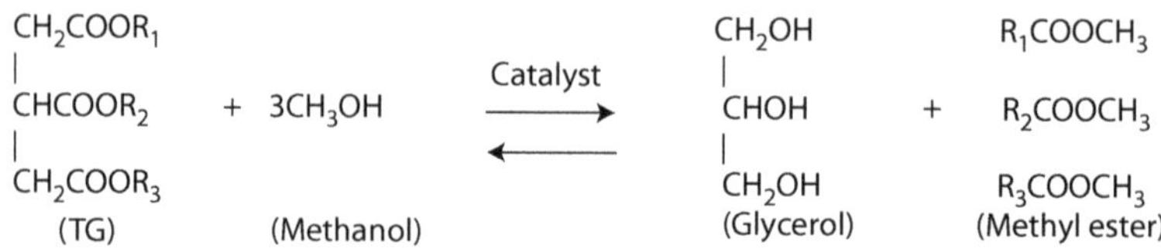

FIGURE 18.2 Transesterification reaction of oil with alcohol in the presence of catalyst.

low emission profiles, and it is environmentally beneficial (Krawczyk, 1996). Biodiesel has very low sulfur content and releases lower CO during combustion in diesel engines. Biodiesel contains oxygen, which leads to complete combustion in internal combustion engines. Biodiesel can be used in the existing diesel engines without any modification and can mix readily with diesel at any blending level (Krawczyk, 1996).

The comparison of important fuel properties of date seed oil with other fuels is furnished in Table 18.1. Cetane number (CN) can be used for determining the ignition quality of diesel. The CN for conventional diesel fuel ranges from 40 to 55. Based on the values of iodine and saponification values of date seed oil, the predicted value of CN is found to be 60.34, which is within the upper limit of CN (65) specified in U.S. biodiesel standard (ASTM PS 121–99). The biodiesel produced from date seed oil is also possible and it can be used as an alternate fuel for conventional diesel fuel. The steps involved in biodiesel production from date seed oil is presented in Figure 18.3.

THERMOCHEMICAL CONVERSION OF DATE SEEDS

The pyrolysis products are divided into a volatile fraction consisting of gases, vapors, and tar components and a carbon-rich solid residue. The pyrolysis process consists of a very complex set of reactions involving the formation of radicals (Demirbas, 2009). Synthetic diesel fuel cannot yet be produced directly by the pyrolysis process from any organic material. But diesel-like liquid called bio-oil can be produced through pyrolysis after removal of some valuable biochemicals that can be used in food additives or pharmaceuticals. Bio-oil can be used as diesel fuel in vehicles. Fluidized bed reactors with a char separation system are used for pyrolysis of date seed oil cake. Phenol, phenolic compounds, and pyrolytic lignin are also obtained during pyrolysis.

Charcoal is obtained by heating the whole date seed until its complete pyrolysis occurs, leaving only carbon and inorganic ash (Figure 18.4). The heat generated by the burning part of the date seed

TABLE 18.1
Comparison of Selected Properties of Date Seed Oil with Other Fuels

S. No.	Particulars	Date Seed Oil	Jatropha Oil	Diesel[c]
1	Specific gravity	0.9174 at 20°C	–	836–850 at 30°C
2	Iodine value	54.8[a]	93.0[b]	–
3	Saponification value	207[a]	202.6[b]	–
4	Cetane number	60.34[d]	52.31[d]	40–55

[a] El-Shurafa, M. Y., H. S. Ahmed, and S. E. Abu-Naji. 1982. Organic and inorganic constituent of dates palm pit (seeds). *Date Palm J.*, 1:75.

[b] Bringi, N. V. 1987. *Non-Traditional Oilseeds and Oils of India*. New Delhi: Oxford & IBH Publishing Co. Pvt. Ltd.

[c] Pramanik, K. 2003. Properties and use of *Jatropha curcas* oil and diesel fuel blends in compression ignition engine. *Renew. Energy*, 28:239–248.

[d] Calculated by using an equation developed by Krisnangkura, K. 1986. A simple method for estimation of Cetane index of vegetable oil methyl esters. *JAOCS*, 63:552–553.

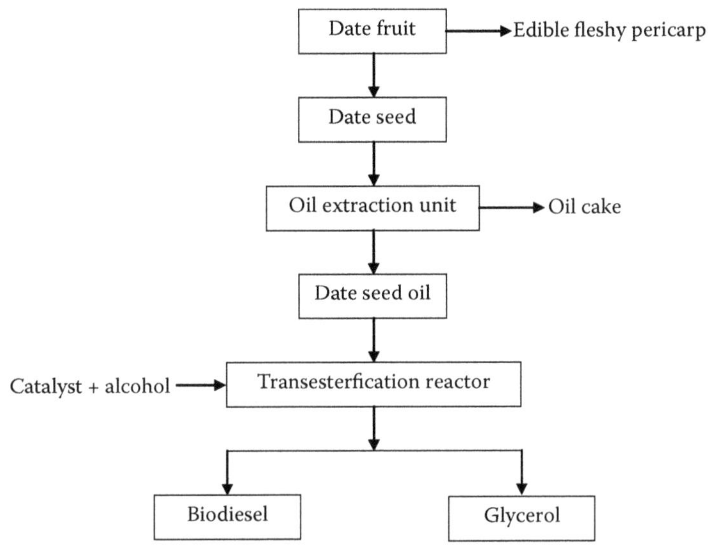

FIGURE 18.3 Biodiesel production from date seed oil.

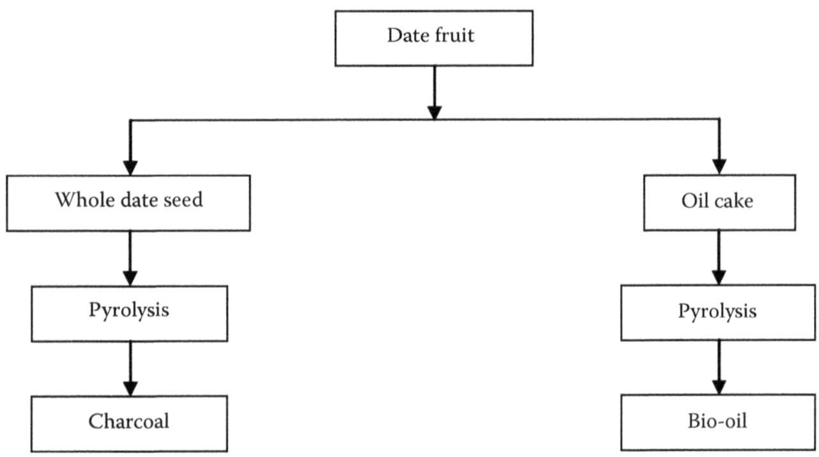

FIGURE 18.4 Thermo-chemical conversion route for production of charcoal and bio-oil from date seed and oil cake.

and volatile by-products pyrolyzes the rest of the pile. The composition of end product produced by dry distillation of date pits is presented in Table 18.2. The best reaction temperature for charcoal production from date seed ranges from 300°C to 400°C (Barreveld, 1993). The limited supply of oxygen to this process prevents the charcoal from burning.

BIOGAS PRODUCTION FROM DATE SEED MEAL

The anaerobic digestion of any organic matter in the absence of air yields gas, which comprises gases like methane (CH_4), carbon dioxide (CO_2), hydrogen (H_2), and so on. The hydraulic retention time varies from 10 days to few weeks depending upon the feedstock used for anaerobic digestion. The composition of the biogas generated depends upon feedstock compositions. The main uses of biogas are thermal applications, running the dual-fuel engines, and fuel cells.

TABLE 18.2
Compositions of Dry Distillation of Date Pits

S. No.	Components	Composition (%)
1	Carbon	27
2	Tar, crude	13–14
3	Acetic acid	2–3
4	Methanol	2

Source: Adapted from Barreveld, W. H. 1993. *Date Palm Products*, Agricultural Services Bulletin No. 101, 216 pp, FAO, Rome.

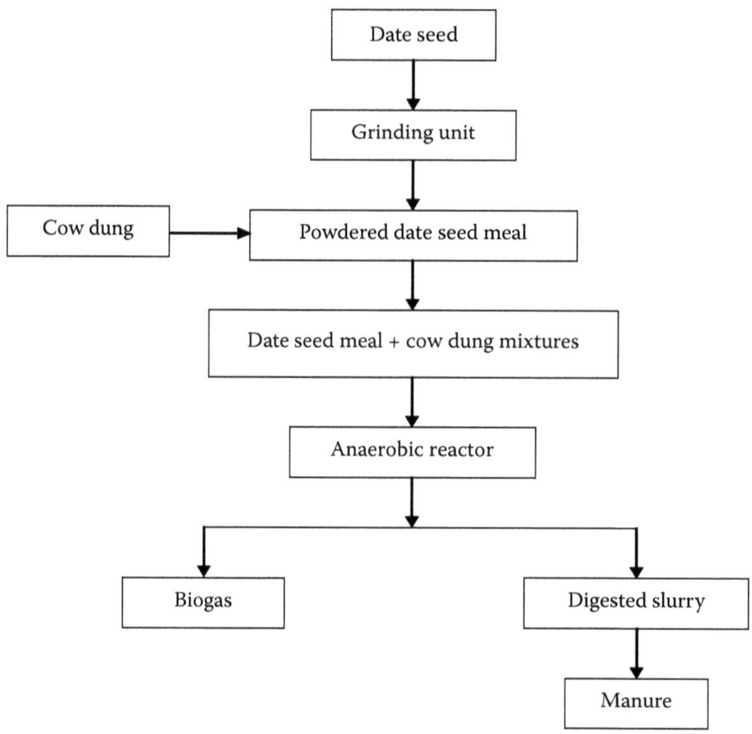

FIGURE 18.5 Biogas production from date seed meal.

The date seeds generated from the date-processing industry are rigid and hard in nature, which cannot be digested in biogas plant. The whole seed can be crushed before feeding into the biogas digester (Figure 18.5). Al-Turki et al. (2004) studied biogas production from different combination date seed meal and cow dung mixtures. They found that the combination of 90% of cattle dung and 10% of date pits meal yielded the highest biogas production.

PRODUCTION OF ETHYL ALCOHOL

Ethanol is an alcohol and is fermented from different feedstock such as sugars, starches, or lignocellulosic biomass. Most of the commercial production of ethanol is from crops like sugar cane or sugar beet, as starches and cellulosic biomass. It is possible that wood, straw, and even household wastes may be economically converted into bioethanol. Carbohydrates (hemicelluloses and cellulose) in

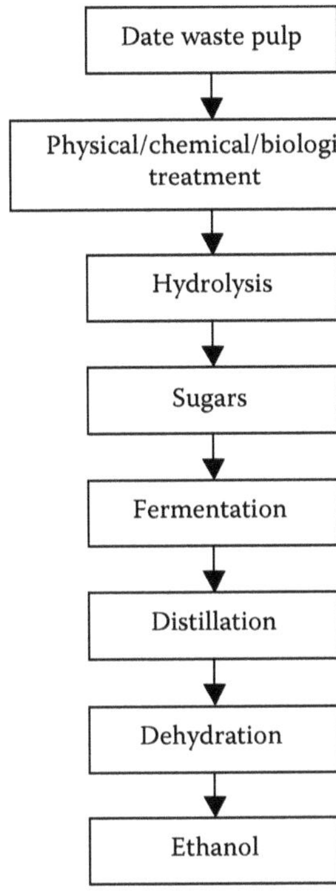

FIGURE 18.6 Bioethanol production from date waste pulp.

plant materials can be converted into sugars by hydrolysis. Fermentation is an anaerobic biological process in which sugars are converted into ethyl alcohol by the action of fermentative microorganisms, usually yeast like *Saccharomyces* or bacteria like *Zymomonas mobilis*. The value of any particular type of biomass as feedstock for fermentation depends on the ease with which it can be converted into sugars.

Residuals of date factories are among the best food stuffs exposed to fermentation because they are rich in sugars, vitamins, and mineral ingredients in addition to other nonorganic substances. By virtue of the composition of date waste pulp, it serves as an excellent growth medium for the alcoholic fermentative microflora. Date waste pulp can be chopped or crushed, boiled at suitable concentration, and filtered. After sterilization, the extract (wort) can be added with suitable starter inocula for hydrolysis (Figure 18.6). Hydrolyzed and filtered extracts are fermented in retorts at ambient temperature and the slurry is kept under agitation for fermentation. After fermentation, the biomass can be separated and the slurry can be distilled for recovery of fuel ethanol. Ethanol concentration increases with higher fruit glucose concentration. Hence, to enhance the yield of ethanol, suitable sugar additives or adjuncts may be explored.

CONCLUSION

Date palm can be used for sustainable biodiesel production, not only from the preextracted oil but also directly from the wastes of dates. Date biodiesel quality meets the biodiesel international standards. The approach suggested addresses the production of cost-effective biodiesel using a neglected but highly adapted plant and utilization of the wastes accrued from the plant for sustainable and environmentally safe biofuel, which ultimately helps in the use of renewable energy for the future. Further, use of 5% of bioethanol as fossil fuels will reduce the emission of carbon monoxide into the atmosphere by 25%–30%. More importantly, creating such an insight will call for more palm tree plantations that will act as a true carbon sink to offset carbon dioxide that is released into the atmosphere.

REFERENCES

Ahmed, A. I., A. W. K. Ahmed, and R. K. Robinson. 1995. Chemical composition of date varieties as influenced by the stage of ripening. *Food Chem.*, 54:305–309.

Al-Farsi, M. A. and C. Y. Lee. 2008. Nutritional and functional properties of dates: A review. *Crit. Rev. Food Sci. Nutr.*, 48:877–887.

Al-Hooti, S., J.S. Sidhu, and H. Qabazard. 1998. Chemical composition of seeds date fruit cultivars of United Arab Emirates. *J. Food Sci. Technol.*, 35:44–46.

Almana, H. A. and R. M. Mahmoud. 1994. Palm date seeds as an alternative source of dietary fibre in Saudi bread. *Ecol. Food Nutr.*, 32:261–270.

Al-Qaessi, F. A. H. 2010. Production of activated carbon from date stones by using zinc chloride. *Energy Sour., Part A*, 32:917–930.

Al-Shahib, W. and R. J. Marshall. 2003. The fruit of the date palm: Its possible use as the best food for the future? *Int. J. Food Sci. Nutr.*, 54(4):247–259.

Al-Showiman, S. S. 1990. Chemical composition of some date palm seeds (*Phoenix dactylifera* L.) in Saudi Arabia. *Arab Gulf J. Sci. Res.*, 8:15–24.

Al-Turki, A., Y. M. El-Hadidi, and S. A. Al-Yahya. 2004. Utilization of date pits as a potential source of biogas and organic fertilizer. *J. Food. Agric. Environ.*, 2:369–374.

Bagby, M. O. 1987. *Vegetable Oils for Diesel Fuel: Opportunities for Development*. ASAE Paper No. 87–158. ASAE, St. Joseph, MI.

Barreveld, W. H. 1993. *Date Palm Products*, Agricultural Services Bulletin No. 101, 216 pp, FAO, Rome.

Besbes, S., C. Blecker, C. Deroanne, N. E. Drira, and H. Attia. 2004. Date seeds: Chemical composition and characteristic profiles of the lipid fraction. *Food Chem.*, 84(4):577–584.

Besbes, S., S. Cheikh Rouhou, C. Blecker, C. Deroanne, G. Lognay, N. E. Drira, and H. Attia. 2006. Voies de valorisation des sous produits de dattes: Valorisation de la pulpe. *Microbiol. Hyg. Alim.*, 18:3–7.

Besbes, S., R. Ghorbel, R. Salah, M. Masmoudi, F. Jedidi, H. Attia, and C. Blecker. 2010. Date fiber concentrate: Chemical compositions, functional properties and effect on quality characteristics of beef burgers. *J. Food Drug Anal.*, 18(1):8–14.

Bringi, N. V. 1987. *Non-Traditional Oilseeds and Oils of India*. New Delhi: Oxford & IBH Publishing Co. Pvt. Ltd.

Burnie, G., S. Forrester, D. Greig, and S. Guest. 2006. *Botanica-Encyclopédie de botanique et d' horticulture*, 1st Edn. Place Des victoires Eds, Paris.

Demirbas, A. 2009. Biofuels from agricultural biomass. *Energy Sour., Part A*, 31:1573–1582.

Devshony, S., E. Eteshola, and A. Shani. 1992. Characteristics and some potential applications of date palm seeds and seed oil, *JAOCS*, 69(6):595–597.

El-Shurafa, M. Y., H. S. Ahmed, and S. E. Abu-Naji. 1982. Organic and inorganic constituent of dates palm pit (seeds). *Date Palm J.*, 1:75.

Fayadh, J. M. and S. S. Al-Showiman. 1990. Chemical composition of date palm (*Phoenix dactylifera* L.). *J. Chem. Soc. Pak.*, 12:84–103.

Matteson, G. C. and B. M. Jenkins. 2007. Food and processing residues in California: Resource assessment and potential for power generation. *Bioresour. Technol.*, 98:3098–3105.

Ghazanfari, A., S. Panigrahi, and L. Tabil Jr. 2005. *Experiments on Production of Bio-Composite Plates from Pistachio Shells, Date Pits and HDPE*. Paper No. 05–079. Paper presented at the CSAE/SCGR 2005 Meeting, Winnipeg, Manitoba during June 26–29, 2005.

Goering, C. E., A. W. Schwab, M. J. Daugherty, E. H. Pryde, and A. J. Heakin. 1982. Fuel properties of eleven vegetable oils. *Trans. ASAE*, 25(6):1472–1477.

Hamada, J., I. Hashim, and F. Sharif. 2002. Preliminary analysis of potential uses of date pits in foods. *Food Chem.*, 76:135–137.

Hui, Y. H. (Ed.), J. Barta, M. Pilar Cano, T. Gusek, J. S. Sidhu, and N. Sinha (Assoc. Eds.). 2006. *Handbook of Fruits and Fruit-Processing*. Ames, Iowa: Blackwell Publishing Professional.

Hussein, A. S., G. A. Alhadrami, and Y. H. Khalil. 1998. The use of dates and date pits in broiler starter and finisher diets. *Bioresour. Technol.*, 66:219–223.

Juneja, N. N. and S. R. Singh. 1995. Petroleum conservation through use of vegetable oil (rape seed oil) as diesel engine supplementary fuel. *Agric. Eng. Today*, 19(3–4):8–15.

Kalam, M. A. and H. H. Masjuki. 2002. Biodiesel from palm oil-an analysis of its properties and potential. *Biomass Bioenergy*, 23:471–479.

Katikaneni, S. P. R., J. D. Adjaye, and N. N Bakhshi. 1995. Catalytic conversion of canola oil to fuels and chemicals over various cracking catalyst. *Can. J. Chem. Eng.*, 73:384–397.

Kim, S. and B. E. Dale. 2005. Life cycle assessment of various cropping systems utilized for producing: Bioethanol and biodiesel. *Biomass Bioenergy*, 29:426–439.

Korus, R. A., T. L. Mousetis, and L. Lloyd. 1982. Polymerization of vegetable oils in vegetable oil fuels. *Proc. Int. Conf. Plant and Vegetable Oils as Fuels*, Furgo, ND, 2–4 August:218–223.

Krawczyk, T. 1996. Biodiesel—Alternative fuel makes in roads but hurdles remain. *INFORM*, 7:801–829.

Krisnangkura, K. 1986. A simple method for estimation of Cetane index of vegetable oil methyl esters. *JAOCS*, 63:552–553.

Mazaheri Assadi, M. and M. Nikkhah. 2002. Production of citric acid from date pulp by solid state fermentation. *J. Agric. Sci. Technol.*, 4:119–125.

Mossa, J. S., M. S. Hifnawy, and A. G. Mekkawi. 1986. Photochemical and biological investigation on date seeds (*Phoenix dactylifera* L.) produced in Saudi Arabia. *Arab Gulf J. Sci. Res.*, 4:495.

Perkins, L. A., C. L. Peterson, and D. L. Auld. 1991. Durability testing of transesterified winter rape oil (*Brassica napis* L.) as fuel in small bore, multi cylinder DI, CI engines. SAE Paper No. 911764. Warrendale, MI: SAE.

Peterson, C. L., D. L. Auld, and R. A. Korus. 1983. Winter rape oil fuel for diesel engines: Recovery and utilization. *JAOCS*, 60(8):1579–1587.

Pramanik, K. 2003. Properties and use of *Jatropha curcas* oil and diesel fuel blends in compression ignition engine. *Renew. Energy*, 28:239–248.

Rygg, G. L. 1977. *Date Development, Handling and Packing in the United States*, Agricultural Handbook No. 482. USDA, Washington, DC, pp. 25–41.

Schwab, A. W., M. O. Bagby, and B. Freedman. 1987. Preparation and properties of diesel fuels from vegetable oils. *Fuel*, 66(10):1372–1378.

Sheehan, J., V. Cambreco, J. Duffield, M. Garboski, and H. Shapouri. 1998. *An Overview of Biodiesel and Petroleum Diesel Life Cycles.* A report by US Department of Agriculture and Energy, Washington, D.C., pp. 1–35.

Shinwari, M. A. 1993. Date palm. In: Macrae R., Robinson R.K., and Sadler M. (Eds.). *Encyclopedia of Food Science*, Food Technology and Nutrition. London, UK: Academic Press Ltd, pp. 1300–1305.

Zaid, A. (Ed.). 1999. *Date Palm Cultivation.* Rome: United Nations FAO Plant Production and Protection Paper.

Zhang, R. H. and Z. Zhang. 2002. Biogasification of solid wastes by anaerobic phased solids digester system. US Patent 6,342,378.

Part III

Food

19 Physical Characteristics and Chemical Composition of Date Palm Fruits

Nejib Guizani and Vandita Singh

CONTENTS

INTRODUCTION

The fruits of the date palm (*Phoenix dactylifera* L.) are one of the important agricultural commodities in North-African countries, the Middle East, and Asia. The fruit of date palm has served as a staple nutritious food and source of wealth for many years (Khan et al., 2008). In addition to their use as a dietary supplement, dates are of commercial importance and sold worldwide as a commodity or processed into various products (Omezzine et al., 1997). Dates are produced in 35 countries worldwide and cultivated on about 2.9 million acres of land. The world production of dates has increased considerably during the last 30 years, from ~2,289,511 tonnes in 1974 to ~6.9 million tonnes in 2004 (FAOSTAT, 2007). Edible dates pass through four distinct stages of ripening termed in Arabic—Kimri, Khalal, Rutab, and Tamr—and they represent, respectively, the immature green, the mature full colored, the soft brown, and the hard raisin-like stages of development (Ahmed et al., 1995). Following harvesting, dates at Tamr stage are dried typically by air or solar methods to different levels of moisture content for preservation.

The dates could be consumed as fresh fruits at Khalal and Rutab stages (short shelf life), or at Tamr stage (good storability), or may be processed into various products such as date paste, syrup or powder which are used as ingredients in cookies or cake manufacturing. The largest consumption of dates is at Tamr stage due to their good storability and availability all year around. Some date varieties, such as Khasab and Hilali Pakistan, do not produce Tamr stage (Ahmed et al., 1995), and are thus consumed at Khalal and Rutab stages.

Date composition can be strongly influenced by the variety and degree of ripeness. The physical and compositional characteristic of date fruit is of prime importance. Its physical properties are good indicators of maturity, degree of ripeness, and predicted shelf life. On the other hand, the chemical composition relates to nutritional significance and health benefits. Consideration of these properties is also essential for sorting and grading as well as handling of products for quality control (Khan et al., 2008).

This chapter deals with the physical characteristics and chemical composition of dates at the Tamr stage in which they are mostly traded and consumed.

Typically, dates at the Tamr stage contain water (7%–38%), carbohydrate (total sugars, 44%–88%), fat (0.1%–3.25%), protein (1.5%–5.4%), dietary fiber (6.4%–11.5%), minerals, vitamins (such as vitamin C, B_1, B_2, A, riboflavin, and niacin), and phenolic compounds.

MORPHOLOGICAL CHARACTERISTICS

The date palm fruit is a berry consisting of pericarp and seed. The pericarp, which constitutes between 85% and 90% of date fruit weight (Hussein et al., 1998), is composed of exocarp, fleshy mesocarp, and papery endocarp.

The morphological characteristics of different date varieties are shown in Table 19.1. Significant differences in the weight, length, and width are observed among the varieties. The weight of individual fruit ranged from 4.25 g for Zaidi to 10.00 g for Deglet Nour. The length ranged from 28.55 mm for Ruzeiz to 60.00 mm for Deglet Nour and the width ranged from 16.00 mm for Allig to 23.35 for Barhee. If the dates show no apparent abnormality and visual damage, and had a total weight, a pulp weight, and a length greater than 6 g, 5 g, and 35 mm respectively, then it is considered as acceptable morphological quality (Açouren, 2001).

TABLE 19.1
Morphological Characteristics of Dates

Variety	Average Length (mm)	Average Width (mm)	Average Fruit Weight (g)	Reference
Deglet Nour	40.00	18.00	10.40	Bellagha et al. (2008)
	60.00	18.00	12.00	Maatalah (1970)
Khalas	35.00	18.00	8.50	Rahman and Al-Farsi (2005)
	31.65	21.50	10.32	Ismail et al. (2006)
Fard	31.95	20.48	7.55	Ismail et al. (2006)
Allig	44.00	16.00	11.9	Bellagha et al. (2008)
Kentichi	37.00	17.00	6.70	Bellagha et al. (2008)
Barni	34.00	17.00	6.00	Kasapis et al. (2000)
Aseel	40.00	18.30	7.18	Khan et al. (2008)
Basra	38.00	17.00	4.76	Khan et al. (2008)
Begun	32.01	15.26	6.16	Khan et al. (2008)
Zaidi	34.01	17.40	4.25	Khan et al. (2008)
Barhee	29.76	21.50	8.78	Ismail et al. (2006)
Boumaan	31.20	23.35	9.24	Ismail et al. (2006)
Ruzeiz	28.55	20.82	7.86	Ismail et al. (2006)
Ghars	40.00	18.00	9.00	Belguedj (2002)

PHYSICOCHEMICAL CHARACTERISTICS

pH

The pH of dates is slightly acidic; it varies between 5 and 6. This pH range is not favorable for bacterial growth; however, it is favorable for mold growth (Reynes et al., 1994), and for the preservation of some predominant B vitamins found in dates such as B_1, B_2, B_5, B_9, and B_{12} (Bourgeois, 2003). The pH is related to the degree of maturity of dates. The pH of dates increases from Kimri to Tamr stage. It increased for Deglet Nour from pH 5.0 at the Kimri stage to 6.2 of the Tamr stage (Maatalah, 1970).

Acidity

Dates have in general a low acidity, varying between 2.02 and 6.3 g of acid/kg of fresh dates (Rygg et al., 1953). Malic and acetic were the two main organic acids responsible for the acidity in the Deglet Nour variety (Maatalah, 1970). These organic acids hinder the growth of microorganisms and influence the sensory characteristics of fruits (Jadhav and Andrew, 1977; Siebert, 1999). However, high acidity is usually associated with low-quality dates.

Moisture Content

The moisture content of dates at Tamr stage varies between 7% and 38%. Table 19.2 shows the moisture content of various date varieties grown in different parts of the world.

The moisture content of mature dates is very important to their stability during storage. Traditional dry dates in the Tamr stage are microbiologically stable when their water activity (a_w) is reduced below 0.6 (Beuchat, 1981). This critical value is achieved by removing water from the fresh fruit. The relationship between a_w and moisture content (at a constant temperature) is described practically and theoretically by a moisture sorption isotherm. Determination of the moisture sorption isotherm of dates can be used directly to solve food-processing design problems, predict energy requirements, determine proper storage conditions, and aid in food product formulations (Myhara et al., 1998a). Desorption isotherms of the different varieties of dates showed a less pronounced sigmoidal shapes. This shape is characteristic of high-sugar foods (Saravacos et al., 1986). At low relative humidity, dates adsorbed a low amount of water because of the water sorption of high molecular components. However, for $a_w > 0.5$, a larger amount of moisture is adsorbed because of the dominant sorption of water. A slight difference in the sorption behavior for the different varieties was also reported. This difference may be due to differences in the composition of dates and more specifically due to the total sugar content and the type of sugar present. For instance, Kentichi has the lowest sugar content (0.68 g/g) (Bellagha et al., 2008) compared to Allig (0.86 g/g) (Bellagha et al., 2008) and to Deglet Nour (0.741 g/g) (Guizani et al., 2010). For the same water activity, Kentichi has thus the lowest water content compared to Deglet Nour and Allig dates. The type of sugar present also plays a role in the reduction of water activity. Fructose has a higher water activity reduction capacity than sucrose and glucose due to its high water-holding capacity. The moisture content of the dates contributes to its texture through interaction with carbohydrates and to a lesser extent with proteins, lipids, and salts. In general, the moisture content is used to classify the dates as dry, semidry or semisoft, and soft (Espiard, 2002).

As an illustration, three main cultivar groups of date were reported:

1. *Soft*: Barhee, Halawy, Khalas, and Qassimand Khalas Kharj
2. *Semisoft or Semidry*: Deglet Nour, Zahidi, Ajwa, Anbara, Mabroom, Medjool, and Barni
3. *Dry*: Thoory, Ruthana, Nabtat Ali, Segae, and Sukkari

TABLE 19.2
Proximate Analysis (g/100 g Fresh Fruit) of Some Important Commercial Varieties of Dates at Tamr Stage

Variety	Moisture	Protein	Lipid	Ash	Sugars[a]	Reference
Deglet Nour	21.43	2.47	0.31	1.68	74.11	Guizani et al. (2010)
Mazafati	37.50	3.70	0.54	1.25	57.01	Sahari et al. (2007)
Kabkab	31.00	3.70	0.30	1.66	63.34	Sahari et al. (2007)
Zahedi	30.90	5.00	0.28	1.50	62.32	Sahari et al. (2007)
Khanizi	25.70	5.00	0.37	1.62	67.31	Sahari et al. (2007)
Zark	19.50	3.70	0.49	1.88	74.43	Sahari et al. (2007)
Khalas	22.30	2.10	0.1	1.40	74.1	Ahmed et al. (1995)
	12.60	1.70	0.50	1.80	83.4	Al-Farsi et al. (2005)
Fard	27.70	2.10	0.1	1.80	68.3	Ahmed et al. (1995)
	18.50	1.50	1.40	1.50	77.1	Al-Farsi et al. (2005)
Khasab	16.50	1.60	1.00	1.60	79.3	Al-Farsi et al. (2005)
Naghal	9.20	2.70	0.20	0.70	87.2	Ahmed et al. (1995)
Khunaizy	18.00	2.20	0.20	1.50	78.1	Ahmed et al. (1995)
Barhi	29.50	2.30	0.10	1.50	66.6	Ahmed et al. (1995)
Hallawi	07.30	2.10	0.50	1.80	88.3	Yousif et al. (1982)
Sayer	07.50	2.60	0.30	1.70	87.9	Yousif et al. (1982)
Khadrawi	09.50	2.20	0.40	1.90	86	Yousif et al. (1982)
Zahdi	08.30	2.00	0.40	1.70	87.6	Yousif et al. (1982)
Boushibal	16.30	1.90	0.10	1.80	79.9	Al-Hooti et al. (1997)
Gash Gaafar	07.20	2.20	0.10	1.90	88.6	Al-Hooti et al. (1997)
Gash Habash	12.80	1.70	0.10	1.60	83.8	Al-Hooti et al. (1997)
Lulu	13.10	2.50	0.10	1.60	82.7	Al-Hooti et al. (1997)
Shahla	12.60	2.10	0.20	1.90	83.2	Al-Hooti et al. (1997)
Shahal	17.52	1.10	2.20	1.84	77.34	Al-Farsi et al. (2007)
Um-Sellah	09.73	1.79	2.04	1.99	84.45	Al-Farsi et al. (2007)
Mabseela	14.58	1.15	3.25	1.41	79.61	Al-Farsi et al. (2007)

[a] Calculated by difference.

Carbohydrates

Carbohydrates are the most important constituents of dates. Dates provide a good source of rapid energy (sugars) and a good nutritional value, based on their dietary fiber.

The sugar composition of different varieties of dates is presented in Table 19.3. The sugar content varied between 35% and 88%. The sugars consisted mainly of sucrose, fructose, and glucose. The relative amounts of sucrose, glucose, and fructose are determined mainly by varietal characteristics. Most of the date varieties belong to the invert sugar type where sucrose is totally inverted into glucose and fructose at the mature edible Tamr stage. The presence of a high content of fructose and glucose is the result of a relatively important invertase activity in these varieties, which would considerably reduce their content in sucrose (Fayadh and Al-Showiman, 1990). Deglet

TABLE 19.3
Sugar Content (g/100 g Fresh Weight) of Some Important Commercial Varieties of Dates at Tamr Stage

Variety	Fructose	Glucose	Sucrose	Total Sugars	Reference
Deglet Nour	22.26	23.75	24.33	74.11	Guizani et al. (2010)
	09.53	10.36	39.84	59.73	Elleuch et al. (2008)
Medjool	33.73	32.67	nd	66.40	Rock et al. (2009)
Hallawi	36.43	37.17	2.21	75.81	Rock et al. (2009)
Khalas	31.90	30.30	nd	62.20	Al-Farsi et al. (2005)
	21.90	29.40	nd	51.30	Rahman et al. (2005)
	28.59	27.34	nd	55.93	Ismail et al. (2006)
	26.5	30.30	nd	57.00	Ahmed et al. (1995)
Fard	28.20	28.50	nd	56.70	Al-Farsi et al. (2005)
	28.34	26.86	nd	55.20	Ismail et al. (2006)
	29.80	29.80	nd	59.50	Ahmed et al. (1995)
Khasab	27.40	28.70	nd	56.10	Al-Farsi et al. (2005)
Allig	21.20	21.86	10.16	53.22	Elleuch et al. (2008)
Aseel	39.20	39.20	nd	78.40	Khan et al. (2008)
Basra	37.20	40.60	nd	77.80	Khan et al. (2008)
Begun	39.60	39.60	nd	78.20	Khan et al. (2008)
Zaidi	39.40	39.40	nd	78.60	Khan et al. (2008)
Barhee	29.05	32.60	nd	61.65	Ismail et al. (2006)
Boumaan	29.50	26.80	nd	56.30	Ismail et al. (2006)
Ruzeiz	29.40	24.60	nd	54.00	Ismail et al. (2006)
Khunaizy	28.50	25.40	nd	53.90	Ahmed et al. (1995)
Barhi	27.60	29.70	nd	57.20	Ahmed et al. (1995)
Mazafati	21.70	35.90	nd	57.60	Sahari et al. (2007)
Zahidi	35.23	47.57	nd	82.80	Sahari et al. (2007)
Khanizi	37.73	49.81	nd	87.54	Sahari et al. (2007)
Zark	12.62	16.41	28.9	57.93	Sahari et al. (2007)
Boushibal	33.40	33.00	nd	66.40	Al-Hooti et al. (1997)
Gash Gaafar	36.80	36.80	nd	73.60	Al-Hooti et al. (1997)
Gash Habash	34.40	33.50	nd	67.90	Al-Hooti et al. (1997)
Lulu	39.95	38.55	nd	78.50	Al-Hooti et al. (1997)
Shahla	38.58	40.04	nd	78.62	Al-Hooti et al. (1997)
Smiti	17.83	17.75	nd	35.58	Chaira et al. (2009)
Kenta	21.36	22.04	nd	43.40	Chaira et al. (2009)
Mermella	32.39	33.67	nd	66.06	Chaira et al. (2009)
Garn Ghzel	29.55	30.91	nd	59.86	Chaira et al. (2009)

Note: nd, not detected.

Nour, an important variety in North Africa and California, is a good example of dates where sucrose is dominant as a result of its partial inversion at commercial maturity.

Dates are generally classified into three commercial subdivisions based on their external qualities of texture and pliability at the Tamr stages (Barreveld, 1993). This classification seems to be correlated to the sucrose content. With the exception of Deglet Nour and Zahdi varieties, the degree of dryness increases with an increased sucrose content. Moisture and fiber contents also play a role in determining this classification.

Dates are a good source of dietary fibers, mainly the insoluble materials; they provide 6.3–10.9 g total fiber/100 g of date flesh, depending on variety, degree of ripeness, growth location, and the method of analysis used. The content of dietary fiber was between 1.8% and 2.5% for Iraqi dates (Yousif et al., 1982), 7.9 and 7.0 g/100 g fresh dates for Medjool and Hallawi, respectively (Rock et al., 2009), 5.94, 8.72, and 8.46 g/100 g fresh dates for Mabseeli, Um-Sellah, and Shahal, respectively (Al-Farsi, 2007), and 8.1%, 9.7%, 10.5%, 11.8%, and 12.75% for Safawy, Mabroom, Bamy, Shalaby, and Lobanah Masery, respectively (Al-Shahib and Marshall, 2002). Dates and their by-products serve as good sources of dietary fiber compared to syrups and other fresh and dried fruits.

Proteins

Proteins are present in dates in small amounts and vary from 2.3% to 5.6%. The protein contents of several date varieties are shown in Table 19.2. In spite of their low-protein content, dates may contribute to the human diet with high quality of some essential amino acids (Salem and Hegazi, 1971). Table 19.4 gives the amino acid pattern of some varieties of dates.

Lipids

The lipid content of dates is generally low, 0.1%–3.25% of the fresh weight, and is mainly concentrated in the skin. Table 19.2 gives the lipid content of some date varieties grown in different countries. The major free fatty acids identified in the date flesh are caprylic acid followed by linoleic, lauric, pelargonic, myristic acids, and so on.

Minerals

Dates contain low percentage of ash, between 0.70 and 1.99 g/100 g dates (Table 19.2). Table 19.5 shows that among the minerals studied, potassium was the most abundant, with a concentration of 413–1164 mg/100 g, followed by calcium (15–89.9 mg/100 g), sodium (2.4–287 mg/100 g), and phosphorus (59–104 mg/100 g). Dates were also found to be rich sources of micronutrients such as iron (0.3–6.03 mg/100 g), zinc (0.20–2.41 mg/100 g), magnesium (47–89.6 mg/100 g), copper (0.10–2.38 mg/100 g), and manganese (02–1.20 mg/100 g).

Dates are also good sources of copper, zinc, and manganese. These microminerals intervene in several therapeutic aspects such as normal functioning of immune system (zinc) and nausea and headache diseases treatments (iron).

Vitamins

Dates are not an important source of vitamins, in particular the fat-soluble vitamins. The only vitamins that are relatively important in dates belong to water-soluble vitamins of the B group and C. In general, dates are a moderate source of vitamins B_1, B_2, and B_9. Vitamins C and A are found in relatively low concentrations in dried dates. Yousif et al. (1982) reported that Hallawi dates contain 92 μg/100 g of vitamin B_1, 160 μg/100 g of vitamin B_2, 53 μg/100 g of vitamin B_9, and 3300 μg/100 g of vitamin C; Sayer dates contain 120 μg/100 g of vitamin B_1, 125 μg/100 g of vitamin B_2, 65 μg/100 g of vitamin B_9, and 16,000 μg/100 g of vitamin C content. Khadrawi dates contain 85 μg/100 g of vitamin B_1, 135 μg/100 g of vitamin B_2, 39 μg/100 g of vitamin B_9, and 2900 μg/100 g of vitamin C content while Zahidi dates contain 73 μg/100 g of vitamin B_1, 153 μg/100 g of vitamin B_2, 58 μg/100 g of vitamin B_9, and 2200 μg/100 g of vitamin C content.

Carotenoids and Phenolics

Table 19.6 shows the total carotenoids and total phenolics in dried dates grown worldwide. Carotenoids content varied significantly between date varieties from 5.13 to 2900 μg/100 g date

TABLE 19.4
Amino Acid Content of Dried Dates (mg/100 g Wet Basis)

Varieties	Ala	Arg	Asp	Cys	Glu	Gly	His	Iso	Leu	Lys	Met	Phe	Pro	Ser	Thr	Try	Tyr	Val	Reference
Khastawi	92.7	80.9	161.2	58.7	226.8	120.1	36.8	61.7	114.6	83.8	15.9	62.0	126.6	67.4	62.8	—	36.6	85.6	Auda et al. (1976)
Khadhrawi	116.2	62.3	159.9	40.0	257.5	113.4	0.1	60.8	110.8	66.5	22.6	62.2	108.1	69.0	55.0	—	43.8	80.9	Auda et al. (1976)
Zahdi	133	67	117	53	231	106	24	43	76	85	11	49	69	58	41	—	25	—	Auda et al. (1976)
Zahdi	63	81	67	—	200	45	46	37	61	42	24	45	94	45	55	—	37	—	Al-Aswad (1971)
Hallawi	105	39	129	32	158	98	21	43	84	51	18	53	110	64	51	46	15	—	Al-Rawi et al. (1967)
Hillawi	70	91	59	—	225	64	44	44	72	46	28	66	81	53	44	—	49	—	Al-Aswad (1971)
Sayir	79	45	119	35	183	92	19	41	78	51	12	43	100	58	45	46	15	—	Al-Rawi et al. (1967)
Sayer	117	95	148	—	248	105	29	50	105	52	62	67	148	74	57	—	52	—	Al-Aswad (1971)
Deglet Nour	83	136	213	67	359	101	32	49	84	66	22	50	130	57	43	12	15	—	USDA (2007)
Deglet Nour	—	—	—	31.85	—	—	—	41.95	86.25	64.50	39.35	55.10	—	—	76.35	19.50	46.35	91.10	Açouren (2001)
Algerian	130	68	174	51	258	130	36	64	103	72	25	70	144	88	69	66	26	88	Favier et al. (1993)

TABLE 19.5
Mineral Content of Date Flesh (mg/100 Dry Weight)

Variety	K	Na	Mg	Ca	Fe	Zn	Mn	P	Cu	Se	Reference
Naghal	788	287	47	15	1.2	0.2	0.7	—	0.2	—	Ahmed et al. (1995)
Khuneizi	704	197	59	15	1.5	0.2	0.4	—	0.1	—	Ahmed et al. (1995)
Bahri	855	75	82	12	0.3	0.1	—	—	0.2	—	Ahmed et al. (1995)
Khalas	630	82	62	16	1.7	0.4	0.3	—	0.4	—	Ahmed et al. (1995)
	1164	85	54.1	—	1.6	0.25	0.47	—	—	—	Ismail et al. (2006)
	742	3.6	76	85	0.6	0.5	0.3	74	0.6	0.28	Al-Farsi et al. (2005)
Khasab	603	2.4	67	55	1.1	0.6	0.2	63	0.6	0.4	Al-Farsi et al. (2005)
Ruzeiz	413	55	41	—	1.4	0.38	0.4	—	—	0.24	Ismail et al. (2006)
Fard	914	141	63	14	1.2	0.2	0.39	—	0.4	—	Ahmed et al. (1995)
	624	3.5	61	82	0.6	0.5	0.3	59	0.8	0.4	Al-Farsi et al. (2005)
	664	9.6	52.3	—	1.32	0.27	0.5	—	—	—	Ismail et al. (2006)
Barhree	789	59.9	62.3	—	—	0.6	0.45	—	—	—	Ismail et al. (2006)
Aseel	519	9.68	—	27.4	1.31	0.55	—	—	0.13	—	Khan et al. (2008)
Basra	556	9.02	—	40.6	1.16	0.71	—	—	0.14	—	Khan et al. (2008)
Begun	527	5.83	—	41	1.87	0.61	—	—	0.14	—	Khan et al. (2008)
Smiti	740	11.1	84	71.1	6.03	2.40	1.06	82.3	2.07	—	Chaira et al. (2009)
Kenta	471	9.15	61.8	89.2	4.47	2.06	1.20	66.6	2.38	—	Chaira et al. (2009)
Mermella	668	12.9	75.3	89.9	5.20	2.41	1.06	69.1	2.57	—	Chaira et al. (2009)
Garn Ghzel	422	9.31	89.6	68.3	4.62	2.08	1.04	63.2	1.98	—	Chaira et al. (2009)
Deglet Nour	863	10.2	41.6	47.7	2.5	—	—	101	—	—	Elleuch et al. (2008)
Allig	823	10.1	44.1	63	2.0	—	—	104	—	—	Elleuch et al. (2008)
Medjool	539	—	50	54	0.83	0.58	0.36	—	—	—	Rock et al. (2009)
Hallawi	525	—	75	50	1.14	0.61	0.50	—	—	—	Rock et al. (2009)

flesh. The major components of the carotenoids present in dried dates are β-carotene, lutein, and neoxanthin (Boudries et al., 2007; Al-Farsi et al., 2005; Gross, 1991). Similar to carotenoids, phenolics content varied significantly between date varieties from 3.87 to 661 mg/100 g date flesh. The total phenolic content of seven varieties of ripe date fruits grown in Algeria ranged from 2.49 to 8.36 mg gallic acid equivalent/100 g of fresh weight (Mansouri et al., 2005). For dates grown in Oman, total phenolic content ranged from 134 to 280 and from 217 to 343 mg of ferullic acid equivalent/100 of fresh and sundried date varieties, respectively (Al-Farsi et al., 2005). Wu et al. (2004) reported much higher contents of total phenolics of two varieties of dates grown in California with values of 572 and 661 mg of gallic acid equivalent/100 g of fresh weight in Deglet Nour and Medjool varieties, respectively. Many factors may be responsible for these wide variations, including the use of different acid standards, variety, growing conditions, maturity, season, geographic origin, storage conditions, soil type, fertilizer, and amount of sunlight received (Al-Farsi, 2005).

Phenolic compounds in fruit tissues include a wide range of compounds and broad spectrum of functional activities. Phenolic compounds are comprised mainly of free and bound phenolic acids. At least eight phenolic acids were present in dried Omani and Tunisian dates (Al-Farsi et al., 2005, Regnault-Roger et al., 1987). Five phenolic acids consisted of hydroxylated derivatives of benzoic acid (gallic acid, protochatechuic acid, *p*-hydroxybenzoic acid, vanillic acid, and syringic acid) and three were cinnamic acid derivatives (caffeic acid, *p*-coumaric acid, and ferulic acid).

Fruits typically show a decline in phenolic compounds with ripening, and an increase in response to stress like bruising and fungal infection. The extent and magnitude of these changes varies widely

TABLE 19.6
Total Carotenoids and Phenolics

Variety	Total Carotenoids (μg/100 g Date Flesh)	Total Phenolics (mg/100 g Date Flesh)	Reference
Deglet Nour	64.3	nd	Boudries et al. (2007)
	nd	661	Wu et al. (2004)
	920	nd	Gross et al. (1985)
	nd	6.73	Mansouri et al. (2005)
	nd	230.90	Saafi et al. (2009)
Hayany	1260	nd	Gross (1991)
Hamraya	51.3	nd	Boudries et al. (2007)
Barhee	1320	nd	Gross (1991)
Frad	1200	343	Al-Farsi et al. (2005)
Khalas	900	217	Al-Farsi et al. (2005)
Khasab	2900	339	Al-Farsi et al. (2005)
Tazerzait	nd	3.91	Mansouri et al. (2005)
Tafiziouine	nd	4.59	Mansouri et al. (2005)
Medjool	nd	572	Wu et al. (2004)
	nd	517	Rock et al. (2009)
Hallawi	nd	619	Rock et al. (2009)
Mabseela	nd	246	Al-Farsi et al. (2007)
Um-Sellah	nd	186	Al-Farsi et al. (2007)
Shahal	nd	172	Al-Farsi et al. (2007)
Allig	nd	447.73	Saafi et al. (2009)
Khouet Kenta	nd	236.27	Saafi et al. (2009)
Kentichi	nd	209.42	Saafi et al. (2009)
Smiti	nd	9.71	Chaira et al. (2009)
Kenta	nd	7.27	Chaira et al. (2009)
Mermella	nd	3.87	Chaira et al. (2009)
Garn Ghzel	nd	6.82	Chaira et al. (2009)

Note: nd, not determined.

depending on the plant material and the conditions of storage (Haard and Chism, 1996). Shahidi and Naczk (2004) reported that drying, in general, is regarded as unfavorable due to the possibility of inducing oxidative decomposition either enzymatically by polyphenol oxidase and glucosidase or by thermal degradation of phenolic compounds. However, Al-Farsi (2005) reported an increase in total phenolics contents after sun drying of dates grown in Oman.

CONCLUSION

Physicochemical characteristics of date palm fruits suggested that DPF (date palm fruit) can be considered as an important source of energy and nutritious food. DPF contains high amount of sugars, mostly in the invert sugar form. Although DPF is a low source of fat and protein, it can be regarded as a good source of minerals like potassium, calcium, and selenium. DPF is slightly acidic in nature and is a good source of dietary fibers, mainly the insoluble materials which help in reducing many diseases. It is considered as a moderate source of vitamins like B-complex and C. Furthermore, dates can be a good source of polyphenolics and antioxidants which can reduce oxidative stress and help in combating many diseases.

REFERENCES

Açouren, S. 2001. Caractérisation, evaluations de la qualité de la date et identification des cultivars rares de palmier de la region Ziban. *Revue de l'INRA*. pp. 21–39.

Ahmed, I.A., Ahmed, A.W.K., and Robinson, R.K. 1995. Chemical composition of date varieties as influenced by the stage of ripening. *Food Chemistry*. 54:305–309.

Al-Aswad, M.D. 1971. The amino acid content of some Iraqi dates. *Journal of Food Science*. 36:1019–1020.

Al-Farsi, M., Alasalvar, C., Al-Abid, M., Al-Shoaily, K., Al-Amry, M., and Al-Rawahy, F. 2007. Compositional and functional characteristics of dates, syrups, and their by-products. *Food Chemistry*. 104:943–947.

Al-Farsi, M., Alasalvar, C., Morris, A., Baron, M., and Shahidi, F. 2005. Comparison of antioxidant activity, anthocyanins, carotenoids, and phenolics of three native fresh and sun-dried date (*Phoenix dactylifera* L.) varieties grown in Oman. *Journal of Agricultural and Food Chemistry*. 53:7592–7599.

Al-Hooti, S., Sidhu, J.S., and Qabazard, H. 1997. Physicochemical characteristics of five date fruit cultivars grown in the United Arab Emirates. *Plant Foods for Human Nutrition*. 50:101–113.

Al-Rawi, N., Markakis, P., and Bauer, D.H. 1967. Amino acid composition of Iraqi dates. *Journal of the Science of Food and Agriculture*. 18:1–2.

Al-Shahib, W. and Marshall, R.J. 2002. Dietary fibre content of dates from 13 varieties of date palm *Phoenix dactylifera* L. *International Journal of Food Science & Technology*. 37:719–721.

Auda, H., Al-Wandawi, H., and Al-Adhami, L. 1976. Protein and amino acid composition of three varieties of Iraqi dates at different stages of development. *Journal of Agricultural and Food Chemistry*. 24:365–367.

Barreveld, W.H. 1993. *Date Palm Products*. FAO, Agricultural Services Bulletin No. 101, FAO, Rome, 211p.

Belguedj, M. 2002. Les ressources génétiques du palmier dattier. Caractéristiques des cultivars de dattiers du Sud-Est du Sahara algérien. Dossiers-Documents-Débats No. 1. Alger: éditions INRAA, 2002.

Bellagha, S., Salhi, A., and Farhat, A. 2008. Desorption isotherms and isostatic heat of three Tunisian date cultivars. *Food Bioprocess Technology*. 1:270–275.

Beuchat, L.R. 1981. Microbial stability as affected by water activity. *Cereal Foods World*. 26:345–349.

Boudries, H., Kefalas, P., and Hornero-Méndez, D. 2007. Carotenoid composition of Algerian date varieties (*Phoenix dactylifera*) at different edible maturation stages. *Food Chemistry*. 101:1372–1377.

Bourgeois, C. 2003. *Les Vitamines Dans les Industries Agroalimentaires*. Ed. Tech & Doc, Paris, 483p.

Chaira, N., Mrabet, A., and Ferchichi, A. 2009. Evaluation of antioxidant activity, phenolics, sugar and mineral contents in date palm fruits. *Journal of Food Biochemistry*. 33:390–403.

Elleuch, M., Besbes, S., Roiseux, O., Blecker, C., Deroanne, C., Drirar, N., and Attia, H. 2008. Date flesh: Chemical composition and characteristics of the dietary fibre. *Food Chemistry*. 111:676–682.

Espiard, E. 2002. *Introduction à la transformation industrielle des fruits*. Ed. Lavoisier, pp. 147–155.

FAOSTAT. 2007. Statistical Databases, http://faostat.fao.org, accessed June 2, 2007.

Fayadh, J.M. and Al-Showiman, S.S. 1990. Chemical composition of date palm (*Phoenix dactylifera* L.). *Journal of the Chemical Society of Pakistan*. 12:84–103.

Guizani, N., Al-Saidi, G.S., Rahman, M.S., Bornaz, S., and Al-Alawi, A.A. 2010. State diagram of dates: Glass transition, freezing curve and maximal-freeze-concentration condition. *Journal of Food Engineering*. 99:92–97.

Gross, J. 1991. *Pigments in Vegetables: Chlorophylls and Carotenoids*. Van Nostrand Reinhold, New York.

Haard, N.F. and Chism, G.W. 1996. Characteristics of edible plant tissues. In: *Food Chemistry*, 3rd edition. Fennema, O.R. (ed), Marcel Dekker, Inc., New York, pp. 943–1011.

Hussein, A.S., Alhadrami, G.A., and Khalil, Y.H. 1998. The use of dates and date pits in broiler starter and finisher diets. *Bioresource Technology*. 66:219–223.

Ismail, B., Haffar, I., Baalbaki, R., Mechref, Y., and Henry, J. 2006. Physico-chemical characteristics and total quality of five date varieties grown in the United Arab Emirates. *International Journal of Food Science and Technology*. 41:919–926.

Jadhav, S.J. and Andrew, W.T. 1977. Effects of cultivars and fertilizers on nonvolatile organic acids in potato tubers. *Canadian Institute of Food Science and Technology Journal*. 10:13–21.

Kasapis, S., Rahman, M.S., Guizani, N., and Al-Amri, M. 2000. State diagram of temperature vs. date solids obtained from the mature fruit. *Journal of Agricultural and Food Chemistry*. 48:3779–3784.

Khan, M.N., Sarwar, A., Wahahb, F., and Haleem, R. 2008. Physico-chemical characterization of date varieties using multivariate analysis. *Journal of the Science of Food and Agriculture*. 88:1051–1059.

Maatalah, S. 1970. Contribution à la valorization de la date algèrienne. Mémoire d'ingéniorat en agronomie. *INA Alger*. 120.

Mansouri, A., Guendez, E., Kokkalouc, E., and Panagiotis, K. 2005. Phenolic profile and antioxidant activity of the Algerian ripe date palm fruit (*Phoenix dactylifera*). *Food Chemistry*. 89:411–420.

Myhara, R.M., Sablani, S.S., Al-Alawi, S.M., and Taylor, M.S. 1998. Water sorption isotherms of dates: Modeling using GAB equation and artificial neural network approaches. *Lebensmittel-Wissenschaft und-Technologie*. 31:699–706.

Myhara, R.M., Taylor, M.S., Slominski, B.A., and Al-Bulushi, I. 1998. Moisture sorption isotherms and chemical composition of Omani dates. *Journal of Food Engineering*. 37:471–479.

Omezzine, A., El Mardi, O., El Kharbotly, A., Ampratwum, D., and Myhara, R. 1997. Introduction. In *Dates Duality Improvement in the Sultanate of Oman—A Comprehensive Document*; Sultan Qaboos University Publications, Muscat, pp. 1–4.

Rahman, M.S. and Al-Farsi, S.A. 2005. Instrumental texture profile analysis (TPA) of date flesh as a function of moisture content. *Journal of Food Engineering*. 66:505–511.

Regnault-Roger, C., Hadidane, R., Biard, J.F., and Boukef, K. 1987. High performance liquid and thin-layer chromatography determi-nation of phenolic acids in palm (*Phoenix dactylifera*). *Food Chemistry*. 25:61–71.

Reynes, M., Bouabidi, H., Piombo, G., and Risterucci, A.M. 1994. Caractérisation des principales variétés de dattes cultivées dans la région des Djerid en Tunisie. *Fruit*. 49:289–298.

Rock, W., Rosenblat, M., Borochov-Neori, H., Volkova, N., Judeinstein, S., Elias, M., and Aviram, M. 2009. Effects of date (*Phoenix dactylifera* L., Medjool or hallawi variety) consumption by healthy subjects on serum glucose and lipid levels and on serum oxidative status: A pilot study. *Journal of Agricultural and Food Chemistry*. 57:8010–8017.

Rygg, G.L., Furr, J.R., Nixon, R.W., and Armstrong, W.W. 1953. Factors affecting the spoilage of dates at room temperature. *Date Growers' Institute Report*. 30:10–14.

Saafi, E.B., El Arem, A., Issaoui, M., Hammami, M., and Achour, L. 2009. Phenolic content and antioxidant activity of four date palm (*Phoenix dactylifera* L.) fruit varieties grown in Tunisia. *International Journal of Food Science and Technology*. 44:2314–2319.

Sahari, M.A., Barzegar, M., and Radfar, R. 2007. Effects of varieties on the composition of dates (*Phoenix dactylifera* L.)—Note. *Food Science and Technology International*. 13:269–275.

Salem, S.A. and Hegazi, S.M. 1971. Chemical composition of the Egyptian dry dates. *Journal of the Science of Food and Agriculture*. 22:632–633.

Saravacos, G.D., Tsiourvas, D.A., and Tsami, E. 1986. Effect of temperature on the water adsorption isotherms of sultana raisins. *Journal of Food Science*. 51:381–383,387.

Shahidi, F. and Naczk, M. 2004. *Phenolics in Food and Nutraceuticals*. CRC Press, Boca Raton, FL.

Siebert, K.J. 1999. Modeling the flavor thresholds of organic acids in beer as a function of their molecular properties. *Food Quality and Preformance*. 10:129–137.

Wu, X., Beecher, G., Holden, J., Haytowitz, D., Gebhardt, S., and Prior, R. 2004. Lipophilic and hydrophilic antioxidant capacities of common foods in the United States. *Journal of Agricultural and Food Chemistry*. 52:4026–4403.

Yousif, A.K., Benjamin, N.D., Kado, A., Alddin, S.M., and Ali, S.M. 1982. Chemical composition of four Iraqi date cultivars. *Date Palm Journal*. 1:285–294.

20 Sorption and Structural Characteristics of Date Palm Fruit

Mohammad Shafiur Rahman

CONTENTS

INTRODUCTION

Dates are generally consumed as fresh fruits or processed into various products such as paste, syrup, and powder which are used as ingredients in cookies or cakes manufacturing. Knowledge on the physico-chemical, mechanical, structural, textural, and sensory properties is necessary for its processing, storage stability, and consumer acceptability. The rank of the most preferred sensory quality attributes of date at *Tamr* stage of maturity were as follows: high (color, appearance, sweetness), medium (fruit size, chewiness, solubility, flesh thickness), and low (elasticity, mouth feel, and shear force) (Ismail et al., 2001). Considering subjects within the UAE population, it was found that the highest preference variety of dates was Khalas, medium preference was Boumaan, Barhee, and Fard; and low preference variety was Ruzeiz, respectively (Ismail et al. 2001). In this chapter, the sorption characteristics, structural properties related to the state diagram, and instrumental textural characteristics of dates are presented. The basic understanding of the above-mentioned parameters is presented in the following sections for general readers.

Drying is one of the oldest or traditional methods of food preservation. Traditionally, dates are dried to reduce its moisture content so that it could be stored at room temperature for a longer time. Drying reduces water content (i.e., water activity), thus preserving foods by avoiding microbial growth and deteriorative chemical reactions. Water activity or moisture isotherm plays an important role in designing, operation, and control of drying processes, as well as in determining the food's stability during storage.

It has been mentioned in the literature that foods can be considered very stable at the glassy state, since below glass temperature the compounds involved in the deterioration reactions take many months or even years to diffuse over molecular distances and approach each other to react (Slade and Levine, 1991). Similarly, mechanical and transport properties could also be related with glass transition.

The rules of glass transition concept are (1) the food is the most stable at and below its glass transition, and (2) higher the T–T_g (i.e., above glass transition), higher the deterioration or reaction rates. The glassy state in foods could be achieved in two ways. One by reducing the water content to such a low level so that its glass transition is above the room temperature, thus foods are in the glassy state at room temperature and are stable. This is the reason why the dried products are stable at room temperature for many months or years depending on the types of foods. The lowering of water could be achieved by drying or adding solids to the foods. Another way to achieve a glassy state in food is by reducing the storage temperature below or close to the glass transition (T_g'). This is the reason why frozen products are generally stable for months or years in the freezer (i.e., commercial freezer at −18°C). The physico-chemical properties of food materials change at its glass–rubber transition; thus, different properties (such as specific heat, diffusivity, volume, mechanical properties) are measured as a function of temperature and any change or shift in the property is used to trace the glass transition. More details of the measurement methods of glass transition are available in Rahman (2009a).

The glass transition concept provided a basis for developing state diagram. The main purposes of drawing a stability map, such as state diagram, are to help in understanding the complex changes when food's water content and temperature are changed. It also assists in identifying the food's stability during storage as well as selecting a suitable temperature and moisture content for processing. The glass transition temperature, freezing point, and maximal-freeze-concentration conditions are mainly needed to draw the basic maps. Many applications of the state diagram in food processing and preservation are presented by Rahman (2010).

Sensory properties of a food are the ultimate considerations used by the consumers for accepting a product. Sensory properties are mainly measured by expert-panel, trained panel, and consumer panel. In many instances, measurement techniques and data analysis are complex, and costly. In addition, variability could be higher since it is a subjective method. Considering this perspective, it is commonly applied to instrumental methods to measure the different physico-chemical parameters of foods and these parameters are correlated with sensory attributes measured by subjective methods. The advantages of the instrumental methods are its reproducibility, fast and could be measured in the food-processing line, thus very helpful for quality control of a food product. Instrumental Texture Profile Analysis (TPA) is widely used technique to measure the textural attributes of food products since these parameters could be correlated with sensory texture measured by sensory panels.

WATER ACTIVITY AND MOISTURE SORPTION ISOTHERM

Water is an important constituent of all foods. In the middle of the twentieth century, scientists began to discover the existence of a relationship between the water contained in a food and its relative tendency to spoil. The preservation of foods by drying is based on the water activity concept. The water activity concept considers that the active water could be much more important to the stability of a food than the total amount of water present (Scott, 1953). Scott (1957) clearly stated that the water activity of a medium correlated with the deterioration of food stability due to the growth of microorganisms. Thus, it is possible to develop generalized rules or limits for the stability of foods using water activity. For example, there is a critical water activity level below which no microorganisms can grow. Pathogenic bacteria cannot grow below a water activity of 0.85, whereas yeasts and molds are more tolerant to reduced water activity, but usually no growth occurs below a water activity of about 0.6. It is also possible to identify the specific critical limits for a group of specific types of microflora. This was the main reason why food scientists started to emphasize water activity along with water content. Since then, the scientific community has explored the great significance of water activity in determining the physical characteristics, processes, shelf life, sensory properties of foods. Recently, Rahman and Labuza (2007) and Rahman (2009b) have presented a detailed review on this aspect of water activity. Water activity, a thermodynamic property, is defined as the ratio of the vapor pressure of water in a system to the vapor pressure of pure water at the same temperature, or the equilibrium relative humidity of the air surrounding the system at the same temperature. Thus, water activity can be expressed as follows:

$$a_w = \frac{(P_w^v)_{sy}}{P_w^v} = \text{ERH} \tag{20.1}$$

where a_w is water activity (fraction) at T (°C), $(P_w^v)_{sy}$, P_w^v are vapor pressures of water in the system and pure water at T°C (Pa), and ERH is equilibrium relative humidity of air at T°C.

The moisture sorption isotherm is the dependence of moisture content on the water activity of one of the samples at a specified temperature. It is usually presented in graphical form or as an equation. The moisture sorption isotherm is an extremely valuable tool for food scientists because of its usefulness in predicting the food stability. Most foods there have a critical moisture content below which the rate of quality loss is negligible. The difference in the equilibrium moisture content between the adsorption and desorption curves is called hysteresis and shown in Figure 20.1. BET and GAB equations are mainly used to model isotherm of food materials. The monolayer value can be determined from BET isotherm and is widely used to determine the stability of foods. The BET equation can be derived on kinetic, statistical mechanics, or thermodynamic considerations. The equation can be written as

$$\frac{a_w}{M_w(1-a_w)} = \left[\frac{D-1}{M_m}\right] a_w + \frac{1}{M_m D} \tag{20.2}$$

where a_w is the water activity, M_m is the BET-monolayer (g/g dry solids), M_w is the total moisture content (g/g dry solids), and D is the temperature dependence for sorption excess enthalpy. The value of D indicates how strongly water is bound to the polar sites of the solid matrix and can be related with temperature as

$$D = \alpha\left[\exp\left(\frac{Q_s}{RT}\right)\right] \tag{20.3}$$

where Q_s is the excess heat of sorption (kJ/kg mol), R is the gas constant (8.314×10^{-3} J/kg mol K), T is the temperature (K), and α is the preexponent factor. The monolayer can be estimated from

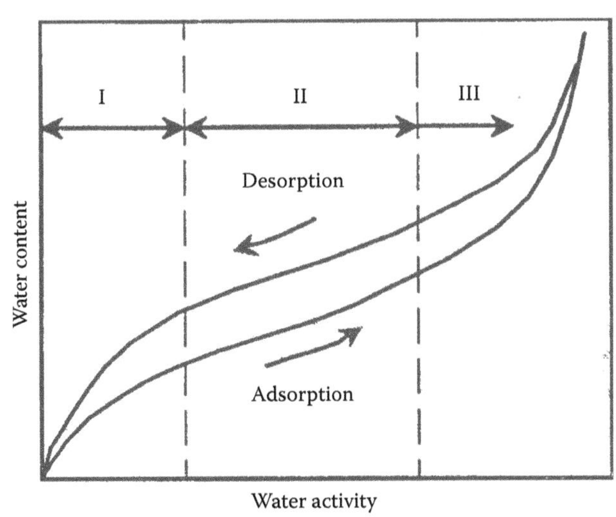

FIGURE 20.1 Sorption isotherm for typical food product showing the hysteresis.

the slope of the linear line of the plot $a_w/M_w(1-a_w)$ versus a_w. The BET equation is valid only within 0.05–0.50 water activity thus values within that range should be used to estimate the monolayer value. The monolayer value is generally at around water activity of 0.2–0.4 (Labuza, 1984). In addition, the BET monolayer calculation is an effective method for estimating the amount of bound water to specific polar sites in dehydrated food systems (McLaren and Rowen, 1952). The BET monolayer values usually varied from 0.01 to 0.14 g/g dry solids (dry basis) in the case of foods and food components. Macromolecules, such as starch, protein, and agar usually have higher BET monolayer whereas high-fat content foods, such as avocado, peanuts, and whole milk showed lower monolayer. Iglesias and Chirife (1976) found that monolayer values decreased significantly with increasing temperature after studying 100 foods and food components. This may be due to the thermodynamics where higher temperatures increase the escaping tendencies of gas molecules.

In recent years, the most widely accepted and represented model for sorption isotherms for foods has been the GAB. This is mainly due to its accuracy and its validity over a wide range of water activities from 0.1 to 0.9. The GAB isotherm was developed by Gugenheim, Anderson, and De Boer and can be written as

$$M_w = \frac{M_{gm}CKa_w}{(1-Ka_w)(1-Ka_w+CKa_w)} \tag{20.4}$$

where C and K are the model parameters and are related to the temperature. The GAB isotherm equation is an extension of the two-constant BET model and takes into account the modified properties of the sorbate in the multilayer region and bulk liquid properties through the introduction of a third constant K. It is important to point that BET monolayer has more physical meaning and acceptability to be used for food stability compared to the GAB monolayer although GAB provided better mathematical prediction of isotherm over the wide range of water activity (Rahman and Al-Belushi, 2006). The BET and GAB models' parameters for different date varieties are given in Table 20.1. The BET monolayer values of date flesh were varied from 8.88 to 12.43 g/g dry solids, while date pits showed a lower value as 4.27 g/g dry solids. Myhara et al. (1998) also used an artificial neural network modeling technique to predict the moisture sorption isotherm of date as a function of its 12 chemical compositions including fructose, glucose, sucrose, protein, ash, fibers (total, arabinose, xylose, mannose, galactose, glucose, uronic acids), and temperature. They observed that chemical composition showed more impact upon the water sorption isotherm.

A food product is most stable at its monolayer moisture content, which varies with the chemical composition, structure and environmental conditions, such as temperature. One of the earlier food stability maps based on the water activity concepts considers growth of microorganisms and different types of biochemical reactions (Labuza et al., 1970, 1972). The updated food stability map is presented in Figure 20.2 (Rahman, 2009b). In the present map, the trends of microbial growth, biochemical reactions, and mechanical characteristics are presented in the three zones of water activity.

STATE DIAGRAM AND MACRO–MICRO REGION CONCEPT IN FOODS

State Diagram

Glass transition temperature is the first component of the state diagram. The state diagram is stability map of different states of a food as a function of water or solids content and temperature (Rahman, 2004). Levine and Slade (1986) presented a state diagram of polyvinyl pyrrolidone (PVP) containing the glass line, freezing curve, and the intersection of these lines as T_g'. (In this chapter it is defined as T_g''.) This is most probably the first state diagram in the food science literature. Many of their publications later presented the state diagram of starch by quoting the reference of van den Berg (1986)

TABLE 20.1
GAB and BET Parameters for Different Varieties of Date at *Tamr* Stage of Maturity

	GAB Parameter			BET Parameter			
Varieties	$M_g \times 100$	K	C	$M_b \times 100$	D	$M_{su} \times 100$	**Reference**
Barhi	9.01	1.03	7.76	10.22	6.26	89.97	Belarbi et al. (2000)
Fard	9.59	1.00	6.62	9.02	10.40	88.55	Belarbi et al. (2000)
Fard	9.28	1.01	94.87	—	—	—	Myhara et al. (1998)
Kadrawi	8.83	1.01	11.67	8.88	11.34	82.31	Belarbi et al. (2000)
Khalas	7.61	1.06	12.17	11.00	6.39	92.52	Belarbi et al. (2000)
Khalas	12.10	0.92	2289.10	—	—	—	Ahmed et al. (2005)
Khalas	9.23	1.01	103.31	—	—	—	Myhara et al. (1998)
Khalas (freeze-dried)	10.2	1.00	43.05	—	—	—	Ahmed et al. (2005)
Madjol	9.13	1.04	6.94	9.80	8.61	80.23	Belarbi et al. (2000)
Madjhoul	7.90	1.08	16.37	9.77	9.56	86.33	Belarbi et al. (2000)
Rizaz	9.11	1.00	12.47	9.32	10.30	81.35	Belarbi et al. (2000)
Haloua	10.51	1.01	2.78	11.58	2.05	81.74	Belarbi et al. (2000)
Kentichi	9.89	1.01	14.99	11.05	6.75	62.97	Belarbi et al. (2000)
Mouche-Deglat	5.97	1.12	24.89	10.23	2.44	74.82	Belarbi et al. (2000)
Deglet-Nour (pulp)	8.73	1.02	5.75	12.43	1.03	70.34	Belarbi et al. (2000)
Deglet-Nour (kernel)	4.89	0.93	3.76	4.27	52.18	—	Belarbi et al. (2000)
Deglet-Nour (whole fruit)	5.94	1.06	2.25	10.06	2.13	—	Belarbi et al. (2000)

Note: M_g, GAB monolayer water (g/g dry solid); M_b, BET monolayer water (g/g dry solids); M_{su}, sugar content (g/g dry solids).

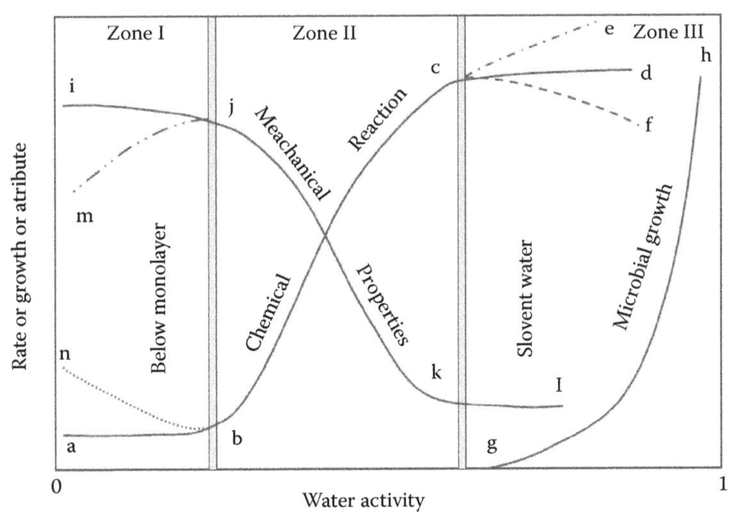

FIGURE 20.2 Stability diagram based on the water activity concepts. gh: microbial growth trend; oa, ab, nb: chemical reaction trends below BET-monolayer; ab, nb, bc: chemical reaction trends in the adsorbed water; ce, cd, cf: chemical reaction trends in the solvent water region; ij, mj: mechanical properties trends below BET-monolayer; jk: mechanical properties trend in the adsorbed water region; kl: mechanical properties trend in the solvent water region. (From Rahman, *Int. J. Food Properties*, 12(4), 726–740, 2009b. With permission.)

(Slade and Levine, 1988a). Later, they presented another state diagram including glass line, freezing curve, melting line, eutectic point, T_g', and vapor line (Levine and Slade, 1993). The main advantages of drawing a map are to help in understanding the complex changes that occur when the water content and temperature of a food are changed (Slade and Levine, 1988b, 1994, 1995; Slade et al. 1993; Kokini et al. 1994, 1995; Roos 1995; Roos et al. 1996). It also assists in identifying food's stability during storage as well as selecting a suitable condition of temperature and moisture content for processing (Roos and Karel, 1991a; Cuq et al., 2003; Rahman, 2006, 2009b, 2010). The regions of drying and freezing processes can be easily visualized in the diagram, and product stability could be assessed based on moisture content and temperature. Figure 20.3 shows a state diagram indicating different states as a function of temperature and solids mass fraction (Rahman, 2009b).

Macro–Micro Region Concept

Rahman (2009b) used the state diagram to hypothesized macro–micro regions for determining the food stability. The state diagram as shown in Figure 20.3 indicates the freezing line (ABC) and solubility line (BD) are shown in relation to the glass transition line (EQFS). The point F (X_s' and T_g') lower than T_m' (point C) is a characteristic transition (maximal-freeze-concentration condition) in the state diagram defined as the intersection of the vertical line from T_m' to the glass line EQFS (Rahman, 2006). The water content at point F or C is considered as the unfreezable water ($1–X_s'$). The unfreezable water mass fraction is the amount of water remaining unfrozen even at a very low temperature. It includes both uncrystallized free water and bound water attached to the solids matrix. The point Q is denoted as T_g'' and X_s'' and defined as the intersection of the freezing curve to the glass line by maintaining a similar curvature of the freezing curve ABC.

Point R is defined as T_g''' the glass transition of the solids matrix in the frozen sample, which is determined by DSC below T_m'. This is due to the formation of similar solid matrix associated with unfreezable water and transformation of all free water into ice although the sample contains different

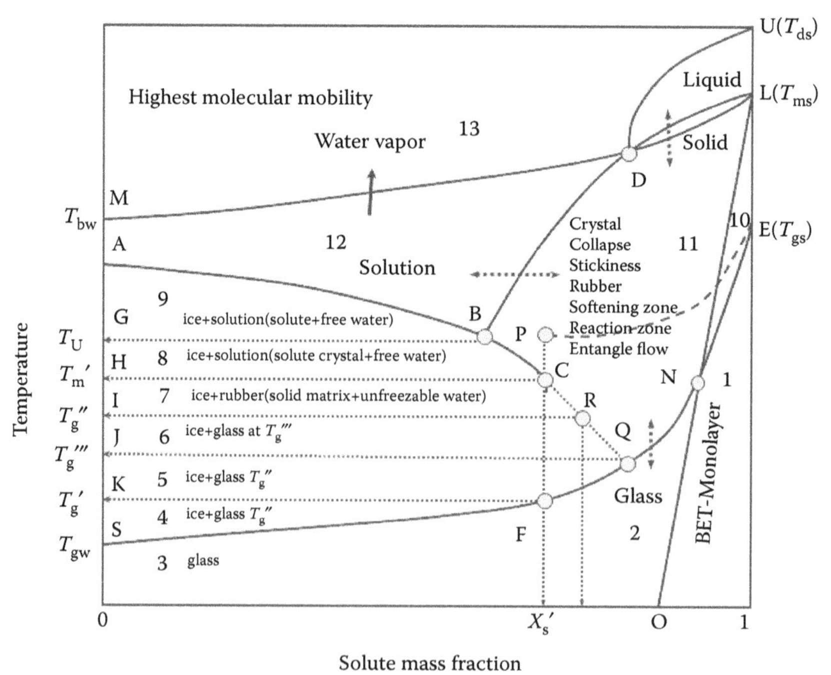

FIGURE 20.3 State diagram showing different regions and state of foods. (From Rahman, *Int. J. Food Properties*, 12(4), 726–740, 2009b. With permission.)

levels of total water before the start of DSC scanning (Rahman et al., 2005). The values of T_g''' decreased as the solids content increased falling to a minimum value. In the region AHB, the phases present are ice and solution. Below point B, first crystallization of solute occurs, and transforming GBCH region to three states: ice, solution, and solute crystal. No free water (i.e., able to form ice) exists to the right side of point C (T_m', end point of freezing with maximal freeze-concentration-condition) and then the very concentrated solution is transformed to a rubber state. The maximal-freeze-concentration condition could be achieved using optimum conditions by slow cooling and/or annealing of the samples. The region HCRI contains ice, rubber, and solute crystal. The point F is the T_g', below this point a portion of the rubber state is transformed to the glass state, thus region KFS contains glass, ice, and solute crystal. The rate of cooling can shift the points B, C, R, Q, and F. More detailed effects of cooling on the shift are discussed by Rahman (2004). The glass line should follow the line EQFS if maximum, ideal plasticization occurs in the solid matrix with the addition of water. The line EQF could be shifted upward if water causes different degrees of plasticization (as shown in curve EP). The vertical line from F will intersect at point P and the temperature is defined as T_g^{iv} (i.e., characteristic temperature) as defined by Rahman (2009b). The region BQEL is important in food processing and preservation, many characteristics such as crystallization, stickiness, collapsed phenomenon are observed in this region (Roos and Karel, 1991b; Roos, 1995).

The line BDL is the melting line which is important when products are exposed to high temperatures during processing, such as frying, baking, roasting, and extrusion cooking. In the case of a multicomponent mixture, such as food, a clear melting point is difficult to observe at a high temperature due the reactions or interactions between the components. In this case, Rahman (2004) and others defined melting as the decomposition temperature. Line MDL is the boiling/evaporation line for water evaporation from the liquid phase (line MD) and solid matrix with a degree of saturation with water (line DL).

It is important to combine water activity and glass transition concepts so that food stability rules from both concepts could be used for determining the stability. Rahman (2006) attempted to combine both concepts by plotting the BET-monolayer value as the LNO line in the state diagram shown in Figure 20.3. It intersects at point N with the glass line EQFS, which shows that at least in one location (point N), glass and water activity concepts provide the same stability criterion. This approach forms more microregions, which could give different degrees of stability in the state diagram. More studies regarding the stability need to be done on the left (above and below glass) and right sides (above and below glass) of the line LNO.

It is evident from the literature that stability below or above glass transition varied based on the specific cases indicating that developing the stability rule based on only the glass transition temperature could be inadequate. Samples having freezable water are more complex and four temperatures are defined as $T_g^{iv} > T_m' > T_g''' > T_g'' > T_g'$ (Rahman et al., 2005; Rahman, 2009b). In addition, the stability below or above $T_g^{iv} > T_m' > T_g''' > T_g'' > T_g'$ should also be explored. There are only a few references available including all five characteristic temperatures with their moisture content. It is important to know how these temperatures affect the stability of foods. It would be interesting to explore the differences in stability of various products within these different ranges (Rahman, 2006, 2009b).

The BET-monolayer line shown as LNO intersecting glass line in the state diagram (Figure 20.3) makes four regions: below BET-monolayer, one above and one below; and above BET-monolayer, one above and another below (Rahman, 2006). This approach forms more microregions, which could give different stability in the state diagram and could explain the limitations of each concept. The present hypothesis proposed by Rahman (2009b) traces 13 microregions having the highest to the lowest stability based on the location from the glass and BET-monolayer lines. For example, region-1 (relatively nonreacting zone, below the BET-monolayer line and glass line) is the most stable and region-13 (highly reacting zone, far from BET-monolayer line and glass line) is the least stable. The stability decreased as the zone number increased. Each microregion could be studied for specific attributes separately considering the different characteristics of a medium. The microregions could also be subdivided further to explain the variability of stability, for example regions between 3 and 8 based on T_g^{iv}, T_m', T_g''', T_g'', or T_g'.

CHARACTERISTIC PARAMETERS OF DATES FOR DETERMINING STATE DIAGRAM

The required data for the state diagram of dates were measured by Kasapis et al. (2000), Rahman (2004), and Guizani et al. (2010). They have measured freezing point, glass transition, and maximal-freeze-concentration condition. The theoretical Clausius–Clapeyron equation is usually used to estimate the freezing point and the equation can be written as

$$\delta = -\frac{\beta}{\lambda_w}\ln\left[\frac{1 - X_s^o}{1 - X_s^o + EX_s^o}\right] \tag{20.5}$$

where δ is the freezing point depression (T_w–T_F), T_F is the freezing point of food (°C), T_w is the freezing point of water (°C), β is the molar freezing point constant of water (1860 kg K/kg mol), λ_w is the molecular weight of water, X_s^o is the initial solids fraction, and E is the molecular weight ratio of water and solids (λ_w/λ_s). The model parameter E was found as 0.147 for Khalas variety of date flesh (Rahman, 2004). The Clausius–Clapeyron equation is limited to the ideal solution (i.e., for a very dilute solution). Theoretical models can be improved by introducing parameters for nonideal behavior when fraction of total water is unavailable for forming ice. The unfreezable water content B can be defined as the ratio of unfrozen water even at a very low temperature to total solids. Equation 20.2 can be modified based on this concept as (Chen, 1986)

$$\delta = -\frac{\beta}{\lambda_w}\ln\left[\frac{1 - X_s^o - BX_s^o}{1 - X_s^o - BX_s^o + EX_s^o}\right] \tag{20.6}$$

Table 20.2 shows the compiled freezing points and glass transition temperatures of date flesh as a function of solids content. The model parameters E and B of Equation 20.6 were estimated as 0.129 and 0.053 g/g dry solids for Khalas variety dates (Rahman, 2004). The state diagram of date flesh was presented for *deglet nour* variety (Guizani et al., 2010) and Khalas variety (Rahman, 2004) by incorporating freezing curve, glass transition, and maximal-freeze-concentration. The ultimate maximal-freeze-concentration condition for Deglet Nour variety date was determined T_m' equal to −38.2°C and T_g' as −48.0°C with unfreezable water as 0.22 g/g sample (X_s': 0.78), respectively (Guizani et al., 2010). Similarly, the T_g'' and X_s'' were found as −42.0°C and 0.82, respectively (Guizani et al., 2010). In the case of Khalas variety Rahman (2004) determined T_m' equal to −43.6°C and T_g' as −46.4°C with unfreezable water as 0.24 g/g sample (X_s': 0.76), respectively.

Table 20.3 shows the compiled data of the glass transition of date flesh containing unfreezable water content as a function of water activity and solids content. The glass transition temperature of foods and biological materials is commonly modeled by the Gordon and Taylor (1952) equation as

$$T_g = \frac{X_s T_{gs} + kX_w T_{gw}}{X_s + kX_w} \tag{20.7}$$

where T_g used here is the T_{ge}; and T_{gs} and T_{gw} are the glass transition temperature of mixture, solids and water, respectively; k is the Gordon–Taylor parameter and X_w and X_s are the mass fraction of water and solids (wet basis). The parameters T_{gs} and k usually were determined by considering that glass transition of pure water as −135° C (Johari et al., 1987). The values of T_{gs} and k were deter-

TABLE 20.2
Freezing Point and Glass Transition of Date Flesh as a Function of Solids Content

							Transition 1			Transition 2 (Close to Ice Melting)		
Sample	r_c	X_s	T_F^a (°C)	T_F^b (°C)	$(\Delta H)_{sa}$ (kJ/kg)	$(T_m')^c$ (°C)	T_{gi}(°C)	T_{gp}(°C)	T_{ge}(°C)	T_{gi} (T_g''') (°C)	T_{gp}(°C)	T_{ge} (T_m') (°C)
Date[1]	5	0.220	−1.8	−4.1	172	−26.1	−61.4	−57.6	−54.5	−44.2	−42.6	−41.5
	5	0.298	−3.2	−5.5	151	−34.9	−60.6	−58.4	−56.0	−44.3	−42.2	−40.0
	5	0.360	−4.9	−7.2	122	−31.0	−60.7	−58.4	−54.9	−44.7	−42.9	−40.7
	5	0.414	−7.8	−10.8	87	−40.9	−60.9	−57.1	−54.8	−46.3	−43.3	−41.4
	1	0.511	−12.6	−15.2	59	−37.1	−59.7	−56.9	−54.4	−46.4	−44.3	42.4
	1	0.541	−18.3	−20.2	55	−38.9	−58.5	−55.3	−54.0	−44.1	−39.7	−39.4
	1	0.589	−20.3	−23.4	41	−37.9	−62.6	−58.6	−56.1	−44.8	−42.6	−39.9
	1	0.611	−22.0	−25.6	27	−43.6	−63.4	−59.3	−57.0	−46.4	−45.4	43.8
Date[2]	10	0.100	−2.5	—	196	−8.2	N	N	N	−11.7	—	−8.2
	10	0.200	−4.7	—	150	−11.0	N	N	N	−17.4	—	−11.0
	10	0.300	−6.8	—	111	−15.3	N	N	N	−18.8	—	−15.3
	10	0.400	−6.2	—	125	−14.0	N	N	N	−19.2	—	−14.0
	10	0.500	−11.2	—	97	−21.8	−39.5	−37.4	−35.9	−25.7	—	−21.8
	10	0.600	−18.9	—	62	−31.3	−55.7	−51.2	−50.5	−42.6	—	−31.3
	10	0.700	−33.3	—	29	−35.6	−59.0	−53.5	−49.6	−37.1	—	−35.6
	10	0.820	−27.2	—	39	−38.5	−60.8	−54.2	−51.9	−41.4	—	−38.5
Date[3]		0.117	−1.9									
		0.236	−3.6									
		0.288	−5.1									
		0.377	−9.4									
		0.444	−13.6									
		0.544	−24.5									
		0.589	−31.7									
		0.640	−47.0									

Source: [1] Adapted from Rahman, M. S. 2004. *International Journal of Food Properties* 7(3): 407–428.
[2] Adapted from Guizani, N. et al., 2010. *Journal of Food Engineering* 99: 92–97.
[3] Adapted from Kasapis, S. et al., 2000. *Journal of Agricultural and Food Chemistry* 48: 3779–3784.

Note: r_c, cooling rate (C/min); X_s, solids content (g/g sample, wet basis), T_F, freezing point, $(\Delta H)_{sa}$ latent heat of fusion (kJ/kg).
T_g''' and T_m' are the onset and end of glass transition before melting of ice.
T_{gi}, T_{gp}, T_{ge} are the onset, mid, and end of glass transition temperature.
[1] DSC method, annealing time are the determined optimum conditions
[a] Maximum peak.
[b] Maximum slope.
[c] At optimum annealing condition (30 min at T_m'–1).
[2] DSC method, Deglet Nour variety, r_c (cooling rate, °C/min) and AT (annealing time) are the determined optimum conditions.
[a] Maximum peak.
[b] Maximum slope.
[c] At optimum annealing condition (30° min at t_m'–1).
[3] Cooling curve method for freezing point.

TABLE 20.3
Glass Transition Temperature above Unfrozen Water Content When There Is No Formation of Ice

Date Flesh[a]					Date Flesh[b]			Date Flesh[b]			Date Flesh[c]					Date Flesh[d]	
a_w	X_s	T_{gi} (°C)	T_{gp}(°C)	T_{ge}(°C)	a_w	X_s	T_{gp} (°C)	a_w	X_s	T_{gp} (°C)	a_w	X_s	T_{gi} (°C)	T_{gp} (°C)	T_{ge} (°C)	X_s	T_{gi} (°C)
0.755	0.675	−71.5	−68.3	−65.4	0.851	0.644	−78.9	0.851	0.600	−81.4	0.113	0.790	−48.7	−42.5	−38.5	0.708	−42.5
0.717	0.716	−51.3	−54.7	−58.1	0.755	0.714	−69.0	0.755	0.667	−72.8	0.225	0.810	−44.8	−38.6	−34.8	0.781	−32.0
0.570	0.782	−43.7	−46.4	−49.9	0.544	0.828	−36.8	0.544	0.818	−41.8	0.331	0.820	−39.8	−28.9	−24.2	0.802	−26.0
0.544	0.802	−34.6	−38.2	−42.6	0.331	0.836	−23.4	0.331	0.875	−18.6	0.432	0.830	−33.8	−25.7	−21.1	0.838	−19.0
0.432	0.821	−26.3	−29.8	−34.3	0.113	0.861	−13.4	0.113	0.901	−6.7	0.570	0.840	−32.0	−21.5	−12.9	0.881	0.0
—	0.832	−23.7	−29.3	−35.6	0.070	0.893	−9.1	0.070	0.922	−2.0	0.708	0.850	−28.4	−18.9	−16.8	0.907	10.0
0.331	0.880	−13.8	−17.2	−32.6							—	0.910	−13.8	−7.7	−2.0	0.923	23.5
0.231	0.851	−13.9	−20.3	−24.9												1.000	57.4
0.113	0.862	−19.6	−24.3	−27.8													
—	0.889	−17.2	−23.3	−28.4													
—	0.927	29.2	32.7	35.6													
—	0.932	29.8	33.2	35.9													
—	0.969	33.9	36.4	38.5													
0.000	1.000	62.8	69.2	76.2													

Note: a_w, Water activity; X_s, Solids content (g/g sample, wet basis). T_{gi}, T_{gp}, T_{ge} are the onset, mid, and end of glass transition temperature.

[a] Rahman (2004) (without annealing and cooling rate 5°C/min and glass transition from heating scan).

[b] Ahmed et al. (2005) (cooling rate 5°C/min).

[c] Guizani et al. (2010) (Deglet Nour) DSC heating rate 10°C/min.

[d] Kasapis et al. (2000), Oscillation method (rheological).

mined as 9.7°C and 2.6 for Deglet Nour variety (Guizani et al., 2010), 63.4°C and 4.0 for Barni variety (Kasapis et al., 2000), 57.4°C and 3.2 for Khalas variety (Rahman, 2004).

MECHANICAL AND TEXTURAL CHARACTERISTICS

Instrumental TPA is the extension of the one cycle imitative tests. Instrumental TPA is one of the major breakthroughs in the imitative test (Bourne, 1978). In many cases in the literature it is reported as TPA. However, it is scientifically sound to use instrumental TPA since human senses are not used in these tests. Instrumental TPA involves two or more cycles of compression of a sample between two parallel surfaces. The first attempt to imitate mastication by instrumental means was the denture tenderometer (Proctor et al., 1955). The major breakthrough in instrumental TPA came with the development of the General Foods Texturometer (Bourne, 1978). In this attempt to simulate mastication, a group at General Foods, led by Dr. A. S. Szczesniak, developed the Texturometer (Friedman et al., 1963). The equipment, which was a modification of the original MIT Denture Tenderometer, was designed to simulate the masticating action of the human mouth and can be described as a unit composed of a plate supported by a flexible arm attached to a strain gauge, and a plunger which acted upon the food sample, usually a bite-size piece, compressing it twice in a reciprocating motion that imitated the biting action of jaw. The strain gauge detected the force generated, which was recorded on a strip chart recorder. The plunger was originally constructed in five sizes and was made of three different materials. The sizes usually varied from 2 to 5 cm in diameter. All the plungers were 2.5 cm in height. The materials used were lucite, aluminum, and brass (Pons and Fiszman, 1996). This device usually compressed a bite size piece (i.e., a cube approximately 1.3 cm) of food twice and generated a force–time curve, which was then used to define some of the textural characteristics (Figure 20.4). This device was then followed by Instron, which produces similar output. The Texturometer uses a small flat-faced cylinder to compress to 25% of its original height (75% compression) two times in a reciprocating motion that imitated the biting action of the jaw by back-and-forth movement. A plot of force–time curve by strip-chart recorder portrays the force history of the simulated mastication action. Original analyses of the force–time curve led to the extraction of seven textural parameters or attributes: five measured and two calculated from the measured parameters (Friedman et al., 1963). A typical curve is shown in Figure 20.4. The original parameters were defined from the curve given by Friedman et al. (1963) and a discussion on their improvements is given below.

Fracturability (*FR*) is defined as the force at first significant break in the curve shown as a small peak in Figure 20.4. This is originally called *brittleness*. The unit is Newton. Hardness (*HA*) is defined as the peak force (i.e., force necessary to attain a given deformation) of area A_1 during the first compression cycle envisaged as the first bite. The unit is Newton. Cohesiveness (*CO*) (the strength of the internal bonds making up the body) is defined as the ratio of positive force area during the second compression to that during the first compression ($CO1 = A_2/A_1$). The unit is dimensionless. It is one of the complex attributes since it does not correspond to a single sensation such as hardness and brittleness. The area of integration is a measure of the power needed to impose that force, assuming that the head moves at constant speed. The ratio does imply some measure of the ability of the food to recover and the process to be repeated with a different resistance or reaction. Samples that are very cohesive (high values) are perceived as tough and difficult to break in the mouth. It was modified as the positive areas of compression and excluded the negative areas of the decompression for each cycle and calculated as $CO2 = (A_2 - A_4)/(A_1 - A_3)$ (Drake, 1966; Bourne, 1978). Adhesiveness (*AD*) is defined as the negative force area of A_3 to the first bite, representing the work necessary to pull the compressing plunger away from the sample. If the material is at all sticky or adhesive, the force becomes negative. The unit is N m. Actual adhesiveness is usually measured by touching the plunger without compression of the sample and then withdrawing it. In the case of TPA, the negative area is not actual adhesiveness since additional compression is applied and a pull

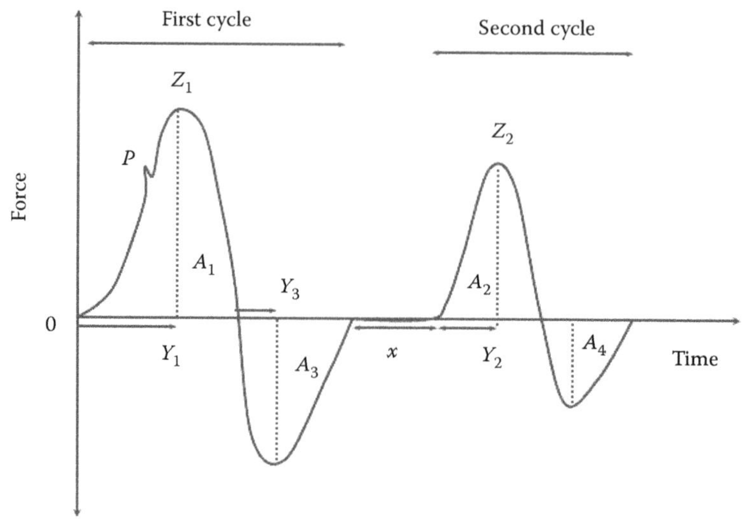

FIGURE 20.4 A typical instrumental TPA graph showing first and second cycles of compression–decompression.

or push action from the sample like spring could affect the area A_3 (Rahman and Al-Farsi, 2005). Thus, this adhesiveness is a function of surface adhesion and elastic–plastic nature of the sample.

Springiness (*SP*) (originally called elasticity) is a measure of the recovery after the first compression. It is an indication of recovery during the time that elapses between the end of first bite and start of the second bite. The unit is second or m. It is a measure of how much the original structure of the sample was broken down by the initial compression. The spring back is measured at the down stroke of the second penetration, so the wait time between two strokes can be relatively important. In some cases, an excessively long wait time allows a product to spring back more than it might under conditions being researched (e.g., we would not wait 60 s between chews) (Ma et al., 1998). The original definition of springiness (Type 1) used the length 2 ($SP1 = Y_2$) only and units are m or s. Springiness was modified typically as the distance of detected height of the product on the second penetration as divided by the original penetration ($SP2 = Y_2/Y_1$) (Type 2). Resilience (*RE*) is the instant springiness since it is measured on the withdrawal of the first penetration before the waiting period started. The calculation is the area during the withdrawal of the first penetration divided by the area of the first penetration (A_3/A_1). Gumminess (*GU*) is defined as the product of hardness and cohesiveness. The unit is Newton. Gumminess only applies to semisolid products. Gumminess is a measure of the force to disintegrate the particles ready for swallowing; thus, higher the hardness higher the gumminess. Chewiness (*CH*) is related to the primary parameters of hardness, cohesiveness, and elasticity. It is defined as the product of gumminess and springiness (=hardness × cohesiveness × springiness). The unit is N s or N m (when Type 1 springiness used), and N (when Type 2 springiness used). Chewiness is the energy for masticating the food until it is ready for swallowing. It is incorrect to quantify and report chewiness and gumminess in TPA of solid or semisolid products. Gumminess refers to semisolid foods and chewiness to solid foods and therefore chewiness and gumminess should not be reported both for the same product (Szczesniak and Bourne, 1995). However, many products may transform from solid to semisolid state in the course processing, storage, and consumption.

The variations of chemical composition in dates produce large effect upon their structural, sensory, and textural characteristics. Rahman and Al-Farsi (2005) measured instrumental TPA of date flesh (*Tamr* stage of maturity) as a function of moisture content (Table 20.4). They observed that hardness, resilience, and chewiness increased reverse exponentially with the decrease in water content, whereas cohesiveness, adhesiveness, springiness showed a peak at around 21.5% water content.

TABLE 20.4
Instrumental TPA of Date (*Tamr* Stage of Maturity) as a Function of Water Content

X_w	a_w	Hardness (N)	Cohesiveness	Adhesiveness (J) × 10^{-3}	Springiness	Resilience	Chewiness (N)
0.171	0.300	98	0.20	4.60	0.55	−29.54	10.15
0.172	0.113	171	0.20	8.42	0.50	−39.37	19.07
0.177	0.231	118	0.19	7.18	0.43	−26.64	11.80
0.186	0.331	119	0.24	9.50	0.61	−22.73	16.59
0.215	0.432	60	0.40	13.93	0.82	−5.73	15.76
0.246	0.544	36	0.28	4.76	0.71	−7.86	5.62
0.256	0.570	27	0.25	3.26	0.71	−8.25	4.87
0.345	0.755	6	0.13	0.79	0.59	−7.48	0.43
0.435	0.851	4	0.20	0.68	0.54	−6.09	0.47
0.582	0.976	3	0.20	0.55	0.34	−6.91	0.25

Source: Adapted from Rahman, M. S., Al-Farsi, S. A. 2005. *Journal of Food Engineering* 66: 505–511.
Note: a_w, Water activity; X_w, water content (g/g sample, wet basis).
Sample shape: rectangular slab of 1.0 × 1.0 × 0.4 cm.
Crosshead speed during compression: 0.5 mm/s.
Total deformation: 75%.

The trends or curvature above and below the peaks were different for cohesiveness, adhesiveness, and springiness. The principal component and factorial analysis showed two factors: the first factor indicated the "elastic nature" (characteristics of deformation in first compression: hardness, adhesiveness, and chewiness), and the second factor indicated the "plastic nature" (fighting back to regain its original shape after first compression: cohesiveness, resilience, and springiness).

CONCLUSION

This chapter presents the moisture sorption isotherm, parameters related to the state diagram, and instrumental TPA of dates. These parameters are needed to determine processing and storage stability as well as quality assessment of the dried date at *Tamr* stage of maturity. The BET monolayer values (i.e., stability moisture content) of date flesh varied from 8.88 to 12.43 g/g dry solids, while date pits showed a lower value as 4.27 g/g dry solids. The glass transition temperature of dried date flesh (i.e., zero moisture content) varied from 57°C to 63°C, much higher than room temperature at 20°C. The characteristics ultimate-freeze-concentration conditions T_m' and T_g' varied from −38°C to −44°C, and −46°C to −48°C; while X_s' from 0.76 to 0.78, respectively. This implies that date flesh in the frozen state will be the most stable below −38°C and unfreezable water content is 24 g/100 g date flesh. The model parameters of GAB, Clausius–Clapeyron equation, and Gordon–Taylor equation are presented and it could be used to predict the isotherm, freezing point, and glass transition of date as a function of its water content. The TPA analysis showed that two major parameters, elastic, and plastic nature are mainly affecting the textural characteristics of date flesh.

ACKNOWLEDGMENT

Dr. Rahman would like to acknowledge the financial support of Sultan Qaboos University toward this research in the area of food structure and stability.

REFERENCES

Ahmed, J., Ramaswamy, H. S., Khan, A. R. 2005. Effect of water activity on glass transitions of date pastes. *Journal of Food Engineering* 66: 253–258.

Belarbi, A., Aymard, C., Meot, J. M., Themelin, A., Reynes, M. 2000. Water desorption isotherms for eleven varieties of dates. *Journal of Food Engineering* 43: 103–107.

Bourne, M. C. 1978. Texture profile analysis. *Food Technology* 32(7): 62–66 and 77.

Chen, C. S. 1986. Effective molecular weight of aqueous solutions and liquid foods calculated from the freezing point depression. *Journal of Food Science* 51(6): 1537–1553.

Cuq, B., Abecassis, J., Guilbert, S. 2003. State diagrams to help describe wheat bread processing. *International Journal of Food Science and Technology* 38: 759–766.

Drake, B. 1966. *Advances in the Determination of Texture and Consistency of Foodstuffs*. SIK Report No. 207.

Friedman, H. H., Whitney, J. E., Szczesniak, A. S. 1963. The texturometer: A new instrument for objective texture measurement. *Journal of Food Science* 28: 390–396.

Gordon, M., Taylor, J. S. 1952. Ideal copolymers and the second order transitions of synthetic rubbers. I. Noncrystalline copolymers. *Journal of Applied Chemistry* 2: 493–500.

Guizani, N., Al-Saidi, G. S., Rahman, M. S., Al-Alawi, A. A. 2010. State diagram of dates: Glass transition, freezing curve and maximal-freeze-concentration condition. *Journal of Food Engineering* 99: 92–97.

Iglesias, H. A., Chirife, J. 1976. On the local isotherm concept and modes of moisture binding in food products. *Journal of Agricultural and Food Chemistry* 24(1): 77–79.

Ismail, B., Haffar, I., Baalbaki, R., Henry, J. 2001. Development of a total quality scoring system based on consumer preference weightings and sensory profiles: Application to fruit dates (*Tamr*). *Food Quality and Preference* 12: 499–506.

Johari, G. P., Hallbrucker, A., Mayer, E. 1987. The glass–liquid transition of hyperquenched water. *Nature* 330: 552–553.

Kasapis, S., Rahman, M. S., Guizani, N., Al-Aamri, M. 2000. State diagram of temperature vs date solids obtained from the mature fruit. *Journal of Agricultural and Food Chemistry* 48: 3779–3784.

Kokini, J. L., Cocero, A. M., Madeka, H. 1995. State diagrams help predict theology of cereal proteins. *Food Technology* 49(10): 74–82.

Kokini, J. L., Cocero, A. M., Madeka, H., De Graaf, E. 1994. The development of state diagrams for cereal proteins. *Trends in Food Science and Technology* 5: 281–288.

Labuza, T. P. 1984. *Moisture Sorptions: Practical Aspects of Isotherm Measurement and Use*. Am Assoc. Cereal Chemists, St. Paul, MN.

Labuza, T. P., McNally, L., Gallagher, D., Hawkes, J., Hurtado, F. 1972. Stability of intermediate moisture foods. I. Lipid oxidation. *Journal of Food Science* 37: 154–159.

Labuza, T. P., Tannenbaum, S. R., Karel, M. 1970. Water content and stability of low-moisture and intermediate-moisture foods. *Food Technology* 24(5): 543–544.

Levine, H., Slade, L. 1986. A polymer physico-chemical approach to the study of commercial starch hydrolysis products (SHPs). *Carbohydrate Polymer* 6: 213–244.

Levine, H., Slade, L. 1993. The glassy state in applications for the food industry, with an emphasis on cookie and cracker production. In *The Science and Technology of the Glassy State in Foods*. eds. Blanshard, J. M. V., Lillford, P. J., pp. 333–373. Nottingham: University of Nottingham Press.

Ma, L., Davis, D. C., Obaldo, L. G., Barbosa-Canovas, G. V. 1998. *Engineering Properties of Foods and other Biological Materials*. American Society of Agricultural Engineers, St. Joseph, MI.

McLaren, A. D., Rowen, J. W. 1952. Sorption of water vapor by proteins and polymers: A review. *Journal of Polymer Science* 7(2/3): 289–324.

Myhara, R. M., Sablani, S. S., Al-Alawi, S. M., Taylor, M. S. 1998. Water sorption isotherm of dates: modeling using GAB equation and artificial neural network approaches. *Food Science and Technology* 31: 699–706.

Pons, M., Fiszman, S. M. 1996. Instrumental texture profile analysis with practical reference to gelled systems. *Journal of Texture Studies* 27: 597–624.

Proctor, B. E., Davison, S., Malecki, G. J., Welch, M. 1955. A recording strain-gage denture tenderometer for foods. I. Instrument evaluation and initial tests. *Food Technology* 9: 471.

Rahman, M. S. 2004. State diagram of date flesh using differential scanning calorimetry (DSC). International *Journal of Food Properties* 7(3): 407–428.

Rahman, M. S. 2006. State diagram of foods: Its potential use in food processing and product stability. *Trends in Food Science and Technology* 17: 129–141.

Rahman, M. S. 2009a. Glass transition data and models of foods. In *Food Properties Handbook*. ed. Rahman, M. S. Boca Raton, pp. 247–285. FL: CRC Press, Boca Raton, FL.

Rahman, M. S. 2009b. Food Stability beyond water activity and glass transition: Macro–micro region concept in the state diagram. *International Journal of Food Properties* 12(4): 726–740.

Rahman, M. S. 2010. Food stability determination by macro–micro region concept in the state diagram and by defining a critical temperature. *Journal of Food Engineering* 99(4): 402–416.

Rahman, M. S., Al-Belushi, R. H. 2006. Dynamic isopiestic method (DIM): Measuring moisture sorption isotherm of freeze-dried garlic powder and other potential uses of DIM. *International Journal of Food Properties* 9(3): 421–437.

Rahman, M. S., Al-Farsi, S. A. 2005. Instrumental texture profile analysis (TPA) of date flesh as a function of moisture content. *Journal of Food Engineering* 66: 505–511.

Rahman, M. S., Labuza, T. P. 2007. Water activity and food preservation. In *Handbook of Food Preservation*. 2nd Edition. ed. M. S. Rahman, pp. 447–476. Boca Raton, FL: CRC Press.

Rahman, M. S., Sablani, S. S., Al-Habsi, N., Al-Maskri, S., Al-Belushi, R. 2005. State diagram of freeze-dried garlic powder by differential scanning calorimetry and cooling curve methods. *Journal of Food Science* 70(2): E135–E141.

Roos, Y. 1995. Characterization of food polymers using state diagrams. *Journal of Food Engineering* 24: 339–360.

Roos, Y., Karel, M. 1991a. Applying state diagrams to food processing and development. *Food Technology* 44(12): 66–71.

Roos, Y., Karel, M. 1991b. Plasticizing effect of water on thermal behavior and crystallization of amorphous food models. *Journal of Food Science* 56: 38–43.

Roos, Y. H., Karel, M., Kokini, J. L. 1996. Glass transitions in low moisture and frozen foods: Effects on shelf life and quality. *Food Technology* 50(11): 95–108.

Scott, W. J. 1957. Water relations of food spoilage microorganisms. *Advances in Food Research* 72: 83

Scott, W. J. 1953. Water relations of *Staphylococcus aureus* at 30°C. *Australian Journal of Biological Science* 6: 549–564.

Slade, L., Levine, H. 1988a. Non-equilibrium behavior of small carbohydrate–water systems. *Pure Applied Chemistry* 60: 1841–1864.

Slade, L., Levine, H. 1988b. Structure stability of intermediate moisture foods—A new understanding? In *Food Structure—Its Creation and Evaluation*. eds. J. M. V. Blanshard, J. R. Mitchell. pp. 115–147. London: Butterworths.

Slade, L., Levine, H. 1994. Water and the glass transition-dependence of the glass transition on composition and chemical structure: Special implications for flour functionality in cookie baking. *Journal of Food Engineering* 22: 143–188.

Slade, L., Levine, H. 1995. Glass transitions and water–food structure interactions. *Advances in Food and Nutrition Research* 38: 103–179.

Slade, L., Levine, H., Ievolella, J., Wang, M. 1993. The glassy state phenomenon in applications for the food industry: Application of the food polymer science approach to structure–function relationships of sucrose in cookie and cracker systems. *Journal of the Science and Food Agriculture* 63(2): 133–176.

Slade, L., Levine, L. 1991. A food polymer science approach to structure property relationships in aqueous food systems: Nonequilibrium behavior of carbohydrate–water systems. In *Water Relationships in Food*. eds. H. Levine, L. Slade, New York: Plenum Press.

Szczesniak, A. S., Bourne, M. C. 1995. Letters: Texture profile analysis—methodology interpretation clarified. *Journal of Food Science* 60(6): vii.

Van den Berg, C. 1986. Water activity. In *Concentration and Drying of Foods*. ed. D. MacCarthy, pp. 11–36. London: Elsevier Applied Science.

21 Fermentative Products Using Dates as a Substrate

Nallusamy Sivakumar

CONTENTS

INTRODUCTION

The date palm tree (*Phoenix dactylifera* L.) is grown extensively in arid and semiarid regions of the world like North Africa, Arabian Peninsula, and South Asian countries. Botanically, date fruits are one-seeded berries consisting of fleshy mesocarp covered by a thin epicarp and a hard endocarp surrounding the seed (Munier, 1973). The date palm is cultivated over 1.18 million ha all over the world (FAOSTAT, 2008). The world production of dates has increased considerably during the last three decades and estimated more than 6.7 million tonnes. Dates have always played a significant role in the economy of the cultivated countries. Date fruits are rich in carbohydrates, minerals, and vitamins (Morton, 1987). Fresh sap from date contains sugars as the major components (92%–95% dry matter basis), 2.7%–5% of proteins, 0.1%–0.5% fats (Hasnaoui et al., 2010), and 2.3%–2.6% of minerals (Ben Thabet et al., 2010). Since date fruit is a good source of sugar, it could be a cheaper raw material for microbial fermentation processes. Due to the high sugar content, it could also be used to produce microbial biomass. Another economical significance is that date does not require any special treatment like acid hydrolysis, steam explosion, or any other enzymatic pretreatment to

FIGURE 21.1 Wasted fruits left uncollected in a date farm.

release sugars in fermentable form. Ben Thabet et al. (2010) observed that even the date sap, after 4 h of latent period, the pH values started to decrease while the turbidity of the sap rose remarkably representing the potential of easily fermentable product. In view of this fact, date can be utilized as a raw material for biomass production and synthesis of microbial products.

The date production process is accompanied by an undesirable increase of loss during picking, storage, commercialization, and conditioning process (Besbes et al., 2005). These lost dates could amount to more than 2,000,000 tonnes/year worldwide. These lost dates are not consumed by human beings because of inadequate texture, insect infection, and poor quality. In addition, date fruits dropped from date palm trees are also not utilized for human consumption (Figure 21.1). At present, these lost dates are simply discarded or seldom used as animal feed (Besbes et al., 2006). To date, emphasis is on biological conversion of agricultural wastes into value-added products. For this purpose, different agro-industrial residues such as grape pomace, apple pomace, banana, sugar beet, okara, jack fruit carpel, kiwi fruit peel, and banana peel were investigated as a substrate for microbial conversion (Hang and Woodams, 1995; Luciana et al., 2000; Shojaosadati and Babaeipour, 2002; Sassi et al., 1991; Khare et al., 1995; Angumeenal and Venkappayya, 2005; Karthikeyan and Sivakumar, 2010).

At present research on date by-products has not been reflecting the importance and prospective of the crop. In this chapter, different products obtained from date using microorganisms are discussed. The future possibilities of utilizing date as a substrate for biomass production and other economically valuable products are also evaluated.

WINE

Wine is a beverage in which all or part of sugar is converted into alcohol by a simple fermentation process. The alcohol content of wine is about 11%–13% (v/v). All the constituents of a fruit and the newly formed components during fermentation play a decisive role in the taste, color, flavor, and thickness of wine. The selection of raw materials, fermentation conditions, and aging are very essential parameters in the production of high-grade wines. Wine is mostly produced from grapes. Other wines are named by the fruit from which they are fermented. For example, elderberry wine, peach wine, and so on. Fruits like dates may also be used for wine production. Although dates lack natural acidity, typical flavor, and the slight acerbity found in grapes, research on date wine production has been continuing over the last few decades.

The production process of date wine involves the selection of different yeast strains, fermentation conditions, and the influence of additional nutrients (Ali and Dirar, 1984; Hassan et al., 1980; Ogaidi et al., 1984; Sachde et al., 1981). *Saccharomyces uvarum, S. cerevisiae, S. carlbergensis*, and *S. ellipsoideus* were tested for their abilities for wine production. The best biochemically active strain, *S. ellipsoideus*, was used for fermentation along with baker's yeast (*S. cerevisiae*) to produce wine from grapefruit, orange, and pineapple juices. *S. ellipsoideus* was found to be the best wine-producing strain with the highest acceptable score of 7.41 from orange. Ndip et al. (2001) revealed the possibility of producing wine from the locally available fruits using simple, cheap, and adaptable technology with biochemically characterized yeast strains. So date could be a good alternative source for wine production because of its high sugar content.

Fermentation of Date Fruit

A variety of vessels may be used as the fermentation tank for date wine production. The low-cost and commonly used vessel is a 32-gallon, plastic garbage can. Date juice could be used as a medium for wine production. Once the date juice is in the fermentation tank, it should be inoculated with preferred yeast strain. This step is not needed when grape juice is used as a fermentation medium because the skin of the grapes already has yeasts on them. The date juice is allowed to ferment for 20–30 days. During the fermentation, the dead yeast cells and other debris settle at the bottom. If all the sugars in date juice have been metabolized, the wine is said to be dry. But the fermentation is stopped before the complete utilization of sugars by the yeast, the final product is a sweet wine. In beer, the percent of alcohol will be 14%–18%. However, the percentage is lower in fruit wines because the amount of sugars in other fruits is generally lower. Even with the addition of sugar, fruit wines are generally 5%–7% alcohol. It could be possible to get more percentage of alcohol in wine by using dates because of its high sugar content. Once the fermentation is complete, the date wine is separated from the sediment and transferred to an aging tank. During the aging process also, more sedimentation occurs and the wine is frequently transferred across a series of tanks. This process is called racking. During different stages of aging, usually the wine is sampled and judged by a wine master. The date wine may be bottled after aging and sold immediately or again kept for aging in the bottles.

Early references are available on date wine and the use of date as a strengthening and flavoring agents in beer making (Darby et al., 1977; Popenoe, 1973). Zahdi dates of Iraq were used to produce light-colored, good-quality wine. But for darker and sweeter wine, date syrup is suitable. It is reported that the addition of 0.025% urea improved the fermentation and also the aging improved the quality of wines (Benjamin et al., 1975). The Sudanese date wines, sherbote, and nabit prepared from date syrup are produced by a quick fermentation process taking about 36–48 h. The fermentation of the third wine, dakkai, prepared from whole date fruits, is a slow process taking about 96 h. A small quantity of spices and sorghum malt was used as additives in the production of sherbote and nabit, respectively. In dakkai fermentation, the inoculum comes from the natural flora of the date fruit itself. The most important genera of yeast, acetic, and lactic acid bacteria were identified during the fermentation (Ali and Dirar, 1984). It is imperative that the knowledge gained in wine fermentation using grapes should be employed profitably to improve the production of wine using date fruits as raw materials.

ALCOHOL

Alcohol fermentation is a biochemical process in which sugars are converted into ethanol and carbon dioxide by the enzyme produced from *S. cerevisiae*. Alcohol is produced in different purities to be used in beverages, as a fuel and also for some medicinal purposes. In addition, in industries, alcohol is used as a precursor material for the production of acetaldehyde and ethylene. Acetaldehyde is further used in the production of acetic acid and butanol and ethylene is the base to manufacture polyethylene, polystyrene, and poly vinyl chloride. By utilizing date fruits as a substrate for alcohol production, it is possible to get all the above-mentioned components. Although alcohol is produced as an integral part of the final product in wine making, for large-scale production, it can also be produced by distillation and rectification of mother liquor. Although alcohol is a versatile chemical, it needs a well-developed downstream process. The production of 95% (v/v) alcohol from dates involves three steps, (i) extraction of date juice, (ii) fermentation, and (iii) distillation and rectification.

Extraction of Date Juice

The date fruits can be kept in a vessel for a series of extractions with juices of decreasing Brix content until the sugars are removed from the dates. A batch-wise extraction of dates is practiced to get the juice which may have some nonsolubles. As yeast cells will grow better in solutions with 20 Brix content, the interference by the nonsolubles in date extract is of less importance. During extraction, the dates can be kept in a vessel followed by series of 3–4 vessels to get the extraction liquid from the dates. Steam can be supplied to maintain the extraction temperature at the same level. Once the dates in the vessel are exhausted, it can be replenished by a new batch of dates. The extract with high sugar content can be used as a medium to produce alcohol. A semicontinuous extraction system could be used to get a better fermentation liquid with a residual left over sugar in dates. The vessels can be loaded and unloaded intermittently with the help of valves. In addition, date juice can also be recirculated into the same vessel until the equilibrium and passed to another vessel containing dates with high Brix content. The advantage here is in three to four extractions we can get an extract suitable for fermentation and in stationary position dates release very little fiber and acting as a natural filter for circulating the date juices. A counter current continuous system may also be used to get fermentation juice from dates but it should be filtered before fermentation (FAO, 1993).

Fermentation

It is safe to maintain the temperature of date juice around 60°C to avoid contamination. But before inoculation of yeast cultures, the medium should be cooled because the fermentation temperature will be maintained around 32°C (Ogaidi et al., 1988). At least 2 days of fermentation is needed for date juice to get ~9% of alcohol in the fermented medium. After fermentation, the fermented liquid containing alcohol and other mixtures are distilled and collected at different boiling temperatures to remove the undesired materials. We can get alcohol strength of 90% by repeating this process at least for four times.

Distillation and Rectification

In industrial scale, a continuous distillation followed by rectification gives a pure form of alcohol. Distillation results in the formation of a drink with 25%–30% alcohol content and flavors contributed by the raw materials which distilled along with alcohol such as brandy (grapes), gin (juniper), grappa (grapes), kirsch (cherries), rum (molasses), schlivovitch (plums), and tequila (cactus) (Barreveld, 1971). In this way, we can produce a pure alcohol using the date extract as a substrate. In date-producing countries, the production of alcohol using date fruits is restricted to medicinal purpose and household use.

XANTHAN GUM

Xanthan gum is a biopolymer produced by *Xanthomonas campestris*, a bacterium found on cabbage plants. This was the first industrially produced biopolymer used in the food industry. After the authorization by Food and Drug Authority (FDA) the demand for xanthan gum has increased about 5%–10% every year (Salah et al., 2010). Xanthan molecular weight is heavily affected by the composition of the medium used. Several studies have been conducted on the influence of different nutrients, nitrogen, and carbon in particular, with glucose and sucrose as the most frequently used carbon sources (Garcia-Ochoa and Gomez, 1998; Garcia-Ochoa et al., 2000). On the other hand, the cost of fermentation medium plays a significant role in commercial production of xanthan. Recent research in this field focuses on cheaper natural alternatives. In this scenario, the waste date gains substantial importance as a substrate for xanthan gum production.

Preparation of Date Juice

The waste dates or dates by-products can be used as a raw material for the production of xanthan gum. After removal of pits the date flesh can be collected, washed, and air dried for at least 12 h. From these dried samples date paste can be prepared by simple grinding. By adding water to the date paste (3:1 v/w) we can get the date juice (Youssif et al., 1990). Then boil the juice gently for 15 min and mix it continuously. Now centrifuge the liquid for 45 min at 10,000 rpm. In this way, we can get about 2.5 Lof date juice from 1 kg date. Usually, the Brix content of date juice will be around 20.

Composition of Date Medium

The following medium can be used for the production of xanthan gum (Salah et al., 2010) (g/L): Date juice: D-glucose, 84.68; KH_2PO_4, 5; $MgSO_4$, 0.2; $(NH_4)_2SO_4$, 2.7; citric acid, 2; H_3BO_3, 0.006; $ZnCl_2$, 0.006; $FeCl_3$, 0.0024; and $CaCO_3$, 0.02. Adjust the pH to 7 by the addition of HCl (1 M). Sterilize the medium at 121°C for 20 min. The production media will be inoculated with 5% inoculum.

Optimum Conditions for Xanthan Production Using Date Medium

Salah et al. (2010) standardized the optimum conditions for the xanthan production by palm date juice using response surface methodology. They found that carbon source had a more positive influence on xanthan production than the nitrogen. But temperature had an effect 131.4 times less than the carbon source. The xanthan production increased proportionally with increasing concentration of carbon source. So it is clear that during xanthan fermentation the carbon source in the date fruits not only plays a major role in building cellular constituents, but is also used in the synthesis of growth-associated polysaccharides. High yield of xanthan gum was obtained when the date juice was in the temperature range of 30–33°C (Shu and Yang, 1990; Esgalhado et al., 1995). In contrast to the xanthan production, carbon source in date fruits exerted a negative influence on biomass production but nitrogen source had a more positive effect than temperature.

Xanthan yield of 43.35 g/L was obtained at 30.1°C. This yield was 50% higher than the yield using date juice without central composite orthogonal design and those obtained using other agro-industrial residues such as carob extract, 17 g/L (Roseiro et al., 1992); citrus waste, 14.5 g/L (Green et al., 1994); olive mill waste waters, 7 g/L (Lopez et al., 2001); ram horn hydrolysate, 25.6 g/L (Basaran-Kurbanoglu and Izzet-Kurbanoglu, 2007); and cheese way, 26.35 g/L (Fernandez-Silva et al., 2009).

Economical Importance

It is noteworthy that the cost of fermentation medium is one of the deserving factors to determine the viability of xanthan gum production. The date juice offers a relatively low price and could be a promising alternative for the currently used substrates like glucose and sucrose. Pragmatically, 1 kg of xanthan produced using date juice costs around \$2.4. On the other hand, current commercial xanthan costs around \$10. As date juice reduces the cost of xanthan production fourfold, admittedly this could be a better alternative source for xanthan gum production.

VINEGAR

Acetic acid is an important feed stock for various industrial products and also used in food and pharmaceutical industries. Due to depleting natural resources, the production of acetic acid by microbial fermentation gains more significance than the chemosynthesis. More emphasis is given to the use of cheap and abundantly available biomass as the substrate (Slapack et al., 1985). A large number of organic acids can be derived from sugars using microorganisms. But dates have been used as a raw material for a very few organic acids production because of market limitations and economy of production. One among them is acetic acid. The production of vinegar (acetic acid) is possibly one of the oldest organic acid fermentations known. Strains of *Acetobacter* are generally involved in vinegar production (Sokollek et al., 1998). The vinegar production from dates may involve two stages. In the first stage, sugar in date juice will be converted into alcohol by *Saccharomyces* sp. and in the second stage alcohol will be converted into acetic acid by *Acetobacter* sp. The commonly used organisms for acetic acid production are *S. cerevisiae*, *S. ellipsoideus*, *Acetobacter xylinum*, and *A. cycendens*.

PRODUCTION OF VINEGAR USING DATE SYRUP

The manufacturing principle is the same like that of wine. Both products can be produced at the cottage level incorporating the flavors of dates extract and secondary fermentation products. Any raw material that will undergo an alcoholic fermentation is a basic requirement for vinegar production. It can be produced from any type of fermentable carbohydrate source such as fruits, vegetables, syrups, and wine. Hydrolyzed starches, cereals, syrups, beer, wine, honey, pears, grapes, and dates are ideal substrates. Date syrup supplemented with some nutrients (malt/urea) can be converted into acetic acid by oxidative fermentation. The rate and final amount of acetic acid produced depends greatly on the oxygen supply, the initial alcohol concentration and the strain of the bacterium. Thus, for increasing efficiency of acetic acid production various types of vinegar generators are used. In one type of generator, wooden shavings packed in a column are saturated with old vinegar containing acetic bacteria (mother vinegar) and circulated with fresh-fermented alcoholic solution. Acetic acid bacteria develop as a thin film on the wooden shavings and forms acetic acid from ethanol by oxidation. Solution with 5%–6% alcohol is converted into vinegar in <5 days. The desirable acetic acid level can be achieved in few hours by the use of stirred tank fermentors or column fermentors with vigorous aeration. Vinegar is produced industrially by two main methods, a slow process involving static surface acetic acid fermentation and a fast submerged fermentation process. Direct conversion of date syrup into acetic acid by a single fermentative organism is economical.

LACTIC ACID

Lactic acid has a wide range of applications in food, pharmaceutical, and cosmetic industries (Hujanan and Linko, 1994). It is also used as acidulant, preservative for beverages and food stuffs, in textile, leather, and laundry industries. Calcium lactate is used in baking powders, as a supplement in animal feed and a source of calcium in pharmaceutical preparations (Litchfield, 1996). New applications like degradable plastics made from polylactic acid have the prospective to increase the market for lactic acid, if more economical processes could be developed (Cheng et al., 1991). The economics of lactic acid fermentation is still a problem in many points of view and composition of production media plays a pivotal role in the improvement of such process. So the fermentative production of lactic acid using waste dates remains attractive. Lactic acid production has been carried out with date juice by fermentation using *Lactobacillus casei* sub sp. *rhamnosus* (Nancib et al., 2001) and *Lactobacillus* sp. KCP01 (Chauhan et al., 2007). These homofermentative lactic acid-producing bacteria can be grown in deMan, Rogosa, and Sharpe (MRS) medium and maintained in 20%–50% glycerol and stored below 0°C.

Date Juice Preparation

Date juice can be prepared by mincing 40 g of thoroughly washed seedless dates in distilled water and make up the final volume to 300 mL. Boil the juice for 10 min to flocculate the undissolved suspended matter that settled rapidly upon cooling to room temperature. Then filter this to obtain a clear date juice (Nancib et al., 2005; Samuel and Lee, 1980). This crude date juice of dates was further diluted with distilled water to get final concentration of 5% reducing sugar. Sometimes water–date mixture (2:1 w/w) is heated at 80°C for 2 h with uninterrupted stirring. Then the extract is centrifuged at 5000 rpm for 30 min to remove the cellulose debris and the supernatant serves as a carbon source in the fermentation medium (Nancib et al., 2001).

Date Medium for Lactic Acid Production

The date juice alone can be used as production medium or with the nutrient supplement of MRS medium, except for the recommended concentration of sugar source. The MRS medium used for lactic acid production contained (g/L): peptone, 10; beef extract, 10; yeast extract, 5; K_2HPO_4, 2; sodium acetate, 5; tri-sodium citrate, 2.15; $MgSO_4 \cdot 7H_2O$, 0.2; $MnSO_4 \cdot 4H_2O$, 0.05; Tween-80, 1; and date juice to give 5% reducing sugar. The pH of the medium is adjusted to 7.0 and sterilized by autoclaving at 121°C for 15 min. After inoculation, incubate the flasks at 40°C under static condition. Samples can be collected at every 24 h up to 2 days for further studies. Harvest the cells by centrifugation at 9000*g* for 10 min and clear supernatant will be subjected to lactic acid estimation (Chauhan et al., 2007).

Factors Influencing the Production of Lactic Acid from Date Juice

Nancib et al. (2001) conducted an experiment to demonstrate the effect of date juice as the only carbon and nitrogen source on the final concentration of lactic acid produced. Maximum amount of lactic acid was obtained when the medium contains 60 g/L of glucose concentration. But it was less than the yield produced in the MRS medium. This could be due to the deficiency of nitrogen sources in the date juice (Abou-Zeid and Khoja, 1993). Lactic acid bacteria are reported to be fastidious and need complex nutrients for growth and production (Fitzpatrick and O'Keeffe, 2001). Chauhan et al. (2007) suggested that even though the palm dates are rich in easily extractable sugar source compared to other raw materials, they need a prior treatment with enzyme or other agents to release the sugar in utilizable and fermentable form by the organism. As dates have low protein content (1.5%–3%, w/w), the date juice in salt solution was supplemented with yeast extract which increased the lactic acid production. Although yeast extract was found as the best nitrogen source, its cost has a negative impact on the economy of industrial production. The effect of supplementing date juice with different nitrogen sources such as peptone, urea, ammonium sulfate, and corn steep liquor was compared. Among the different nitrogen sources, ammonium sulfate was found to be a feasible and inexpensive alternative substrate (Nancib et al., 2001). For the optimum lactic acid production, the addition of vitamin B is needed in addition to ammonium sulfate (Nancib et al., 2005). This again gives a negative impact on the cost of lactic acid production. Hence, peptone could be a better option for the production of lactic acid using date juice. Addition of sodium acetate and potassium phosphate to the date juice also increases the lactic acid production significantly (Chauhan et al., 2007). It was also observed that biomass and lactic acid production was higher when the date juice was supplemented with both ammonium sulfate and yeast extract. When the ratio of ammonium sulfate to yeast extract was 4:1, the lactic acid production was found comparable to date juice supplemented with 20 g/L yeast extract. This would reduce the cost of nitrogen source by 80% (Nancib et al., 2001). Controlling the pH, by adding calcium carbonate to neutralize the lactic acid, significantly increased the lactic acid production (45.59 g/L) as compared to the control (16.5 g/L) but it affects the cell growth.

BIOMASS PRODUCTION

There are numerous reports on utilizing waste and surplus products into high-protein food and feed in the form of microbial biomass using a variety of microorganisms, yeasts, fungi, and algae in particular. Earlier, in large-scale productions of beer, wine, and pure alcohol, the yeasts are recuperated, purified and sold, but the yield compared to alcohol production is small. If *S. cervesiae* strains are provided with appropriate nutrients and the condition is changed to aerobic from anaerobic, the process can be directed toward biomass production. A well-known microbial product is baker's yeast available as pressed fresh yeast (27%–29% dry substance) or as active dry yeast (90%–92% dry matter), the later often imported and used as a leavening agent in bread making in the date-producing countries of the Old World (FAO, 1993).

The conditions needed for the growth of baker's yeast in date juice medium were studied by Mudhaffer (1978). Date extract and date molasses were successfully used as a media for the production of baker's yeast (Mohammed et al., 1986). This research was mainly aimed to test the suitability of date extracts and molasses as the raw materials for yeast production. The outcome of the research showed that there does not seem any technical limitation in using date extract which would be of special interest to date-producing countries which have no beet or cane sugar production. Apart from the experimental work on using date extracts, feed yeast production using dates has reached the industrial level in Iraq (Ogaidi et al., 1981; Ogaidi and Al-Khafaji, 1983). The optimum concentration of micro- and macronutrients in date juice for the production of baker's yeast was also standardized (Obaidi et al., 1985a, 1987). Date extract was also used to produce fungal protein using *Aspergillus niger* (Obaidi et al., 1985b) while the same organism was also tested on date extract for the production of amylase (Tai et al., 1988). The date extract, date-coat sugar extract, and date-seed hydrolysate proved as good carbon and nitrogen sources suitable for biomass production by *S. cerevisiae*. Promising results were obtained in the natural medium containing only date extracts, which are very economical (Nancib et al., 1997).

CITRIC ACID

In addition to date fruit, the date pit can also be used as a source for microbial conversion. Date pits were odorless and had light-to-dark-brown color. It has 7.1%–10.3% moisture, 5.0%–6.3% protein, 9.9%–13.3% fat, 65%–69% neutral detergent fiber, and 1.0%–1.8% ash. The date pit hydrolysate and whey were utilized as the main constituents of a medium for the formation of citric acid by *Candida lipolytica*. Maximum amount of citric acid is formed at 72–96 h of fermentation. The supplement of glucose to the fermentation medium increases the citric acid production because the carbon source in natural medium is not sufficient. The addition of date seed ash to the fermentation medium increased the output of citric acid, signifying the role of certain elements such as calcium, iron, magnesium, manganese, zinc, and nickel in the formation of citric acid (Abou-Zeid et al., 1983). Pretreatment of date syrup with sulfuric acid, tricalcium phosphate, tricalcium phosphate with HCl, potassium ferricyanide, and EDTA to increase the production of citric acid was also investigated (Roukas and Kotzekidou, 1997).

CONCLUSION

It is necessary to search some alternative organic substrates for microbial growth and production because of the existing economic recession and increasing cost of substrates. Waste date fruit has a greater feasibility to be utilized as a substrate for the production of some economically important microbial products. Any material that contains this many nutrients could serve as a possible alternative substrate for microbial fermentation. Although very few fermentation studies carried out using date as a substrate, this nutrient-rich fruit has a great potential as a substrate for the microbial fermentation products. Further research should focus on the development of new progressive fermentation techniques to achieve high substrate conversion and high production yields.

REFERENCES

Abou-Zeid, A.A. and Khoja, S.M., 1993. Utilization of dates in the fermentative formation of citric acid by *Yarrowia lipolytica. Zentrabl. Mikrobiol.*, 148:213–221.

Abou-Zeid, A.A., Baghlaf, A.O., Khan, J.A., and Makhashin, S.S. 1983. Utilization of date seeds and cheese whey in production of citric acid by *Candida lipolytica. Agric. Wastes*, 8:131–142.

Ali, M.Z. and Dirar, H.A. 1984. A microbiological study of Sudanese date wines. *J. Food Sci. (USA)*, 49:459–460.

Angumeenal, Ar. and Venkappayya, D., 2005. *Atrocarpus heterophyllus*—A potential substrate for citric acid biosynthesis using *Aspergillus niger. LWT. Food Sci. Technol.*, 38:89–93.

Barreveld, W.H. 1971. Industrial use of secondary quality dates. *World Crop.*, 23:209–210.

Basaran-Kurbanoglu, E. and Izzet-Kurbanoglu, N. 2007. Ram horn hydrolysate as enhancer of xanthan production in batch culture of *Xanthomonas campestris* EBK-4 isolate. *Process Biochem.*, 42:1146–1149.

Benjamin, N.D., Zubair, H., and Al-Tai, W. 1975. Zahdi date and syrup wine. Paper presented at the *Third FAO Int. Palm and Dates Conference*, Baghdad, Iraq.

Ben Thabet, I., Francis, F., Pauw, E., Besbes, S., Attia, H., Deroanne, C., and Blecker, C. 2010. Characterisation of proteins from date palm sap (*Phoenix dactylifera* L.) by a proteomic approach. *Food Chem.*, 123:765–770.

Besbes, S., Cheikh Rouhou, S., Blecker, C., Deroanne, C., Lognay, G., and Drira, N.E. 2006. Voies de valorisation des sous produits de dattes: Valorisation de la pulpe. *Microbiologie Hygiène Alimentaire*, 18:3–7.

Besbes, S., Hentati, B., Blecker, C., Deroanne, C., Lognay, G., and Drira, N.E. 2005. Voies de valorisation des sous produits de dattes: Valorisation du noyau. *Microbiologie Hygiène Alimentaire*, 18:3–11.

Chauhan, K., Trivedi, U., and Patel, K.C. 2007. Statistical screening of medium components by Plackett–Burman design for lactic acid production by *Lactobacillus* sp. KCP01 using date juice. *Bioresour. Technol.*, 98:98–103.

Cheng, P., Muller, R.E., Jaeger, S., Bajpai, R., and Iannoti, E.L. 1991. Lactic acid production from enzyme-thinned corn starch using *Lactobacillus amylovorus. J. Ind. Microbiol.*, 7:27–34.

Darby, J., Ghalioungui, P., and Grivetti, L. 1977. *Food: The Gift of Osiris*, Vol. 2. Academic Press, London.

Esgalhado, M.E., Roseiro, J.C., and Amaral, C.M.T. 1995. Interactive effects of pH and temperature on cell growth and polymer production by *Xanthomonas campestris. Process Biochem.*, 30:667–671.

FAO (Food and Agricultural Organization of the United States). 1993. *Date Palm Products*. FAO Agricultural Services Bulletin–101.

FAOSTAT (The FAO Statistical Database). 2008. *Agro-Statistics Database*. Rome: Food and Agriculture Organization of the United Nations.

Fernandez-Silva, M., Fornari, R.C.G., Mazutti, M.A., De Olivera, D., Ferreira-Padhila, F., Jose Cichoski, A., Luis-Cansian, R., Di Luccio, M., and Helen, T. 2009. Production and characterization of xantham gum by *Xanthomonas campestris* using cheese whey as sole carbon source. *J. Food Eng.*, 90:119–123.

Fitzpatrick, J.J. and O'Keeffe, U. 2001. Influence of whey protein hydrolysate addition to whey permeate batch fermentation for producing lactic acid. *Process Biochem.*, 37:183–186.

Garcia-Ochoa, F. and Gomez, E. 1998. Mass transfer coefficient in stirred reactors for xanthan gum solutions. *Biochem. Eng.*, 1:1–10.

Garcia-Ochoa, F., Santos, V.E., Casas, J.A., and Gomez, E. 2000. Xanthan gum: Production, recovery, and properties. *Biotech. Adv.*, 18:549–579.

Green, M., Shelef, G., and Bilanovic, D. 1994. The effect of various citrus waste fractions on xanthan fermentation. *Chem. Eng.*, 56:37–41.

Hang, Y.D. and Woodams, E.E., 1995. Grape pomace: A novel substrate for microbial production of citric acid. *Biotechnol. Lett.*, 7:253–254.

Hasnaoui, A., Elhoumaizi, A., Asehraou, A., Sindic, M., Deroanne, C., and Hakkou, A. 2010. Chemical composition and microbial quality of dates grown in Figuig Oasis of Moracco. *Int. J. Agric. Boil.*, 12:311–314.

Hassan, H.K., Al-Tai, W.F., and Hamando, M.M. 1980. Studies on the production of date wine. I: Selection of active strains for date juice fermentation. Palm and Dates Research Centre, Baghdad, Iraq, Technical Bulletin No. 6.

Hujanan, M. and Linko, Y.Y. 1994. Effect of temperature and various nitrogen sources of L(+)-lactic acid production by *Lactobacillus casei. Appl. Microbiol. Biotechnol.*, 45:307–313.

Karthikeyan, A. and Sivakumar, N. 2010. Citric acid production by Koji fermentation using banana peel as a novel substrate. *Bioresour. Technol.*, 101:5552–5556.

Khare, S.K., Krishna, J., and Gandhi, A.P., 1995. Citric acid production from Okara (soy residue) by solid state fermentation. *Bioresour. Technol.*, 54:323–325.

Litchfield, J.H. 1996. Microbial production of lactic acid. *Adv. Appl. Microbiol.*, 42:48–95.

Lopez, M.J., Moreno, J., and Ramos-Cormenzana, A. 2001. *Xanthomonas campestris* strain selection for xanthan production from olive mill wastewaters. *Water Res.*, 35:1828–1830.

Luciana, P.S., Vandenberghe, L.P.S., Saccol, C.R., Pandey, A., and Lebeault, J.M., 2000. Solid state fermentation for the synthesis of citric acid by *Aspergillus niger. Bioresour. Technol.*, 74:175–178.

Mohammed, N.A., Al-Obaidi, Z.S., Hasson, N.A., and Jassem, M.A. 1986. Semi-industrial production of baker's yeast using date extract and molasses (Arabic). *J. Agric. Water Resour. Res. (Iraq)*, 5(1):147–162.

Morton, J. 1987. Date. p. 5–11. In: Julia F. Morton, ed., *Fruits of Warm Climates*. 5--11. Miami, FL.

Mudhaffer, S. 1978. Studies on the conditions of growth of baker's yeast, grown in solution of molasses or debis (date juice). *Nahrung*, 22(4):285–289.

Munier, P. 1973. *Le Plamier Dattier*. Maisonneuvre et Larose, Paris, France, 220.

Nancib, A., Nancib, N., Meziane-Cherif, D., Boubendir, A., Fick, M., and Boudrant, J. 2005. Joint effect of nitrogen sources and B vitamin supplementation of date juice on lactic acid production by *Lactobacillus casei* subsp. *rhamnosus. Biores. Technol.*, 96:63–67.

Nancib, N., Nancib, A., and Boudrant, J. 1997. Use of waste products in the fermentative formation of baker's yeast biomass by *Saccharomyces cerevisiae. Bioresour. Technol.*, 60:67–71.

Nancib, N., Nancib, A., Boudjelal, A., Benslimane, C., Blanchard, F., and Boudrant, J. 2001. The effect of supplementation by different nitrogen sources on the production of lactic acid from date juice by *Lactobacillus casei* subsp. *rhamnosus. Biores. Technol.*, 78:149–153.

Ndip, R.N., Akoachere, F.K.T., Dopgima, L.L., and Ndip, L.M., 2001. Characterization of yeast strains for wine production. Effect of fermentation variables on quality of wine produced. *Appl. Biochem. Biotechnol.*, 95:209–220.

Obaidi, Al-, Z.S., Al-Hilli, M.A., and Al-Hakkak, Th. S. 1985a. Optimization of propagation medium for baker's yeast using date extract and molasses. 1. Determination of the optimum concentration of the macro elements. *J. Agric. Water Resources Res. Iraq*, 5(2):97–108.

Obaidi, Al-, Z.S., Aziz, G.M., Al-Hakkak, T.S., and Al-Hilli, M.A. 1987. Optimization of propagation medium for baker's yeast using date extract and molasses. Determination of the optimum concentration of microelements and vitamins. *Date Palm J.*, 5(1):64–78.

Obaidi, Al-, Z.S., Hassan, H.S., Jassem, M.A., and Jaffer, T.S. 1985b. Production of fungal protein using date extract by *Aspergillus niger* (in Iraq) (Arabic). *J. Agric. Water Resour. Res.*, 4(1).

Ogaidi, Al-, H.K., Al-Hammando, M.M., and Al-Hakkak, T.J. 1981. Production of single cell protein from yeast growing in date juice (variety Zahdi). I. Selection of yeast strains with high protein yield (Arabic). Palm and Dates Research Centre, Baghdad, Iraq. Technical Bulletin No. 1/81.

Ogaidi, AL-, H.K., AL-Nagash, S., Mutlak, H.H., and Al-Bradiw, G. 1984. The effect of some factors on wine produced from Sayer date juice (in Iraq) (Arabic). *J. Agri. Water Resour. Res.*, 3(2):111–119.

Ogaidi, Al-, H.K.H. and Al-Khafaji, N.A. 1983. The effect of hydrogenion concentration and temperature on biomass production of *Candida* sp. using date extract (Arabic). *Proc. First Symp. Date Palm*, Saudi Arabia.

Ogaidi, Al-, H.K.H., Muslh, R., and Ali, N.M. 1988. Isolation and identification of some Iraqi local ethanol producing yeasts grown on date juice. *Date Palm J.*, 6(2):433–446.

Popenoe, P. 1973. *The Date Palm*. Edited by Henry Field. Field Research Projects, Coconut Grove, Miami, FL.

Roseiro, J.C., Costa, D.C., and Collaco, M.T.A. 1992. Batch and fed-batch cultivation of *Xanthomonas campestris* in carob extracts. *Lebensmittel-Wissenschaft und-Technologie-Food Sci. Technol.*, 25:289–293.

Roukas, T. and Kotzekidou, P. 1997. Pretreatment of date syrup to increase citric acid production. *Enzyme Microbiol. Technol.* 21:273–276.

Sachde, A.G., Al-Kiasy, A.M., and Norris, R.A.K. 1981. A study of the possibility of producing quality wines from some commercial varieties of Iraqi dates. *Mesopotamia J. Agric.*, 16(1):93–106.

Salah, R.B., Chaari, K., Besbes, S., Ktari, N., Blecker, C., Deroanne, C., and Attia, H. 2010. Optimisation of xanthan gum production by palm date (*Phoenix dactylifera* L.) juice by-products using response surface methodology. *Food Chem.*, 121:627–633.

Samuel, W.A. and Lee, Y.Y. 1980. Lactic acid fermentation of crude sorghum extract. *Biotechnol. Bioeng.*, 22:757–777.

Sassi, G., Ruggeri, B., Specchia, V., and Gianetto, A., 1991. Citric acid production by *Aspergillus niger* with banana extract. *Bioresour. Technol.*, 37:259–269.

Shojaosadati, S.S. and Babaeipour, V. 2002. Citric acid production from apple pomace in multilayer packed bed solid-state bioreactor. *Process Biochem.*, 37:909–914.

Shu, C.H. and Yang, S.T. 1990. Effects of temperature on cell growth and xanthan production in batch cultures of *Xanthomonas campestris. Biotechnol. Bioeng.*, 35:454–468.

Slapack, G.E., Russell, I., and Stewart, G.G. 1985. Thermophilic bacteria and thermotolerant yeasts for ethanol production, NRCC No. 24410. Project Report submitted to Division of Energy, NRC, Ottawa, pp. 1–404.

Sokollek, S.J., Hertel, C., and Hammes, W.P. 1998. Cultivation and preservation of vinegar bacteria. *J. Biotechnol.*, 60:195–206.

Tai, Al-, W.F., Al-Obaidi, Z.S., Al-Hakkak, Th.S., and Al-Hilli, M.A.1988. Production of *Aspergillus niger* amylase using date extract. *Date Palm J.*, 6(2):447–456.

Youssif, A.K., Abou Ali, M., and Abou Idreese, A. 1990. Processing, evalution and storability of date jelly. *J. Food Sci. Technol.*, 27:264–266.

22 Dates as Potential Substitute for Added Sugar in Food

A. Manickavasagan

CONTENTS

INTRODUCTION

Sugar has been used in human food for a long time because of its organoleptic and preservative properties (Mann 2004). In the perspective of human health, sugars can be broadly classified into two main groups: (1) sugar incorporated within the structure of intact fruits and vegetables (intrinsic sugars) and (2) sugar added to foods and drinks by the manufacturer, cook, or consumer (added sugar). Total sugar includes all sugar in diet. The dietary guidelines from health authorities do not recommend any restriction for intrinsic sugars (such as fruit sugar [fructose] and milk sugar [lactose]); however, added sugars and concentrated sugars (also referred to as free sugar or refined sugar or sucrose) should be limited in the diet by considering their untoward effects (Mann 2004).

Sugar is added to fruit drinks, soft drinks, desserts, snacks, cereals, coffee, tea, and many other regional foods. The consumption of added sugar varies from person to person and region to region based on their dietary behavior. In some regions, the intake of added sugar becomes very high during festival seasons and special occasions. The food guide from the United States Department of Agriculture (USDA) recommends consumption of added sugar in the range between 6% and 10% of total energy, and the World Health Organization (WHO) recommends to limit added sugar intake to <10% of total energy (Kranz et al. 2005; Welsh et al. 1993; World Health Organization 2003). The American Heart Association recommends limiting the daily added sugar intake to 100 cal for women and 150 cal for men (Johnson and Yon 2010).

In spite of these recommendations, it was estimated from a survey (1996–2006) that added sugar intake from a representative sample of 6113 US adults was 359 cal, which represented 15.8% of total daily energy intake (Johnson and Yon 2010; Welsh et al. 2010). Although added sugar intake is higher in many other regions, information for per capita intake is not available in the public domain.

ADDED SUGAR CONSUMPTION AND HUMAN HEALTH

The scientific evidence demonstrates that added sugar increases the risk of overweight and obesity as well as other risk factors for cardiovascular disease including dyslipidemia and high blood pressure (Johnson and Yon 2010). Sugar-sweetened beverage is the primary source of added sugar intake in the United States (Johnson and Yon 2010) and many other countries. The reason for weight gain by the consumption of sugar-sweetened beverages is partly due to incomplete compensation for energy at subsequent meals following the intake of liquid calories. This may also increase the risk of type 2 diabetes and cardiovascular disease as a contributor to a high dietary glycemic load leading to inflammation, insulin resistance, and impaired beta-cell function (Hu and Malik 2010).

Apart from causing obesity, type 2 diabetes, and cardiovascular disease, diets containing a high proportion of added sugars are lower in micronutrients than diets containing a moderate portion of added sugar (Gibson 2007). Kranz et al. (2005) investigated the effect of added sugar intake levels on nutrients and food group consumption by preschoolers (2–5 years old) in the United States. About 11%–12% of children consumed more than 25% of energy from added sugar. Around 50% of the added sugar consumption was contributed by top three items: fruit juice, high-fat desserts, and regular soft drinks. It was determined that increasing the added sugar consumption was paralleled by the decreased intake of nutrients, minerals (including calcium), and other food groups such as fruits and vegetables.

The combined effect of high sugar and saturated fat intake on human health has been widely studied. The diets rich in saturated fat and refined sugar can contribute to cognitive decline in aging, and can accelerate the course of dementia in Alzheimer's disease (Knopman et al. 2001; Kalmijn et al. 1997, 2000; Molteni et al. 2002).

In addition to the quantity of added sugar intake, the mode of consumption seems to play a role in determining their health functionalities. Baschetti (2004) hypothesized that the sugar can predispose to type 2 diabetes and can cause unhealthy changes in blood lipids if it is consumed in solid forms or in dense solution containing more than 250 g/L, whereas sugar is harmless if it is consumed in more dilute concentrations.

DATES AS A SWEETENER IN FOOD

The health authorities suggest using fruits in food preparation whenever sweet taste is required. The functionality of fruit sugar is different from the added sugar in human health. The sugar in fruits is mainly fructose, whereas in refined sugar, it is sucrose. The glycemic (GI) index of fructose is 26% whereas it is 83% for sucrose when compared with standard white bread (Anderson 1997; Wolver 1990).

Like other fruits, dates contain dietary fiber, phenolic compounds, minerals, vitamins, and antioxidant compounds (Al-Farsi et al. 2007). Al-Shahib and Marshall (2002) reported that the dietary fiber content of dates ranged from 8.1% to 12.7% (w/w). It has been determined that date fruits have potent antioxidant and antimutagenic activities (Vayalil 2002). The GI of several date varieties has been in the range of medium-to-low category. For Omani date varieties (Khalas, Khasab, and Fard), the GI was in the range of 48–58 (Ali et al. 2009). The GI values of UAE dates (Fard, Lulu, Bomaan, Dabbas, and Khalas) were in the range of 46–55 (Alkaabi et al. 2011). Miller et al. (2003) determined the GI of Khalas variety (dates alone and in mixed meals with plain full-milk yogurt) with healthy male and female subjects (aged 32–67 years). The mean GI of the dates were 47, 45, 36, 37, 29 for rutab, traditionally stored, commercial, rutab/yogurt, and commercial tamer/yogurt preparations, respectively. It was concluded that Khalas dates, when eaten alone or in mixed meals with plain yogurt, have low GI and it may be of benefit in the glycemic and lipid control of diabetic patients.

DATE SYRUP

In terms of application of dates as sweetener in foods, date syrup has been widely studied for its suitability. Aboubacar et al. (2010) determined potential uses of date syrup as a substitute sweetener for sucrose in food using muffin as a model product. Five types of muffins were prepared with date syrup replacing 25%, 50%, 75%, and 100% of sucrose, control (100% sucrose), and 50% sucrose replacement by high-fructose corn syrup (HFCS). The physicochemical and sensory attributes were evaluated for the developed products. Control muffin had lighter color and the lowest red and yellow colors compared to muffin with syrups. Date syrup used in that study (Al-Dhafra, Kingdom and Al-Saad) had very low sucrose content (0–0.37 g/100 g).

Muffins made with date syrup had significantly lower hedonic rate for appearance and color compared to the control muffin. Products containing up to 50% date syrup and control had similar sweetness, flavor, texture, and acceptance rating. But the muffin containing 50% HCFS had the lowest rating for flavor, texture, and acceptance. It was concluded that date syrup can be used just as HFCS in food industry. However, color appears to be the major sensory attribute affecting consumer acceptability while adding more than 50% of date syrup.

Gouhari et al. (2005) investigated the possibilities of using date syrup to replace sucrose in ice cream. Five types of ice creams were prepared: 0% (control), 25%, 50%, 75%, and 100% replacement of sucrose with date syrup. The viscosity of ice cream increased significantly as the level of substitution increased. Replacement with date syrup decreased the freezing point of ice cream. The sensory properties of the modified ice cream were not affected up to 50% replacement and samples prepared with 100% replacement were rejected by panelists.

Gad et al. (2010) used date syrup as part of water (2%, 4%, 6%, 8%, and 10% v/v) in reconstituting skim milk powder during the processing of yogurt with 14% total solids. Yogurt enriched with 10% date syrup had the highest rating for sweetness, taste, total phenolic content, antioxidant values, HCl-soluble minerals, and folate concentrations.

Sidhu et al. (2003) used date syrup from two varieties in Kuwait (Birhi and Safari) to replace sucrose in pan bread formulations. Five types of pan breads were prepared by replacing sucrose (0% (control), 50%, and 100% replacement with Birhi syrup and 50% and 100% replacement with Safari syrup). The use of concentrated date syrup in pan bread formulations did not adversely affect the baking loss, but rather gave a significantly higher-specific loaf volume (4.32–4.35 cm^3/g) for 100% replacement level compared with the control (4.09 cm^3/g). The lower compression force (358.8–368.5 g) at 100% replacement of sucrose with date syrup also indicated a softer texture of test bread samples than the control (479.6 g). Even though the crumb color of date syrup pan bread scored lower in sensory evaluation, there were no significant differences in texture, flavor, and overall acceptability of control and date syrup pan breads.

Date Sugar and Date Flour

The advantages of dates over other fruits in replacing added sugar are its higher sugar content. Date sugar can be prepared by grinding dried dates into coarse powder. Since it is the product of whole fruit, it contains considerable amount of fiber, minerals, and vitamins. However, the utilization of date sugar might be limited to specific baking application as it contains 30%–50% of nonsugar compounds which will not provide the desired physical properties needed in many food products and dissolve in liquids (Phillips et al. 2009; Kohli 2011). Another method to extract various sugars (fructose, glucose, and sucrose) from date flesh has been explained in a patent by Silver et al. (1998).

The recent dietary recommendations are to increase the intake of antioxidant-rich foods rather than supplement-specific nutrients (Anonymous 2000; Phillips et al. 2009). Unrefined sweeteners are expected to contain higher levels of antioxidants than refined sugars. Phillips et al. (2009) investigated the antioxidant content of various alternatives to refined sugar including date sugar. It was determined that the antioxidant capacity of date sugar ranged from 2.24 to 4.59 FRAP (ferric-reducing ability of plasma) mean mmol/100 g whereas the granulated white sugar contained only 0.004–0.017 FRAP mean mmol/100 g. Figure 22.1 explains the estimated antioxidant contribution from various sweeteners used as a substitute for refined sugar in a standard cake recipe.

Harrak and Jaouan (2010) prepared date flour from the flesh of six date varieties (Ademou, Bouijjou, Bouskri, Bouslikhene, Jihel, and Oum-N'hal) in Morocco, and determined their nutritional,

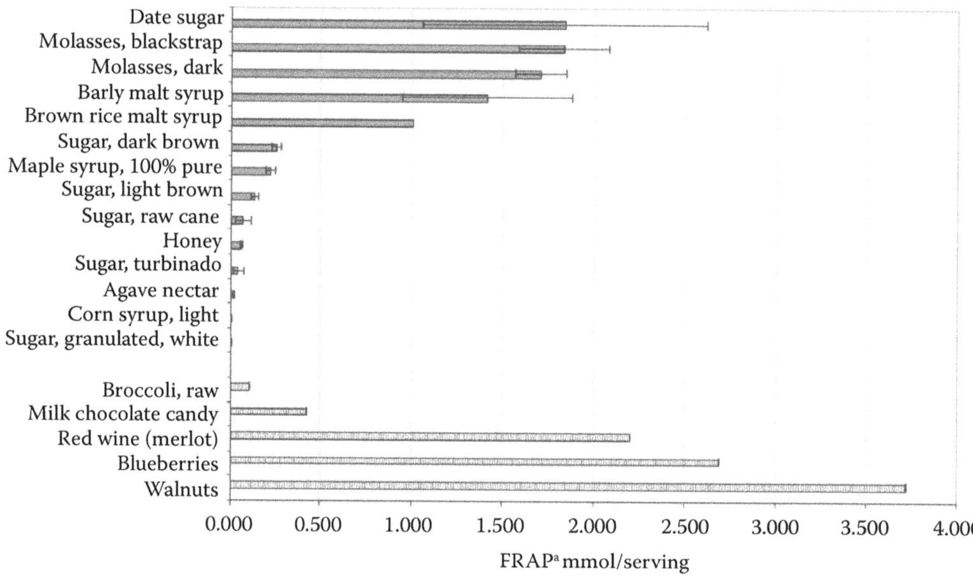

FIGURE 22.1 Estimated antioxidant contribution from sweeteners used as a substitute for refined sugar in a standard cake recipe containing 1.5 cups (350 g) granulated sugar and yielding nine servings, compared to one serving of selected foods. [a]FRAP = ferric-reducing ability of plasma. (Adapted from Phillips, K.M., Carlsen, M.H., and R. Blomhoff. 2009. *Journal of the American Dietetic Association* 109:64–71.)

organoleptic, and storage qualities. It was reported that Bouskri and Bouijjou varieties were suitable for processing into flour. Besides a higher processing yield (up to 75%), the date flours showed interesting nutritional and organoleptic qualities. The water activity (<0.41) indicated the microbial and enzymatic stability of date flours. The storage of date flour at 25°C storage for 3 months did not affect the quality; however, the color was changed when it was stored for up to 6 months. The sensory evaluation demonstrated the suitability of date flour in enriching various food products and the potential to blend with the ingredients of biscuits, cakes, infant foods, and other similar products.

Although people across the world use dates as sweetener for many traditional and modern foods, most of the details are not documented. Ready-to-eat cereals are popular breakfast globally, and "dried date slices" with good rehydration properties would be a great choice to sweeten the cereals. Parfitt and Valentine (2008) have published a book containing several recipes with dates. Banana and date pudding, cinnamon date soufflé, date ice cream, date and walnut ice cream, date pan cake, rhubarb and date fool, baked apples with dates, date muffin, date balls, oaty date nut cookies, date cake, sugar-free fruit cake, and wholemeal date and caramel cake are some of the recipes in which dates can be used as sweetener for added sugar.

CONCLUSIONS

Globally, consumption of added sugar is much higher than the dietary recommendation. It has been scientifically proven that increased intake of added sugar causes various diseases. To minimize the intake of added sugar, and enjoy the delicious sweet taste, the natural sugar from the fruits should be used in the recipes. Dates in its various forms have the potential to blend with the ingredients of many products. There are ample opportunities to use date syrup, date sugar, date flour, and date slices in various food preparations to replace added sugar. Research must be carried out in various aspects of manufacturing date powder and date sugar. These products should be made available commercially in the retail market. In certain products, it may not be possible for a complete substitution of date-based sweeteners for refined sugars because of the changes in product quality; however, partial replacements up to acceptable levels on a routine basis might provide desired health benefits.

ACKNOWLEDGMENT

This work was supported by SQU—Internal Grant No. IG/AGR/ SWAE/11/02 (Nutrification of traditional foods with Omani dates).

REFERENCES

Aboubacar, A., Hashim, I.B., and H.S. Afifi. 2010. Quality characteristics of muffin containing date syrup as sweetener. Poster presented in *Fourth International Date Palm Conference (FIDPC)*, March 15–17. Abu Dhabi, UAE.

Al-Farsi, M., Alasalvar, C., Al-Abid, M., Al-Shoaily, K., Al-Amry, M., and F. Al-Rawahy. 2007. Compositional and functional characteristics of dates, syrups, and their by-products. *Food Chemistry* 104:934–947.

Ali, A., Al-Kindi, Y.S.M., and F. Al-Said. 2009. Chemical composition and glycemic index of three varieties of Omani dates. *International Journal of Food Sciences and Nutrition* 60:51–61.

Alkaabi1, J.M., Al-Dabbagh, B., Ahmad, S., Saadi, H.F., Gariballa, S., and M. Al-Ghazali. 2011. Glycemic indices of five varieties of dates in healthy and diabetic subjects. *Nutrition Journal* 10:1–26 (doi:10.1186/1475-2891-10-59).

Al-Shahib, W. and R.J. Marshall. 2002. Dietary fibre content of dates from 13 varieties of date palm *Phoenix dactylifera* L. *International Journal of Food Science and Technology* 37:719–721.

Anderson, G.H. 1997. Sugar and health: A review. *Nutritional Research* 17:1485–1498.

Anonymous. 2000. Dietary guidelines for chronic disease prevention. *Southern Medical Journal* 93:1157–1161.

Baschetti, M.D.R. 2004. Evolutionary legacy: Form of ingestion, not quantity, is the key factor in producing the effects of sugar on human health. *Medical Hypotheses* 63:933–938.

Gad, A.S., Kholif, A.M., and A.F. Sayed. 2010. Evaluation of the nutritional value of functional yogurt resulting from a combination of date palm syrup and skim milk. In *Proceedings of the Fourth International Date Palm Conference*, Ed. Zaid, A. and G. Alhadrami. Abu Dhabi, United Arab Emirates, March 15–17: International Society for Horticultural Science, pp. 583–592.

Gibson, S.A. 2007. Dietary sugars intake and micronutrient adequacy: A systematic review of the evidence. *Nutrition Research Review* 20:121–131.

Gouhari, A.A., Habibinajafi, M.B., and K.M.H. Hadad. 2005. Effect of date syrup as a substitute for sugar on the physicochemical and sensory properties of soft ice cream. *Iranian Food Science and Technology Research Journal* 1:23–32.

Harrak, H. and F. Jaouan. 2010. Processing dates of low market value into flour: Evaluation of quality and storage stability. In *Proceedings of the Fourth International Date Palm Conference*, Ed. Zaid, A. and G. Alhadrami. Abu Dhabi, United Arab Emirates March 15–17: International Society for Horticultural Science, pp. 593–602.

Hu, F.B. and V.S. Malik. 2010. Sugar-sweetened beverages and risk of obesity and type 2 diabetes: Epidemiologic evidence. *Physiology & Behavior* 100:47–54.

Johnson, R.K. and B.A. Yon. 2010. Weighing in on added sugars and health. *Journal of the American Dietetic Association* 110:1296–1299.

Kalmijn, S., Foley, D., White, L., Burchfiel, C.M., Curb, J.D., Petrovich, H., Ross, G.W., Havlik, R.J., and L.J. Launere. 2000. Metabolic cardiovascular syndrome and risk of dementia in Japanese-American elderly men: The Honolulu-Asia aging study. *Arterioscler. Thromb. Vasc. Boil.* 20:2255–2260.

Kalmijn, S., Launere. L.j., Ott. A., Witteman, J.C., Hofman, A., and M.M. Breteler. 1997. Dietary fat intake intake and the risk of incident dementia in the Rotterdam study. *Ann. Neurol.* 42:776–782.

Knopman, D., Boland, L.L., Mosley, T., Howard, G., Liao, D., Szklo, M., McGovern, P., and A.R. Folsom. 2001. Cardiovascular risk factors and cognitive decline in middle-aged adults. *Neurology* 56:42–48.

Kohli, M. 2011. Healthy ways to sweeten your food. *The Times of India* dated May 08, 2011. Available at: http://timesofindia.indiatimes.com/life-style/health-fitness/diet/Healthy-ways-to-sweeten-your-food/articleshow/8195147.cms. Accessed on June 07, 2011.

Kranz, S., Smiciklas-Wright, H., Siega-Riz, A.M., and D. Mitchell. 2005. Adverse effect of high added sugar consumption on dietary intake in American preschoolers. *The Journal of Pediatrics* 146:105–111.

Mann, J. 2004. Free sugars and human health: Sufficient evidence for action? *Lancet* 363:1068–1070.

Miller, C.J., Dunn, E.V., and I. B. Hashim. 2003. The glycaemic index of dates and date/yoghurt mixed meals. Are dates 'the candy that grows on trees'? *European Journal of Clinical Nutrition* 57:427–430.

Molteni, R., Barnard, R.J., Ying, Z., Roberts, C.K., and F. Gomez-Pinilla. 2002. A high-fat, refined sugar diet reduces hippocampal brain-derived neurotrophic factor, neuronal plasticity, and learning. *Neuroscience* 112:803–814.

Parfitt, J. and S. Valentine. 2008. *Cooking with Arabia's Richest Fruit Dates*. Zodiac Publishing UK limited, UK.

Phillips, K.M., Carlsen, M.H., and R. Blomhoff. 2009. Total antioxidant content of alternatives to refined sugar. *Journal of the American Dietetic Association* 109:64–71.

Sidhu, J.S., Al-Saqer, J.M., Al-Hooti, S.N., and A. Al-Othman. 2003. Quality of pan bread made by replacing sucrose with date syrup produced by using pectinase/cellulase enzymes. *Plant Foods for Human Nutrition* 58:1–8.

Silver, B.S. 1998. *Process for Extracting Sugars from Dates, and for Making Novel Food Products*. US Patent No. 5853488, Publication date December 29, 1998.

Vayalil, P.K. 2002. Antioxidant and antimutagenic properties of aqueous extract of date fruit (*Phoenix dactylifera* L. Arecaceae). *Journal of Agriculture and Food Chemistry* 50:610–617.

Welsh, J.A., Sharma, A., Abramson, J.L., Vaccarino, V., Gillespie, C., and M.B. Vos. 2010. Caloric sweetener consumption and dyslipdemia among US adults. *JAMA* 303:1490–1497.

Welsh, S., Davis, C., and A. Shaw. 1993. *USDA Food Guide: Background and Development*. Hyattsville, MD: United States Department of Agriculture. Report No. 1514.

Wolver, T.M.S. 1990. The glycemic index. In: *Aspects of some Vitamins, Minerals and Enzymes in Health and Disease. World Review of Nutrition and Dietitics*. Ed. Bourne, G.H. Basel: Karager, pp. 120–185.

World Health Organization. 2003. Food and Agriculture Organization of the United Nations Expert Consultation. Diet, nutrition and the prevention of chronic diseases. *Technical Support Series 916*. World Health Organization, Geneva, Switzerland.

23 Use of Date Palm By-Products in Feeding Livestock

O. Mahgoub, I. T. Kadim, and W. Al-Marzooqi

CONTENTS

INTRODUCTION

A major problem that limits livestock production in the arid region of the world is the high cost of animal feed which may reach up to 70% of the total production cost. This is mainly due to an acute shortage in fresh water that limits the utilization of arable land for the production of food for human and animal consumption. Droughts and progressive salinization of water and soils are also limiting factors for fodder production around the world. Any reduction in the cost of livestock feeding will result in higher revenues for livestock owners. Date palm (*Phoenix dactylifera*) belongs to the *Phoenix* genus which includes about 17 palm species and is a member of the Palmae family (Nehdi et al., 2010). Date palm is a major crop in many countries in the Middle East, North Africa, and Western Asia. Date by-products in the form of low-quality dates, date pits, fronds, and date by-products such as date syrup by-product (DSBP) (date fiber) and syrup are available in large quantities in these countries. Date fruits, a cheap nutritious source full of energy, are available in many parts of the arid region. About 20% of the total date production is unsuitable for human consumption and is wasted (Belal et al., 1999). Wasted dates are defined as 1-year-old dates and/or fresh dates that are infested, which are generally fed with no processing to livestock, including sheep, cattle, and camels. In Saudi Arabia, more than half-a-million tons of date palm fruits may be produced annually (Al-Dabeeb, 2005). Dates may also be used as a binding material for pelleting livestock feeds instead of molasses (Belal and Al-Jasser, 1997).

Date by-products had been included into ruminant animals and poultry feeds either as a mix or as pelleted feed. Dates and their by-products such as date pits may positively contribute to the expanding poultry industry in the Arabian Gulf region where an overpopulation of dates occurs (Kamel et al., 1981). Date produces date waste without date pits that may be used in broiler diets for partial substitution of maize or other cereal grains (Al-Hiti and Rous, 1978).

This chapter reviews available literature on the use of date palm by-products for feeding ruminant animals, poultry, and fish.

USES OF DATES AND DATE BY-PRODUCTS FOR FEEDING LIVESTOCK

Whole dates and date palm by-products that may be used for feeding livestock include fruits, pits, pulp, leaves (rachis), and industrial products (syrup and fiber). The date-processing industry generates a sizable amount of by-products mainly in the form of discarded dates and pits. This results in large amounts of date palm products annually. For instance, 20% of dates produced in Saudi Arabia are not used for human consumption and may be used as a high-energy animal supplementary feed (Al-Dabeeb, 2005). The date industry also produces surplus amounts of syrup and DSBP which may be used as a source of fiber for ruminant nutrition.

DATE FRUITS

Date fruits are mainly used for human consumption. They come in many varieties and types ranging from expensive delicate table dates to low-grade dates unsuitable for human consumption. In many cases, there is a surplus of fruits from those meant for human consumption that may be used for livestock feeding. These include low grade and rejected, and infested or damaged during harvesting or processing. They could be pitted or unpitted.

Whole Date Fruit Composition

Whole dates are a good nutritional source for livestock. They vary in chemical composition, especially dry matter (DM) content according to the variety and stage of ripening, or being pitted or unpitted. Table 23.1 summarizes the chemical composition of date fruits reported by several workers who used dates for feeding livestock. The DM in dates generally ranges between 80% and 90%. The moisture content of date fruits has an important consideration in harvesting time, storage, and processing to incorporate in animal feeds (Vandepopuliere et al., 1995). Immature dates contain higher levels of moisture than mature ones (70% and 23%, respectively). Their crude protein (CP) (2%–4.0% of DM) and crude fiber are low (3%–7%). The fat content of date fruits is low, ranging between 1% and 3% DM and ash ranges between 2% and 4% DM. They contain high levels of nitrogen-free extract (NFE) (86% DM) and Ca and P (0.2% and 0.07%, respectively). The carbohydrate in the date fruits includes 21.4% neutral detergent fiber (NDF), 29.3% acid-detergent fiber (ADF), 7.2% permanganate lignin, and 21.3% cellulose (Elhag and Elkhanjari, 1992). Pitted dates (date flesh) contain 133 g/kg moisture and 30.2, 3.0, 3.4, 39.2, and 54.9 g/kg CP, ether extract (EE), ash, NDF, and sucrose, respectively (Belal et al., 1999). The lower values of fiber (NDF) and ash in the pitted dates indicate that most of the fiber in the whole date fruits is found in the date pits.

Date fruit is a good source of sugars as about 60%–73% of the total dry weight of date flesh is made of total sugars (Chaira et al., 2007). Saafi et al. (2008) reported that Tunisian dates (Khalti

TABLE 23.1
Dry Matter Content of Whole Dates and Chemical Composition (% DM) of Whole Date Fruits Used for Feeding Livestock

Nutrient	Elhag and Elkhanjari (1992)	Alhadrami (1996)	Kamel et al. (1981)	Yousif et al. (1996)	Genin et al. (2001)
DM (%)	84.4	89.9	85.28	80.0	84.2
CP	4.0	4.1	2.75	3.0	4.5
CF	5.7	5.0	4.1	3.0	7.4
EE	1.6	0.9	2.8	2.0	
Ash	2.6	3.3	2.6	2.0	4.4
NFE	86.2	86.7			
Ca	0.2		0.1		
P	0.07		0.12		
NDF	21.4				22.2
ADF	29.3				18.0
Lignin	7.2				9.3
Cellulose	21.3				
Sugar				75–80	

variety) contained 63.4, 51.6, 11.8, 3.86, and 0.26% total sugars, reducing sugars, sucrose, protein, and oil, respectively (on DM basis). Dates are a good source of K (0.7% DM) and contain Mg, Ca, P, Na, Zn, Cu, Fe, and Mn (Chaira et al., 2007). Sahari et al. (2007) analyzed 34 Iranian date varieties from the start of the tamr stage. The mean percentage of moisture, protein, lipid, and ash were 29.4, 3.3, 0.42, and 2.25 g/100 g on fresh weight basis, respectively. The predominant sugars were fructose (12.6%–43.3%) and glucose (16.4%–54.2%) on fresh weight basis while sucrose was not detected in most varieties. The minerals varied greatly and values detected in the tamr were 4.5–47.7, 18.4–79.4, 203.6–983, and 23.2–73.9 mg/100 g (DM) for Na, Mg, K, and Ca, respectively.

Dates have some resemblance to molasses in composition and uses. Both contain simple carbohydrates but dates contain mainly monosaccharides (glucose, fructose 1:1) (73%) and a very small percentage of disaccharides sucrose (0.5%) as compared to molasses that contain a high percentage of sucrose (46%) and small percentage of monosaccharides (Belal et al., 1999). Therefore, dates would serve as a better preservative for animal feeds as the monosaccharides are better preservatives against bacteria and fungi than disaccharides in molasses. Dates could be used in higher proportions in poultry, fish, and ruminant's feeds as they contain less amounts of potassium (which has a laxative effect) than molasses (Belal et al., 1999).

Belal et al. (1999) evaluated pitted date mixtures with various animal feed ingredients. Pitted dates were mixed separately in a semiliquid form with yellow maize and wheat bran and dried and ground in proportions of 50:50 to 0:100 and mixtures were stored at room temperature for 3 months. The quality of diet ingredients increased with increasing level of date pits in terms of keeping quality and binding ability.

Use of Whole Dates in Ruminants' Diets

Whole dates had been used as animal feed for centuries but in recent years there has been a surge of interest in research using dates for feeding livestock. Al-Dabeeb (2005) used two levels of low-quality date flesh for fattening lambs in Saudi Arabia. Although diets containing 10% and 20% date were less consumed than controls, they had similar feed conversion ratio and there was no significant difference in growth rates of lambs. However, the digestibility of organic matter (OM), CP, NFE, and CF significantly decreased with increasing dates in the diet as well as a similar pattern for total digestible nutrients (TDN) and digestible crude protein (DCP). The authors concluded that it is economically feasible to fatten lambs on diet containing up to 20% date flesh without adverse effects. Elhag and Elkhanjari (1992) used whole dates to replace barley at ~30% for feeding 6–8-month, 51–83 kg Omani calves. Consumption of date-containing diets was higher than barley-based ones. Feed conversion and daily gain were better for calves on the date-based diets. The cost of feeding was reduced by 8.4% by using dates.

Whole unpitted dates were used to replace 12.5% of a commercial pelleted concentrate in goat diets in the United Arab Emirates (Alhadrami, 1996). Animals fed the date-replaced diet gained less body weight and had lower feed intake but did not lose weight. Therefore, diets with date's inclusion may be used as maintenance diets in arid areas where animal feed scarcity is common, especially during the dry season.

Rekik et al. (2008) studied the effects of incorporating waste dates on the reproductive traits and digestion of prolific D'man ewes in Tunisia. Animals were fed 500 g/day of dates to replace barley with 800 g of oat hay for 90 days. In another experiment, ewes were fed 320, 650, and 970 g/day of dates plus oat hay. Partial and total replacement of barley gain with dates had no effect on relative feed intake, OM digestibility, digestible OM intake, urinary excretion of allantoin, and microbial N supply. It was concluded that when animal's daily intake of wasted dates was equal or above 42 g DM/kg $LW^{0.75}$, reproductive performances of prolific ewes were likely to be lowered. However, it appears that this effect may be restored with higher date allowances.

Feeding date fruits which contain high sugar content may affect rumen chemistry in ruminants. Ilian et al. (1988) evaluated the effects of feeding whole dates and other nonconventional feeds (NCFs) on rumen and blood chemistry of sheep. The rumen fluid of animals fed diets containing whole dates showed significantly lower acetate to propionate ratio than diets without whole dates, whereas butyrate and isobutyrate levels were not affected by dietary treatments.

Utilization of date fruits by animals may vary according to animal species. Genin et al. (2001) studied the digestibility of date by-products by ruminant animals. The wasted Deglet Nour dates digestibility coefficients were 77%, 76%, and 81% in sheep, goat, and camel, respectively. It was much higher than that of Deglet Nour stones which were 32%, 60%, and 35% for sheep, goat, and camel, respectively. Therefore, the dromedary appears to be more efficient in using date by-products compared to other ruminants.

Use of Whole Dates in Poultry Feeds

Kamel et al. (1981) studied the feeding value of whole dates for broiler chicken. Dates were included in the diets of broilers at levels of 0%, 5%, 10%, 30%, and 47.7%, replacing corn as an energy source. Up to 30% inclusions, whole dates supported chicken growth as control diets but inclusion of dates at 47.7% (sole energy source) slightly decreased growth rate and feed conversion ratio. Hussein et al. (1998) included whole dates at a rate of 10% in starter broiler diets. This improved daily weight gain and feed conversion. They concluded that whole dates can be included at 10% level in broiler diets to support and enhance the growth performance in areas with an abundance of palm trees such as in the Middle East to reduce the poultry feed cost.

Vandepopuliere et al. (1995) included date flesh at rates ranging from 8% to 43% and date meat (16%–43%) in broiler's starting diets and 10%–30% dates and 8%–24% date meat in quail breeder's diet. The diets also contained date pits (5%–27% and 5%–15%). The date ingredient diets supported broiler weights and feed conversion comparable to or better than the control diets. Quail breeder feed consumption decreased at the 30% level but 24% date diet were equal to those of controls.

Al-Hiti and Rous (1978) used date waste without stones at levels of 0, 50, 100, or 150 g/kg to substitute shredded maize in broiler diets. Supplementary date waste resulted in small increases in weight gain and feed consumption and improved feed conversion in comparison with control diet.

Hussein et al. (1998) used dates and date pits in broiler starter and finisher diets. Addition of dates and date pits to the diets significantly improved the body weight of chicks compared to controls after the first 2 weeks of the trial but there was no significant difference between experimental groups after 4 weeks.

There were some attempts to improve the nutritive value of dates for feeding poultry. For instance, Zangiabadi and Torki (2010) studied the use of ß-mannanase-based enzyme on growth performance and immune response of broilers fed dates. The diets containing 175 or 350 g/kg whole dates produced body weight gain similar to those obtained without dates. Enzyme addition improved weight gain and feed conversion throughout the experimental period. They concluded that whole dates can be included in broiler diets at levels up to 350 g/kg.

Use of Whole Dates in Fish Feeds

Dates and date by-products had been used in feeds of aquaculture, mainly to minimize the feed cost. Yousif et al. (1996) included dates in tilapia feeds at a rate of 15% plus other ingredients such as fish meal, soybean meal, and wheat flour. The inclusion of dates in the experimental diets had no apparent effect on growth performance or feed utilization efficiency of the fish.

DATE PITS

Date pits, also known as date kernel, seeds, or stones, are a readily available source of animal feed in countries such as those in the Middle East with date-packing and -processing industry, and could generate large amounts of material that may be used for feeding livestock. For instance, date pits yield in the United Arab Emirates (UAE) was estimated at over 50,000 tons annually. The date pits constitute about 10%–15% of the fruit weight (Al-Bowait and Al-Sultan, 2006; Almana and Mahmoud, 1994; Hussein et al., 1998; Kamel et al., 1981). Farmers have used date pits for centuries as animal feed. However, one of the limiting factors of using them as animal feed is the difficulty to reduce their size for more efficient ruminal digestion. Farmers usually break them manually, but if well dried they may be hammer-milled.

Chemical Composition of Date Pits

Date pits were reported to contain high levels of fiber as well as substantial amounts of tannins and resistant starch (Hamada et al., 2002). In particular, they contain high levels of cell wall (75%) which is of low nutritive value for animals, especially monogastrics (Al-Yousef et al., 1986). Date pits vary in composition according to variety. Date pits from Oman contain 89.5% DM and 6.9, 20, 8.5, 1.9, 62.7, 0.14, and 0.13% CP, CF, EE, ash, NFE, Ca, and P in the DM, respectively (Elhag and Elkhanjari, 1992). They also contained 67.9, 52.8, 5.2, 47.2, and 15.1% DM NDF, ADF, lignin, cellulose, and hemicellulose, respectively (Elhag and Elkhanjari, 1992). Hamada et al. (2002) reported that date pits from the UAE contained 7.1%–10.3% moisture, 5.0%–6.3% protein, 9.9%–13.5% fat, 46%–51% ADF, 65%–69% NDF, and 1.0%–1.8% ash in the DM. Besbes et al. (2004) reported that pits from Tunisian dates, Deglet Nour and Allig, contained 5.56% and 5.17% CP, 10.2% and 12.7% oil, 1.15% and 1.12% ash, and 83.1% and 81.0% total carbohydrates. One of the earliest reports on date pits composition was by El-Shazly et al. (1963). They reported that date seeds contained 89.6% DM and 2.0%, 1.4%, 44.3%, 8.9%, and 33.0% CP, EE, CF, ash, and NFE. El-Jabouri (1994) reported that date pits contained 94.5% DM and 6.8% CP, 10.2% crude fat, 20.3% crude fiber, and 61.6% carbohydrates. Proximate analyses indicated that date pits contained 71.5% NFE in which 3% was starch (Aldhaheri et al., 2004). However, when the analysis was expressed on dietary fiber, NFE was only 26.7% in which 21% was mannose. Saafi et al. (2008) reported that Tunisian dates (Khalti variety) contained 8.12, 6.63, 1.49, 5.31, and 8.3 total sugars, reducing sugars, sucrose, protein, and oil (on % DM basis).

Devshony et al. (1992) reported values of 5.6% protein, 8.15% oil, 16.13% fiber, and 1.13% ash for four date varieties. Ash contained 1.55%, 0.97%, 8.07%, 27.6%, 0.73%, 0.13%, and 0.08% Ca,

Na, Mg, K, Fe, Cu, and Mn, respectively. Ali-Mohamed and Khamis (2004) evaluated the mineral content of date pits from six Bahraini date cultivars. These included (μg/g) K (4620–4906), Na (217–261), Ca (64–113), Mg (613–696), Fe (28–60), Mn (12.7–17), Zn (10–14.5), Cu (3.8–6.3), Ni (0.3–2.1), Co (0.03–1.3), Cr (0.05–0.06), Pb (0.4–2.5), and Cd (0.01–0.09). Date pits contained lower levels of minerals and trace elements than the date flesh (Chaira et al., 2007).

However, the composition of date pits may vary greatly (Table 23.2) possibly due to the date variety or the method of processing of the date. Table 23.2 shows the chemical composition of date pits from various varieties of dates (Kamel et al., 1981).

Seventeen amino acids were detected in date seeds with glutamic acid as the major amino acid (16%–18%) of total amino acids (Bouaziz et al., 2008). Date pits contained a majority of essential amino acids (lysine, isoleucine, leucine, methionine, threonine, valine, and phenylalanine). They also contained a number of proteins with molecular weights ranging from 22 to 70 kDa (Bouaziz et al., 2008).

The date pits contain significant amounts of oil which may be extracted for cosmetic and pharmaceutical purposes (Devshony et al., 1992). Extraction of oil from date pit will produce a date palm pit meal. However, the low-protein content coupled with the relatively high-fiber content will seriously limit the use of the defatted date seed meal in feed without significant supplementation. The major unsaturated fatty acid in the pits was oleic acid (42.3%) and the major saturated fatty acid was lauric (21.8%) with myristic, palmitic, and linoleic acids contributing 10.9%, 9.6%, and 13.7%, respectively. Besbes et al. (2004) reported that the major unsaturated fatty acid was oleic acid (41.3%–47.7%) while the main saturated acid was lauric acid (17.8%) in pits from Tunisian dates, Deglet Nour, and palmitic acid (15%) for Allig. Capric, myristic, myristoleic, palmitoleic, stearic, linoleic, and linolenic acids were also detected.

Date pits are reported to contain measurable levels of hormones which may affect the performance and physiology of farm animals when used as a feed component. Aqueous extract of date pits induced *in vitro* uterus contraction and increased uterine weight in immature rats, in a pattern similar to that of estrogens (Elgasim et al., 1995). Ali et al. (1999) added raw and acid-treated powdered date pits to diets of male rats at levels of 7% and 14% for 28 days. Inclusion of untreated date pits at 7% and 14% levels significantly increased concentrations of plasma testosterone and acid-treated pits at 14% increased plasma levels of luteinizing hormone (LH) but there was no effect of date pits on estrogen levels in male rat plasma. Treatment of female rats with lyophilized extracts of date pits for 10 days or with a polar or nonpolar fraction at oral and intraperitoneal dose of 500 mg/kg did not significantly affect body or uterine weights or degree of vaginal orifice opening. Aldhaheri et al. (2004) also fed date pits to rats to evaluate their effects on reproductive hormonal status. There was no effect on testosterone levels in male rats. In female rats, estradiol levels were decreased with increasing level of date pits. This was attributed to the estrogenic effect of date pits that may cause

TABLE 23.2
Moisture (%) and Chemical Composition of Date Pits Obtained from Various Varieties (% of DM)

Date Variety	Moisture	Protein (CP)	Fat (EE)	Crude Fiber	Ash	NFE	Ca	P
Deari	10.0	5.23	4.75	21.5	1.72	56.8	0.60	0.21
Halawee	8.45	5.35	4.48	47.0	1.22	33.5	0.25	0.20
Sumoran	9.27	6.11	5.00	28.0	1.29	50.33	0.0	0.20
Berbee	9.81	5.62	6.07	43.2	1.16	34.14	0.30	0.21
Khadrawee	9.62	5.50	3.85	36.5	1.34	43.19	0.20	0.20
Zahidi	6.51	10.62	6.18	26.5	2.45	47.74	0.57	0.31

Source: Kamel, B.S. et al. 1981. *Poultry Science*, 60, 1005–1011.

reduced fertility. Ali et al. (1999) reported that the inclusion of untreated powdered date pits as 14% in rat diets significantly increased body weight. However, enhancing effects of date pits are not easy to explain. Their composition indicated that they contained low levels of protein, fat, and sugars which exclude additional nutritive value of date pits (Ali et al., 1999). Inclusion of date pits in rat diets did not improve feed intake and there was no pharmacological or endocrinological evidence for estrogen-like action in date pits but there was some evidence of testosterone enhancement (Ali et al., 1999). The latter effect may be the cause of improved body weight growth through its anabolic effects. Treatment of date pits with acids to remove their fibrous coat resulted in losing their enhancing effects on body weight growth, testes weight, and plasma testosterone levels (Ali et al., 1999). Treatment of date pits with solutions reduced estrogen levels in female rat plasma as phytoestrogens may act as antiestrogens.

Date pits were subjected to chemical treatment to reduce their fiber content. Al-Yousef et al. (1986) treated ground date pits with 2.2%, 4.8%, and 9.6% NaOH solution (w/v). Increasing NaOH concentrations tended to solubilize the NDF, ADF, hemicellulose, cellulose, and ash, but not lignin. Sodium hydroxide treatment increased the potentially digestible cell wall and increased the rate of *in vitro* digestibility by solubilizing some of the unavailable fiber component of the cell wall. Sulfuric acid (70% concentrate) treatment had no effect on amino acid profile (Hussein et al., 1998).

Use of Date Pits in Ruminants' Feeds

Date pits had been used for centuries for feeding livestock. However, their effects on animal performance are somewhat ambiguous (Ali et al., 1999). For instance, inclusion of date pits increased body weight gain, improved feed efficiency, and enhanced meat palatability in sheep (Rashid and Alwash, 1976). Date flesh and pits increased body weight gain in the deposition of back fat of sheep (Elgasim et al., 1995). The authors attributed these effects to the presence of estrogens in date flesh and pits.

El-Shazly et al. (1963) used date pits plus a source of protein such as urea or cotton seed cake in sheep diets. They reported that date seeds have a higher TDN value than oil-extracted rice bran and it contains a relatively high percent of oil with a high coefficient of digestibility. Consequently, date seeds may be utilized for feeding sheep if a good protein source supplement such as cotton seed cake is added. The digestibility of date pits in ruminant's feeds was affected by adding other nutrients. For instance, El-Shazly et al. (1963) reported that TDN and apparent DM digestibility for ground date seeds were improved by addition of 15 g urea whereas the digestibility of fat and crude fiber was depressed with the addition of urea.

Elhag and Elkhanjari (1992) used complete diets containing 20% date pits and 20% date leaves for feeding Omani goats. Nutrient composition of both date leaves and pits was similar to that of Rhodesgrass hay but they contained lower CF and higher NFE. Goats fed the date-based diets had higher intakes (absolute and per kg body weight [BWT]) and gained more body weight than those fed barley-based diets.

Use of Date Pits in Poultry Feeds

In spite of their high fiber contents, date pits were reported to improve the growth rates of broiler chicken and results were encouraging. Kamel et al. (1981) evaluated the feeding value of date pits for broiler chicken. Ground date pits were included in the diets of broilers at levels of 5%, 10%, and 15%, replacing wheat bran, corn, and alfalfa, respectively. The date pits supported chick growth as efficiently as control diets at all dietary levels tested. Date pits were included in the diets of broilers at levels ranging from 5% to 27% (Vandepopuliere et al., 1995). Birds on diets containing date pits had similar growth rates and feed conversion as controls. However, there are some reports of less enhancement of broilers performance with date pits feeding. Jumah et al. (1973) reported a gradual reduction in broiler's body weight gain with feeding 5% and 15% date pits due to their high fiber contents.

There were some attempts to reduce the fiber content of date pits by treating them with acids or alkalis. Hussein et al. (1998) reported that diets containing date pits treated with sulfuric acid or diets left nontreated significantly improved body weights of chicks compared to those fed control diets. The authors concluded that adding date pits to the finisher diets improved body weight gain and feed conversion of broiler chicks. Replacement of corn with NaOH-treated date pits as high as 62.4% in broiler diets did not produce serious adverse effects on performance, serum biochemistry, carcass, or eating quality of broiler chicken (Al-Bowait and Al-Sultan, 2006).

Some workers attempted to improve the feeding value of date pits by adding enzymes or alkali or by using acids. Hussein and Alhadrami (2003) fed diets containing 10% of cooked or uncooked date pits with or without enzyme supplementation to broiler chicken. Body weight gains of chicks receiving diets containing 10% uncooked date pits were significantly higher than those of the control group. The addition of enzyme to diets did not affect broiler performance. In another experiment, date pits given at four levels did not affect body weight gain or feed conversion in broiler chicken. However, adding 0.1% enzyme to the starter diets significantly increased body weight gain but had no effect on feed intake or conversion. They concluded that adding date pits with or without enzyme to the broiler diets had no effect on broiler performance.

Use of Date Pits in Fish Feeds

Dates pits had been used in feeds of aquaculture mainly to minimize the feed cost. Yousif et al. (1996) included date pits in tilapia feeds at a rate of 15% plus other ingredients such as fish meal, soybean meal, and wheat flour. The inclusion of date pits in the experimental diets had no apparent effect on growth performance or feed utilization efficiency of the fish. However, in the groups fed diets in which blood meal was the sole animal protein source, the addition of date pits resulted in better performance compared to other groups supplemented with dates. This was attributed to the higher content of simple sugars in the dates compared to the higher levels of fiber in the pits which was more suited to tilapia feeding.

DATE PALM FRONDS (RACHIS PLUS LEAVES)

Date palm trees need to be pruned annually to allow for growth regeneration. This yields large amounts of DM which is usually burnt by farmers. This has serious environmental implications. A date palm tree annually produces ~20 kg of leaves (Pascual et al., 2000). El Hag (1995) estimated the yield of palm fronds in Oman at 30,000 tons of DM/year. Using this readily available cheap material as animal feed will help the environment and provide a high-energy source for ruminants. Frond of other palm species had been widely used for feeding livestock. Extensive research has been carried out on using oil palm (*Elaeis guineensis* Jacq.) fronds for feeding animals in South East Asian countries (Dahlan et al., 2000). They concluded that this readily available cheap fibrous source may support maintenance requirements of milking goats, rabbits, and sheep.

Date Palm Frond and Leaf Composition

The date palm fronds are made up of a central rachis with leaflets originating from it. Date palm fronds are generally fibrous and deficient in CP and soluble carbohydrates (El Hag, 1995). The date leaves contain 90%, 5.6%, 33%, 2%, 10.5%, 48.9%, 0.7%, and 0.01% CP, CF, EE, ash, NFE, Ca, and P in the DM (Elhag and Elkhanjari, 1992). They also contain 73.5%, 54.2%, 8.7%, 32.5%, and 19.3% DM NDF, ADF, lignin, cellulose, and hemicellulose, respectively (Elhag and Elkhanjari, 1992). Pascual et al. (2000) analyzed the leaflets alone and leaflets plus rachis (Table 23.3). The leaflets alone had higher CP, CF, and gross energy but lower hemicellulose and lignin than leaflets plus rachis or raceme without dates.

TABLE 23.3
Chemical Composition and Energy Content of Date Palm Leaves and Raceme

Content	Leaflets	Leaflet + Rachis	Raceme (without Dates)
DM (g/kg)	930.2	914.5	931.6
Organic matter (g/kg DM)	930.5	923.8	934.6
Crude protein (g/kg DM)	16.9	6.4	12.6
Ether extract (g/kg DM)	44.1	21.3	21.2
Crude fiber (g/kg DM)	331.6	246.4	405.8
Nitrogen-free extract (g/kg DM)	537.9	649.7	495.0
NDF (g/kg DM)	544.3	539.4	682.2
ADF (g/kg DM)	400.7	359.2	475.3
Hemicellulose (g/kg DM)	143.6	180.2	206.9
Acid-detergent lignin (g/kg DM)	69.1	89.3	102.1
Cellulose (g/kg DM)	331.6	269.9	363.2
Gross energy (MJ/kg DM)	19.4	16.9	18.5

Source: Pascual, J.J. et al. 2000. *Journal of Arid Environment*, 45, 183–189.

Use of Date Palm Fronds in Feeding Ruminants

Palm fronds have been used as a source of roughage in diets of ruminants for a long time. El Hag and El Shargi (1996) used a complete diet containing 25% date leaves with other ingredients, including 7.5% DSBP and fish sardines as a source of protein. More than half of the diet was made of barley grain. Intake of the date leaves diet was higher than the control and these animals gained weight with a rate equal to 1.5 times the controls due to higher feed conversion ratio. Date frond-fed animals had similar carcass and noncarcass components and dressing percentage as controls. The cost of gain and cost of meat/kg was significantly lower for animals-fed palm frond diets. El Hag and Al Shargi (1998) also experimented with date by-products where they used 30% of palm leaves in a total mixed diet with DSBP (9.5%) and barley (50%). Sheep and goat on this diet gained at a higher rate (122 g/day) than those on other diets.

El Hag and El Shargi (1998) studied the effects of adding urea (4%) or urea and date syrup to date palm fronds as well as studying the effect of storage period (0, 1, 2, or 3 months). Both chemical treatment and storage period had a significant effect on CP and ash content of chopped palm frond but there was no effect on nutritive value or crude fiber although there was high retention value (>80%) for nitrogen at zero storage period but decreased thereafter. In comparison with Rhodesgrass hay, chopped palm fronds treated with urea alone or urea plus date syrup depressed sheep and goat performance possibly due to reduced feed intake. The authors concluded that untreated chopped palm frond has a good potential as cheaply available roughage which may be used as an emergency feed in arid regions.

Bahman et al. (1997) used shredded date palm leaves for feeding Friesian and Holstein dairy cows in comparison with barley straw. Milk yields, milk composition, live weight gains, or feed intakes of cows offered either date palm leaflets or barley straw did not differ significantly. They concluded that date palm leaves, a waste by-product, may be an alternative to barley straw in date-producing countries such as Kuwait.

Mahgoub et al. (1998) fed Omani sheep two diets which contained 40% of either ground palm fronds or Rhodesgrass hay. Each diet contained 38% barley grain and 11.3% soybean and had a 15% CP. Animals were slaughtered at an average body weight of 28 kg. Animals-fed diets based on ground palm frond had a slower growth rate (122 g/day) and needed more feed (7.19 kg/body weight) than those fed diets based on Rhodesgrass hay (154 g/day; 6.09 food conversion ratio [FCR]).

Animals-fed Rhodesgrass-based diet had heavier empty body weight (EBW) hot carcass weight, cold carcass weight, and higher dressing out percentage than those fed diet based on palm frond. Lambs-fed palm frond-based diet had higher proportions of head and gut fill but lower proportions of liver, kidney, plus pelvic fat than those fed Rhodesgrass-based diet. There were no major effects of diet on weight of wholesale cuts, and the physical or chemical composition of the rack cut. This study indicated that while palm fronds may not equal Rhodesgrass in feeding value for native sheep, it may still offer a source of roughage for feeding local ruminants given the costs of irrigated forage in Oman. Myhara et al. (2000) included ground date palm fronds at a level of 42% in Omani sheep diets. They concluded that although the inclusion of milled palm fronds did reduce the daily growth rate of lambs, it did not adversely affect meat quality.

Pascual et al. (2000) used three fractions of date by-products leaflets, leaflets plus rachis, and racemes (clusters without dates) for feeding goats. Feed intake, especially protein, and energy intake of animals fed the leaflets were higher than those fed the leaflets plus rachis or the raceme. They concluded that despite their low nutritive value, date palm leaf fractions could be used as an acceptable alternative to barley straw in goat diets.

Processing of Palm Fronds

Date palm frond contains high levels of fiber which is less digestible by ruminant animals. There were some attempts to physically, chemically, or biologically process the palm fronds to improve their digestibility. An experiment was carried out by Mahgoub et al. (2007) to ensile palm fronds and feed it to sheep with a concentrate made of local agricultural by-products. Shredded palm fronds (600 kg) were weighed and loaded in the Reel Augie mixer. The urea solution was sprayed on the shredded palm fronds using a sprinkler and mixed thoroughly using the Reel Augie mixer. The fronds–urea mix was then transferred into a 600-gallon plastic tank and manually pressed as hard as possible to reduce air. The tank was tightly sealed to avoid air entrance to the silage to allow anaerobic processing of fronds and urea mixture. The silage was kept in the tank for 2 months at the end of which it was emptied and analyzed before feeding to animals. A feeding trial was designed with two roughages and two concentrates. The roughages were palm frond urea-treated silage (UTPF) and Rhodesgrass hay (RGH). The concentrates were commercial general ruminant (GR) and a concentrate made from local by-products (LBP). Animals were fed for 132 days. The concentrate was made from 25% ground date fronds, 25% wheat bran, 12% dried sardines, and 15% barley plus other vitamin and mineral additives. All experimental animals including those fed the palm by-products were in excellent health condition throughout the trial. The animals-fed palm frond silage and by-product concentrate and those fed commercial concentrate (CC) gained weight at 20 and 30 g/day, respectively. Those fed the Rhodesgrass hay with commercial or by-product concentrate gained at 70 and 50 g/day, respectively. There was no effect of by-products feed on carcass and meat quality. This experiment indicated that date palm by-products may be used for feeding sheep for maintenance or low body weight gain during times of feed shortage.

Khorchani et al. (2000) made two silages from 45% chopped palm fronds or date stems plus 35% wasted dates, 15% wheat bran, and 5% urea in hermetic bags for 2 months. Chemical analysis of the silages indicated that they contained 52.2 and 60 DM and 22.2 and 22.4, 22.4 and 21.7, 33.2 and 33.1, 6.1 and 9.2, 6.4 and 8.5, and 50.3 and 46.5% for DM; CP, CF, ADF, ADL, ash, and ash-free NDF, respectively. The *in vitro* DM digestibility for the stem and leaves silages were 55.5 and 50.3, 48.6 and 50.1, and 53.0 and 53.4% in sheep, goats, and camels, respectively. The *in vivo* digestibility for the stem and frond silage was 51.5% and 47.9% for DM; 54.9% and 58.1% for OM; 76.8% and 71.0% for CP; and 26.4% and 27.7% for NDF with higher feed intake in kids on the frond silage (65.5 vs. 46.0 g/DM/kg $LW^{0.75}$). The authors claimed that this technique of silage making is appropriate for date palm by-products conservation and enhancement use.

Chopped palm fronds were ensiled in plastic containers for 60 days with a protein source (urea) and energy sources (beet pulp, molasses, or citrus pulp) by Ziaei and Sharifi Hosseini (2009).

Supplementation of date palm leaves with energy sources significantly increased *in vitro* dry matter digestibility (IVDMD) and improved feeding value of the fronds.

INDUSTRIAL PRODUCTS

Date-processing industries are well established in countries growing date palms on a large scale such as the Middle East and North Africa. Date-processing operations include cleaning, sorting, grading, and packing processes as well as the production of syrup. The industry generates significant by-products, including rejected fruits, pits, and other products. These by-products are cheap and may be used as a good energy source for feeding livestock.

Date Syrup By-Product

Date syrup is an important product of date fruit. It had been produced and used for thousands of years as a nutritious human food. In recent years, syrup production had been mechanized and commercialized. Besides syrup, the industry of syrup extraction produces a by-product known as DSBP or "date fiber." The DSBP is a useful cheap energy source (El Hag and El Shargi, 1996). It contains 84% DM and 5.6, 12.4, 3.4, 3.0, 75.6, 55.0, 48.7, 21.0, 6.3, 5.5, and 71.0% CP, CF, EE, ash, NFE, NDF, ADF, cellulose, hemicellulose, lignin, and TDN in the DM, respectively (it also contained 10.7 MJ/kg DM of metabolizable energy) (El Hag and El Shargi, 1996).

Use of Syrup By-Products in Feeding Ruminants

Mahgoub et al. (2005) fed Omani sheep a concentrate made from LBP including 10% DSBP and 10% date syrup to replace a CC. The findings of the experiment are summarized in Table 23.4. The objective was to replace commercial expensive-pelleted feeds with a local cheap concentrate source in rations for fattening sheep. Animals were divided into three groups of eight animals each. One group was fed a concentrate-pelleted feed made mainly from LBP including *Prosopis* pods, wheat bran, and date syrup and date fiber. The other two groups were fed either a CC or a 50:50 mixture of local and CC. All groups were fed Rhodesgrass hay *ad lib.* The animals fed the LBP concentrate had the same amount of feed intake, grew at the same rate, and had similar feed conversion ratio (13 kg feed/kg body weight) as those fed the other two concentrate rations. There was no effect of diet on hematological parameters or carcass composition. This experiment showed that date by-products might be included up to 20% in sheep diets without causing negative effects on their performance. The animals fed the LBP feed consumed similar amount of feed, grew at a similar rate, and had similar feed conversion efficiency (12.9 kg feed/kg body weight). This experiment indicated that local feed resources, especially date palm by-products, may be used in manufacturing animal feeds to reduce the feed cost without compromising animal performance. This option also offers an environmental-friendly way of disposing agricultural waste.

Date syrup and DSBP may be incorporated in ruminant's feeds in various forms such as whole mixed rations or pelleted concentrates. Ben Salem and Nefzaoui (2003) reported that date syrup, DSBP, ground date frond, and date pulp may be incorporated in feeding blocks at levels of 35, 35, 10, and 25%, respectively.

Use of Syrup By-Products for Feeding Poultry

El Hag and Al Shargi (1995) fed a diet comprised of 15% DSBP, 22% fish sardines, 3% oyster sheep, 57% barley, and 2% ground alfa alfalfa to laying hens in Oman. The performance of birds on the by-product diet was similar to those on the control diet in the form of eggs laid per day, feed conversion, and mortality. By-product diet decreased the cost of feeding per ton by 17% and cost of eggs by 18%.

TABLE 23.4
Feed Intake and Body Weight Parameters of Omani Sheep Fed a Diet Composed of Native Feed Sources and Commercial Diet

	Diet				
	By-Product	Mixed	Commercial	SE	Effect of Diet
Number of animals	8	8	8		
Days of experiment	79	79	79		
Feed intake (kg)					
Total hay intake	54.685	55.300	57.393	1.908	NS
Total concentrate intake	31.600	31.600	31.600	0	NS
Total feed intake	86.285	86.900	88.993	1.908	NS
Daily hay intake	0.692	0.700	0.726	0.024	NS
Daily concentrate intake	0.400	0.400	0.400	0	NS
Daily feed intake	1.092	1.100	1.126	0.024	NS
Intake per body weight (%)	3.298	3.339	3.366	0.118	NS
Concentrate: Hay	41:59	43:57	45:55	0.02	*
Body weight growth (kg)					
Starting body weight	31.475	31.350	31.263	1.095	NS
Final body weight	38.575	39.036	40.238	1.005	NS
Total weight gain	7.100	7.713	8.975	0.768	NS
Average daily gain	0.090	0.098	0.114	0.010	NS
FCE (kg feed/kg BW)	12.940	11.932	10.443	1.016	NS
FCR (kg BW/kg feed)	0.081	0.088	0.101	0.008	NS

Source: Mahgoub, O. et al. 2005. *Animal Feed Science & Technology*, 120, 33–44.
*: $p < 0.05$; NS: $p > 0.05$.

Al-Marzooqi et al. (2000) investigated the potential use of date fiber in poultry diets and its effect on performance. They concluded that date fiber may be used at levels up to 15% in diets of broiler and layer chicken without reducing their performance. They also suggested that date fiber may be used to control body weight and growth rate in broiler breeder pellets because of its low-energy content.

Tabook et al. (2006) used DSBP or date fiber for partial replacement of maize as a source of energy in broiler feeds with an attempt to improve the quality of the product as broiler feed by adding enzymes. Date fiber had lower apparent ME content (8.7 vs. 13.6 MJ/kg) and apparent ileal digestibility coefficients for DM (0.35 vs. 0.83), crude fiber (0.20 vs. 0.54), and amino acids. Diets containing 5%, 10%, and 15% date fiber with or without enzyme supplementation were fed to broiler chicken. Substitution of maize by 10% and 15% date fiber significantly depressed apparent metabolizable energy (AME), increased digestive organs weight, and increased ileal viscosity. Enzyme supplementation did not improve crude fiber digestibility or produce any effect on carcass and meat quality of birds. The authors concluded that date fiber may be used up to 5% in broiler diets without negatively affecting performance or meat quality.

EFFECTS OF DATE AND DATE BY-PRODUCT FEEDS ON ANIMAL HEALTH

Feeding diets containing agricultural by-products may produce negative effects on animal health. Such effects include gastrointestinal problems such as diarrhea or constipation. Animal health problems may be reflected in abnormal blood and serum biochemistry profiles. However, negative effects of feeding NCFs depend on the type and proportion of the by-product, animal species and age, and length of the feeding period. Mahgoub et al. (2005) reported no negative effects of diets containing date palm by-products on sheep for short periods of feeding. Replacement of corn with NaOH-treated date pits as high as 62.4% in broiler diets did not produce serious adverse effects on performance or serum biochemistry of broiler chicken (Al-Bowait and Al-Sultan, 2006).

Various date by-products have different effects on body systems of animals. Date by-products contain sugars, fiber, ash, hormones, and antinutritional factors such as phenols and tannins. Al-Qarawi et al. (2003) studied the effects of water and ethanol extracts of date flesh and date pits on gastrointestinal transit in mice. Compared with controls mice that received ethanol and water extracts of both date flesh and pits emptied, in a dose-dependent manner, more of the gastrointestinal content over a range between 4% and 22%. Addition of date fiber (DSBP) to broiler's diets significantly increased the weights of total digestive tract, pancreas, and cecum. Date fiber increased ileal digesta viscosity and addition of exogenous enzymes decreased the viscosity of the basal diets but not date fiber ones (Tabook et al., 2006).

Experimentally, date extracts increased sperm count in guinea pigs and enhanced spermatogenesis and increased concentration of testosterone, follicle stimulating hormone (FSH), and LH in rats (El-Mougy et al., 1991). Powdered date pits promoted the growth of male rats and produced significant increases in the plasma concentrations of testosterone and LH (Ali et al., 1999).

However, there are some reports indicating that NCFs including date by-products may produce some negative effects on animal health. A study was carried out to investigate the effects of feeding low-quality NCFs containing phenols and condensed tannins on the clinical profiles of sheep (Mahgoub et al., 2008a). Omani sheep were fed one of four diets with two base roughages, urea-treated palm frond (UTPF) and Rhodesgrass hay (RGH) and two concentrates, CC and a by-products concentrate (BC), for 120 days. Hematological, serum biochemical, and urine analyses were used to assess sheep health. Nonconventional feeds (urea-treated palm frond and BC) contained higher levels of polyphenols and condensed tannins than conventional feeds (Rhodesgrass hay and commercial cubes). Animals fed NCF had lower feed intake and lower body gain than those fed conventional ones. They also produced larger volumes of feces which contained higher levels of nitrogen and had lower viscosity values of intestinal content. Rumen liquor of NCF-fed animals had higher pH and lower ammonia-nitrogen levels. Animals-fed urea-treated palm frond plus BC had lower lymphocyte, monocyte, and eosinophil counts by the end of the trial than those fed Rhodesgrass hay-based diets. The urea-treated palm frond and BC-fed animals had lower blood urea nitrogen (BUN) levels, higher alanine aminotransferase (ALT), and lower serum iron than those fed Rhodesgrass hay-based diets. There was a trend of increasing blood, leukocytes, and specific gravity in the urine of NCF-fed animals. This experiment indicated that feeding low-quality NCFs containing antinutritional factors for relatively long periods might produce subtle negative effects on the physiology and chemistry of the digestive system and blood parameters which might negatively affect sheep health and make them more susceptible to diseases. Mahgoub et al. (2008b) investigated the effects of diets containing ensiled palm fronds and ground palm fronds which contained phenols and tannins on sheep health. Villus height and height:crypt depth ratio were significantly lower in animals on NCFs. There were signs of chronic inflammation in the intestine; kidneys showed signs of nephritis and liver showed congestion and hemorrhagic changes. They concluded that feeding diets based on by-products such as date fronds containing phenols and condensed tannins for extended periods may produce subtle negative effects on animal health and production.

CONCLUSION

This chapter reviewed the relevant literature on composition and uses of date by-products for livestock feeding. Date palm by-products used for feeding livestock, poultry, and fish included whole dates (pitted and unpitted), date pits, palm fronds, syrup, and syrup by-product. They are found in large numbers in date palm-growing countries of the Middle East, North Africa, and Western Asia. They vary in chemical composition but they are generally poor in protein and fat but they contain high levels of carbohydrates although some of the latter are in the form of cell wall fiber or lignified carbohydrates. Date fruits contain mainly high levels of sugars and date pits contain fiber and oil. Experimental work indicated that date by-products may be used as partial or full replacement of expensive energy sources such as cereal grains for feeding ruminants, poultry, and aquaculture. There are some reports that feeding livestock on high levels of date by-products may cause some health problems in animals. Utilization of date by-products would reduce feeding cost and consequently increase farmer's revenue and offer an environment-friendly way of waste disposal.

REFERENCES

Al-Bowait, M. and S.I. Al-Sultan. 2006. Aspects of the serum biochemistry, carcass quality and organoleptic characteristics of broilers fed alkali-treated date pits. *International Journal of Poultry Science*, 5, 284–288.

Al-Dabeeb, S.A. 2005. Effect of feeding low quality date palm on growth performance and apparent digestion coefficients in fattening Najdi sheep. *Small Ruminant Research*, 57, 37–42.

Aldhaheri, A., G. Alhadrami, N. Aboalnaga, I. Wasfi, and M. Elridi. 2004. Chemical composition of date pits and reproductive hormonal status of rats fed date pits. *Food Chemistry*, 86, 93–97.

Alhadrami, G. 1996. Effects of date supplementation and crossing with Dhofari goats on the fattening performance of local goats in the United Arab Emirates. *Emirates Journal of Agricultural Science*, 8, 63–72.

Al-Hiti, M.K. and J. Rous. 1978. Date waste without stones in broiler diets. *British Poultry Science*, 19, 17–19.

Ali, B.E., A.K. Bashir, and G. Alhadrami. 1999. Reproductive hormonal status of rats treated with date pits. *Food Chemistry*, 66, 437–441.

Ali-Mohamed, A. and A.S.H. Khamis. 2004. Mineral ion content of the seeds of six cultivars of Bahraini date palm (*Phoenix dactylifera*). *Journal of Agricultural and Food Chemistry*, 52, 6522–6525.

Almana, H. and R.M. Mahmoud. 1994. Palm date seeds as an alternative source of dietary fibre in Saudi bread. *Ecology of Food and Nutrition*, 32, 262–270.

Al-Marzooqi, W., A.S. Al-Halhali, R.S. Al-Maqbaly, A. Ritchie, K. Annamalai, and N.F. Forsberg. 2000. Date fibre as a constituent of broiler starter d. *Agricultural Science*, 5, 59–61.

Al-Qarawi, A.A., B.H. Ali, S.A. Al-Mougy, and H.M. Mousa. 2003. Gastrointestinal transit in mice treated with various extracts of date (*Phoenix dactylifera* L.). *Food and Chemistry Toxicology*, 41, 37–39.

Al-Yousef, Y., R.L. Belyea, and J.M. Vandepopuliere. 1986. Sodium hydroxide treatment of date pits. *Proceedings of II Symposium on Date Palm in Saudi Arabia.* King Faisal University, Al-Hassa, Saudi Arabia, 197–206.

Bahman, A.M., J.H. Topps, and J.A. Rooke. 1997. Use of date palm leaves in high concentrate diets for lactating Friesian and Holstein cows. *Journal of Arid Environment*, 35, 141–146.

Belal, I.E.H. and M.S. Al-Jasser. 1997. Replacing dietary starch with pitted date fruits in Nile Tilapia *Oreochromis niloticus* (L.) feed. *Aquaculture Research*, 28, 101–105.

Belal, I.E.H., M.S. Al-Jasser, I.A. Mustafa, and M.N. Al-Dosari. 1999. Evaluation of date-feed ingredient mixes. *Animal Feed Science & Technology*, 81, 291–298.

Ben Salem, H. and A. Nefzaoui. 2003. Feed blocks as alternative supplements for sheep and goats. *Small Ruminant Research*, 49, 275–288.

Besbes, S., C. Blecker, C. Deroanne, N.-E. Drira and H. Attia. 2004. Date seeds: Chemical composition and characteristic profiles of lipid fraction. *Food Chemistry,* 84, 577–584.

Bouaziz, M.A., S. Besbes, C. Blecker, B. Wathelet, C. Deroanne, and H. Attia. 2008. Protein and amino acid profiles of Tunisian Deglet Nour and Allig date palm fruit seeds. *Fruits*, 63, 37–43.

Chaira, N., A. Ferchichi, A. Marbet, and M. Sghairoun. 2007. Chemical composition of the flesh and the pit of date palm fruit and radical scavenging activity of their extracts. *Pakistan Journal of Biological Sciences*, 13, 2202–2207.

Dahlan, I., M. Islam, and M.A. Rajion. 2000. Nutrient intake and digestibility of fresh, ensiled and pelleted oil palm (*Elaeis guineensis*) frond by goats. *Asian Austral-Asian Journal of Animal Science*, 13, 1407–1413.

Devshony, S., E. Eteshola, and A. Shani. 1992. Characteristics and some potential applications of date palm (*Phoenix dactylifera* L.) seeds and seed oil. *Journal of the American Oil Chemists' Society,* 69, 595–597.

Elgasim, E.A., Y.A. Alyousef, and A.M. Humeida. 1995. Possible hormonal activity of date pits and flesh fed to meat animals. *Food Chemistry*, 52, 149–152.

El Hag, M.G. 1995. Production of animal feeds from dates and date by-products in the Sultanate of Oman. A review. *International Journal of Animal Science,* 10, 195–201.

El Hag, M.G. and K.M. Al Shargi. 1995. Performance of laying hens on concentrate and by-products based diets. *International Journal of Animal Science*, 10, 181–183.

El Hag, M.G. and K.M. Al Shargi. 1998. Comparative performance of goats and sheep fed on high-fiber pelleted diets supplemented with different nitrogen sources. *Journal of Applied Animal Research*, 13, 179–184.

El Hag, M.G. and H.H. Elkhanjari. 1992. Dates and sardines as potential animal feed resources. *World Animal Review*. No. 73, 16–23.

El Hag, M.G. and K.M. El Shargi. 1996. Feedlot performance and carcass characteristics of local (Dhofari) and exotic (Cashmere) goats fed on a high-fiber by-product diet supplemented with fish sardines. *Asian-Australasian Journal of Animal Science*, 9, 389–396.

El Hag, M.G. and K.M. El Shargi. 1998. Chemical composition and feeding value of chopped date fronds (CDF) as affected by urea and date syrup extract treatment. *Asian Australasian Journal of Animal Science*, 11, 685–691.

El-Jabouri, H.J. 1994. The use of palm date trees in the United Arab Emirates. *Agricultural Development in Arab World*, 13, 36–40.

El-Mougy, S.A., S.A. Abdel-Aziz, M. Al-Shanwany, and A. Alexandria. 1991. The gonadotropic activity of Palmae in mature rats. *Journal of Pharmacology Science*, 15, 156–159.

El-Shazly, K., E.A. Ibrahim, and H.A. Karam. 1963. Nutritional value of date seeds for sheep. *Journal of Animal Science*, 22, 894–897.

Genin, D., A. Kadri, T. Khorachani, K. Sakkal, F. Belgacem, and M. Hammadi. 2001. Volarisation of date-palm by-products (DPBP) for livestock feeding in Southern Tunisia. I—Potentialities and traditional utilization. *Options Méditerranéennes. Serie A: Séminaires Méditerranéennes, Numéro*, 59, 221–226.

Hamada, J.S., I.B. Hashim, and F.A. Sharif. 2002. Preliminary analysis and potential uses of date pits in food. *Food Chemistry*, 76, 135–137.

Hamidreza, Z. and M. Torki. 2010. The effect of a ß-mannanase-based enzyme on growth performance and humoral immune response of broiler chicken fed diets containing graded levels of whole dates. *Tropical Animal Health and Production*, 42, 1209–1217.

Hussein, A.S. and G.A. Alhadrami. 2003. Effect of enzyme supplementation and diets containing date pits on growth and feed utilization of broiler chicks. *Agricultural Sciences*, 8, 67–71.

Hussein, A.S., G.A. Alhadrami, and Y.H. Khalil. 1998. The use of dates and date pit in broiler starter and finisher diets. *Bioresource Technology*, 66, 219–223.

Ilian, M.A., M.A. Razzaque, and A.J. Salman. 1988. Unconventional feeds for sheep: Some aspects of rumen and blood biochemistry. *Biological Wastes*, 24, 187–197.

Jumah, H.F., I.I. Al-Azzawi, and S.A. Al-Hashimi. 1973. Some nutritional aspects of feeding ground date pits for broilers. *Mesopotamia Journal of Agriculture*, 8, 139–145.

Kamel, B.S., M.F. Diab, M.A. Ilian, and A.J. Salman. 1981. Nutritional value of whole dates and date pits in broiler rations. *Poultry Science*, 60, 1005–1011.

Khorchani, T., M. Hammadi, D. Genin, K. Sakkal, H. El Jani, and H. Abdouli. 2000. Valorization of date-palm by-products (DPBP) for livestock feeding in Southern Tunisia. II. Characteristics and digestibility of DPBP-based silages. *Options Méditerranéennes. Serie A: Séminaries Méditerranéennes, Numéro*, 59, 227–232.

Mahgoub, O., D.J. Byerley, J.M. Chesworth, and R.M. Myhara. 1998. Effect of testicular status and feeding diets containing palm by-products on growth and composition of the rack cut in Omani sheep. *Small Ruminant Research*, 28, 281–288.

Mahgoub, O., I.T. Kadim, M.H. Al-Busaidi, K. Annamalai, and N.M. Al-Saqri. 2007. Effects of feeding ensiled date palm fronds and a by-product concentrate on performance and meat quality of Omani sheep. *Animal Feed Science & Technology*, 135, 210–221.

Mahgoub, O., I.T. Kadim, E.H. Johnson, A. Srikandakumar, N.M. Al-Saqri, A.S. Al-Abri, and A. Ritchie. 2005. The use of a concentrate containing Meskit (*Prosopis juliflora*) pods and date palm by-products to replace commercial concentrate in diets of Omani sheep. *Animal Feed Science & Technology*, 120, 33–44.

Mahgoub, O., I.T. Kadim, M.H. Tageldin, W.S. Al-Marzooqi, S.Q. Khalaf, and A.A. Ali. 2008a. Clinical profile of sheep fed non-conventional feeds containing phenols and condensed tannins. *Small Ruminant Research*, 78, 115–122.

Mahgoub, O., I.T. Kadim, M.H. Tageldin, W.S. Al-Marzooqi, S.Q. Khalaf, A.A. Ali, and I. Al-Amri. 2008b. Pathological features in sheep fed rations containing phenols and condensed tannins. *Journal of Animal & Veterinary Advances*, 7, 1105–1109.

Myhara, R.M., D.J. Byerley, O. Mahgoub, and J.M. Chesworth. 2000. Effects of testicular status and feeding diets containing date palm by-product on the sensory attributes of Omani lamb meat. *Small Animal Research*, 38, 57–61.

Nehdi, I., S. Omri, M.I. Khalil, and S.I. Al-Resayes. 2010. Characteristics and chemical composition of date palm (*Phoenix canariensis*) seeds and seed oil. *Industrial Crops and Products*, 32, 360–365.

Pascual, J.J., C. Fernández, J.R. Díaz, C. Garcés, and J. Rubert-Alemán. 2000. Voluntary intake and *in vivo* digestibility of different date-palm fractions by Murciano-Granadina (*Capra hircus*). *Journal of Arid Environment*, 45, 183–189.

Rashid, N.H. and A.H. Alwash. 1976. The effect of the proportion of date stones in the diet on its digestion and fermentation in the sheep rumen. *Iraqi Journal of Agriculture Science*, XI, 51–65.

Rekik, M., N. Lassoued, H. Ben Salem, and M. Mahouachi. 2008. Effects of incorporating wasted dates in the diet on reproductive traits and digestion of prolific D'Man ewes. *Animal Feed Science & Technology*, 147, 193–205.

Saafi, E.B., M. Trigui, R. Thabet, M. Hammami, and L. Achour. 2008. Common date palm in Tunisia: Chemical composition of pulp and pits. *International Journal of Food Science & Technology*, 43, 2033–2037.

Sahari, M.A., M. Barzegar, and R. Radfar. 2007. Effect of varieties on composition of dates (*Phoenix dactylifera* L.). *Journal of Food Science & Technology International*, 13, 269–275.

Tabeok, N.M., I.T. Kadim, O. Mahgoub, and W. Al-Marzooqi. 2006. The effect of date fibre supplemented with an exogenous enzyme on the performance and meat quality of broiler chicken. *British Poultry Science*, 47, 73–82.

Vandepopuliere, J.M., Y. Al-Yousef, and J.M. Lyons. 1995. Dates and date pits as ingredients in broiler starting and Coturnix quail breeder diets. *Poultry Science*, 74, 1134–1142.

Yousif, O.M., M.F. Osman, and G.A. Alhadrami. 1996. Evaluation of dates and date pits as dietary ingredients in tilapia (*Orcochromis aureus*) diets differing in protein sources. *Bioresource Technology*, 57, 81–85.

Ziaei, N. and S.M.M. Sharifi Hosseini. 2009. Feeding value and *in vitro* digestibility of date-palm leaves supplemented with different supplementary energy. *Pakistan Journal of Biological Sciences*, 12, 817–820.

Part IV

Medicinal Values

24 Dates
A Fruit from Heaven

Muhammad Qasim and Summar A. Naqvi

CONTENTS

Shake also to thee (Mary) the Palm trunk toward thee, there shall come tumbling upon thee dates fresh and ripe. Eat therefore, and drink, and rejoice.

The Holy Quran, **Maryam 19:25–26**

Symbols made by humans are so close to his nature and feelings that even God the Almighty has chosen to address mankind while using anthropomorphic symbols. Humans have always referred to things around them to express their feelings of love, hate, adoration, sacrifice, and abundance. For instance, we can trace the historical and spiritual reference of date palm tree as being a source of religious, commercial, and nutritious values. Since prehistoric times, the symbol of date palm tree was believed to be an expression of life, love, and prosperity; its archeological and religious reference is in plenty and explains avidly how things around us serve as a source of expression. In support, following are the historical and religious references of date palm tree for candid analysis.

HISTORY

The date palm, the oldest cultivated tree, has a fascinating history. For centuries, the date palm has been an important crop in the desert regions of Asia and North Africa. Out of the 100 million date palm trees worldwide, 62 million are found in the Arab world. It has also been the basic source of survival for many ancient nomads (Mohammed et al., 1983; Morton, 1987). At present, more than 2000 cultivars of date palm are known to exist all over the world, but only a few important ones have been evaluated for their agronomic performance and fruit quality (Al-Hooti et al., 1997).

Figure 24.1 shows the historical references that the most well-spread belief attached with dietes was that it was the fruit given by Eve to Adam for eating, when he was tempted by a serpent in Eden.

FIGURE 24.1 Adam and Eve with a date palm between them. (From Soussa, A. 1969. *Irrigation and Civilisation in Oued Rafi din*. First Part. Adib [ed.]. Baghdad, Iraq.)

DATE PALM IN THE OLD WORLD

The domestication and cultivation of date palm is recorded in ancient history starting 6000–8000 years ago in Mesopotamia (now Iraq), the Nile Valley (Egypt), and the Indus Valley (now Pakistan).

Mesopotamia (Iraq)

Mesopotamia is one of the oldest civilizations, known as cradle of civilization in the world. It is the home of the earliest developments of the Neolithic revolution from around 10,000 BC (Beverley, 2003). The earliest evidence of date palm cultivation was recorded during the Eridu period in Lower Mesopotamia. According to Sumerologist, Benno Landsberger, the Sumerian words for date (zulum) and date palm (gishimmar), adopted from non-Sumerian words, are associated with a pre-Sumerian aboriginal population. So, it is assumed that date was cultivated before the Sumerians and maybe it was a home-grown plant in that region.

There is evidence that the date palm trunks were used for the construction of the temple of the moon God in Ur in Southern Iraq between 4000 and 5000 years ago (Popenoe, 1913, 1973). Date palm was a sacred tree among Sumerians, Akadians, and Babylonians (Zaid and de Wet, 2002), as shown in Figure 24.2.

The Nile Valley (Egypt)

The earliest date palm findings were from the Old Stone Age period, recovered from Egypt and found in Kharga Oasis—western desert. It is believed that the cultivation of date palm in the Nile Valley started thousand of years before the first hieroglyphics marks appeared. According to Darby et al. (1977) and Bircher (1990), different names in hieroglyphic symbols, that is, "Buno," "Phuno," "Bni," "Bnr," "Benrt," "Amt," "Bnrit," and "Bniw," were given to date juice in ancient Egypt. Date palm cultivation in Egypt is divided in pre-Dynastic (3500 BC), Dynastic (2950 BC, 2600, and 2400), Greco-Roman (350 AD), and Roman (500) periods.

Similarly, a mummy covered with date palm leaves was found during the excavation at Ruzikate (Sharkia province), at least in 3500 BC (pre-Dynastic period) (Bircher, 1990), and fleshy part of date palm fruits of pre-Dynastic period (3450 BC) was identified from a beer cocktail, excavated in Hierakonpolis site (upper Egypt) (Amer, 1994).

If we focus on the dynastic period, date palm seeds were excavated among the botanical remains recovered from Abu Sir (Giza) tombs. It is reported that King Down's family used date palm as

FIGURE 24.2 Sacred date palm in Sumerian and Babylonian era. (From Soussa, A. 1969. *Irrigation and Civilisation in Oued Rafi din*. First Part. Adib [ed.]. Baghdad, Iraq.)

mortuary offerings, during the first dynasty (2950 BC) (Annual Report, 1992). Similarly, date palm tree of early second dynasty (2850 BC) was found from the Saqqara tomb. Evidence was found that the Nfer-Mehat tomb of the fourth dynasty (2600 BC) in Midoun was decorated with date palm mentioned with the names in hieroglyphic symbols, that is, Bnrit. Similarly, rocky stones decorated with palm tree were discovered in "Ra Or" tomb in Giza and "Betah Hotop" tomb in Saqqara from the fifth dynasty (2400 BC; Nazir, 1970). In Tel-El Amarana eighteenth dynasty (1580 BC), the garden in the magnificent temple of priest "Meri Ra" was decorated with various types of trees; the most notable ones were date palm (Nazir, 1970).

Date palm seeds were excavated from Douche Necropolis (Kharga Oasis—western desert) in 350 AD, which represents the Greco-Roman period (Wagner, 1982; Barakat, 1986). A large number of date palm seeds were excavated from Abu-Sha'ar site (Red Sea coast); the site dates back to the Roman period (500 AD) (Amer, 1994).

Ancient civilizations revered palm trees as symbols of fertility, peace, and victory. Palm tree images were struck and minted on ageless coins of the Greeks and the Romans. Some ancient coins of Tyre and Sidon bear the image of the date palm. After the banishment of Herod Archelaus, Judaea was ruled by a series of Procurator, appointed by Rome. In 70 AD, a Roman emperor Vespasian issued a coin showing a weeping woman, beneath a date palm tree (Zaid and de Wet, 2002) as shown in Figure 24.3.

Antonius Felix who presided at St. Paul's trial in 52–60 AD also issued a coin showing date palm leaf flanked by a Greek inscription giving the date (Year-5 of Nero, AD 58/9) as shown in Figure 24.4.

FIGURE 24.3 (**See color insert.**) Vespasian coin celebrating the victory over the rebels. The legend says: IVDÆA CAPTA (Judea captured).

FIGURE 24.4 (**See color insert.**) Antonius Felix issued a coin showing date palm leaf.

The Indus Valley (Pakistan)

Phoenix dactylifera most likely grew wild in the Indus Valley and it is considered a natural hybrid of *Phoenix sylvestris* Roxb., where it was appreciated as a wild fruit and probably cultivated as early as the sixth millennia BC. Date seeds at least 5000 years old were found in the storage godowns, excavated at Mohenjo Daro, Sindh, Pakistan (Ahmad and Smead, 2005; Kumar, 2009). Alexander the Great harvested dates from the Ketch valley of Balochistan and fed his soldiers traveling along the Mekran coast in fourth century BC. At the beginning of the eighth century, it was spread by the Muslim conqueror, Muhammad Bin Qasim in Sindh (Ahmad and Tahir, 2005; Dhillon et al., 2005).

P. sylvestris Roxb. (sugar date palm or toddy palm) is a sister species, which still occurs in the wild part of the Indus Valley. The botanical and geographic proximity of the sister species *P. sylvestris* (Indian sugar date palm) reinforces the argument that *P. dactylifera* also originated in the Indus Valley.

DATE PALM IN THE MODERN WORLD

It is said that the Moorish tribes from Arabia brought date palms across North Africa to Spain. Seeds of the date palm eventually spread from Spain to the Caribbean, Brazil, and throughout the Pacific coast up toward northern Mexico and California by Spanish missionaries in the late 1700s. In the twentieth century, the date palm industry took a new twist in the United States after the introduction of superior cultivars. The date palm was first imported on a larger scale into the United States in the 1890s, specifically into the Coachella Valley in California (Morton, 1987). At present the date palm is cultivated on an area of 2185 hectares with production of 18,597 tons in USA (FAO, 2009).

DATE PALM AND RELIGIONS

Besides a long history, date palm has special religious values in ilhami (divine) religions, that is, Judaism, Christianity, and Islam. In these three religions, there are various citations where God mentioned the date palm tree. These references made the date palm the most prominent tree. Hence, some references have been derived from the ilhami literature of these three religions.

Date Palm in the Holy Bible (Old Testament)

Perhaps the most distinctive tree in the Bible land is the date palm with its tall, unbranched trunk, spreading its crown with several meter long fronds. In the Holy Bible, its morphology can be interpreted at various levels; only some prominent features are covered here.

1. *General importance*

> The righteous shall flourish like the palm trees; he shall grow like a cedar in Lebanon. There that be planted in the house of the LORD shall flourish in the courts of our GOD. They shall still bring fresh fruit in old age; they shall be fat and flourishing. (*The Holy Bible*: Psalms 92:12–14)

2. *Plant characteristics*

How fair and how pleasant art thou, O love, for delights! Thy stature is like to a palm tree, and thy breasts to clusters of grapes. I said, I will go up to the palm tree, I will take hold of the boughs thereof. (*The Holy Bible*, Song of Solomon 7:6–8)

The date palm possesses some unique characteristics. Plant is dioecious, male and female flowers are born on separate trees in clusters among the leaves, so, in order to pollinate the female flower, one has to climb up the male tree to get the male flower and then these pollens spread on to the female flower. After fertilization, fruit develops into a hanging cluster.

3. *Multiple uses.* Palm was the only tree that served to fulfill everyday needs of ancient Jews in the desert. It provides dates as food, date juices as wine after fermentation, the trunk for shelter-roof, and leaves for baskets, mats, brooms, beds, ropes, and furniture.

Take boughs of fruit trees laden with fruit, and palm fronds, and the boughs of leafy trees—such as willows that grow by the brooks ... and (build shelters with them.). (*The Holy Bible*, Leviticus 23:40)

4. *New homes.* It is believed that God made promises for a new home for the nation of Israel, and the moving finger on the wall pointed to the palm tree as the perfect supplier of food, shelter, and medicine. This was the promised land of "milk and honey, (palm dates)" the land of Canaan located on the banks of the Jordan River.

And by the river upon the bank thereof, on this side and on that side, shall grow all trees for meat, whose leaf shall not fade, neither shall the fruit thereof be consumed: it shall bring forth new fruit according to his months, because their waters they issued out of the sanctuary: and the fruit thereof shall be for meat, and the leaf thereof for medicine. (*The Holy Bible*, Ezekiel 47:12)

In the scriptures of the Hebrew Bible, Joshua 6–20, the brother of Moses stood at the miraculous crumbled walls at the city of Jericho, "the City of Palms" that was situated in the land of the Canaanites, located along the Mediterranean coast. All these coastal lands were the land of the palms, which is today named Beirut, Lebanon.

And the south, and the plains of the valley of the Jericho, the city of palm trees, unto Zo'ar. (*The Holy Bible*, Deuteronomy 34:3)

The Israelis left Egypt and set up their camp at Elim—one of the oases across the Sinai to the promised land, the further illustration of which is found in the following verses:

They left Marah and went to Elim, where there were twelve springs and seventy palm trees, and they camped there. (*The Holy Bible*, Numbers 33:9)

And they came to Elim, where there were twelve springs and seventy palm trees, and they camped there beside the spring. (*The Holy Bible*, Exodus 15:27)

And the children of the Ken'ite, Moses' father in law, went up out of the city of palm trees with the children of Judah into the wilderness of Judah, which lieth in the south of A'rad; and they went and dwelt among the people. (*The Holy Bible*, Judges 1:16)

And he gathered unto him the children of Ammon and Am'a-lek, and went and smote Israel, and possessed the city of palm trees. (*The Holy Bible*, Judges 3:13)

And the men which were expressed by name rose up, and took the captives, and with the spoil clothed and all were naked among them, and arrayed them, and shod them, and gave them to eat and to drink, and anointed them, and carried all the feeble of them upon asses, and brought them to Jericho, the city of palm trees, to their brethren: then they returned to Sa-ma'ri-a. (*The Holy Bible*, 2 Chronicles 28:15)

5. *Names.* Tamar was the Hebrew word given to palm trees in the Genesis and it was often hoped that her daughters would become like the blessed date: tall, pretty, and

fertile. In Israel and other countries, "Tamar," "Tamara," and "Tamarah" are still used as girl's names, as it is seen in the following verses:

Ab'sa-lom the son of David had a fair sister, whose name was Tamar. (*The Holy Bible*, 2 Samuel 13:1)

He, (Absalom), had three sons and one daughter, Tamar, who was a very beautiful girl. (*The Holy Bible*, 2 Samuel 14:27)

Judah arranged for him to marry a girl named Tamar (Palm Tree). (*The Holy Bible*, Genesis 38:6)

6. *Decoration.* The following verses indicate that palm had been used for the decoration of the temple:

And he carved all the walls of the house with carved figures of cherubim, and palm trees and open flowers, within and without. (*The Holy Bible*, 1 Kings 6:29)

The two doors also were of olive tree; and the carved upon them carving of cherubim and palm trees and open flowers, and overlaid them with gold, and spread gold upon the cherubim, and upon the palm trees. (*The Holy Bible*, 1 Kings 6:32)

And he carved thereon cherubim and palm trees and open flowers; and covered them with gold filled upon the carved work. (*The Holy Bible*, 1 Kings 6:35)

For on the plates of the ledges thereof and on the borders thereof, he graved cherubim, lion, and palm trees. (*The Holy Bible*, 1 Kings 7:36)

And the greater house he ceiled with fir, which he overlaid with fine gold, and set thereon palm trees and chains. (*The Holy Bible*, 2 Chronicles 3:5)

And there were narrow windows to the little chambers and to their posts within the gate round about, and likewise to the arches: and windows were round about inward: and upon each post were palm trees. (*The Holy Bible*, Ezekiel 40:16)

And their windows, and their arches, and their palm trees, were after the measure of the gate that looketh toward the east. (*The Holy Bible*, Ezekiel 40:22)

And there were seven steps to go up to it, and the arches thereof were before them: and it had palm trees, one on this side. (*The Holy Bible*, Ezekiel 40:26)

And the arches thereof were toward the utter court; and palm trees were upon the posts thereof: and the going up to it had eight steps. (*The Holy Bible*, Ezekiel 40:31)

Its portico faced the outer court; palm trees decorated the jambs on either side, and eight steps led up to it. (*The Holy Bible*, Ezekiel 40:34)

And the posts thereof were toward the utter court; and palm trees were upon the posts thereof, on this side, and on that side: and going up to it had eight steps. (*The Holy Bible*, Ezekiel 40:37)

were carved cherubim and palm trees. Palm trees alternated with cherubim. Each cherub had two faces: the face of a man toward the palm tree on one side and the face of a lion toward the palm tree on the other. They were carved all around the whole temple. From the floor to the area above the entrance, cherubim and palm trees were carved on the wall of the outer sanctuary. (*The Holy Bible*, Ezekiel 41:18–20)

And there were made on them, on the doors of the temple, cherubims and palm trees, like as were upon the walls, and there were thick planks upon the face of the porch outside and there were narrow windows and palm trees on the one side and on the other side, on the sides of the porch, and upon the side chambers of the house, and thick planks. (*The Holy Bible*, Ezekiel 41:25–26)

Date Palm in the Holy Bible (New Testament)

In the *Holy Bible* (New Testament) too, date palm is mentioned almost in the same contents, with the same status. However, these references bear some differences, as reproduced below:

1. *Expansion.* The New Testament scriptures note the spreading of Christianity throughout the lands bordering Israel. Tlue lands were known for their palm trees.

> Believers who fled from Jerusalem during the persecution after Stephen's death traveled as far as Phoenicia, land of palms. (*The Holy Bible*, Acts 11:19)
>
> The delegates went on to Jerusalem, stopping along the way in the cities of Phoenicia, the land of the palms. (*The Holy Bible*, Acts 15:3)
>
> The next day, the news that Jesus was on the way to Jerusalem swept through the city, and a huge crown of Passover visitors took palm branches and went down the road to meet him shouting, 'The Savior! God Bless the King of Israel! Hail to God's Ambassador!' (*The Holy Bible*, John 12:12–13)
>
> After this I saw a vast crowd, too great to count, from all nations and provinces and languages, standing in front of the throne and before the Lamb, clothed in white, with palm branches in their hands. (*The Holy Bible*, Revelation 7:9)
>
> Took branches of palm trees, and went forth to meet him, and cried, Hosanna: Blessed is the King of Israel that cometh in the name of the Lord. (*The Holy Bible*, John 12:13)

Date Palm in the Holy Quran

The date palm enjoys equal status, importance, and meanings in the Quran as well. It is in consonance with the respected verses in the Holy Quran that Islam is the evolutionary continuation of the former two religions. However, the meanings are expanded toward other aspects.

The date palm tree is one of the repetitive symbols used in the holy books. It is mentioned 20 times in the Holy Quran. It is highly suggestive in nature. It works on both spiritual and physical levels. Formerly, it is a challenge for the reasoning faculty of human race as a quest for latent meanings of life and a message for rebirth. Latterly, its physical attributes symbolize as an ideal human character that can bear the severity of natural calamity with a unique elegance, perseverance, and strength.

According to the Islamic tradition, the date tree was said to be the "tree of life" in the Garden of Eden, also a tree for shelter (in another place), and a tree of rich food source. All the verses of the Holy Quran are explaining the botanical, physiological, morphological, pharmacological, nutritional, and many other aspects of the date palm.

1. *Surat Maryam.* A famous reference to the date palm can be found in Surat Maryam regarding Maryam's childbirth:

> And the pains of childbirth drove her to the trunk of a palm tree; she cried (in her anguish) "Ah! Would that I had died before this! Would that I been a thing forgotten and out of sight." But (a voice) cried to her from beneath the (palm-tree): Grieve not! for thy Lord hath provided a rivulet beneath thee; and shake towards thyself the trunk of the palm-tree: it will let fall fresh ripe dates upon thee. So eat and drink and cool (thine) eye. (*The Quran*, Mary 19:23–26)

2. *Multifarious importance.* Similar references about the multifarious importance of date palm can be found in the Holy Quran as follows:

> Would any of you like to have a garden of palm-trees and vines, with rivers flowing underneath it, with all kinds of fruit for him therein; and old age hath stricken him and he hath feeble offspring; and a fiery whirlwind striketh it and it is (all) consumed by fire. Thus Allah maketh plain His revelations unto you, in order that ye may give thought. (*The Quran*, Al-Baqara [The Cow], 2:266)
>
> And of the fruits of the date-palm, and grapes, whence ye derive strong drink and (also) good nourishment. Lo! therein is indeed a portent for people who have sense. (*The Quran*, An-Nahl [The Bee], 16:67)

Growing from seed, it might be having two or more than two stems with a single root system. Suckers produced on a stem of date palm are used for further asexual propagation. Male and female flowers are arranged in clusters covered in spathes on separate plants. After fertilization, fruits are developed in hanging clusters.

3. *God's gifts*

He it is Who sendeth down water from the sky, and therewith We bring forth buds of every kind; We bring forth the green blade from which We bring forth the thick-clustered grain; and from the date-palm, from the pollen thereof, spring pendant bunches; and (We bring forth) gardens of grapes, and the olive and the pomegranate, alike and unlike. Look upon the fruit thereof, when they bear fruit, and upon its ripening. Lo! herein verily are portents for a people who believe. (*The Quran*, Al-Anaam [Cattle, Livestock], 6:100)

4. *Diversity*

It is He who produceth gardens, with trellises and without, and dates, and tilleth with produce of all kinds and olives and pomegranates, similar (in kind) and different (in variety): eat of their fruit in their season, but render the dues that are proper on the day that the harvest is gathered. But waste not by excess: for Allah loveth not the wasters. (*The Quran*, Al-Anaam [Cattle, Livestock], 6:141)

And tall date-palms, with ranged clusters. (*The Quran*, Qaf [The Letter Qaf], 50:10)

Therein are fruits, date-palms producing sheathed fruit-stalks (enclosing dates). (*The Quran*, Ar-Rahman [The Beneficent, The Mercy Giving], 55:11)

And tilled fields and heavy-sheathed palm-trees. (*The Quran*, Al-Shuara [The Poets], 26:148)

5. *Message through gifts*

And in the earth are neighbouring tracts, and gardens of vines, and green crops (fields), and date-palms, growing into two or three from a single stem root, or otherwise (one stem root for every palm), watered with the same water, yet some of them We make more excellent than others to eat. Verily, in these things, there are Ayât (proofs, evidences, lessons, signs) for the people who understand. (*The Quran*, Al-Rad [The Thunder], 13:4)

6. *God's blessed fruit.* The following repeated references indicate date palm as an important, heavenly gifted, blessed fruit for mankind:

Therewith He causeth crops to grow for you, and the olive and the date-palm and grapes and all kinds of fruit. Lo! herein is indeed a portent for people who reflect. (*The Quran*, An-Nahl [The Bee], 16:11)

Then we produce for you therewith gardens of date-palms and grapes, wherein is much fruit for you and whereof ye eat. (*The Quran*, Al-Mumenoon [The Believers], 23:19)

Or thou have a garden of date-palms and grapes, and cause rivers to gush forth therein abundantly. (*The Quran*, Al-Isra [ISRA', The Night Journey, Children of Israel], 17:91)

And We have placed therein gardens of the date-palm and grapes, and We have caused springs of water to gush forth therein. (*The Quran*, Ya-Seen [YA-SEEN], 36:34)

Wherein is fruit, the date-palm and pomegranate. (*The Quran*, Ar-Rahman [The Beneficent, The Mercy Giving], 55:68)

And olive-trees and palm-trees. (*The Quran*, Abasa [He Frowned], 80:29)

And put forward to them the example of two men; unto one of them We had given two gardens of grapes, and We had surrounded both with date-palms; and had put between them green crops (cultivated fields). (*The Quran*, Al-Kahf [The Cave], 18:32)

Which He imposed on them for seven long nights and eight long days so that thou mightiest have seen men lying overthrown, as they were hollow trunks of palm-trees. (*The Quran*, Al-Haaqqa [The Reality], 69:7)

Pharaoh said: "Believe you in him (Moses) before I give you permission? Verily! he is your chief who has taught you magic. So I will surely cut off your hands and feet on opposite sides, and I will surely crucify you on the trunks of date-palms, and you shall surely know which of us (I [Pharaoh] or the Lord of Musa [Moses] [Allah]) can give the severe and more lasting torment. (*The Quran*, Ta-Ha [TA-HA], 20:71)

Sweeping men away as though they were uprooted trunks of palm-trees. (*The Quran*, Al-Qamar [The Moon], 54:20)

The reference to man-made symbols in Holy Scriptures speaks volumes of their significance in human history. Some of these ancient symbols like the date palm tree are still believed to have celestial values. In today's world, the date palm tree carries a special connotation to those who believe in hereafter and life after death. They believe it to be the tree of Eden and its presence on earth reminds them of life after death beside those historical and religious date palm trees. Its truth has a nutritional value. It contains carbohydrates, magnesium, calcium, iron, and other nutritional elements.

REFERENCES

Ahmad, S. and P.F. Smead. 2005. Dates in Pakistan. In: Ahmad, S. (ed.). *Date Palm*. Horticulture Foundation of Pakistan, Islamabad, and Pakistan Science Foundation, Islamabad. pp. 71–94.

Ahmad, S. and A. Tahir. 2005. Dates culture. In: Ahmad, S. (ed.). *Date Palm*. Horticulture Foundation of Pakistan, Islamabad, and Pakistan Science Foundation, Islamabad. pp. 41–45.

Al-Hooti, S., J.S. Sidhu, and H. Qabazard. 1997. Physico-chemical characteristics of five date fruit cultivars grown in the United Arab Emirates. *Plant Foods Hum. Nutr*. 50:101–113.

Amer, W.M. 1994. *Taxonomic and Documentary Study of Food Plants in Ancient Egypt*. PhD thesis, Cairo University.

Annual Report of Faculty of Archaeology. 1992. *The Evolution of Abu Sir Area*. Cairo University Press, Egypt, pp. 2–15 (in Arabic).

Barakat, H.N. 1986. *Plant Life in Douch Area: Kharga Oasis, A Comparative Study of the Present and Greaco-Roman Period*. MSc thesis, Cairo University, Egypt.

Beverley, M.E. 2003. Iraq, past, present and future: A thoroughly-modern mandate? [in English]. *History & Policy*. United Kingdom: History & Policy. Retrieved December 9, 2010.

Bircher, W.H. 1990. *The Date Palm; A Boon for Mankind*. Cairo University Herbarium, Egypt. p. 100.

Darby, J.W., P. Ghalioungi, and G. Louis. 1977. *Food the Gift of Osiris*. 2 vols. Academic Press, London.

Dhillon, B.S., R.K. Tyagi, S. Saxena, and G.J. Randhawa. 2005. *Plant Genetic Resources: Horticultural Crops*. Narosa Publishing House, New Delhi, India. pp. 174–176.

FAO, 2009. Food and Agriculture Organization. http://www.fao.org.com

Kumar, R.J. 2009. *New Interpretations on Indus Valley Civilization*. All Right Publishers, Chennai, India.

Mohammed, S., H.R. Shabana, and E.A. Mawlood. 1983. Evaluation and identification of Iraqi date cultivars: Fruit characteristics of 50 cultivars. *Date Palm J*. 2:27–55.

Morton, J. 1987. Date. In: *Fruits of Warm Climates*. Julia F. Morton, Miami, FL. pp. 5–11.

Nazir, W. 1970. *Plants in Ancient Egypt*. Egyptian Co, Cairo. pp. 340 (in Arabic).

Popenoe, P.B. 1973. The date palm. In: H. Field (ed.). *Field Research Projects*. Coconut Grove, Miami, Florida. 247pp.

Popenoe, W. 1913. *Date Growing in the Old and New Worlds*. West India Gardens. Altadena, California. 316pp.

Soussa, A. 1969. *Irrigation and Civilisation in Oued Rafi din*. First Part. Adib (ed.). Baghdad, Iraq. In: Zaid, A. and Arias-Jimenez, E.J. (eds.). *Date Palm Cultivation*. FAO, Rome.

Wagner, G. 1982. *Excavation Report for 1982 Season*. pp. 1–7; Report introduced to Egyptian Antiquity Organization.

Zaid, A. and P.F. de Wet. 2002. Origin, geographical distribution and nutritional values of date palm. In: Zaid, A. and Arias-Jimenez, E.J. (eds.). *Date Palm Cultivation*. FAO, Rome.

25 The Functional Values of Dates

Mohamed Ali Al-Farsi and Chang Yong Lee

CONTENTS

INTRODUCTION

The dates (*Phoenix dactylifera* L.) are produced largely in the hot desert regions of North Africa and Southwest Asia, and are marketed worldwide as a high-value fruit. With the present uncertainty in the world food supply and an expected increase in demand, the date palm is likely to continue to provide a good source of low-cost fruit.

Date flesh consists of 70% carbohydrates, most of which is in the form of sugars. In most varieties, the sugar content is almost entirely invert sugars, which is rapidly absorbed by the human body (Ahmed and Ahmed, 1995; Al-Hooti et al., 1997; Myhara et al., 1999; Al-Farsi et al., 2005). Dates also contain large amounts of dietary fiber, and are thought to be a good source of antioxidants and some minerals. However, relatively few detailed analyses of their composition have been published. In some instances, the variety analyzed has not been identified and, in addition, conflicting results for some constituents have been published.

There is a particular lack of information on functional (bioactive) constituents of dates and their potential value as functional foods. Functional foods are defined as those foods that provide health benefits beyond basic nutrition (IFICF, 1998). Epidemiological studies have consistently shown that there are clear significant positive associations between intake of fruits and vegetables and reduced rate of heart diseases mortality, common cancers, and other degenerative diseases as well as aging (Joseph et al., 1999; Dillard and German, 2000; Prior and Cao, 2000; Wargovich, 2000). This is attributed to the fact that these foods may provide an optimal mix of dietary fiber, natural antioxidants, and other biotic compounds.

The functional constituents of dates include dietary fiber, which is important for the health of the digestive tract. Dietary fiber consists of the edible plant material which is not hydrolyzed by the human digestive tract. Many studies recommend the public to consume adequate amounts of dietary fiber from a variety of plant foods (Institute of Medicine, 2001; Marlett et al., 2002; Mai et al., 2003).

Dates also contain significant quantities of antioxidants. Antioxidants are thought to play an essential role in the prevention of cardiovascular disease (Renaud and De Lorgeril, 1992; Fuhrman et al., 1995), cancers (Wargovich, 2000; Dragsted et al., 1993), neurodegenerative diseases, such as

Parkinson's and Alzheimer's diseases (Joseph et al., 1999; Okuda et al., 1992; Clarke, 1999), as well as inflammation (Joseph et al., 1999; Lietty et al., 1976) and aging (Prior and Cao, 2000; Ames et al., 1993; Gaulejac et al., 1999).

A dietary antioxidant is defined as a substance in foods that significantly decreases the adverse effects of reactive species, such as reactive oxygen and nitrogen, on normal physiological function in humans (Institute of Medicine, 2000). Antioxidants markedly delay or prevent oxidation of the substrate when they are present in foods or in the body at low concentrations (Halliwell, 1999; Shahidi, 2000). Natural antioxidants consist primarily of phenolics, vitamin C, carotenoids, and selenium (Institute of Medicine, 2000; Shahidi and Naczk, 2004). Examples of common plant phenolic antioxidants include flavonoid compounds such as anthocyanins, cinnamic acid derivatives, coumarins, and tocopherols (vitamin E) (Shahidi and Naczk, 2004).

This review examines the functional constituents of dates and discusses their functional properties. In order to compare the composition of fresh and dried dates with other dried fruits, all presented data from cited references have been recalculated on wet weight bases.

SELENIUM

The available data on selenium content of dates ranged between 0.24 mg/100 g in Brahi and 0.4 mg/100 g in Fard and Khasab. The recommended dietary allowance (RDA) of selenium per adult (male and female) per day is 0.055 mg (Whitney and Rolfes, 2002); thus, dates can be regarded as a major source of selenium and this could be used for promotion and marketing strategies. Selenium is a coenzyme for the antioxidant enzyme glutathione peroxidase, and therefore, has a role in the protection of body tissues against oxidative stress, maintenance of defenses against infection, and modulation of growth and development (Institute of Medicine, 2000).

VITAMINS

Dates contain mainly water-soluble vitamins (B-complex and C). They dissolve in water and are not stored in the body; they are eliminated in urine and therefore we need a continuous supply of them in our diets.

Table 25.1 shows the vitamin content of fresh and dried dates with their daily RDA/adequate intakes (AI). Vitamin C content in fresh dates ranged between 2.4 and 3.4 mg/100 g. Dried dates are generally a moderate source of vitamin B6, B9, B2, and B3 as 100 g of dates provides over 9% of the daily RDA/AI for adults. Dried fruits of plums, apricot, figs, raisins, and peaches contain, on average, 0.052 mg B1, 0.136 mg B2, 2.046 mg B3, and 1.980 mg vitamin C per 100 g (USDA, 2007). Therefore, compared to these dried fruits, dates are regarded as a reasonable source of vitamins, particularly vitamin C (3.9 mg/100 g).

Vitamins are essential nutrients found in foods; the daily requirements are small but they perform specific and vital functions essential for maintaining health. B and C vitamins serve as coenzymes that facilitate the work of every cell in our body. Vitamins are active in carbohydrates, fat, protein metabolism, and in the making of DNA of new cells. Vitamin C acts as well as an antioxidant, protects tissues from oxidative stress, and thus may play an important role in preventing the diseases (Whitney and Rolfes, 2002).

DIETARY FIBER

There are several methods used to determine the dietary fibers in foods. The references selected in this review used the same principle for determining the dietary fiber (Association of Official Analytical Chemists [AOAC] method), mainly cellulose, hemicelluloses, pectin, and lignin. Table 25.2 shows the

TABLE 25.1
Vitamin Content of Dates (mg/100 g)[a]

Varieties	B1 Thiamin	B2 Riboflavin	B3 Niacin	B6 Pyridoxal	B9 Folate	C Ascorbic	References
				Fresh Dates			
Berhi						3.400	Allaith (2008)
Hallawi						2.900	Allaith (2008)
Khalas						2.400	Allaith (2008)
Average						2.900	
				Dried Dates			
Medjool	0.050	0.060	1.610	0.249			USDA (2010)
Deglet Noor	0.052	0.066	1.274	0.165		0.400	USDA (2010)
Hallawi	0.092	0.160			0.053	3.300	Yousif et al. (1982)
Sayer	0.120	0.125			0.065		Yousif et al. (1982)
Khadrawi	0.085	0.135			0.039	2.900	Yousif et al. (1982)
Zahdi	0.073	0.153			0.058	2.200	Yousif et al. (1982)
Khudari						1.000	Sawaya et al. (1982)
Sullaj						1.500	Sawaya et al. (1982)
Average	0.079	0.117	1.442	0.207	0.054	1.88	
RDA/AI[b] mg/day	1.200	1.300	16.000	1.300	0.400	90.000	Whitney and Rolfes (2002)

[a] All data are expressed on wet weight basis.
[b] RDA/AI is recommended dietary allowance/adequate intakes per adult male per day.

content of soluble, insoluble, and total dietary fiber of fresh and dried dates. Insoluble dietary fiber is the major fraction of dietary fiber in dates. The total fiber content in fresh dates ranged between 6.9 and 8.6 g/100 g, whereas in dried dates it ranged between 3.6 and 13.5 g/100 g. The average total fiber content in fresh dates increased from 7.5 to 8.6 g/100 g in dried dates.

From the total dietary fiber content in dates and the recommended daily intake of total dietary fiber (25 g/day) (Marlett et al., 2002), dates could be a good source of dietary fiber in the diet, as 100 g of dates provides 34% of the recommended daily intake of dietary fiber. The high content of the insoluble fiber in dates induces satiety, and has a laxative effect due to increased stool weight. It therefore may reduce the risk of serious conditions such as bowel cancer, and diverticular disease (Marlett et al., 2002; Cummings et al., 1992). In comparison with other dried fruits (Table 25.5), their dietary fibers ranged between 3.7 and 9.8 g/100 g; thus, dates are regarded as a rich source of dietary fiber.

CAROTENOIDS

The average total carotenoids content of fresh and dried dates are 0.913 and 0.973 mg/100 g, respectively (Table 25.3). The carotenoids content of dates varied among varieties in the range of 0.033 and 3.03 mg/100 g in fresh dates and between 0.051 and 2.900 mg/100 g in dried dates. The factors causing these variations are probably due to the differences between maturation, methods of drying, and analysis methods. Also, the total carotenoids content in dates varied between the yellow- and red-colored varieties, with lutein, β-carotene, and neoxanthin being the major carotenoids. Fruits that are red usually contain hydrocarbon carotenoids such as lycopene, neurosporene, γ-carotene,

TABLE 25.2
Dietary Fiber of Fresh and Dried Dates[a]

Varieties	Soluble (g/100 g)	Insoluble (g/100 g)	Total (g/100 g)	References
		Fresh Dates		
Khalas			7.1	Myhara et al. (1999)
Fard			8.6	Myhara et al. (1999)
Hayani	0.96	5.9	6.9	El-Zoghbi (1994)
Average	0.96	5.9	7.5	
		Dried Dates		
Fard	1.3	6.7	8.0	Al-Farsi et al. (2005)
Khasab	1.1	7.4	8.4	Al-Farsi et al. (2005)
Khalas	0.4	5.9	6.3	Al-Farsi et al. (2005)
Deglet Noor			8.0	USDA (2010)
Deglet Nour	3.9	7.0	10.9	Elleuch et al. (2008)
Allig	4.9	8.6	13.5	Elleuch et al. (2008)
Medjool			6.7	USDA (2010)
Hayani	0.54	3.0	3.6	El-Zoghbi (1994)
Khalas			10.9	Myhara et al. (1999)
Fard			10.1	Myhara et al. (1999)
Average	2.02	6.4	8.6	

[a] All data are expressed on wet weight basis.

and δ-carotene, whereas yellow-colored fruit contains α-carotene, β-carotene, and a mixture of carotenol fatty acid esters (Fennema, 1996). Drying processes can result in degradation or formation of *cis*-isomers of carotenoids (Chen et al., 1995; Sánchez-Moreno et al., 2003). Al-Farsi et al. (2005) reported that the destruction ranged between 4% and 30% of dates carotenoids during sun drying. This destruction was attributed to the drying temperature (30–50°C) and the duration of the process (7–10 days) (Al-Farsi et al., 2005).

Typical carotenoid concentrations in other dried fruits range from 0.032 mg/100 g for figs to 2.2 mg/100 g in apricot (Table 25.5). Therefore, dates (0.973 mg/100 g) can be considered a moderate source of carotenoids compared to other dried fruits.

ANTHOCYANINS

Table 25.4 shows the total content of anthocyanins, phenolics, and antioxidants of fresh and dried dates. Anthocyanins were found only in fresh date varieties, especially the red-colored varieties, and ranged between 0.2 and 1.5 mg/100 g. The absence of anthocyanins from dried dates is probably due to their destruction during the drying process (Al-Farsi et al., 2005; Markakis, 1982; Shahidi and Naczk, 1995). Al-Farsi et al. (2005) reported 100% destruction of date's anthocyanins during drying. In addition, many other factors such as genetics, light, agronomic factors, and storage are also responsible for the degradation of anthocyanins (Shahidi and Naczk, 2004; Mazza and Miniati, 1993). In comparison with other fruits, anthocyanins present in red grapes ranged from 7.0 to 15.0 mg/100 g (Cantos et al., 2002), whereas in peaches it ranged from 5.4 to 14.3 mg/100 g for white varieties and from 8.6 to 27.4 mg/100 g for the yellow varieties (Tomas-Barberan et al., 2001).

TABLE 25.3
Carotenoid Content of Fresh and Dried Dates (mg/100 g)[a]

Varieties	α-Caro.	β-Caro.	Zeax.	β-Zea.	Lute.	Neox.	Total	References
				Fresh Dates				
Date (unknown)	0.003	0.018	0.033	0.009			0.063	Ben-Amotz and Fishler (1998)
Deglet Nour		0.006			0.156		0.167	Boudries et al. (2007)
Tantebouchte		0.003			0.028		0.033	Boudries et al. (2007)
Hamraya		0.003			0.034		0.037	Boudries et al. (2007)
Hayany		0.116			0.541	0.381	1.270	Gross et al. (1983)
Deglet Noor		0.060			0.461	0.230	0.920	Gross et al. (1983)
Fard							1.390	Al-Farsi et al. (2005)
Khasab							1.310	Al-Farsi et al. (2005)
Khalas							3.030	Al-Farsi et al. (2005)
Average	0.003	0.034	0.033	0.009	0.244	0.306	0.913	
				Dried Dates				
Deglet Nour		0.003			0.060		0.064	Boudries et al. (2007)
Tantebouchte		0.010			0.129		0.145	Boudries et al. (2007)
Hamraya		0.003			0.046		0.051	Boudries et al. (2007)
Hayany		0.146			0.520	0.365	1.260	Gross et al. (1983)
Deglet Noor		0.054			0.485	0.206	0.920	Gross et al. (1983)
Barhee		0.143			0.491	0.184	1.320	Gross et al. (1983)
Fard							1.200	Al-Farsi et al. (2005)
Khasab							0.900	Al-Farsi et al. (2005)
Khalas							2.900	Al-Farsi et al. (2005)
Average		0.060			0.289	0.252	0.973	

[a] All data are expressed on wet weight basis. α-Caro.: α-carotene; β-caro.: β-carotene; zeax.: zeaxanthin; β-zea.: β-zeaxanthin; lute.: lutein; neox.: neoxanthin.

PHENOLICS

The average phenolics content ranged from 150.7 mg/100 g for fresh dates to 353 mg/100 g for dried dates (Table 25.4). The total phenolics of dates varied among dried varieties, although they used the same methodology (Folin Ciocalteu). The use of different phenolic acid standards, such as ferulic acid and gallic acid, makes the quantitative comparison invalid. In general, drying is regarded as unfavorable due to the possibility of inducing oxidative decomposition either enzymatically by polyphenol oxidase and glycosidase or by thermal degradation of phenolic compounds (Shahidi and Naczk, 2004). However, phenolics increased after drying of some varieties. This could be explained by the degradation of tannins by heat and maturation enzymes during the drying process, which leads to the release of phenolic compounds (Maillard and Berset, 1995). According to Maillard and Berset (1995), the linkages between *p*-coumaric acid and lignin and between ferulic acid and arabinoxylans could be broken at a high temperature. In comparison with other dried fruits in Table 25.5, which ranged between 160 and 1065 mg/100 g, dates can be considered to be a good source of total phenolics.

ANTIOXIDANTS

The total antioxidant content of fresh and dried dates were reported by two different methods, oxygen radical absorbance capacity (ORAC) and ferric reducing ability of plasma (FRAP). The

TABLE 25.4
Anthocyanins, Phenolics, and Antioxidants of Fresh and Dried Dates[a]

Varieties	Anthocyanins (mg/100 g)	Phenolics (mg/100 g)	Antioxidants ORAC (μmol/100 g)	Antioxidants FRAP (μmol/100 g)	References
			Fresh Dates		
Fard	0.9	280	1738		Al-Farsi et al. (2005)
Khnaizi		135		0.62	Allaith (2008)
Ashaal		88		0.80	Allaith (2008)
Berhi		100		0.70	Allaith (2008)
Khasab	1.5	167	1169		Al-Farsi et al. (2005)
Khalas	0.2	134	2060	0.73	Al-Farsi et al. (2005), Allaith (2008)
Average	0.87	150.7	1656	0.71	
			Dried Dates		
Fard	nd	343	999		Al-Farsi et al. (2005)
Khasab	nd	217	821		Al-Farsi et al. (2005)
Khalas	nd	339	1254	0.99	Al-Farsi et al. (2005), Allaith (2008)
Ruzaiz		301		1.01	Allaith (2008)
Hallaw		342		0.65	Allaith (2008)
Deglet Noor		661	3895	1.97	Wu et al. (2004), Allaith (2008)
Medjool		572	2387		Al-Farsi et al. (2005)
Kharak		130			Biglari et al. (2008)
Kentichi		272			Besbes et al. (2009)
Average		353	1871	1.16	

[a] All data are expressed on wet weight basis; nd: not detected.

antioxidants by the FRAP method ranged between 0.62 and 0.80 μmol/100 g for fresh dates and between 0.65 and 1.97 μmol/100 g for dried dates. Most data available for antioxidants in dates are reported by the ORAC method. ORAC values of fresh dates Fard, Khasab, and Khalas averaged 1656 μmol trolox/100 g and reduced after drying to average 1025 μmol trolox/100 g (Al-Farsi et al., 2005). The reduction in antioxidants on drying has been reported by Al-Farsi et al. (2005). They found that the antioxidant loss ranged from 29.7% to 42.5% during sun drying of three date varieties.

The antioxidant content of other dried fruits ranged between 1863 μmol/100 g for peaches and 3383 μmol/100 g for figs (Table 25.5). Thus, in comparison with these fruits, dates are a good source of antioxidants. This finding is supported by other studies published on date antioxidants; Vayalil (2002) and Guo et al. (2003). Vayalil (2002) stated that the antioxidant and the antimutagenic activity in dates is potent and implicates the presence of compounds with potent free radical scavenging activity. Guo et al. (2003) reported that dates had the second-highest antioxidant value among 28 fruits commonly consumed in China, and Hawthorn fruit had the highest amount of antioxidants. The variation between samples could be due to varietal, extraction techniques, and instrumental analysis.

CONCLUSION

The data presented in this chapter show that dates are a good source of nutrients and antioxidants that may play an important role in human health. Dates are rich in dietary fiber and natural antioxidants,

TABLE 25.5
Composition of Common Dried Fruits[a]

Fruits	Moisture (g/100 g)	Fiber (g/100 g)	Carotenoids (mg/100 g)	Vitamin C (mg/100 g)	Phenolics (mg/100 g)	Antioxidants[b] (μmol/100 g)	References
Dates	15.2	8.6	0.973	3.900	353	1871	
Plum	30.9	7.1	0.69	0.600	500		USDA (2010), Guo et al. (2003), Vizzotto et al. (2006)
Apricot	30.9	7.3	2.2	1.000	160		USDA (2010), Guo et al. (2003), Ruiz et al. (2006)
Figs	30.1	9.8	0.032	1.200	960	3383	USDA (2010), Wu et al. (2004)
Peaches	31.8	8.2	2.08	4.800	163	1863	USDA (2010), Wu et al. (2004)
Raisins	15.0	3.7		2.300	1065	3037	USDA (2010), Wu et al. (2004)

[a] All data are expressed on a wet weight basis.
[b] Antioxidants were measured by the ORAC method.

such as selenium, phenolics, and carotenoids. Therefore, dates could potentially be used as a supplement of fiber and antioxidants in nutraceutical, pharmaceutical, and medicine industries.

REFERENCES

Ahmed, I.A., and Ahmed, W.K. 1995. Chemical composition of date varieties as influenced by the stage of ripening. *Food Chem.*, **54**:305–309.

Al-Farsi, M., Alasalvar, C., Morris, A., Baron, M., and Shahidi, F. 2005. Comparison of antioxidant activity, anthocyanins, carotenoids, and phenolics of three native fresh and sun-dried date (*Phoenix dactylifera* L.) varieties grown in Oman. *J. Agric. Food Chem.*, **53**:7592–7599.

Al-Hooti, S., Sidhu, J.S., and Qabazard, H. 1997. Physicochemical characteristics of five date fruit cultivars grown in the United Arab Emirates. *Plant Food Hum. Nutr.*, **50**:101–113.

Allaith, A.A.A. 2008. Antioxidant activity of Bahraini date palm (*Phoenix dactylifera* L.) fruit of various cultivars. *Int. J. Food Sci. Tech.*, **43**:1033–1040.

Ames, B.N., Shigenga, M.K., and Hagen, T.M. 1993. Oxidants, antioxidants and degenerative diseases of ageing. *Proceedings of the National Academy of Sciences,* pp. 7915–7922.

Ben-Amotz, A. and Fishler, R. 1998. Analysis of carotenoids with emphasis on 9-*cis*-*â*-carotene in vegetables and fruits commonly consumed in Israel. *Food Chem.*, **62**:515–520.

Besbes, S., Drira, L., Blecker, C., Deroanne, C., and Attia, H. 2009. Adding value to hard date (*Phoenix dactylifera* L.): Compositional, functional and sensory characteristics of date jam. *Food Chem.*, **112**:406–411.

Biglari, F., AlKarkhi, Abbas F.M., and Easa, A.M. 2008. Antioxidant activity and phenolic content of various date palm (*Phoenix dactylifera*) fruits from Iran. *Food Chem.*, **107**:1636–1641.

Boudries, H., Kefalas, P., and Hornero-Méndez, D. 2007. Carotenoid composition of Algerian date varieties (*Phoenix dactylifera*) at different edible maturation stages. *Food Chem.*, **101**:1372–1377.

Cantos, E., Espin, J.C., and Tomas-Barberan, F.A. 2002. Varietal differences among the polyphenol profiles of seven table grape cultivars studied by LC-DAD-MS-MS. *J. Agric. Food Chem.*, **50**:5691–5696.

Chen, B.H., Peng, H.Y., and Chen, H.E. 1995. Changes of carotenoids, color, and vitamin A contents during processing of carrot juice. *J. Agric. Food Chem.*, **43**:1912–1918.

Clarke, M.P. 1999. Can food forestall ageing. *Agricultural Research Magazine*, USA, **47**:15–17.

Cummings, J.H., Bingham, S., Heaton, K.W., and Eastwood, M.A. 1992. Fecal weight, colon cancer risk, and dietary intake of nonstarch polysaccharides (dietary fibre). *Gastroenterology*, **103**:1783–1789.

Dillard, C.J., and German, J.B. 2000. Phytochemicals: Nutraceuticals and human health. *J. Sci. Food Agric.*, **80**:1744–1756.

Dragsted, L.O., Strube, M., and Larsen, J.C. 1993. Cancer-protective factors in fruits and vegetables: Biochemical and biological background. *Pharmacol. Toxicol.*, **72**:116–135.

Elleuch, M., Besbes, S., Roiseux, O., Blecker, C., Deroanne, C., Drira, N.-E., and Attia, H. 2008. Date flesh: Chemical composition and characteristics of the dietary fibre. *Food Chemistry*, **111**:676–682.

El-Zoghbi, M. 1994. Biochemical changes in some tropical fruits during ripening. *Food Chem.*, **49**:33–37.

Fennema, O.R. 1996. *Food Chemistry*, 3rd edition. Dekker: New York.

Fuhrman, B., Lavy, A., and Aviram, M. 1995. Consumption of red wine with meals reduces the susceptibility of human plasma and low-density lipoprotein to lipid peroxidation. *Am. J. Clin. Nutr.*, **61**:549–554.

Gaulejac, N., Vivas, N., and Borgeois, G. 1999. The influence of various phenolics compounds on scavenging activity assessed by an enzymatic method. *J. Sci. Food Agric.*, **79**:1081–1090.

Gross, J., Haber, O., and Ikan, R. 1983. The carotenoid pigments of the date, *Sci. Hort.*, **20**:65–68.

Guo, C., Yang, J., Wei, J., Li, Y., Xu, J., and Jiang, Y. 2003. Antioxidant activities of peel, pulp and seed fractions of common fruits as determined by FRAP assay. *Nutr. Res.*, **23**:1719–1726.

Halliwell, B. 1999. Food-derived antioxidants. Evaluating their importance in food and *in vivo. J. Food Sci. Agric. Chem.*, **1**:67–109.

IFICF. 1998. International Food Information Council Foundation. Backgrounder: Functional foods. In: *Food Insight Media Guide*. Washington.

Institute of Medicine. 2000. *DRI Dietary Reference Intakes for Vitamin C, Vitamin E, Selenium, and Carotenoids*. National Academy Press: Washington, DC.

Institute of Medicine. 2001. *Dietary Reference Intakes: Proposed Definition of Dietary Fiber*. National Academy Press: Washington, DC.

Joseph, J.A., Shukitt-Hale, B., Denisova, N.A., Bielinski, D., Martin, A., McEwen, J.J., and Bickford, P.C. 1999. Reversals of age-related declines in neuronal signal transduction, cognitive, and motor behavioral deficits with blueberry, spinach, or strawberry dietary supplementation. *J. Neurosci.*, **19**:8114–8121.

Lietty, A., Cristoni, A., and Picci, M. 1976. Studies on *Vaccinium myrtillus* anthocyanosides: Vasoprotective and anti-inflammatory activity. *Arzneimittel-Forschung*, **26**:829–832.

Mai, V., Flood, A., Peters, U., Lacey, J., Schairer, C., and Schatzkin, A. 2003. Dietary fiber and risk of colorectal cancer in the Breast Cancer Detection Demonstration Project (BCDDP) follow-up cohort. *Int. J. Epidemiol.*, **32**:234–239.

Maillard, M.N., and Berset, C. 1995. Evolution of antioxidant activity during kilning: Role of insoluble bound phenolic acids of barley and malt. *J. Agric. Food Chem.*, **43**:1789–1793.

Markakis, P. 1982. *Anthocyanins as Food Colors*. Academic Press: New York.

Marlett, J.A., McBurney, M.I., and Slavin, J.L. 2002. Position of the American Dietetic Association: Health implications of dietary fiber. *J. Am. Diet. Assoc.*, **102**:993–1000.

Mazza, G., and Miniati, E. 1993. *Anthocyanins in Fruits, Vegetables, and Grains*. CRC Press: Boca Raton, FL.

Myhara, R.M., Karkalas, J., and Taylor, M.S. 1999. The composition of maturing Omani dates. *J. Sci. Food Agric.*, **79**:1345–1350.

Okuda, T., Yoshida, T., and Hatano, T. 1992. Antioxidant effects of tannins and related phenols. In *Phenolic Compounds in Foods and Their Effect on Health*. Huang, M.T., Ho, C.T., and Lee, C.Y. (eds), pp. 87–97. American Chemical Society: Washington, DC.

Prior, R.L., and Cao, G. 2000. Antioxidant phytochemicals in fruits and vegetables: Diet and health implications. *J. Hort. Sci.*, **35**:588–592.

Renaud, S., and De Lorgeril, M. 1992. Wine, alcohol, platelets, and the French paradox for coronary heart disease. *Lancet*, **339**:1523–1526.

Ruiz, D., Egea, J., Gil, M.I., and Tomas-Barberan, F.A. 2006. Phytonutrient content in new apricot (*Prunus Armeniaca* L.) varieties. *Acta Hort.* (ISHS), **717**:363–368.

Sánchez-Moreno, C., Plaza, L., De Ancos, B., and Cano, M.P. 2003. Vitamin C, provitamin A carotenoids, and other carotenoids in high-pressurized orange juice during refrigerated storage. *J. Agric. Food Chem.*, **51**:647–653.

Sawaya, H.A., Khatchadourian, H., Khalil, J., Safi, W., and Al-Shalhat, A. 1982. Growth and compositional changes during the various developmental stages of some Saudi Arabian date cultivars. *J Food Sci.*, **47**:1489–1492.

Shahidi, F. 2000. Antioxidants in food and food antioxidants. *Nahrung*, **44**:158–163.

Shahidi, F. and Naczk, M. 1995. Antioxidant properties of food phenolics. In *Food Phenolics: Sources, Chemistry, Effects and Applications*. Shahidi, F., and Naczk, M. (eds), pp. 235–277. Technomic Publishing Co.: Lancaster, PA.

Shahidi, F. and Naczk, M. 2004. *Phenolics in Food and Nutraceuticals*. CRC Press: Boca Raton, FL.

Tomas-Barberan, F.A., Gil, M.I., Cremin, P., Waterhouse, A.L., Hess-Pierce, B., and Kader, A.A., 2001. HPLC-DAD-ESIMS analysis of phenolic compounds in nectarines, peaches, and plums. *J. Agric. Food Chem.*, **49**:4748–4760.

USDA. National Nutrient Database for Standard Reference, United States Department of Agriculture; www.nal.usda.gov/fnic/foodcomp/search/, accessed May 15, 2007.

USDA. 2010. USDA National Nutrient Database for Standard Reference. http://www.nal.usda.gov/fnic/foodcomp/search/, accessed June 5, 2010.

Vayalil, P.K. 2002. Antioxidant and antimutagenic properties of aqueous extract of date fruit (*Phoenix dactylifera* L. Arecaceae). *J. Agric. Food Chem.*, **50**:610–617.

Vizzotto, J., Cisneros-Zevallos, L., Byrne, D. H., Ramming, D.W., and Okie,W. R. 2006. Total phenolic, carotenoid, and anthocyanin content and antioxidant activity of peach and plum genotypes. *Acta. Hort.* **713**:453–456.

Wargovich, M.J. 2000. Anticancer properties of fruits and vegetables. *J. Hort. Sci.*, **35**:573–575.

Whitney, E., and Rolfes, S. 2002. *Understanding Nutrition*, 9th edition. Wadsworth/Thomson Learning Belmont: CA, USA.

Wu, X., Beecher, G.R., Holden, J.M., Haytowitz, D.B., Gebhardt, S.E., and Prior, R.L. 2004. Lipophilic and hydrophilic antioxidant capacities of common foods in the United States. *J. Agric. Food Chem.*, **52**:4026–4037.

Yousif, A.K., Benjamin, N.K., Kado, A., Alddin S.M., and Ali, S.M. 1982. Chemical composition of four Iraqi date cultivars. *Date palm J.*, **1**:285–294.

26 Nutritional and Medicinal Value of Date Fruit

Amanat Ali, Mostafa Waly, M. Mohamed Essa, and Sankar Devarajan

CONTENTS

INTRODUCTION

Date, the fruit of date palm, can be considered as an ideal food that provides a wide range of essential nutrients with many potential health benefits. Traditionally, dates have been considered as the staple food in the Arab Gulf regions (Erskine et al., 2004). The important quality criteria for consumers are the appearance including color, size and shape, physical condition and absence of defects, mouth feel or texture, flavor, and nutritional value (Wills et al., 1998). Dates are consumed in a variety of ways. They are mainly consumed as fresh (30%–40%) or in the dried form (60%–70%) at Rutab (semiripe) and Tamar (fully ripe) stages with little or no processing (Al-Hooti et al., 1997). Dates are usually taken as such or with Arabian coffee, milk, or yoghurt. In the processed form, they are consumed as paste, syrup, pickles, jams, jellies, and are used in many bakery or confectionary products together with chocolate, coconut, honey, vinegar, and others (Khatchadourian et al., 1987; El-Shaarawy et al., 1986; Al-Hooti et al., 1997; Besbes et al., 2009). Masmoudi et al. (2010) reported that various types of jellies prepared from date fruit and lemon by-products had less quantity of sugar, decreased pH, and resulted in significantly firmer jellies, with higher adhesiveness, chewiness, cohesiveness, and taste attributes and gave higher sensory evaluation. The average per capita daily consumption of dates varies in the Arab Gulf region. It has been estimated to be 164 and 114 g in Oman and the United Arab Emirates, respectively (MAF, 2005; Ismail et al., 2006). The average per capita consumption of fresh dates (68 g/day) in Bahrain appears to be lower as compared to other Gulf States (Allaith, 2008). This decline in date consumption, particularly among children and teenagers, is attributed to the changing dietary habits and lifestyle (Musaiger, 2009). The consumption of dates, however, reaches its peak in the Muslim's holy month of fasting "Ramadan," when dates are commonly taken to break the fast. Socioeconomic changes are, however, affecting the consumption of dates due to changes in eating habits, improvement in living standards, continued urban drift, as well as to the availability of a wide variety of other fruits all the year round. The consumption of dates by younger population is less as compared to the older population (Ismail et al., 2006).

Date is a delicious fruit with a sweet taste and a fleshy mouth feel. The major component of dates are carbohydrates (mainly the sugars; sucrose, glucose, and fructose), which may constitute about 70%. The sugars in dates are easily digested and can immediately be moved to the blood after consumption and can quickly be metabolized to release energy for various cell activities. Dates are also a good source of fiber, and contain many important vitamins and minerals, including significant amounts of calcium, iron, fluorine, and selenium (Barreveld, 1993; Makki et al., 1998; Myhara et al., 1999; Al-Shahib and Marshall, 2003; Al-Farsi et al., 2005a; Khan et al., 2008). Dates have also been shown to contain antioxidant and antimutagenic properties. Recent studies have shown that dates and their aqueous extracts have demonstrated the free radical scavenging activity, inhibition of free radical-mediated macromolecular damages, antimutagenic, and immunomodulatory activities (Vayalil, 2002; Al-Farsi et al., 2005b; Allaith, 2008; Saafi et al., 2009). Further research is required to explore the health benefits of date fruit and date pits as well as the use of their functional components in the development of various value-added food products and supplements. The market for dates and their products could therefore be increased significantly in the future.

RELIGIOUS SIGNIFICANCE OF DATE PALMS AND DATE FRUIT

All the three major Biblical religions, Judaism, Christianity, and Islam have given high priority and significance to date palm and its fruit because of Prophet Abraham's love and affection for date palm. In Judaism, date is regarded as one of the seven holy foods/seeds, the others being, barely, wheat, lentils, beans, garlic, and onion. In Genesis, the Hebrew word for date palm is "Tamar," which was often used to describe a beautiful woman and was regarded as a Jewish symbol of grace and elegance. Date palm leaves are used as a "lulav" (a closed frond of date palm tree) in the Jewish holiday of "Sukkot" the Feast of Booths. There are numerous references about date palm and its fruit in the Bible, which testify its manifold virtues. The New Testament Scriptures describe the spreading of Christianity throughout the lands of palm trees. Date palm leaves are also used for Palm Sunday in Christian religion. Islam has, however, given the highest priority to dates. The nutritional significance and health benefits of dates have also been mentioned in the Muslim's Holy Book "*Al-Qur'an* (*The Koran*)." Date palm and date fruit have been mentioned in 20 verses of 17 Surahs (Chapters) in *The Holy Qur'an*. Date has been honored as one of the blessings of paradise (Qur'an, 55:68). Allah (SWT) in Qur'an recommends Maryam (mother of Prophet Jesus) to eat this fruit. "*The pains of labour drove her to the trunk of a date palm.* She [Maryam] said, "*Oh if only I had died before this time and was something discarded and forgotten*!" A voice called out to her from under her, "*Do not grieve! Your Lord has placed a small stream at your feet. Shake the trunk of the palm towards you and fresh, ripe dates will drop down onto you. Eat and drink and delight your eyes*" (Al-Qur'an, 19:23–26, Surah Maryam). Prophet Muhammad (peace be upon him) has urged Muslims to eat dates, maintain date orchards, and has mentioned that dates can cure many diseases/disorders. According to a "*Hadith*," the sayings of Holy Prophet Muhammad (peace be upon him), "*he who eats seven Ajwa dates every morning will not be affected by poison or magic on the day he eats them*" (Al-Bukhari, 2008).

CHEMICAL COMPOSITION AND NUTRITIONAL QUALITY OF DATES

Date fruit can be distinguished from most of other fruits as it achieves its botanical maturity at various distinct maturation levels (Barreveld, 1993), which are known throughout the world by their Arabic names Kimri (green, unripe), Khalal (full-size, crunchy), Rutab (ripe, soft), and Tamar (ripe, sun-dried) (Dowson and Aten, 1982; Fayadh and Al-Showiman, 1990). These stages of maturity are traditionally described by changes in color, texture, taste, and flavor (Myhara et al., 1999). The chemical composition of dates can vary depending on cultivar, soil conditions, agronomic practices as well as the ripening stage (Al-Hooti et al., 1997; Al-Kharusi et al., 2009). During the ripening process, the fruit starts losing its moisture content and its sugars are converted to glucose and fructose. Small amounts of mannose and maltose have also been reported to be present at this stage (Shinwari, 1993). Myhara et al. (1999) reported that moisture content of Omani dates in the last stage of maturity (Tamar) was 21.5 and 18.1 g/100 g in the Fardh and Khasab varieties, respectively. The moisture content of 10 fresh date varieties ranged between 37.6 and 50.4 g/100 g (mean value 42.4 g/100 g). Whereas the moisture content of 16 dried date varieties ranged between 7.2 and 29.5 g/100 g with a mean value 15.2 g/100 g (Al-Farsi and Lee, 2008). Ali et al. (2008) observed that the moisture content of three dried Omani dates varieties (Khasab, Khalas, and Fardh) ranged between 18.77 and 23.71 g/100 g of date flesh. The variation in the moisture contents of different varieties of market-ready dates grown in the same country, or the same variety grown in different regions is largely because of differences in the harvest and post-harvest treatments (Aidoo et al., 1996).

Carbohydrates and Sugars

The most important components of dates are the carbohydrates in particular sugars, which can constitute up to 78% (Makki et al., 1998) and provide a readily available source of energy to the human body. Ali et al. (2009) observed that the total carbohydrate contents of three dried Omani date varieties (Khasab, Khalas, and Fardh) ranged between 68.53 and 75.37 g/100 g of date flesh. The highest value 75.37 g/100 g was observed in Khalas variety. Myhara et al. (1999) reported that at the early green stage (Kimri), dates contained only small amounts of glucose and fructose. The carbohydrate contents in fresh and dried dates ranged from 47.8 to 59.4 (mean 54.9) and 66.1 and 88.6 (mean 80.6) g/100 g, respectively. At the intermediate stages (Khalal and Rutab), large amount of sucrose was accumulated but no additional glucose or fructose were detected. However, at the final stage of maturity (Tamar), almost all the sucrose is converted to glucose and fructose. The total sugar contents of three varieties of dried Omani dates (Khasab, Khalas, and Fardh) ranged between 56.1 and 62.2 g/100 g, the lowest in Khasab and highest in Khalas (Al-Farsi et al., 2005a). Whereas, Ali et al. (2009) reported that the total sugar contents of three dried Omani date varieties (Khasab, Khalas, and Fardh) ranged between 52.17 and 59.96 g/100 g. The highest total sugar content was found in Khalas.

The glucose, fructose, and sucrose contents ranged from 29.04 to 34.53 g/100 g, 20.72 to 23.65 g/100 g, and 1.86 to 2.34 g/100 g, respectively. The highest glucose level was found in Khasab whereas the highest fructose was observed in Khalas. The region of production and variety can significantly affect the glucose and total sugar contents of date fruit (Ahmed et al., 1995; Ismail et al., 2006; Ali et al., 2009). Glucose and fructose contained in date flesh are readily absorbed during the digestion and can lead to a rapid elevation of blood sugar (Liu et al., 2000). The sugars are responsible for the sweet taste of foods. The dates contain almost half of the amount of sugars in the form of fructose, which is twice as sweet as glucose and can induce a feeling of satiety. Ali et al.

(2009) observed an overall glucose-to-fructose ratio of 1.3 in three different Omani date varieties. However, Ismail et al. (2006) reported higher fructose concentrations, with a glucose-to-fructose ratio of <1 at Tamar stage, in five different date varieties (Khalas, Barhee, Fard, Boumaan, Ruzeiz) grown in the UAE.

Dietary Fiber

Dates are a good source of dietary fiber and depending upon the variety and stage of ripening, it ranged from 6.4% to 11.5% in 14 different varieties (Al-Shahib and Marshall, 2003). Some of the low-quality dates, which are used for industrial purposes, have been found to contain up to 10% of crude fiber (Barreveld, 1993). Dates contain both the soluble and insoluble fibers. The main fractions are cellulose, hemicelluloses, pectin, lignin, and insoluble proteins. In one particular determination, date flesh was found to contain (on fresh weight basis) 1.55% cellulose, 1.28% hemicellulose, and 2.01% lignin (Barreveld, 1993). Insoluble fiber is the major component of dietary fiber in dates. Soluble pectins, on the other hand keep on accumulating regularly until the date fruit reaches the Rutab stage (Myhara et al., 1999). During the ripening process, the enzymes gradually break down these substances into more soluble compounds, which render the fruit more tender and soft (Fennema, 1996). Myhara et al. (1999) reported that dietary fiber, determined as nonstarch polysaccharides by gas chromatography, decreased from 250 (Kimri) to 50 g kg^{-1} (Tamar) on a dry weight basis.

Dates contain 0.5%–3.9% of pectin (Fayadh and Al-Showiman, 1990). Pectin starts accumulating during the period of fast fruit growth and reaches a maximum at the beginning of sugar accumulation. It, however, drops with the further growth of fruit. El-Zoghbi (1994) observed that the percentage of pectin decreased from stage 1 (1.58/100 g tissue) to 0.54% at stage 4 of the dried dates. The pectin esterase activity in date increases during ripening stages to reach 60.8 units/100 g tissues, which explains the loss of pectin. The hemicellulose content also decreases from 5.25% to 1.29% from stage 1 to stage 4, while the cellulose content also decreases from 3.4% to 1.4%. The lignin content decreases from 3.5% to 0.3% and the total fiber content decreases from 13.7% in green date, stage 1, to 3.6% in the black date, stage 4. The cellulase activity in date increases during ripening and therefore the cellulose, hemicellulose, and lignin contents decrease with the ripening stage (El-Zoghbi, 1994). Dietary fiber content of dates can further contribute to their nutritional significance as dates can be used in the preparation of fiber-based foods and dietary supplements. Daily intake of 100 g of dates can meet the 32% of recommended dietary allowance for dietary fiber (Marlett et al., 2002). Higher content of the insoluble fiber induces satiety and has a laxative effect due to increased stool weight. Dietary fiber exhibits many therapeutic benefits and helps in lowering the blood cholesterol levels and has been shown to reduce the risk of many disease conditions such as diabetes, hypertension, bowel and colon cancers, cardiovascular diseases, and diverticulosis (Marlett et al., 2002; Cummings et al., 1992).

Proteins and Lipids

Protein and lipid occur in small amounts in dates. Proteins occur in the date fruit in the range of 1%–3%. The increased protein and lipid contents after drying are mainly due to loss of moisture. However, their concentrations differ between the varieties due to differences in cultivation, drying conditions, and the analytical methods used for their determination. The average protein content of fresh and dried dates is 1.50 and 2.14 g/100 g, respectively. Little variation was observed in the protein contents of different date varieties (Al-Farsi et al., 2005a; Ismail et al., 2006; Sawaya et al., 1982b, 1983). Ishurd et al. (2004) reported a reduction in the amino acid contents through maturation stages. Within the same stage of maturation, the amino acid content can vary significantly. Proteins also play a role in the nonoxidative browning and the precipitation of tannins during ripening (Barreveld, 1993). Makki et al. (1998) reported that the protein content of the pulp ranged between 1.7% and 2.95% on fresh weight basis, whereas the date seeds on the average contained 5.22%. Although, the amount of protein is too small to be considered as a significant nutritional source, dates contain many essential amino acids and their amino acids pattern is favorable to the human needs.

Lipids are mainly concentrated in the skin and have a higher physiological importance in the protection of fruit than contributing to the nutritional value of the date flesh (Barreveld, 1993). The average lipid content is about 0.14 g/100 g in fresh and 0.38 g/100 g in dried dates. Al-Hooti et al. (1997) reported that the date flesh contained 0.2%–0.5% saponifiable oil, whereas the date seed contained 7.7%–9.7% oil. Seeds usually contain more lipids than the flesh that may reach to about 8.49% on a fresh weight basis (Makki et al., 1998). Date seeds have been reported to contain relatively high amounts of protein (5.1 g/100 g) and fat (9.0 g/100 g) as compared to date flesh (Almana and Mahmoud, 1994; Hussein et al., 1998). Date pulp can contain lipids ranging from 0.31% to 1.9% on fresh weight basis. The date seeds are also a rich source of dietary fiber (73.1 g/100 g), phenolics (3942 mg/100 g), and antioxidants (80,400 μmol/100 g) and therefore could potentially be considered as an inexpensive source of dietary fiber and natural antioxidants (Almana and Mahmoud, 1994; Hussein et al., 1998). The sugars mannose and maltose are only present in the seeds (Al-Showiman, 1990). The seed of date fruit can constitute from 4.6% to 15.2% of the total weight of the date (Mossa et al., 1986; Al-Showiman, 1990).

Fatty Acid Contents of Date Flesh and Seeds

The date flesh and seed both contain a wide range of saturated and unsaturated fatty acids. The saturated fatty acids include capric, lauric, myristic, palmitic, stearic, margaric, arachidic, heneicosanoic, behenic, and tricosanoic acids. Unsaturated fatty acids include palmitoleic, oleic, linoleic, and linolenic acids. The data published so far, however, does not distinguish between α- and γ-linolenic acid contents. Al-Shahib and Marshall (2002, 2003) observed significant variability in the fatty acids content in seeds of 14 date cultivars. They reported that the concentration of oleic acid varied from 41.1% to 58.8% and suggested that date seeds could be used as a source of oleic acid provided the technical problems related to its extraction are overcome. The presence of organic acids and olfactory oils is regarded as taste enhancer and can help improve the sensory characteristics of specialty date products.

Vitamin and Mineral Contents

Dates contain many important vitamins and minerals and their mineral content in dried dates can vary from 0.1 to 916 mg/100 g of date flesh (Barreveld, 1993; Al-Farsi et al., 2005a; Khan et al., 2008). Dates contain high levels of selenium, copper, potassium, and magnesium, moderate concentrations of manganese, iron, phosphorus, and calcium and small quantities of boron (Barreveld, 1993). Date varieties were also found to be an excellent source of selenium (Al-Farsi et al., 2005a). The high potassium and low sodium contents in dates are desirable for people suffering from hypertension. Boron is useful in the treatment of cancer of the brain. Boron and vitamins are also used in the treatment of rheumatism (Al-Showiman, 1998). The date can be used as a practical supplement for iron deficiency without any side effects such as nausea, headache, and anorexia that may occur with iron tablet supplements. At least six vitamins (thiamin, riboflavin, niacin, ascorbic acid, pyridoxin, and vitamin A) have been reported to be present in dates in visible concentrations (Al-Hooti et al., 1995; Al-Shahib and Marshall, 2003). As compared to other dried fruits, dates can therefore be considered a reasonable source of vitamins (USDA, 2011). Fresh dates contain higher concentrations of vitamins as compared to dry dates as the vitamins are lost in the drying process. Dried dates can be regarded as a moderate source of vitamin riboflavin, pyridoxine, niacin, and folic acid. Thiamin, ascorbic acid (vitamin C), and vitamin A are present in relatively low concentrations in dried dates.

Pigments and Antioxidants

Dates contain various types of phytochemicals such as carotenoids (beta-carotene, lycopene, lutein, zeaxanthin, neoxanthin, etc.), the phenolics, mainly the cinnamic acids and their derivatives (ferulic, sinnapic, syringic, vanillic, gallic, caffeic, protocatechuic, coumaric acids, and their derivatives such as dactilyferic acids, etc.), flavonoid glycosides (luteolin, methyl luteolin, quercetin, and methyl quercetin) flavones, flavonols (catechin, epicatechin), flavoxanthin, anthocyanins, and so on (Shahidi and Naczk, 2004; Al-Farsi et al., 2005a; Biglari et al., 2008). The concentration of these phytochemicals, however, decreases with advancing stage of fruit maturity. The astringent taste of date fruit during the early stages of maturity is closely associated with its level of tannins (Barreveld, 1993; Al-Farsi et al., 2005a) that decreases with the ripening of fruit and almost disappears at Rutab stage (Makki et al., 1998; Myhara et al., 1999). The most astringent dates were found to be at the *Kimri* stage of maturity with highest fresh weight tannin content of 2.8 g/kg. The least astringent dates were found to be at *Rutab* stage of maturity with the lowest fresh weight tannin content (1.0 g/kg). The protein content of dates plays an important role in the nonoxidative browning (Maillard reactions) of dates and helps in the precipitation of tannins during the ripening process (Barreveld, 1993; Makki et al., 1998). Tannins in dates are made mainly of polyphenols and to a lesser extent of flavones. These are responsible for the dark color of dates in the postharvest period (Makki et al., 1998).

The carotenoids and phenolic compounds (flavonoids and anthocyanins) can contribute to varying degrees of antioxidant and antimutagenic activity. The contribution of total phenolics toward antioxidant activity in dates is greater than that of ascorbic acid (Shivashankara et al., 2004). The date fruits as well as its water extracts have shown free radical scavenging activities (Ahmed et al., 1995; Al-Shahib and Marshall, 2003; Al-Farsi and Lee, 2008; Al-Farsi et al., 2005a; Elleuch et al., 2008; Vayalil, 2002; Mansouri et al., 2005; Allaith, 2008; Biglari et al., 2008). Most of the antioxidants in dates have been reported to be hydrophilic or water soluble (Al-Farsi et al. 2005a). The data reported on the antioxidant activity of date fruits from different parts of the world such as from Algeria (Mansouri et al., 2005), Kuwait (Vayalil, 2002), Iran (Biglari et al., 2008), Bahrain (Allaith, 2008), Oman (Al-Farsi et al., 2005b), and the United States (Vinson et al., 2005) suggest that dates cannot only be considered as a good source of natural antioxidants, but can also serve to improve the product's flavor and color because of their high contents of active phenolic acids.

Carotenoids

Dates can be considered a moderate source of carotenoids as compared to other dried fruits (Hart and Scott, 1995; Boudries et al., 2007). The level of carotenoids in date fruit can, however, vary depending upon the variety, stage of maturation, drying, and postharvesting conditions. Boudries et al. (2007) reported the total carotenoids content of different date varieties from Algeria from 32.6 to 773 μg/100 g of fresh date weight. Major carotenoids found in dates are lutein, β-carotene, zeaxanthin, and neoxanthin. Although all carotenoids do not act as provitamin A, dates are likely to contribute to the human requirement for vitamin A. The pro-vitamin A value in different Algerian date varieties ranged between 0.4 and 11.7 RE/100 g in fresh dates. The values, however, decreased with the ripening of dates and were lower at Tamar stage (on the average 0.5 RE/100 g). The total carotenoids content in three varieties of fresh Omani dates (Fard, Khasab, and Khalas) were found to be 1.39, 1.31, and 3.03 mg/100 g, respectively (Al-Farsi et al., 2005b). The yellow color date variety (Khalas) showed higher total carotenoids as compared to red-colored date varieties (Khasab and Fardh). The yellow-colored fruits are reported to also contain a complex mixture of carotenol fatty acid esters (Fennema, 1996). The red color dates on the other hand contain hydrocarbon carotenoids such as lycopene, neurosporene, γ-carotene, δ-carotene, α-carotene, β-carotene, phytofluene, and phytoene (reference). The average total carotenoids content of fresh and dried dates have been reported as 913 and 973 μg/100 g, respectively (Al-Farsi et al., 2005b). Freeze-dried dates have been reported to contain much less concentration of carotenoids, 0.18 mg/100 g of fresh weight (Ben-Amotz and Fishler, 1998).

Phenolic Acids

As compared to other fruits, dates can be considered a good source of phenolics (Vayalil, 2002; Guo et al., 2003). Guo et al. (2003) reported that dates had the second-highest antioxidant value out of 28 fruits commonly consumed in China. Date varieties contain different concentrations and patterns of phenolic acids, which are mostly present in the bound form. Al-Farsi et al. (2005b) reported that the average contents of total phenolics in fresh and dried dates were 193.7 and 239.5 mg/100 g, respectively. They also detected a total of nine free and bound phenolic acids of which five consisted of hydroxylated derivatives of benzoic acid (gallic acid, protocatechuic acid [PCA], *p*-hydroxybenzoic acid, vanillic acid, and syringic acid) and four were cinnamic acid derivatives (caffeic acid, *p*-coumaric acid, ferulic acid, and *o*-coumaric acid). Several other unknown compounds associated with both free and bound phenolic acids were also reported in fresh and sun-dried Omani date cultivars (Fard, Khasab, and Khalas). The predominant phenolic acids were PCA, ferulic acid, caffeic acid, *p*-coumaric acid, and *o*-coumaric acid (Al-Farsi and Lee, 2008; Al-Farsi et al., 2005b). The total concentrations of these phenolic acids varied from 2.61 to 12.27 mg/100 g and from 6.06 to 14.77 mg/100 g in fresh and dried dates, respectively. Fard showed the highest contents of free phenolic acids, followed by Khalas and Khasab in both fresh and sun-dried samples. Overall, the total phenolic acids of sun-dried dates were found to be significantly ($p < 0.05$) higher than those of fresh dates (Al-Farsi and Lee, 2008).

The concentration of phenolics in different date varieties varied significantly ($p < 0.05$). The total bound phenolic contents were in the range of 6.84–30.25 mg/100 g in fresh dates and 14.18–49.67 mg/100 g in sun-dried dates. Regnault-Roger et al. (1987) identified eight phenolic acids (gallic acid, PCA, *p*-hydroxybenzoic acid, vanillic acid, caffeic acid, syringic acid, *p*-coumaric acid, and ferulic acid) in dried Tunisian date cultivars. Mansouri et al. (2005) reported the presence of *p*-coumaric acid, ferulic acid, and sinapic acid as well as some derivatives of cinnamic acid in seven different varieties of Algerian dates. Three different isomers of 5-*o*-caffeoylshikimic acid were also detected. The concentration of phenolics in Omani date cultivars has been reported to be much higher than those of Tunisian date varieties (Al-Farsi et al., 2005b). The variations in the reported results about the phenolics may be due to the analytical procedures, in particular the use of different phenolic acid standards, such as ferulic acid and gallic acid, which make it difficult to have the valid quantitative comparisons. It has been reported that caffeic acid, sinapic acid, ferulic acid, and *p*-coumaric acid have more antioxidative potential than PCA, syringic acid, and vanillic acid (Cuvelier et al., 1992). PCA is a polyphenol antioxidant with mixed effects on normal and cancer cells. PCA has been reported to induce apoptosis in human leukemia cells, as well as in malignant HSG1 cells taken from human oral cavities. Depending on the amount of PCA and the time before its application, PCA can reduce or enhance the tumor growth (Babich et al., 2002; Lim et al., 2007; Guan et al., 2009).

Total Anthocyanins

The different colors (blue, red, violet, and purple) in most of the edible plant species and their fruits (berries, grapes, cherries, cherry laurels, plums, etc.) are due to anthocyanins (Mazza and Miniati, 1993; Kaehkoenen et al., 2003; Wrolstad, 2004). Compared to other fruits such as grapes, blueberries, and so on, dates are a poor source of anthocyanins. The anthocyanin contents of different varieties of fresh dates expressed as cyanidin 3-glucoside equivalents can differ significantly (Eong et al., 2006; Al-Farsi et al., 2005a). The differences in the anthocynin contents could be associated with the color of date fruit and red color varieties have usually higher concentration as compared to others (yellow etc.). The anthocyanins were reported to be present only in fresh dates as the drying may destroy anthocyanins (Al-Farsi et al., 2005a). The drying temperature, light, agronomic practices, and storage conditions affect the fruit anthocyanins. Commercial sun-drying of dates at 30–50°C for 7–10 days can completely destroy the anthocyanins due to enzymatic and nonenzymatic browning reactions (Wrolstad, 2004; Al-Farsi et al., 2005a).

The variability in the data about the total antioxidants content of dates could be attributed to different extraction techniques and analytical methods used for their estimation. Higher antioxidant activity has been reported in fresh dates as compared to dried dates, which may be due to decomposition of natural antioxidants during the drying process. Sun-drying can cause a significant loss ranging from 29.7% to 42.5% in the antioxidant activity of dried dates (Vayalil, 2002; Guo et al., 2003; Al-Farsi et al., 2005a). There are, however, few studies which have reported either an increase or no change in antioxidant activity of some fruits such as plums after drying (Piga et al., 2003). Vayalil (2002) stated that the antioxidant and the antimutagenic activity in dates implicate the presence of compounds with potent free radical scavenging activity. Guo et al. (2003) reported that dates had the second-highest antioxidant value of 28 fruits commonly consumed in China, and the Hawthorn fruit showed the highest amount of antioxidants. The role of antioxidants in minimizing or preventing the risk several chronic diseases such as cancer, cardiovascular diseases, neurodegenerative diseases, and inflammation is well recognized (Benzie, 2003; Kris-Etherton et al., 2004; Cooper, 2004). Because of their high fiber content, certain minerals and vitamins as well as the presence of many antioxidant and antimutagenic phytochemicals, dates can potentially be considered as a functional food in human health. Further research is therefore needed to study the various date varieties in terms of their total antioxidant capacity, free radical scavenging activities, reducing power, anti-inflammatory properties, and protection against chronic diseases.

MEDICINAL VALUE OF DATE FRUIT

Antioxidant, Anti-inflammatory, Antimutagenic, and Anticancer Activities

In addition to their significance as an ideal high-energy food, in the folklore, the dates are believed to have many medicinal properties such as to provide strength, fitness, and relief against a number of ailments and pains including fever, stomach disorders, memory disturbances, nervous disorders, as well as aphrodisiac and to boost the immunity. They are also considered to protect against many chronic diseases including cancer and heart diseases (Lambiote, 1982; Duke, 1982; Vyawahare et al., 2009) as they have been shown to contain antioxidant and antimutagenic properties (Al-Farsi et al., 2005b; Vayalil, 2002; Allaith, 2008). Aqueous extracts of dates have also been shown to inhibit the lipid peroxidation and protein oxidation as well as exhibit a potent superoxide and hydroxyl radical scavenging activity (Allaith, 2005). Panahi and Asadi (2009) reported that extract of date fruit were useful in controlling the blood cholesterol levels and also protected the CA1 neurons against oxidative injury. Methanolic and aqueous extracts of date flesh and seeds have also been shown to exhibit anti-inflammatory properties and suppressed the swelling in the foot and adjuvant arthritis (Mohammed and Al-Okbi, 2004).

Dates are a rich source of hydroxyl pope folic acid and therefore eating on dates can increase the body's immunity and resistance to cancers. Ishrud and John (2005) observed that the preparations made from the extracted polysaccharides (glucans) from Lybian dates exhibited a dose-dependent anticancer activity by suppressing the growth of Sarcoma 180 tumor cells in female CD1 mice. It is a common belief in the Middle East that the consumption of dates, particularly in the morning on an empty stomach, can reverse the actions of any toxic material that the subject may have been exposed to (Vyawahare et al., 2009). Date extracts (DE) have been reported to inhibit the benzopyrene-induced mutagenecity in *Salmonella* tester strains TA-98 and TA-100 with metabolic activation (Vayalil, 2002). It has also been suggested that the extracts of date flesh and pits can reverse the carbon tetra chloride (CCl_4)-induced liver damages in rats (Al-Qarawi et al., 2004). The DE have also been shown to reduce the oxidative stress by normalizing the increased hepatic levels of malondialdehyde (MDA) and by increasing the hepatic glutathione levels (Mohammed et al., 2008). DE and products can also be used prophylactically as a dynamic liver support and can counteract the alcohol intoxication. A number of herbal formulations are available in the markets, which have been shown to reduce or prevent the alcohol-induced hangover symptoms in human volunteers as well as protection against liver disorders in acute and chronic alcoholics (Thornfeldt et al., 2006). It is also believed that consuming seven dates daily can protect the children from anxiety and nervous disorders.

Treatment of Abdominal Troubles

Dates are used for the treatment of abdominal troubles. Because of their high phenolic contents, dates are used medicinally as a detersive (cleansing agent) and astringent in intestinal troubles. Water extracts of date flesh have been shown to induce a dose-dependent increase in the gastrointestinal transit time of foods in rats (Al-Qarawi et al., 2003). The aqueous extracts of dates have also been shown to have antidiarrheal effects in rats (Al-Taher, 2008). The ethanolic and aqueous extracts have been reported to have positive protective effects on the ethanol-induced gastric ulceration in rats (Al-Qarawi et al., 2005). Because of its antidysentery and laxative effects, it is believed that a daily dosage of seven dates soaked in water and taken before sleeping can kill *Ascaris* (giant intestinal roundworms). According to folklore medicine to secure a laxative effect, the dates should be immersed in water at night and taken after making them into fine syrup the next morning. Decoction of dates, free from fibrous material together with some table salt, can be used as a remedy to treat the dehydration resulting from vomiting and diarrhea. Water extracts of dates are also considered to help in lowering the stomach hyperacidity as well as blood acidity. For children with sensitive stomachs, the water extracts of dates are sometimes added to milk to help in the digestion of milk because the date fibers are soft and do not irritate a sensitive bowel or stomach. Dates crushed and boiled in milk are very nourishing and restorative especially during convalescence.

Antianalgesic, Antipyretic, and Protection against Colds, Sore Throat, and Fever

Date products in the form of infusion, decoction, syrup, or paste are administered as a treatment for sore throat, colds, bronchial catarrh (inflammation of mucus membranes). Although, no clinical data are available in the literature for such claims, the antioxidant and anti-inflammatory properties of dates (Mohammed and Al-Okbi, 2004; Allaith, 2008; Vyawahare et al., 2009) could be regarded as the main contributing factor for such effects. The traditional recipe that is considered as an effective expectorant consists of a mixture of 50 g dates, 50 g fig, 50 g hibiscus, and 50 g raisins boiled in 1 L of water. This mixture is recommended to be taken three times daily to control throat and chest infections. Decoction of dates and fenugreek is recommended in the treatment of bronchial asthma. Date pastes with margarine are considered to act as analgesic and antipyretic and are therefore applied externally on the abscesses as well as on the poisonous bites to avert the poisonous effect. Pastes prepared from powdered seeds are given to relieve the ague (malarial fever). Date pastes in water are also believed to act as antihistaminic to control allergies when externally applied to skin. Zinc is regarded as an active ingredient inducing allergic inhibition. Soaps prepared with powdered date seeds are used for treatment of skin allergy and Acrodermatitis enteropathica (AE). AE is a rare inherited autosomal metabolic recessive disorder that results from poor absorption or uptake of zinc and can be resolved by oral supplementation of zinc (Park et al., 2010). Daily diets containing dates can produce sedative action to allergy. It is believed that continuous feeding on dates can relieve the hemorrhoids pain and may increase the body's resistance to this disease. However, no clinical data are available to prove this statement. Coffee made from powdered date stones drunk twice daily is believed to relieve the gout pains. The roots of date palm are used against toothache.

Effects on Pregnancy, Lactation, and Reproductive System

The most common medicinal use of dates and its products is as a tonic, especially for women who are close to delivery and at postpartum stage. The consumption of dates by women before and after delivery can act as a tonic to strengthen the uterine muscles. Dates not only help in activating the delivery process but may also prevent the postdelivery bleeding due to the presence of some constricting substances. The typical example given for such use is the case of Virgin Mary's delivery as mentioned in Holy Quran when she was recommended to eat dates to relieve her delivery pains (Al-Qur'an, 19:23–26). The potassium, glycine, and threonine content of dates are thought to activate the production of milk hormone (prolactin). Small quantities of oxytocin have also been detected in dates and therefore dates can act as lactagogue if used in the daily meal plan of a woman during her lactation period.

Dates mixed with milk and cinnamon are considered as aphrodisiac as they can activate sexual desire. In the Middle East, it is common to consume about 10 to 30 dates daily as part of their normal diet. Studies have shown that DE increased the sperm count in guinea pigs and enhanced the spermatogenesis by increasing the concentration of testosterone, follicle-stimulating hormone, and leutinizing hormone in rats (Elgasim et al., 1995). Dates can help solve the infertility problems in males. Date palm pollen grains have been shown to regulate the sexual hormonal balance in rats (Reshod and Al-Shagrawi, 1998). The date pollens yield an estrogenic principle, estrone that has been shown to have a gonadotropic effect in young rats (El-Moughy et al., 1991).

Nephroprotective and Hemolytic Activities

Dates are also considered as renal restorative and their daily consumption can prevent the formation of renal calculi due to its diuretic and anti-inflammatory actions. A cup of hot decoction of seven dates, when taken twice daily for a period of 15 days, is considered as a remedy for lithontriptic and diuretic. A drink made from powdered date stones is also considered as lithontriptic for hepatic and renal calculi. A gum that exudes from wounded trunk of date palm is employed in India for treating diarrhea and a genito-urinary ailment as it works as diuretic (increase in urine output) and demulcent (soothing effects on the irritated mucous membrane). Extracts of date flesh and pits have been shown to reduce the gentamycin-induced nephrotoxicity in rats (Al-Qarawi et al., 2003). It was suggested that vitamin E, ascorbic acid, and melatonin fractions of dates were responsible for nephroprotection. Siddiqi et al. (2009) reported that ingestion of few dates does not cause any significant hyperkalemia in hemodialysis patients without hyperkalemia. They suggested that hemodialysis patients could be allowed to eat few dates on the days prior to their dialysis sessions.

The DE have been shown to neutralize the hemolytic activity of *Streptococcal exotoxin* and *Streptolysin O*. The inhibitory substance has been found to be steroidal in nature and not proteinaceous as deproteinization of DE did not decrease its inhibitory effects (Abuharfeil et al., 1999). DE have also been reported to inhibit the hemolytic activity of snake and scorpion venoms by interfering with their degenerative effects on the fine structure of hepatocytes (Sallal et al., 1995). In a recent study, the immune-modulatory and immune-stimulant activities of date fruit extracts have also been demonstrated (Puri et al., 2000).

Other Medicinal Benefits of Date Fruits

The daily consumption of dates with meals and its decoction as eye-lotion is considered to help in the maintenance of eye hygiene and a remedy for night blindness and ophthalmic disorders. Because of their high-potassium and low-sodium contents, dates can help alleviate potassium deficiency, maintain the sodium–potassium balance in the body and may help in treating the cardiac disorder especially after diarrhea, vomiting or after the use of diuretic medications. Dates have been shown to have low-to-medium glycemic index values (Ali et al., 2009) and therefore can have some beneficial effect in the glycemic and lipid control of diabetic patients (Miller et al., 2003). In folklore traditional medicine, the date seed powder is considered to be the best medicine for diabetic patients. Studies have also reported the antiviral and heavy metal-absorbing properties of date fruit and date pits.

CONCLUSION

Date is a delicious fruit with a sweet taste and a fleshy mouth feel and can be considered as an almost ideal food that provides a wide range of essential nutrients with many potential health benefits. Its major components are carbohydrates (mainly sucrose, glucose, and fructose), which may constitute to about 70%. Dates are a good source of fiber and contain many important vitamins and minerals, including significant amounts of calcium, iron, fluorine, and selenium and low in sodium. Date fruit also contains a number of phytochemicals such as carotenoids, phenolics, flavonoids, anthocyanins, which possess antioxidant and antimutagenic and immune-modulatory properties. In addition to its nutritional significance, in the folklore, the dates are believed to have many medicinal properties such as aphrodisiac, boost the immunity and provide strength, fitness, and relief against pains as well as protection against many disease including cancer and heart diseases. The date published so far has many unanswered questions but has renewed an increased interest to further explore the nutritional and health benefits of date fruit and date pits. Further research is therefore required to systematically explore the health benefits of date fruit and date pits as well as on the use of their functional components in the development of various values-added food products and supplements.

REFERENCES

Abuharfeil, N.M., El-Sukhon, S., Msameh, Y., and Sallal, A.K.J. 1999. Effect of date fruits, *Phoenix dactylifera* L., on the haemolytic activity of *Streptolysin O. Pharmaceutical Biology* 37: 335–339.

Ahmed, I.A., Ahemed, A.W.K., and Robinson, R.K. 1995. Chemical composition of date varieties as influenced by the stage of ripening. *Food Chemistry* 54: 305–309.

Aidoo, K.E., Tester, R.F., Morrison, J.E., and Macfarlane, D. 1996. The composition and microbial quality of pre-packed dates purchases in Greater Glasgow. *International Journal of Food Science and Technology* 31: 433–438.

Al-Bukhari, S. 2008. *Sahihalbukhari, Hadith Nos. 5396 and 5403*, Available on line at http://www.sahihalbukhari.com/sps/sbk/ (last accessed on Sept 9, 2009).

Al-Farsi, M., Alasalvar, C., Morris, A., Baron, M., and Shahidi, F. 2005a. Compositional and sensory characteristics of three native sundried date (*Phoenix dactylifera* L.) varieties grown in Oman. *Journal of Agriculture and Food Chemistry* 53: 7586–7591.

Al-Farsi, M., Alasalvar, C., Morris, A., Baron, M., and Shahidi, F. 2005b. Comparison of antioxidant activity, anthocyanins, carotenoids, and phenolic of three native fresh and sun dried date (*Phoenix dactylifera* L.) varieties grown in Oman. *Journal of Agriculture and Food Chemistry* 53: 7592–7599.

Al-Farsi, M. and Lee, C.Y. 2008. Nutritional and functional properties of dates: A review. *Critical Review in Food Science and Nutrition* 48: 877–884.

Al-Hooti, S., Sidhu, J.S., and Qabazard, H. 1995. Studies on the physico-chemical characteristics of date fruits of five UAE cultivars at different stages of maturity. *Arab Gulf Journal of Scientific research* 13: 553–569.

Al-Hooti, S., Sidhu, J.S., and Qabazard, H. 1997. Physiochemical characteristics of five date fruit cultivars grown in United Arab Emirates. *Plant Foods for Human Nutrition* 50: 101–113.

Ali, A. Yusra, M. Al-Kindi, and Al-Said, F. 2009. Chemical composition and glycemic index of 3 varieties of Omani dates. *International Journal of Food Science and Nutrition* 60(S4): 51–62.

Al-Kharusi, L.M., El-Mardi, M.O., Ali, A., Al-Said, A.F., and Kadhir, M. 2009. Effect of minerals and organic fertilizers on the chemical characteristics and quality of date fruits. *International Journal of Agriculture and Biology* 11(3): 290–296.

Allaith, Abdull, A.A. 2005. *In-vitro* evaluation of antioxidant activity of different extracts of *Phoenix dactylifera* L. fruit as functional foods. *Deutsche Lebensmittel Rundschau* 101: 305–308.

Allaith, A.A.A. 2008. Antioxidant activity of Bahreini date palm (*Phoenix dactylifera*) fruit of various cultivars. *International Journal of Food Science and Technology* 43(6): 1033–1040.

Almana, H.A. and Mahmoud, R.M. 1994. Date-palm seeds as an alternative source of dietary fibre in Saudi bread. *Ecology Food Nutrition* 32: 261–270.

Al-Qarawi, A.A., Abdel-Rehman, H., Ali, B.E.H., Mousa, H.M., and El-Mougy, S.A. 2005. The ameliorative effect of dates (*Phoenix dactylifera*) on ethanol induced gastric ulcer in rats. *Journal Ethnopharmacology* 98: 313–317.

Al-Qarawi, A.A., Abdel-Rehman, H., Mousa, H.M., Ali, B.H., and El-Moughy, S.A. 2008. Nephroprtective action of *Phoenix dactylifera* in gentamicin-induced nephrotoxicity. *Pharmaceutical Biology* 4: 227–230.

Al-Qarawi, A.A., Ali, B.H., El-Mougy, S.A., and Mousa, H.M. 2003. Gastrointestinal transit in mice treated with various extracts of date (*Phoenix dactylifera* L.). *Food and Chemical Toxicology* 41:37–39.

Al-Qarawi, A.A., Mousa, H.M., Ali, B.E.H., Abdel-Rehman, H., and El-Mougy, S.A. 2004. Protective effect of extracts of dates (*Phoenix dactylifera* L.) on carbon tetrachloride-induced hepatotoxicity in rats. *International Journal of Applied Research and Veterinary Medicine* 2: 176–180.

Al-Shahib, W. and Marshall, R.J. 2002. Dietary fibre content of dates from 13 varieties of 9 date palm *Phoenix dactylifera* L. *International Journal of Food Science and Technology* 37: 719–722.

Al-Shahib, W. and Marshall, R.J. 2003. The fruit of date palm: its possible uses as best food for the future. *International Journal of Foods Science and Nutrition* 54:247–259.

Al-Showiman, S.S. 1990. Chemical composition of date palm seeds (*Phoenix dactylifera* L.) in Saudi Arabia. *Journal of the Chemical Society* 12: 15–24.

Al-Showiman, S.S. 1998. *Al Tamr, Ghetha was Saha (date, Food and Health)*. Saudi Arabia: Dar Al-Khareji Press.

Al-Taher, A.Y. 2008. Possible antidiarrheal effect of date-palm (*Phoenix dactylifera* L.) aqueous extracts in rats. *Scientific Journal of King Faisal University (Basic and Applied Sciences)* 9: 1429–1435.

Babich, H., Sedletcaia, A., and Kenigsberg, B. 2002. *In-vitro* cytotoxicity of protocatechuic acid to cultured human cells from oral tissue: Involvement in oxidative stress. *Pharmacology Toxicology* 91: 245–253.

Barreveld, W.H. 1993. *Date-Palm Products*. Bulletin No 101. Rome, Italy: Food and Agriculture Organization of the United Nations.

Ben-Amotz, A. and Fishler, R. 1998. Analysis of carotenoids with emphasis on 9-*cis*-ß-carotenes of vegetables and fruits commonly consumed in UK. *Food Chemistry* 62: 515–520.

Benzie, I.F. 2003. Evolution of dietary antioxidants. *Comparative Biochemistry and Physiology* 136: 113–126.

Benzie, I.F. and Strain, J.J. 1996. The ferric reducing ability of plasma (FRAP) as a measure of "antioxidant power": The FRAP assay. *Analytical Biochemistry* 239: 70–76.

Besbes, S., Drira, L., Blecker, C., Deroanne, C., and Attia, H. 2009. Adding value to hard date (*Phoenix dactylifera* L.): Compositional, functional and sensory characteristics of date jam. *Food Chemistry* 112: 406–411.

Biglari, F., Abbas, F.M., Alkarkhi, F.M., and Azahar, M.E. 2008. Antioxidant activity and phenolic content of various date palm (*Phoenix dactylifera*) fruits from Iran. *Food Chemistry* 107: 1636–1641.

Boudries, H., Kefalas, P., and Hornero-Mendez, D. 2007. Carotenoid composition of Algerian date varieties (*Phoenix dactylifera*) at different edible maturation stages. *Food Chemistry* 101: 1372–1377.

Cooper, D.A. 2004. Carotenoids in health and disease: Recent scientific evaluations. Research, recommendation and the consumer. *Journal of Nutrition* 134: 221s–224s.

Cummings, J.H., Bingham, S., Heaton, K.W., and Eastwood, M.A. 1992. Fecal weight, colon cancer risk, and dietary intake of non-starch polysaccharides (dietary fibre). *Gastroenterology* 103: 1783–1789.

Cuvelier, M.E., Richard, S., and Berset, C. 1992. Comparison of the antioxidant activity of some acid phenols: Structure–activity relationship. *Bioscience, Biochemistry and Biotechnology* 56: 324–325.

Dowson, V.H.M. and Aten, A. 1982. *Composition Et Maturation. Recolte Et Conditionnement Des Datttes. Collectoiion FAO*. Rome, Italy: Cahier 72, pp. 1–392.

Duke, J.A. 1982. *Handbook of Physiochemical Characteristics of GRAS Herbs and Other Economic Plants*. Boca Raton, FL: CRC Press.

Elgasim, E.A., Alyousif, Y.A., and Homeida, A.M. 1995. Possible hormonal activity of date pits and fleshed of meat animals. *Food Chemistry* 52: 149–150.

Elleuch, M., Basbes, S., Roiseux, O., Blecler, C., Deroenne, N., Driera, E., and Attia, H. 2008. Date flesh: Chemical composition and characteristics of dietary fibre. *Food Chemistry* 111: 676–682.

El-Moughy, S.A., Abdel-Aziz, S.A., Al-Shanawany, M., and Omar, A. 1991. The gonadotropic activity of palmae in mature rats. *Alexandria Journal of Pharmaceutical Research* 5: 156–159.

El-Shaarawy, M.I., Mesalam, A.S., El-Nakhal, H.M., and Wahdan, A.N. 1986. Studies on extraction of dates. *Proceedings of the 2nd Symposium on the Date Palm in Saudi Arabia*. II, pp. 259–270, 3–6 March, 1986.

El-Zoghbi, M. 1994. Biochemical changes in some tropical fruits during ripening. *Food Chemistry* 49:33–37.

Eong, Y.J., Hong, F.A., Tomas-Barberan, A., Kader, S., and Alyson, E. 2006. The flavonoid glycosideds and procynidine composition of Deglet Noor dates (*Phoenix dactylifera*). *Journal of Agriculture and Food Chemistry* 54: 2405–2411(s).

Erskine, W., Mostafa, A.T., and Osman, A.E. 2004. Date palm in GCC countries of the Arabian Peninsula. *International center for Agriculture Research in Dry Area (ICARADA)*, htpp://www.icarda.org/APRP/ Date palm/Introduction/intro-body.htm. (accessed on Oct 10, 2006).

Fayadh, J.M. and Al-Showiman, S.S. 1990. Chemical composition of date palm (*Phoenix dactylifera* L.). *Journal of the Chemical Society of Pakistan* 12: 84–103.

Fennema, O.R. 1996. *Food Chemistry*, 3rd edition. New York: Dekker.

Guan, S., Ge, D., Liu, T.Q., Ma, X.H., and Cui, Z.F. 2009. Protocatechuic acid promotes cell proliferation and reduces basal apoptosis in cultured neural stem cells. *Toxicology In-vitro: An International Journal Published in Association with BIBRA* 23(2): 201–208.

Guo, C., Yang, J., Wei, J., Li, Y., Xu, J., and Jing, Y. 2003. Antioxidant activities of peel, pulp and seed fractions of common fruits as determined by FRAP assay. *Nutrition Research* 23: 1719–1726.

Hart, D. and Scott, K.J. 1995. Development and evaluation of HPLC method for the analysis of carotenoids in fruits, and the measurement of the carotenoid content of vegetables and fruits commonly consumed in UK. *Food Chemistry* 54: 101–111.

Hussein, A.S., Alhadrami, G.A., and Khalil, Y.H. 1998. The use of dates and date pits in broiler starter and finisher diets. *Bioresource Technology* 66: 219–223.

Ishruda, O. and John, F.K. 2005. The anticancer activity of polysaccharide prepared from Libyan dates (*Phoenix dactylifera* L.). *Carbohydrate Polymers* 59: 531–535(s).

Ishurd, O., Zahid, M., Xiao, P., and Pan, Y. 2004. Protein and amino acid contents of Libyan dates at three stages of development. *Journal of Science, Food and Agriculture* 84: 481–484.

Ismail, B., Haffar, I., Baalbaki, R., Mechref, Y., and Henry, J. 2006. Physico-chemical characteristics and total quality of five date varieties grown in United Arab Emirates. *International Food Science and Technology* 41: 919–926.

Kaehkoenen, M.P., Heinamaki, J., Olliainemn, V., and Heinonen, M. 2003. Berry anthocyanin: Isolation, identification, and antioxidant activity. *Journal of Science Food and Agriculture* 83: 1403–1411.

Khan, M., Sarwar, A., Wahab, M., and Haleem, R. 2008. Physio-chemical characterization of date varieties using multivariate analysis. *Journal of Food Agriculture* 88: 1051–1059.

Khatchadourian, H.A., Sawaya, W.N., Ayaz, M., and Al-Mohemmed, M.M. 1987. Processing date varieties into pickles. *International Journal of Food Science and Technology* 22: 243–247.

Kris-Etherton, P.M., Lichtenstein, A.H., Howard, B.V., Steinberg, D., and Witztum, J.L. 2004. Antioxidant vitamin supplements and cardiovascular diseases. *Circulation* 110: 637–641.

Lambiote, B. 1982. Some aspects of the role of dates in human nutrition: In *Proceedings of the First International Symposium on Date Palm*. King Faisal University, Saudi Arabia, 23–25 March, 1982, p.16.

Lim, Y.Y., Lim, T.T., and Tee, J.J. 2007. Antioxidant properties of several tropical fruits: A comparative study. *Food Chemistry* 103(3): 1003–1008.

Liu, S., Willett, W.C., Stampfer, M.J., Hu, F.B., Franz, M., Sampson, L., Hennekens, C.H., and Manson, J.A.A. 2000. Prospective study of dietary glycemic load, carbohydrate intake and risk of coronary heart diseases in U.S women. *American Journal Clinical Nutrition* 71: 1455–1461.

MAF. 2005. *Agriculture and Fisheries Research Report*. Ministry of Agriculture and Fisheries, Muscat, Sultanate of Oman.

Makki, M., Hamooda, A., and Al-Abri, A. 1998. *The Date Palm, Culture, Operation and Maintenance*. Muscat, Oman: Modern Color Publishers.

Mansouri, A., Embarek, G., Kokkalou, E., and Kefalas, P. 2005. Phenolic profile and antioxidant activity of the Algerian ripe date palm fruit (*Phoenix dactylifera*.). *Food Chemistry* 89: 411–420.

Marlett, J.A., Mc Burney, M.I., and Slavin, J. 2002. Position of the American Diabetic Association: Health implications of dietary fiber. *Journal American Diabetic Association* 102: 993–1000.

Masmoudi, M., Besbes, S., Blecker, C., and Attia, H. 2010. Preparation and characterization of jellies with reduced sugar content from date (*Phoenix dactylifera* L.) and lemon (*Citrus limio* L.) by-products. *Fruits* 65: 21–29.

Mazza, G. and Miniati, E. 1993. *Anthocyanins in Fruits. Vegetables and Grains*. Boca Raton, FL: CRC Press.

Miller, C.J., Dunn, E.V., and Hashim, I.B. 2003. The glycemic index of dates and date/yoghurt mixed meals: Are date the candies that grow on plants? *European Journal of Clinical Nutrition* 57: 427–430.

Mohammed, B.A., Nabil, A.H., and Hanan, A.H. 2008. Protective effects of extracts from dates (*Phoenix dactylifera* L.) and ascorbic acid on thioacetamide-induced hepatotoxicity in rats. *Iranian Journal of Pharmaceutical Research* 7: 193–201.

Mohammed, D.A. and Al-Okbi, S.Y. 2004. *In-vivo* evaluation of antioxidant and anti-inflammatory activity of different extracts of date fruits in adjuvant arthritis. *Polish Journal of Food and Nutrition Science* 13: 397–402.

Mossa, J.S., Hifnawy, M.S., and Mekkawi, A.G. 1986. Phytochemical and biological investigation on date seed (*Phoenix dactylifera* L.) produced in Saudi Arabia. *Arab Gulf Journal of Science and Research* 4: 495–501.

Musaiger, A.R. 2009. http://www.bcsr.gov.bh/BCSR/En/News/2 + Dec + 2009 + - + 2.htm (last accessed on 22 Jan, 2011).

Myhara, H.M., Karkala, J., and Taylor, M.S. 1999. The composition of maturing Omani dates. *Journal of the Science of Food and Agriculture* 47: 471–479.

Panahi, A. and Asadi, M. 2009. Cholesterol lowering and protective effects of date fruit extracts: An *in-vivo* study. *Toxicology Letters* 189S: S57–S273(F25).

Park, C.-H., Lee, M.J., Kim, H.-J., Lee, G., Park, J.-W., and Cinn, Y.-W. 2010. Congenital zinc deficiency from mutations of the SLC39A4 gene as the genetic background of Acrodermatitis enteropathica. *Journal of the Korean Medical Science* 25: 1816–1820.

Piga, A., Del Caro, A., and Corda, G. 2003. From plums to prunes: Influence of drying parameters on polyphenols and antioxidant activity. *Journal of Agriculture and Food Chemistry* 51: 3675–3681.

Puri, A., Sahai, R., Kiran, L.S., Saxena, R.P., Tandon, J.S., and Saxena, K.C. 2000. Immunostimulant activity of dry fruits and plant materials used in Indian traditional medical system for mothers after child birth and invalids. *Journal of Ethnopharmacology* 71(1–2): 89–92.

Regnault-Roger, C., Hadidan, R., Biard, J.F., and Boukef, K. 1987. High performance liquid and thin chromatographic determination of phenolic acids in palm (*Phoenix dactylifera* L.). *Food Chemistry* 25: 61–71.

Reshod, A. and Al-Shagrawi, M. 1998. Enzyme activities, lipid functions, and fatty acid composition in male rat fed palm pollen grains (*Phoenix dactylifera*). *Research Bulletin* 79: 5–18.

Saafi, E.B., El-Arem, A., Issaoui, M., Hammi, H., and Achour, L. 2009. Phenolic potent and antioxidant activity of four date palm (*Phoenix dactylifera* L.) fruit varieties grown in Tunesia. *International Journal of Food Science and Technology* 44: 2314–2319.

Sallal, A., Amr, Z., and Al-Bagdadi, F. 1995. The effect of dated on the ultrastructure of hepatocytes in mice inoculated with *Cerastes cerastes* venom. *Proceeding Microscopy and Microanalysis* 79: 870–871.

Sawaya, W.N., Khachadourian, H.A., Khalil, J.K., Sail, W.M., and Al-Shalhat, A., 1982. Growth and compositional changes during the various development stages of some Saudi Arabian date cultivars. *Journal of Food Science* 47: 1489–1493.

Sawaya, W.N., Khalil, J.K., Safi, W.N., and Al-Shalhat, A. 1983. Physical and chemical characterization of three Saudi date cultivars at various stages of development. *Journal Canadian Institute of Good Science and Technology* 16(2): 87–91.

Shahidi, F. and Naczk, M. 2004. *Phenolics in Foods and Nutraceuticals.* Boca, Rotan, FL: CRC Press.

Shinwari, M.A. 1993. Date palm. In *Encyclopedia of Food Science. Food Technology and Nutrition*, Macrae, R., Robinson, R.K., and Sadler, M. (eds). London, UK: Academic Press Limited, pp. 1300–1305.

Shivashankara, K.S., Isobes, S., Al-Haq, M.I., Takenaka, M., and Shinha, T. 2004. Fruit antioxidant activity, ascorbic acid, total phenol, quercetin, and carotene of Irwin mango fruits stored at low temperature after high electric field treatment. *Journal of Agriculture and Food Chemistry* 52:1281–1286.

Siddiqi, N., El Shahat, O., Bokhari, E., Haytham Roujouleh, H., Hamid, H., Sheikh, I., El Sayed, H., Saleh, A., and Seddik, A. 2009. The effect of use of dates on serum potassium in no diabetic hemodialysis patients. *Saudi Journal of Kidney Diseases and Transplantation* 20(6): 1018–1022.

Thornfeldt, C.R., Paster, E.K., and Inventors, D., Shaver and Nipper, LLP, US, Assignee. 2006. *Treatment of Muco-Cutaneous Disorders through Reversing Chronic Inflammation and Barrier Disruption*, US patent Al.

USDA National database for Standard Reference, United States Development of Agriculture: http://www.nal.usda.gov/fnic/foodcomp/cgi-bin/list_nut_edit.pl (last accessed on Jan 26, 2011).

Vayalil, P.K. 2002. Antioxidant and antimutagenic properties of aqueous extract of date fruit (*Phoenix dactylifera* L. Arecaceae). *Journal of Agriculture and Food Chemistry* 50: 610–617(s).

Vinson, J.A., Zubik, L., Bose, P., Samman, N., and Proch, J. 2005. Dried fruits: Excellent in vitro and in vivo antioxidants. *Journal of the American College of Nutrition* 24: 44–50.

Vyawahare, N., Pujari, R., Khsirsagar, A., Ingawale, D., Partil, M., and Kagathara, V. 2009. *Phoenix dactylifera*: An update of its indigenous uses, phytochemistry and pharmacology. *Internet Journal of Pharmacology* 7: 1.

Wills, R., McGladdon, B., Ghramdadi, D., and Joyce, D. 1998. *Post-Harvest: An Introduction to the Physiology and Handling of Vegetables and Ornamentals*, 4th edition. Australia: University of New South Wales Press, Ltd.

Wrolstad, R.E. 2004. Interaction of natural colors with other ingredients: Anthocynin pigments—Bioactivity and coloring properties. *Journal Food Science* 669: 419–425.

27 Uses of Date Palm in Ayurveda

M. Shanmugapriya and Kishor Patwardhan

CONTENTS

AYURVEDA

Ayurveda is the native Indian system of healthcare that is currently used by millions of people in India, Nepal, and Sri Lanka for their healthcare needs. *Vedas*, especially *Rigveda* and *Atharvaveda*, the earliest documented ancient Indian knowledge dated roughly to 2000 BC, have references on health and diseases. Ayurveda texts like *Charaka Samhita* and *Sushruta Samhita* were possibly started to be composed about 1000 BC (Patwardhan et al., 2009). The term "Ayurveda" means "the science of life." This science deals elaborately with measures for healthy living during the entire span of one's life. Besides dealing with the principles for maintenance of health, it also delves upon a range of preventive and therapeutic measures to fight diseases.

Ayurveda derives its medicines mainly from the medicinal plants, though minerals and metals are also used sometimes. Since ancient times, Indian scholars had an extensive knowledge of phyto-medicine, which was due to their thriving passion for the study of nature, including medicinal plants. Ayurveda, the science of a long and healthy life, is a traditional system of medicine and healing that is based on natural resources.

Ayurveda has a holistic approach to disease. According to Ayurveda, disease occurs when there is a deviation from the state of normalcy at the physical or psychological level. Ayurveda lays stress on some prime factors that are to be considered for maintaining a healthy and long life. These factors include a balanced and nutritious diet, wholesome lifestyle, adequate sleep, judicious sexual activity, and physical exercise.

In Ayurveda, it is said that all matter, living and nonliving, is made up of five basic elements (*Panchamahabhutas*), namely, earth (*Prithvi*), water (*Jala*), fire (*Teja*), air (*Vayu*), and space (*Akasha*). These five basic elements are expressed in the living body as three *Doshas*—*Vata*, *Pitta*, and *Kapha*. The state of equilibrium between these three *Doshas* constitutes health and an imbalance leads to disease. In fact, the complete description of human physiology in Ayurveda is based primarily on this theory. The "homeostatic mechanisms" as perceived in the contemporary biomedicine have a very close similarity with this theory. On the basis of this understanding, the three *Doshas*—*Vata*, *Pitta*, and *Kapha*—can be equated with the nervous, endocrine, and immune mechanisms, respectively. This is because, *Vata* controls all the sensory and motor activities, *Pitta* regulates the digestion and metabolism, and *Kapha* confers immunity (Tripathi et al., 2011).

According to Ayurveda, the ultimate goal in the treatment of all diseases is to bring these *Doshas* back to their state of equilibrium or to "restore the homeostasis." The Ayurveda formulations used for such treatment purposes are manufactured based on the procedures given in the ancient texts like *Charaka Samhita*, *Susruta Samhita*, *Sharangadhara Samhita*, *Ashtanga Hridayam*, and so forth. The formulations are prepared out of a variety of plant sources such as stems, roots, leaves, fruits, barks, fruit rinds, seeds, resins, rhizomes, animal products, and minerals. As these natural products also are composed of the five basic elements mentioned earlier, one needs to replenish these elements through these natural resources in order to sustain the healthy body.

DATE PALM AND AYURVEDA

Among a large variety of Ayurveda preparations, dates occupy a considerably significant position and have a wide range of therapeutic uses. The Ayurveda name for date palm is *Kharjura*. Another variety of dates used in Ayurveda is *Pinda Kharjura*. In Ayurveda, date is mentioned as one among the "*Trimadhuras*," which means "the three predominant drugs having extensive sweet taste and add to the body's general vitality." The parts of the plant used in Ayurveda formulations include the leaves, flower, fruits, roots, and seed.

PROPERTIES OF DATE FRUIT

Guna (attributes): viscous, heavy, and anabolic
Rasa (taste): sweet
Vipaka (effect of metabolic transformation): sweet (anabolic)
Virya (potency): cold
Prabhava (impact): blood purifier
Effect on *Doshas*: has a suppressant action on *Vata* and *Pitta*

The fruit is sweet (*Madhura*); cooling (*Sheetha*); tonic; fattening; aphrodisiac; and useful in skin diseases, excessive thirst, asthma, bronchitis, fatigue, tuberculosis, abdominal complaints, fevers, vomiting, psychotic disorders, lymphatic and circulatory system disorders, kidney diseases, and nervous system disorders (Bhavamishra, 1600; Govindadasa, 1700; Shastry, 2008).

The toddy derived out of date palm is intoxicating, fattening, and aphrodisiac, and it improves taste and is useful in bronchitis and *Vata* disorders. The leaves are aphrodisiac and good for the liver. The flower is bitter, purgative, expectorant, tonic to the liver, and is useful in fever and blood dyscrasias. The dry fruit is sweet, diuretic, and aphrodisiac, enriches the blood, and is useful in paralysis. The seed is applied to wounds since it reduces inflammation. The gum is useful in diarrhea and diseases of the genitourinary system. The root is a good analgesic due to its heavy (*Guru*) property. The juice is very effective for relieving toothache. The powder is very effective in various abdominal disorders and also is a stimulant for the digestive system. This improves circulation and also increases sexual potency (The Wealth of India, 1952; Nadkarni, 1976; Vagbhata, 700; Sharngadhara 1300).

THERAPEUTIC USES OF DATES

Ayurveda physicians use date fruit as a remedy for teeth and gum problems, urinary tract infections, upper respiratory and lower respiratory tract infections, general weakness, sciatica, dehydration, asthma, hiccups, cystitis, tuberculosis, nervous disorders, skin problems, leprosy, gonorrhea, anemia, decreased sperm count, purpura, edema, sepsis, cognitive dysfunction, anxiety, psychosis, muscular paralysis, cardiovascular disorders, kidney and liver disorders, and microbial and viral infections. It is also used to counteract alcohol intoxication (The Wealth of India, 1952; Nadkarni, 1976).

Rasayana

Dates have a very important role as a *Rasayana* (rejuvenation). *Rasa*, in Ayurveda terminology, means a common nutrient pool that supplies the nutrients to different tissues as it circulates in the body. All drugs and procedures that enhance the quality of this *Rasa* are collectively known as "*Rasayana*" and this branch of science formed one of the eight specialties of Ayurveda in ancient India. *Rasayana* has been classified as follows (Agnivesha, 1000 BC; Susrutha, 1000 BC):

1. Based on the major effect of *Rasayana*, it is classified into *Kamya Rasayana* (health promoting) and *Naimittika Rasayana* (as treatment in specific diseases). *Kamya Rasayana* helps in promoting the normal health by boosting up energy levels and general health. This is further divided into *Pranakamya* (vitality enhancing), *Medhakamya* (intellect enhancing), and *Srikamya* (enhancing the physical attractiveness). *Naimittika Rasayana* helps in fighting specific diseases.
2. Based on the methods used, *Rasayana* is classified into *Kuti Praveshika* (indoor procedures) and *Vatatapika* (outdoor procedures).
3. These are also *Rasayanas* that are based on *Aushadha* (drug), *Ahara* (diet), or *Achara* (lifestyle modification).

Benefits of Rasayana

In general terms, the *Rasayana* may have different effects depending upon the clinical context. These effects may include adaptogenic, immunomodulatory, antidegenerative, longevity enhancing, or cognition enhancing. The main purpose of *Rasayana* therapy in general is to delay the aging process and to minimize the degenerative process in the body like arthritis and neurodegenerative disorders like dementia. Those specific to brain tissue are called *Medhya Rasayana*. Such *Rasayanas* retard brain aging and help in the regeneration of neural tissues besides producing antistress, adaptogenic, and memory-enhancing effect. The neuronutrient impact of *Rasayana* therapy in brain aging has also been proposed recently (Singh et al., 2008). In total, the *Rasayanas* improves the overall metabolic status, quality of body tissues, and also the microcirculation. The date fruit (*Kharjura*) has been included under *Kamya Rasayana* and *Ahara Rasayana* in Ayurveda.

There are several scientific evidences for the analgesic (Vyawahare et al., 2009a; Mohamed and Al Okbi; 2004), anti-inflammatory (Mohamed and Al Okbi, 2004), antioxidant (Mohamed DA et al., 2004; Vayalil, 2002), anticancer (Ishurda and John, 2005; Vayalil, 2002), antimutagenic (Vayalil, 2002), antiulcer, (Al-Qarawi et al., 2005) nephroprotective (Al-Qarawi et al., 2008), hepatoprotective (Al-Qarawi et al., 2004; El-Desoky et al., 1995), anti-hyperlipidemic (El-Desoky et al., 1995), spermatogenic (Bahmanpour et al., 2006; El-Desoky et al., 1995), antipsychotic, nootropic, anxiolytic (Vyawahare et al., 2009b), antidiarrheal (Al-Taher, 2008), and antiviral (Jassim and Naji, 2007) activities of date palm.

Medicinal Uses of Date Palm in Ayurveda

Nadkarni has enlisted the following medicinal uses of date fruit in his book titled *Indian Matria Medica*.

After soaking fresh dates in clean water for 2–3 h, this mixture can be used to relieve alcohol intoxication. The infusion made of milk and dates is a very nourishing and restorative drink during convalescence from fevers and small pox. The sweet pulp of the date fruit is useful in dysentery. The date fruits are used as an ingredient in various aphrodisiac and tonic confections. They are useful in asthma, tuberculosis, and exhibit antispasmodic activity. Paste made from ground seeds is said to be useful in the treatment of corneal opacity and also in the treatment of headache and hemicrania. The smoke produced from burning of the date seeds in powder form is a useful fumigatory for piles. Dates promote expectoration, soothe the chest, and prevent constipation. Dried fruit pounded and mixed with almonds, quince seeds, and spices like cardamom, ginger, and sugar forms a nutritious formulation. A fine paste of the seed of the date fruit and the roots of *Achyranthes aspera* applied to betel leaves and made into small packets with clove, cardamom, catechu, and betel nut powder is used for the prevention of intermittent fevers, such as malaria. A gum or juice obtained from the stem and named "laghi" (*Khajurnidaru*) is used as a demulcent, diuretic, and a coolant in genitourinary affections (Nadkarni, 1976).

In *Sandhivata*, the Ayurveda name for osteoarthritis, remedies include the use of *Rasayana* which are herbs that improve the joint function by increasing the *Kapha* content of the body in general and at the joints in particular. Dates occupy a prime place for this purpose. Ayurveda preparations made from date palm can be used in therapeutic doses to mitigate the excessive burning sensation, thirst, urinary infections, neuralgia, neuropathy, and nephropathy that are commonly associated with *Madhumeha*—Ayurveda synonym for diabetes mellitus. The date fruits are extensively used in Ayurveda as palliative treatment for *Pitta* and *Rakta* disorders like hemoptysis, hematemisis, chronic idiopathic urticaria, anemia, bleeding disorders, skin diseases, and various arthropathies gout. Date fruit is a potent therapeutic as well as dietetic agent in *Vata* disorders like hemiplegia, paraplegia, motor neuron diseases, convulsions, and many of the nervous system disorders. Chronic and recurrent infections and inflammatory conditions like abscess, boils, and ulcers can also be controlled by a regular use of date fruits and also with external applications of various Ayurveda preparations made of dates (Sastry, J.L.N., 2008; The Wealth of India, 1952).

A Few Important Traditional Preparations That Use Dates as an Ingredient

Most of the preparations mentioned subsequently are quoted from a text called *Sahasra Yogam*, meaning thousand formulations. *Sahasra Yogam* is a compiled text that contains formulations from different classical text books. The name of the author of this work is unknown as it is not mentioned anywhere in the text. This text can be traced back to 1800 AD (Anonymous, 1800).

1. *Kharjuraasava* (Sahasrayogam): Indicated in the treatment of tuberculosis, chicken pox, boils, skin eruptions, edema, bronchitis, hemorrhoids, intermittent fevers, and pyrexia of unknown origin. This preparation can be used for all ages and is tasty. It acts as an appetizer and promotes strength and vitality. This can be used as an adjuvant in diseases of cardiovascular, respiratory, and nervous system. All the ingredients of this formulation are given in Table 27.1.
2. *Draksha-Kashmaryadi Kashaya* (Sahasrayogam): Indicated in pacifying the pain and burning sensation in eruptions caused due to *Rakta* and *Pitta* such as chickenpox, smallpox, herpes simplex, and herpes zoster. All the ingredients of this formulation are given in Table 27.2.

TABLE 27.1
Kharjurasava[a]

Musta (*Cyperus rotundus*)	Trivruth (*Ipomea turpethum*)
Amalaki (*Phyllanthus emblica*)	Lajjalu (*Mimosa pudica*)
Draksha (*Vitis vinifera*)	Rohini (*Picorrhiza kurroa*)
Hareetaki (*Terminalia chebula*)	Jatamamsi (*Nordostachys jatamamamsi*)
Pugaphala (*Acacia catechu*)	Pippali (*Piper longum*)
Paata (*Cissampelos pareira*)	Lavanga (*Syzygyium aromaticum*)
Bharangi (*Clerodendron serratum*)	Jaatiphala (*Myristica fragrans*)
Shati (*Curcuma zedoaria*)	Chandana (*Santalum album*)
Kushta (*Sasurrea lappa*)	Loha churna (purified iron)
Ajamoda (*Carum roxbhurghianum*)	Dhataki pushpa (*Woodfordia fructosa*)
Pippalimoola (*Piper nigrum*)	Nagabala (*Sida veronicaefolia*)
Guggulu (*Commiphora mukul*)	Kachoora (*Mucuna prureins*)
Punarnava (*Boerhavia diffusa*)	Kharjura (*Phoenix dactylifera*)

[a] This is a fermented liquid preparation made out of the following ingredients that contains self-generated alcohol.

TABLE 27.2
Draksha-Kashmaryadi Kashaya[a]

Draksha (*Vitis vinifera*)	Laja (*Oryza sativum*)
Kashmarya (*Gmelina arborea*)	Amalaka (*Phyllanthus emblica*)
Patola (*Trichosanthes cucurmina*)	Dusparsha (*Mucuna prureins*)
Arishta (*Azadyracta indica*)	Kharjura (*Phoenix dactylifera*)
Vasa (*Adathoda vasica*)	Sharkara (sugar)

[a] The water-soluble and heat-stable constituents of the ingredients are extracted to prepare a Kashaya.

3. *Drakshadi Churna* (Sahasrayogam): Indicated in dyspepsia, cough, hepatitis, vomiting, menorrhagia, and leucorrhea. All the ingredients of this formulation are given in Table 27.3.
4. *Eladi Gutika* (Sahasrayogam): Indicated in bronchitis, cough, asthma, vomiting, hiccough, giddiness, fever, hemoptysis, anorexia, edema, disorders of spleen and liver, gout, hoarseness of voice, and pulmonary tuberculosis. All the ingredients of this formulation are given in Table 27.4.
5. *Shiva Gutika* (Sahasrayogam): Used as an aphrodisiac and a spermatogenic agent. Improves general body condition and vitality. All the ingredients of this formulation are given in Table 27.5.
6. *Kharjuradi Mantha* (Sharangadhara Samhita): Indicated in vomiting, hemoptysis, burning sensation, excessive thirst, diarrhea, and very effective to relieve alcohol intoxication. All the ingredients of this formulation are given in Table 27.6.
7. *Mahakushmandaka Ghrita* (Sahasrayogam): Indicated to boost one's immunity and vitality, especially for menorrahgia, metrorrhagia, dysuria, dysmenorrhea, bleeding hemorrhoids, hematemesis, leucorrhea, anemia, jaundice, fever, and all debilitating disorders. It is very efficient to improve the fertility chances in women and is also beneficial to pregnant women, thereby decreasing the chances of miscarriage. All the ingredients of this formulation are given in Table 27.7.
8. *Mahamayuradi Ghrita* (Caraka Samhita): Indicated as a Rasayana therapy and also very effective in diseases of the head and disorders of the female reproductive system. It is also indicated for oligospermia and azoospermia. All the ingredients of this formulation are given in Table 27.8.

TABLE 27.3
Drakshadi Churna[a]

Draksha (*Vitis vinifera*)	Tejopatra (*Cinnamomum tamala*)
Lajjalu (*Mimosa pudica*)	Nagakesara (*Messia ferrea*)
Sitopala (sugar candy)	Gokshura (*Tribulus terrestris*)
Madhuka (*Glycyrrhiza glabra*)	Dhanyaka (*Coriandrum sativum*)
Tavakshira (*Bambusa arundinaceae*)	Shunti (*Zingiber officinale*)
Hrivera (*Pavonia odorata*)	Shati (*Curcuma zedoaria*)
Amalaki (*Phyllanthus emblica*)	Talisa (*Abies webbiana*)
Musta (*Cyperus rotundus*)	Ela (*Elattaria cardamomum*)
Chandana (*Santalum album*)	Amragandhi (*Curcuma amada*)
Kana (*Piper longum*)	Kharjura (*Phoenix dactylifera*)
Usheera (*Vettiveria zizonoides*)	Twak (*Cinnamomum zeylanicum*)

[a] It is a compound formulation prepared by powdering the dried ingredients in a specific proportion.

TABLE 27.4
Eladi Gutika[a]

Ela (*Elettaria cardamomum*)	Sita (sugar)
Patra (*Cinnamomum tamala*)	Madhuka (*Glycyrrhiza glabra*)
Twak (*Cinnamum cassia*)	Kharjura (*Phoenix dactylifera*)
Pippali (*Piper longum*)	Mrudhika (*Vitis vinifera*)
Madhu (honey)	

[a] The fine powder of the ingredients is mixed with equal quantity of honey and rolled into pills.

TABLE 27.5
Shiva Gutika[a]

Shilajit (asphaltum)	Munditaka (*Spharanthus indicus*)
Bilwa (*Aegle marmelos*)	Jeeraka (*Cuminum cyminum*)
Agnimantha (*Clerodendron phlomoides*)	Rasna (*Vanda roxburghii*)
Shyonaka (*Uroxylum indicum*)	Pushkara (*Inula racemosa*)
Kashmari (*Gmelina arborea*)	Chitraka (*Plumbago zeylanica*)
Patala (*Stereospermum suoveolens*)	Danti (*Baliospermum montanum*)
Shaliparni (*Desmodium gangeticum*)	Kalinga (*Holarhenna antidysentrica*)
Prishniparni (*Uraria picta*)	Chavya (*Piper chaba*)
Brihati (*Solanum indicum*)	Abda (*Cyperus rotundus*)
Kantakari (*Solanum xanthocarpum*)	Katuki (*Picorrhiza kurroa*)
Gokshura (*Tribulus terrestris*)	Talisa (*Abies webbiana*)
Guduchi (*Tinospora cardifolia*)	Sita (sugar)
Haritaki (*Terminalia chebula*)	Makshika (honey)
Amalaki (*Phyllanthus emblica*)	Tilataila (gingely oil)
Vibhitaki (*Terminalia belerica*)	Tavakshira (*Bambusa arundinaceae*)
Bala (*Sida cardifolia*)	Patra (*Cinnamomum tamala*)
Madhuka (*Glycyrrhiza glabra*)	Twak (*Cinnamomum cassia*)
Patola (*Trichosanthes cucurmina*)	Ela (*Elatteria cardamomum*)
Kakoli (*Lilium polyphylum*)	Lavanga (*Syzygium aromaticum*)
Ksheerakakaoli (*Frittilaria royle*)	Jatipushpa (*Myristica fragrans*)
Medha (*Litsea sebiteria*)	Ksheera (milk)
Mahamedha (*Polygonatum cirrhifollium*)	Dadima (*Punica granatum*)
Vidari (*Ipomoea digitata*)	Sariba (*Hemidesmus indicus*)
Shatavari (*Asperagus racemosa*)	Shringi (*Rhus succedenea*)
Draksha (*Vitis vinifera*)	Kharjura (*Phoenix dactylifera*)
Ridhi (*Habenaria intermedia*)	Rishabaka (*Microstylis wallachi*)

[a] The fine powder of the ingredients is cooked in sugar, cooled, mixed with honey and rolled into pills.

TABLE 27.6
Kharjuradi Mantha[a]

Kharjura (*Phoenix dactylifera*)	Amalaki (*Phyllanthus emblica*)
Mridvika (*Vitis vinifera*)	Parushaka (*Grewia asiatica*)
Vrikshamala (*Garcinia cambogia*)	Dadima (*Punica granatum*)

[a] The filtrate obtained from the combination of dry powder of the ingredients and cold water in the ratio of 1:4.

TABLE 27.7
Mahakushmandaka Ghrita[a]

Kushmanda (*Benincasa hispida*)	Jeeraka (*Cuminum cyminum*)
Vidari (*Ipomea digitata*)	Chandana (*Santalum album*)
Ikshu (*Sacchurum officinarum*)	Yashti (*Glycyrrhiza glabra*)
Vaasa (*Adathoda vasica*)	Twak (*Cinnamomum cassia*)
Amalaki (*Phyllanthus emblica*)	Tamalapatra (*Cinnamomum tamala*)
Shatavari (*Asperagus racemosa*)	Ela (*Elattaria cardamomum*)
Kadali (*Musa paradisiaca*)	Musta (*Cyperus rotundus*)
Gokshura (*Tribulus terrestris*)	Chabaka (*Piper chaba*)
Jivanti (*Leptadania reticulata*)	Chitraka (*Plumbago zeylanica*)
Ashwagandha (*Withania somnifera*)	Shunti (*Zingiber officinale*)
Mahabala (*Sida cardifolia*)	Pippali (*Piper longum*)
Saariba (*Hemidesmus indicus*)	Pippalimoola (*Piper nigrum*)
Usheera (*Vettiveria zizonoides*)	Lavanga (*Cinnamomum zeylanicum*)
Draksha (*Vitis vinifera*)	Vidanga (*Embelia ribes*)
Dadima (*Punica granatum*)	Padmaka (*Caesalpania sappa*)
Narikela (*Cocus nucifera*)	Haridra (*Curcuma longa*)
Amrita (*Tinospora cardifolia*)	Amalaki (*Phyllanthus emblica*)
Dhaanyaka (*Coriandrum sativum*)	Vibhitaki (*Terminalia belerica*)
Kushta (*Sassurea lappa*)	Haritaki (*Terminalia chebula*)
Nagakesara (*Messua ferrea*)	Avalguja bheeja (*Psoralea corylifolia*)
Kharjura (*Phoenix dactylifera*)	Dadima (*Punica granatum*)

[a] This is a dosage form prepared in clarified butter by cooking the ingredients in it.

TABLE 27.8
Mahamayuradi Ghrita[a]

Bilwa (*Aegle marmelos*)	Shati (*Curcuma zedoaria*)
Shyonaka (*Uroxylum indicum*)	Ela (*Elettaria cardamomum*)
Agnimantha (*Clerodendron phlomoides*)	Tamala (*Cinnamomum tamala*)
Kashmari (*Gmelina arborea*)	Ikshu (*Saccharum officinarum*)
Patala (*Stereospermum souveolens*)	Shatavari (*Asperagus racemosa*)
Shaliparni (*Desmodium gangeticum*)	Bisha (*Nelumbo nucifera*)
Prishniparni (*Uraria picta*)	Shringi (*Rhus succedenea*)
Brihati (*Solanum indicum*)	Shukanasa (*Oroxylum indicum*)
Kantakari (*Solanum xanthocarpum*)	Kakoli (*Lilium polyphylum*)
Gokushura (*Tribulus terrestris*)	Punarnava (*Boerhavia diffusa*)
Atmagupta (*Mucuna prureins*)	Rishabaka (*Microstylis wallachi*)
Bala (*Sida cardifolia*)	Kapikachu (*Mucuna prureins*)
Madhuka (*Glycyrrhiza glabra*)	Badam (*Terminalia cattappa*)
Amalaki (*Phyllanthus emblica*)	Palasha (*Butea monosperma*)
Haritaki (*Terminalia chebula*)	Lajjalu (*Mimosa pudica*)
Vibhitaki (*Terminalia belerica*)	Durva (*Cynodon dactylon*)
Mahamedha (*Polygonatum cirrhifollium*)	Tugakshiri (*Bambosa arundinacea*)
Rasna (*Vanda roxburghii*)	Kaseruka (*Seirpus grossus*)
Pushkaramoola (*Inula racemosa*)	Mayur (purified flesh of peacock)
Bharangi (*Clerodendron indicum*)	Kharjura (*Phoneix dactylifera*)
Jivanthi (*Leptadania reticulata*)	Usheera (*Vettiveria zizonoides*)
Chavya (*Piper chaba*)	Manjishta (*Rubia cardifolia*)
Medha (*Litsea sebiteria*)	Ridhi (*Habenaria intermedia*)
Mridvika (*Vitis vinifera*)	

[a] This is a dosage form prepared in clarified butter by cooking the ingredients in it.

Clinical Trials on the Effect of Date Palm in Ayurveda Preparations

There have been several clinical trials that prove the efficacy of the Ayurveda preparations containing dates as their ingredients. A herbal formulation containing *Phoenix dactylifera* powder as one of the ingredients was found to significantly reduce the mean hangover score along with the blood levels of alcohol and acetaldehyde without producing any clinically significant adverse effects. This result proved that this herbal formulation can be used as a safe and effective novel herbal composition for the prevention of hangover syndromes and liver disorders in acute and chronic alcoholics (Shalini et al., 2005).

A preparation containing alcoholic extract of *P. dactylifera* fruit and seeds showed a significant reduction in wrinkles, toughness, fine lines, mottled hyperpigmentation, inflammatory papules and pustules, lesions in the skins of various groups of patients with mucocutaneous skin disorders such as visible cheek nonhypertrophic actinic keratosis, rosacea, telangectasia, sebaceous hyperplasia, actinic/senile comedones, premalignant keratosis, and acne vulgaris (Dammak et al., 2007).

An extract from the kernel of *P. dactylifera* fruit has been clinically proven for its antiwrinkle and free-radical-scavenging properties. The extract is used as an ingredient in a formulation to protect the skin from environmental sources of aging and wrinkling (Thornfeldt et al., 2006).

An Ayurveda preparation with dates as an ingredient has also been proven for its efficacy due to its antistress and adaptogenic (a drug which improves the overall resistance of the organism to nonspecific stressors) properties that retard the degenerative changes and accelerate cellular regeneration and repair. It is prescribed as a daily health tonic, in generalized anxiety disorders, geriatric stress, in senile pruritis, and in menopausal stress as an immune-booster and as an adjuvant during prolonged illness and convalescence (Singh et al., 1994).

CONCLUSION

Date fruit forms an important part of therapeutics in Ayurveda. The formulations containing date fruit have been used traditionally as *Rasayanas* and also as antidotes for alcoholic intoxication. Its anabolic property makes it a good food supplement. The sweet taste of the date fruit is an added advantage that enhances the palatability of the formulations containing it as an ingredient. Ayurveda formulations that contain date fruit as an ingredient have been shown in recent years to have significant antianxiety, cognition-enhancing, and adaptogenic properties. Further scientific evaluation is required to investigate many more possible medicinal and nutritive uses of date palm.

REFERENCES

Agnivesha. 1000 BC. *Caraka Samhita.* Sharma PV (Editor). 2008. Chaukhmbha Orientalia, Varanasi.

Al-Qarawi, A., Abdel-Rahman, H., and Ali, B.H. 2005. The ameliorative effect of dates on ethanol induced gastric ulcer in rats. *J Ethanopharmacol* 98:313–317.

Al-Qarawi, A.A., Abdel-Rahman, H., Mousa, H.M., Ali, B.H., and El-Mougy, S.A. 2008. Nephroprotective action of *Phoenix dactylifera* in gentamicin-induced nephrotoxicity. *Pharmaceutical Biol* 4:227–230.

Al-Qarawi, A.A., Mousa, H.M., Ali, B.H., and Abdel-Rahman, H. 2004. Protective effects of extracts from Dates (*Phoenix dactylifera*) on carbon tetrachloride induced hepatotoxicity in rats. *Int J Appl Res Vet Med* 2:176–180.

Al-Taher, A.Y. 2008. Possible anti-diahorreal effect of the date palm spathe aqueos extract in rats. *Sci J King Faisal Univ (Basic and Applied Sciences)* 9:1429–1435.

Anonymous 1800 AD. *Sahasra Yogam.* Ramanivasa Sharma (Editor). 1989. Daksina Prakasana, Hyderabad, India.

Bahmanpour, S., Panjeh Shahin, M.R., Talaei, T. et al. 2006. Effect of *Phoenix dactylifera* pollen on sperm parameters and reproductive system of adult male rats. *Iranian J Med Sci* 31:4.

Bhavamishra. 1600 AD. *Bhava Prakasha Nighantu.* Pandeya GS (Editor), Chunekar KC (Commentator). 2010. Chaukhambha Bharati Academy, Varanasi. ISBN: 978-81-909872-0-2.

Dammak, I., Abdallah, F.B., Boudaya, S., and Hentati, B. 2007. Date seed oil limit oxidative injuries induced by hydrogen peroxide in human skin organ culture. *Biofactors* 29(2–3):137–145.

El-Desoky, G.E., Ragab, A.A., Ismail, S.A., and Kamal, A.E. 1995. Effect of palm-pollen grains (*Phoenix dactylifera*) on sex harmones, proteins, lipids and liver functions. *J Agric Sci Mansoura Univ* 20:4249–4268.

Govinda Dasa. 1700 AD. *Bhaishajya Ratnavali*. Shastri, A. (Commentator), Shastri, R.D. (Editor). 2010. Chaukhambha Prakashan, Varanasi, 20th Edition.

Ishurda, O. and John, F.K. 2005. The anti-cancer activity of polysaccharide prepared from Libyan dates (*Phoenix dactylifera*). *Carbohydrate Polym* 59:531–531.

Jassim, S.A. and Naji, M.A. 2007. *In vitro* evaluation of the antiviral activity of an extract of date palm (*Phoenix dactylifera*), pits on a pseudomonas phage. *Evid Based Complement Alternat Med* 15:1–6.

Mohamed, D.A. and Al Okbi, S.Y. 2004. *In vivo* evaluation of antioxidant and anti-inflammatory activity of different extracts of date fruits in adjuvant arthritis. *Pol J Food Nutr Sci* 13:397–402.

Nadkarni, K.M. (Editor). 1976. *Indian Matria Medica*, Vol 1, Bombay Popular Prakashan, Mumbai, India.

Patwardhan, K., Gehlot, S., Singh, G., and Rathore, H.C.S. 2011. The Ayurveda education in India: How well are the graduates exposed to basic clinical skills? *Evid Based Complement Alternat Med* 2011:6, Article ID 197391, doi:10.1093/ecam/nep113.

Sastry, J.L.N. 2008. *Illustrated Dravyaguna Vijnana*. Vol. 2. Chaukhambha Orientalia, Varanasi. 3rd edition.

Shalini, Ramakrishna, Manu, M.B., and Kolhapure, S.A. 2005. Evaluation of the efficacy and safety of "PartySmart" in the prevention of alcohol-induced hangover: A prospective, randaomized, doubleblind, comparative, crossover, Phase III clinical trial. *Medicine Update* 12:46–55.

Sharngadhara. 1300 AD. *Sharngadhara Samhita*. Murthy, K.R.S. (Editor). 2009. Chaukhambha Orientalia, Varanasi.

Singh, B., Sharma, S.P., and Ritu, G. 1994. Evaluation of Geriforte, an herbal geriatric tonic, an anti-oxidant defense system in wistar rats. *Ann the NY Acad Sci* 717, 170.

Singh, R.H., Narasimhamurthy, K., and Singh, G. 2008. Neuronutrient impact of Ayurvedic Rasayana therapy in brain aging. *Biogerontology* 9:369–374.

Sushruta. 1000 BC. *Sushruta Samhita*. Murthy, K.R.S. (Editor). 2010. Chaukhambha Orientalia, Varanasi. 4th Edition.

Thornfeldt, C.R., Pastor, E.K., and Inventors; Dykas, Shaver and Nipper. LLP US, Asignee. Treatment of mucocutaneous disorders through reversing chronic inflammation and barrier disruption. US patent Al, 20060233739. 2006 October 19.w.

Tripathi, P.K., Patwardhan, K., and Singh, G. 2011. The basic cardiovascular responses to postural changes, exercise, and cold pressor test: Do they vary in accordance with the dual constitutional types of Ayurveda? *Evid Based Complement Alternat Med* 2011, Article ID 251850, 10. doi:10.1155/2011/251850.

Vagbhata. 700 AD. *Ashtanga Hridaya*. In: Kunte, A.M. and Navare, K.S. (Editors). 2005. Chaukhambha Orientalia, Varanasi. 9th Edition.

Vayalil, P.K. 2002. Antioxidant and antimutagenic properties of aqueous extract of date fruit. *J Agric Food Chem* 50:610–617.

Vyawahare, N., Pujari, R., Khisirsagar, A., Ingawale, D., Patil, M., and Kagathara, V. 2009a. *Phoenix dactylifera*: An update of its indigenous uses, phytochemistry and pharmacology. *Internet J Pharmacol* 7:1. Available at: http://www.ispub.com/journal/the-internet-journal-of-pharmacology/volume-7-number-1/phoenix-dactylifera-an-update-of-its-indegenous-uses-phytochemistry-and-pharmacology.html (accessed January 21, 2010).

Vyawahare, N.S., Pujari, R.R., and Rajendran, R. 2009b. Neurobehavioral effects of *Phoenix dactylifera* in mice. *J Young Pharm* 1:225–232.

The Wealth of India (A dictionary of Indian raw materials and industrial products) Raw Materials, Sastri, B.N. (Editor). 1952. 1st ed., Vol 3, Council of Scientific and Industrial Research, New Delhi.

28 Dates in Indigenous Medicines of India

Ethirajan Sukumar

CONTENTS

INTRODUCTION

India is home to countless varieties of flora and fauna. Due to its favorable climate, variations in soil, and availability of water resources, a large number of plants grow even in rough terrains. Many virgin forests add natural wealth to the environment that accommodates rare plant species but many still remain unexplored and unidentified. In addition to food and commercial crops, many exotic plants are also cultivated in some parts of the subcontinent. The popular indigenous medical systems and folklore practices flourished over the years utilizing the plant kingdom as the major armamentarium to treat diseases.

The ancient systems of medicine widely known as "Indian systems of medicine" (ISM) have in their fold not only the ones that originated in India, viz. Ayurveda and Siddha, but also from outside and got adopted in due course of time in the soil named Unani medicine. These systems have become part of tradition and culture of the country despite the phenomenal growth of modern medicine. In fact, they were the only systems of treatment available to a large section of people till the beginning of the nineteenth century.

Among the systems under ISM, Ayurveda and Siddha are the two oldest health practices, while Unani medicine came much later (tenth century AD). A sizable population depends on these systems in India even today due to their easy availability and affordability, besides the belief that these medicines are generally free from harmful reactions and after effects (Dharani, 2006).

THE INDIAN *PHOENIX*

Eight species of the genus *Phoenix* Linn. (family: Palmae) are known to occur in India and most of them find use as food as well as medicine in the indigenous medical practices. Among the Indian species, *Phoenix sylvestris* Roxb. is tapped for its sap that in turn is converted into palm jaggery and sugar. *Phoenix dactylifera* Linn. has long been introduced in the country and also grown to a small extent in some parts. Large quantities of some popular varieties are imported while attempts are being made for the commercial cultivation to meet the demand (Anonymous, 2005).

PHOENIX ACAULIS ROXB.

It is a short, handsome palm mostly found in the outer Himalayas and sub-Himalayan tracts from Kumaun eastward, Khasi and Naga hills in Assam, Bihar, Orissa, Central India, and parts of Deccan Peninsula, up to an altitude of 1500 m. This palm is common and occurs in open sal forests and grasslands, especially in clayey soil. In the outer Himalayas, it ascends into pine forests.

The pith of the stem is the source of sago and is consumed by locals. The fruits and tender peduncles are also edible. The terminal leaf bud is a good vegetable. The leaves are used for thatching and making mats, brooms, ropes, and so on. They are also used as fodder. In indigenous medicine, the roots and pith of the young stem are used as aperients.

PHOENIX DACTYLIFERA LINN.

Phoenix dactylifera or date palm is known to exist from prehistoric times in the warm dry zone, extending from North-West Africa to South-West Asian countries and cultivated throughout this region. It is believed to be originally indigenous to countries around the Persian Gulf and is widely cultivated in West Pakistan since early Mughal invasion and is an important fruit crop of the region. Although there are no commercial plantations of dates in India at present, quite a number of trees are, however, found wild or cultivated mostly in the dry districts of Gujarat, Rajasthan, Punjab, Uttar Pradesh, Madhya Pradesh, Andhra Pradesh, and Karnataka which generally yield fruit of inferior quality. Of late, some varieties of dates have been introduced for trial cultivation at Abohar in Punjab and Khedio in Gujarat. Successful varieties are proposed to be introduced in places suitable for cultivation.

Date palm plantations are interplanted with a variety of crops. During the early stages, wheat, barley, vegetables, leguminous fodders, and pulses that do not damage the palm by overshading can be intercropped. Later, when the palms have grown sufficiently tall, low-growing fruit crops such as grapes, pomegranates, figs, peaches, apricots, oranges, etc. are interplanted.

Considerable quantities of dates are consumed in India that are imported from Iraq, Oman, Iran, and Saudi Arabia. They are sold in two forms in the market, viz. soft fleshy dates and the hard dry dates. These are eaten as restorative or added to sweet dishes. Dates are rich in sugars and eaten fresh or dried. They are widely used in bakery and confectionery. Sometimes they are made into jams or added in cakes and dishes with milk, butter, meat, and so on; a type of honey is also prepared from them. Dates are considered demulcent, expectorant, and laxative and are used in respiratory diseases and fever. They are also fed to cattle, sheep, and goats.

Date seeds are ground by soaking in water and used for feeding camels, goats, sheep, and horses; they are also used as a substitute for barley in poultry feed. In South India, a paste made from the seeds by trituration with water is applied to cure eye diseases. Roasted seeds are also used as an adulterant of coffee while small bits of seeds are adulterated in areca-nut bits.

Date palm is tapped for its sugary sap that is usually done by making a circular cut around the base of the exposed terminal bud and running the exuding sap through a spout into the collecting vessel. The yield of sap per palm is estimated at about 10 L/day. The sap has a sweet taste, pleasing flavor, and good nutritive value besides containing about 10% sucrose content. It is also cooling and acts as a laxative. The sap is also used in the preparation of jaggery, sugar, and vinegar. Fermentation of sap leads to the formation of 4%–5% alcohol in about 6–8 h. The fermented liquor (toddy) is slightly acidic and has a pleasant tang.

Date palm is one of the most useful trees and almost all parts are used for various purposes. The leaves are used for thatching, making mats, fans, baskets, ropes, and so on. From the petioles, light walking sticks, crates, boxes, and fishing floats are made. The leaves, especially the petioles, yield a fiber which together with fillers may be used in the production of insulating boards. The fiber extracted from the covering of the sheathing bases of the petiole is made into cordage.

Green spathes also yield a fiber used for making ropes. The terminal leaf bud is eaten raw or cooked as a vegetable. From the fresh spathes, a pleasantly aromatic liquid is obtained by steam distillation and used for making *sherbet*. The light and fairly durable wood is used for house construction, bridges, and water conduits. Pruned leaves and fruit stalks are used as manure.

Fatty oil is extracted from the dried pollen; the nonsaponifiable fraction of the fatty oil yielded an estrogenic substance. The pollen exhibited a gonadotropic activity in immature rats. The presence of cholesterol and estrone was also reported from pollen.

Phoenix humilis Royle.

It is a handsome palm, locally known as dwarf date palm or hill date palm, that grows up to 3 m in height, with bulbiferous stems and is found almost throughout India mainly in hilly tracts up to an altitude of 2000 m. This plant yields sago and the pith from the top of the stem is eaten. The fruits are sweet and contain a small amount of pulp. The leaves are used as cordage and for making mats, baskets, fancy hand-bags, and so on. The wood is light brown and somewhat hard.

Three varieties of this palm are reported to occur in India. They are *Phoenix humilis* var. *humilis* (found in the Himalayas, Assam, and Northern and Central India); *Phoenix humilis* var. *pedunculata* Becc. (distributed mainly in the hills of Deccan Peninsula); and *Phoenix humilis* var. *loureirii* (Kunth) Becc. (occurs in Khasi hills and Manipur).

Phoenix paludosa Roxb.

It is a handsome palm found in the coastal swamps of West Bengal, Orissa, and Andaman Islands. Leaves spread up to 3 m long with fruits of ovoid berry shape and black purple color when ripe. The leaves are used for thatching and making ropes and fences. The stems of the smaller plants are used as walking sticks while larger ones serve as rafters. The fruits possess cooling and anti-inflammatory actions. The pith of the stem is also eaten.

Phoenix pusilla Gaertn. (Syn: *P. farinifera* Roxb.)

This is a shrubby species, widely known as "small wild date palm," and is found on the Coromandel coast and in the dry forests of Kerala, Karnataka, and Tamil Nadu at low elevations (Warriar et al.,

1995). The stem, without leaves and outer woody fibrous covering, gives a farinaceous material when pieces of it are ground in a mortar. This is sifted, boiled into a thick gruel, and consumed in times of famine. The young terminal leaf bud is also eaten and the edible fruits contain a small amount of sweet pulp. The latter are prescribed in respiratory diseases and fever. The fruits are also considered to be laxative, cardiotonic, aphrodisiac, carminative, and useful in general debility. The seed paste is triturated with water and used in treating the inflammation of eyes. Fresh sap is cooling and laxative while the gum exuded from the tree is used in diarrhea and urinary disorders (Caius, 1934). The leaves are used for thatching and making coarse mats and as a fuel for potteries. The petioles are split and woven into baskets.

Phoenix sylvestris Roxb.

It is a graceful and widely growing tree popularly called wild date palm or date sugar palm. It grows up to 10–16 m high with a large crown and is found throughout India up to an altitude of 1500 m. This is commonly seen in many parts of India, either wild or cultivated, and it may be amazing to know that there are about 29 million trees spread all over the country. They are generally found growing on the borders of the fields. The tree flowers in the beginning of the hot season and the fruits ripen in September–October. It does not produce offshoot or suckers (Davis, 1967).

Wild date palm is tapped for its sap that is converted into palm jaggery and sugar. The yield of sap per palm varies from 70 to 310 kg in a season giving 10%–13% of its weight as jaggery. It is also used to make vinegar to a smaller extent. Fresh sap contains protein, sugars, phosphorus, calcium (trace), iron, thiamine, riboflavin, ascorbic acid, nicotinic acid, *iso*-nicotinic acid hydrazide, and organic acids. Fresh unfermented sap known as *Nira* is a refreshing sweet drink and is also used to prepare jaggery and sugar. *Nira* is a rich source of vitamins of B group and also contains a good amount of vitamin C. The concentration of the latter ranges from 7.1 to 12.5 mg/100 mL.

The sap is converted into a common beverage known as *toddy* after spontaneous fermentation for 8–10 h. Toddy contains sucrose, reducing sugar, alcohol, glycerol, nitrogen, and phosphorus. It possesses a higher growth-promoting ability and nourishes microorganisms that synthesize and secrete some vitamins during growth. It is a nutritious drink mainly for those suffering from thiamine deficiency.

Palm jaggery is more nutritious than cane sugar. The former contains vitamins such as thiamine, riboflavin, nicotinic acid, and ascorbic acid and minerals such as calcium, phosphorus, iron, sodium, potassium, copper, cobalt, nickel, magnesium, and molybdenum. The major portion of the total nitrogen in the palm jaggery is present in a nonprotein fraction that is a good source of amino acids such as lysine, phenyl alanine, leucine, threonine, *iso*-leucine, tyrosine, and cystine.

The fruits are edible and are considered to be restorative. The flesh constitutes about 60% of the whole fruit and contains proteins, fat, carbohydrates, fiber, calcium, and phosphorus. The fruits are also used for the preparation of vinegar (100 kg of fresh fruits yield about 135 L of vinegar).

The leaves of wild date palm are widely used for thatching, making mats, fans, baskets, bags, brooms, fishing nets, and so on. They also give a soft fiber and the petioles are beaten and made into ropes. The leaf bud and the inner portion of the stem are eaten and the leaves are also lopped for fodder. The tree yields a light-brown durable wood which is used for temporary construction, bridges, and piers, and for tent pegs. Trunks removed from pith form excellent water conduits. The tree also yields a gum and the bark contains tannin. The roots are reported to relieve toothache.

Phoenix robusta Hook. F.

This is a palm with a trunk up to 6 m in height and about 0.5 m in girth, found in Bihar and parts of Deccan Peninsula up to an altitude of 1700 m. The fruit is black, sweet, and edible.

Phoenix rupicola T. Anders.

It is a graceful and slender palm with trunk up to 7 m in height and about 20 cm in diameter, found in eastern Himalayas and Mishmi hills in Assam, and also cultivated in gardens. The pith of the stem is farinaceous and eaten.

In addition to the above species of *Phoenix*, the following ones that are native of other countries also grow in Indian soil.

Phoenix canariensis Chabaud.

This tree, a native of Canary Islands, is a fast-growing ornamental palm. It grows up to 15 m high and is also cultivated in Indian gardens. It is a stately tree suitable for growing along avenues and borders of parks and lakes.

Phoenix reclinata Jacq. (Syn: *P. spinosa* Thonn.)

It is a native of tropical Africa and is cultivated in Indian gardens as an ornamental tree. It grows to a height of 12 m with several stems and feathery leaves. The terminal leaf bud is consumed as a vegetable. The sap obtained by tapping near the roots becomes intoxicating on fermentation.

Phoenix zeylanica Trimen

This palm is a native of Sri Lanka and grows up to 7 m height. It is cultivated in Indian gardens and bears a sweet and edible fruit. In times of scarcity, the pith of the stem is eaten.

INDIGENOUS MEDICINES OF INDIA

The indigenous medical systems form part of Indian culture and tradition. The early Indians were experts in 64 arts that include medicine besides mathematics, astronomy, astrology, architecture, and so on. The presence of rich and unlimited wealth of medicinal herbs and minerals in the country proved exceptionally favorable to the growth and development of medical science. In the course of time, this became a teaching subject in the universities of Takshashila and Nalanda and students from distant South Asian countries also learnt this science (Dharani, 2006). Recent researches in the Western universities further confirm the validity of the principles and drugs used in these systems in the treatment of various diseases (Anonymous, 2011a).

The Indian medical systems enjoyed high prestige and respect from the rulers and elite who patronized them for their growth and development. During this time, they enjoyed universal acceptance as they were the only available recourse for care and cure of the health problems of the rulers and the people alike.

AYURVEDA

Ayurveda is a great treasure-house of medical wisdom compiled with profound experience and knowledge accumulated over thousands of years. During the very early period of this phase, Ayurveda came out of its original magico-religious status into a rational science through the application of continuous systematic observation and empirical analysis.

The term "Ayurveda" means "science of life." It describes in detail the measures to be undertaken during the entire lifespan as well as in its various phases. Apart from dealing with the principles for maintenance of health, it has also developed a wide range of therapeutic measures to cure illness. These principles of health-promoting measures relate to physical, mental, and spiritual welfare of human beings. Thus, Ayurveda becomes one of the oldest systems of medicine dealing with both preventive and curative aspect of life in a most comprehensive way.

The most important and massive ancient compilation of the school of medicine is known as *Charka Samhita*. It contains several chapters dealing at length with therapeutics. About 600 drugs of herbal, mineral, and animal origin are described in it. In addition, this compendium also deals with other branches of Ayurveda such as anatomy, physiology, etiology, prognosis, pathology, treatment, and medicine.

Treatment of the disease generally consists in avoiding causative factors. Administration of suitable medicines and proper diet are recommended for restoring the balance and strengthening the body mechanism. While treating the disease, the whole body is taken into account for a thorough investigation and hence it is said that Ayurveda treatment is done on the basis of a holistic approach. The treatment part in Ayurveda is individualized, and in addition to giving appropriate drugs, it also considers other factors such as diet, sleep, work, rest pattern, environment, age, sex, appetite, strength, and vitality.

Dates in Ayurveda

The date palm plant is mentioned in Ayurveda as *Kharjuri* while the fruit is denoted by the name *Kharjura*, and *Kharjura Mastaka* or *Kharjura Majja* is the edible part of the stem. Ayurveda also recognizes *Phoenix acaulis*, *P. dactylifera*, *P. sylvestris*, and *P. umilis* as different varieties of Kharjura available in India (Sastry, 1984).

Various anatomical parts of the date palm are mentioned to have medicinal values in Ayurvedic literature (Warriar et al., 1995):

Leaves: aphrodisiac, antihepatotoxic
Flowers: purgative, expectorant, febrifuge
Fruits: cooling, aphrodisiac, tonic, diuretic, bronchitis, cough
Seeds: paste in opacity of cornea, cephalalgia, inflammation, wounds
Toddy: aphrodisiac, digestive, tonic, appetizer

The fruits of date palm are also used in the formulation of some Ayurvedic compound drugs which are mostly prescribed as general tonic and in nervous debility (Anonymous, 2000a):

Dhraaksha Arishta: restorative, cardiotonic, appetizer, digestive, alterative, hemorrhoids, mild laxative
Amrutaprasa Ghrita: restorative, nervous debility, anemia, convalescence

SIDDHA MEDICINE

Siddha system is another oldest system of medicine of India. This system was developed by saintly figures known as *Siddhars* who achieved eternal bliss by their acts known as *Siddhis*. This system was developed in the Dravidian culture which is of the pre-Vedic period. The aim of Siddha medicine is to make the body perfect and promote longevity. The emphasis is on positive health and the object of medicine is to prevent by careful dieting and relaxation of the body and the mind.

There is supposed to be a relationship between the composition of the drugs, the human body, and the universe. Like the universe and drugs, the human body is also a manifestation of five fundamental elements and the latter guides the physician in his choice of appropriate treatment suited to each disease.

In Siddha medicine, naturally available materials such as metals, minerals, animal products, and plants are used to treat diseases. In search of medicine that could offer eternal youthfulness, Siddhars did extensive research on naturally available materials. They aimed at evolving drugs that would arrest decay of the body and in the process ended up in discovering formulations that do not lose their potency with time and could be administered in small and convenient doses, besides being available all seasons and preserved for a long time.

Dates in Siddha Medicine

The fruits and other anatomical parts of both *Phoenix sylvestris* and *Phoenix farinifera* are used in Siddha medicine either as such or in the preparation of drug formulations in combination with other plant products (Mudaliar, 1988).

Phoenix dactylifera

The fruits of this species are used in Siddha medicine under the name *Karchur kaay*. Besides fruits, its resin, seed, toddy, shoot, jaggery, and sugar are mentioned to have various medicinal properties in the Siddha literature (Pillai, 1978):

Ripe fruits: febrifuge, refrigerant, expectorant, laxative, and aphrodisiac; daily intake of fruits soaked in honey prevents chest pain. Also used to treat fever, thirst, constipation, diabetes, tastelessness, and jaundice
Unripe fruits: stomachic, carminative, antidiarrheal, and aphrodisiac; controls thirst, excess secretion of bile, saliva, and urinary output
Terminal shoot: ulcer, dyspepsia, vomiting, and gushing of abdomen

The fruits also enter into the preparation of the following Siddha formulations that are used to treat various disease conditions (Anonymous, 2000b; Anonymous, 2011b):

Sukku Tailam: sinusitis (external use)
Drakshadhi Kashayam: general tonic and antitussive
Pitta Sura Kudineer: antipyretic
Gandhaka Rasayanam: skin disorders
Parangi Rasayanam: chest pain, syphilis, skin diseases, leucorrhea, and gonorrhea
Kumari Lehyam: aphrodisiac and uterine disorders
Narasimha Lehyam: arthritis and neuromuscular disorders
Pericham Ney: urogenital discharge, ascities, and aphrodisiac
Thanneer Vittaan Ney: nervous debility, urinary calculi, inflammation in genitourinary tract, leucorrhea, urethritis, etc.
Maha Vilvadi Lehyam: upper respiratory diseases, anemia, gastrointestinal disorders, and cachexia
Poornadi Lehyam: piles, hepatoprotective, and antipyretic
Drakshadhi Churanam: cough, bronchitis, tastelessness, diabetes, fever, vomiting, gonorrhea, and tuberculosis

Phoenix farinifera (Syn: *P. pusilla*)

This species is mentioned in the Siddha literature as "small date" and besides its fruit, the resin, seeds, toddy, shoot, jaggery, and sugar find use in medicare (Mudaliar, 1988):

Fruits: diuretic and laxative
Resin: suspension in water used to treat diarrhea and urinary problems
Seed: paste made with water or milk used in eye infection
Toddy: ulcer, urinary disorders, and inflammation

UNANI MEDICINE

The Unani system of medicine originated in Greece between 460 and 377 BC. Later, it was extensively adopted and developed by the Arabs. It reached India around the tenth century AD with Mughal rulers and became indigenous to the country. From the history, it is understood that almost all rulers of the Sultanate period were practicing Unani medicine for the benefit of the people.

The Unani system of medicine is based on the humoral theory according to which there are four humors. It laid down a set of essential prerequisites for the prevention of diseases. They are collectively known as *Asbab-e-sitta zarooriya* which are air, drink, food, body movement, mental activity, sleep, wakefulness, excretion, and retention.

There are various types of treatments in Unani systems:

Regimental therapy
Diet therapy
Pharmacotherapy
Surgery

Diet therapy aims at treating certain ailments by the administration of specific diets while pharmacotherapy deals with the use naturally occurring drugs, mostly plant based.

DATES IN UNANI MEDICINE

In Unani medicine, the date plant is known by the names *Khajore*, *Khurma*, and *Chuhara*. Four types of dates are used in the system, viz. *Kalima Khajore*, *Ajwa Khajore*, *Amber Khajore*, and *Kala Khajore*. The parts used in the medicine are resin (gum), fruit pulp, seeds, flowers, heart wood, and the root. They are mentioned to have different medicinal properties as mentioned below (Anonymous, 2006):

Resin and fruit pulp: anti-inflammatory, carminative, aphrodisiac, liver tonic, antiobesity, and hematinic
Seeds: general tonic, antidiarrheal, and hemostatic. Powder of seeds cleanses the wounds, possesses antiulcer activity and is used as tooth powder
Flowers: stomachic, antidiarrheal, refrigerant, hemostatic, quenches thirst, appetizer
Heartwood: chronic gonorrhea
Roots: used as natural toothbrush; fine powder or decoction used in toothaches

The fruits of date palm also enter into the following compound drug formulations that are prescribed in the treatment of different diseases (Anonymous, 2004):

Majoon-e-Aard Qurma: aphrodisiac
Majoon-e-Laboob: aphrodisiac
Halwa-e-Magze sare Kanjeshk: aphrodisiac and spermatogenesis
Majoon-e-Suparipak: aphrodisiac, leucorrhea, and uterine disorders
Halwa-e-Gheekwar: lumbago and pain
Majoon-e-Muraveehul Arvah: aphrodisiac and liver tonic

NATUROPATHY

Naturopathy is mainly a way of life and not a system of treatment. It is based mainly on the application of simple laws of nature. Naturopathy also forms part of Indian systems of medicine and its fundamental principles are similar to Ayurveda. It pays particular attention to eating and living habits, adoption of purificatory measures, use of hydrotherapy, cold packs, mud packs, baths, massages, and so on.

Dates in Naturopathy

Naturopathy advocates the consumption of natural materials like plants, fruits, and so on in their fresh form without subjecting them to any type of processing such as frying, cooking, and so on. Limited intake of dry fruits and nuts are recommended for diet to keep the body active and free from diseases. Among dry fruits, dates form an important naturopathic food especially to break the fasting together with fresh fruit juice.

CONCLUSION

The genus *Phoenix* plays a significant role in the indigenous medical practices of the Indian subcontinent since many centuries, disproving the notion that date species are popular only in Arab countries. The date palm that is considered to be the fruit from heaven and gift to man by God. Probably in the early days, the plant might have been widely grown in many parts of the world and due to changing climatic conditions, natural calamities, and environmental degradation, it is now confined in very large numbers to the Middle East and parts of a few more countries. The nations embarking on agricultural revolution with the help of technological advancements to meet the food demand may come forward in future to take up large-scale cultivation of date palm for its multipronged uses.

ACKNOWLEDGMENTS

The author thanks Dr. T. Anandan, assistant director and head, Dr. T. Apparanantham, former research officer, Central Research Institute for Siddha Medicine, Chennai, and Dr. (Mrs.) Chand Usmani, vice-principal, Government Unani Medical College, Chennai for useful discussions and Dr. S. Viswanathan, professor of Pharmacology, Meenakshi Medical College and Research Institute, Kanchipuram for critically reviewing the manuscript.

REFERENCES

Anonymous, 2000a. *Vaidya Yoga Ratnavali (Formulary of Ayurvedic Medicine)*, Chennai: Indian Medical Practitioners' Co-operative Pharmacy and Stores.

Anonymous, 2000b. *Formulary of Siddha Medicines*, Chennai: Indian Medical Practitioners' Co-operative Pharmacy and Stores.

Anonymous, 2004. *Formulary of Unani Medicine*, Chennai: Indian Medical Practitioners' Co-operative Pharmacy and Stores.

Anonymous, 2005. *The Wealth of India—Raw Materials*, Vol. VIII, New Delhi: Council of Scientific and Industrial Research.

Anonymous, 2006. *National Formulary of Unani Medicine*, Part IV, New Delhi: Ministry of Health and Family Welfare, Government of India.

Anonymous, 2011a. *The Hindu*, July 20.

Anonymous, 2011b. *The Siddha Formulary of India*, Part II, New Delhi: Ministry of Health and Family Welfare, Government of India.

Caius, P. 1934. The medicinal and poisonous palms of India. *Journal of Bombay Natural History Society*, 37:915–941.

Davis, R. 1967. *Science Reporter*, 4:70.

Dharani, D. 2006. *Historical Perspectives of the Practice of Indigenous Medicines in Tamil Nadu with Special Reference to Siddha Medicine*, PhD dissertation, Chennai: University of Madras.

Mudaliar, M. 1988. *Siddha Materia Medica*, Part I, 4th Ed, Madras: Directorate of Indian Medicine, Government of Tamil Nadu.

Pillai, T.V.S. 1978. *Tamil–English Dictionary of Medicine, Chemistry, Botany and Allied Sciences (Based on Indian Medical Science)*, Vol. 5, Madras: Government of Tamil Nadu.

Sastry, J.L.N. 1984. *Dravyaguna Vijnana—Study of Essential Medicinal Plants in Ayurveda*, Varanasi: Chaukhambha Orientalia.

Warriar, P.K., Nambiar, V.P.K., and Kutty, C.R. 1995. *Indian Medicinal Plants—A Compendium of 500 Species*, Part 4, Hyderabad: Orient Longman Limited.

29 Antiamyloidogenic Effect of Dates Grown in Oman with Reference to Their Possible Protection against Alzheimer's Disease

M. Mohamed Essa, Gilles J. Guillemin, Samir Al-Adawi, Abdullah Al-Asmi, Ragini Vaishnav, Nandhagopal Ramachandiran, and Mustaq A. Memon

CONTENTS

INTRODUCTION

Date palm tree (*Phoenix dactylifera* L.) is an important crop, which is cultivated in many countries extending from North Africa to the Middle East including many of the Arabian Gulf Cooperation Countries (GCC) [1]. The production of dates has been increased worldwide from 2.8 million tons in 1985 to 6.8 million tons in 2003 [2]. Oman produced about 240,000 metric tons in 2005 which contributed to around 3.5% of date global production [3]. Dates are frequently consumed among the Arabian people due to the awareness of its high nutritive value and are chemically composed of sugars (81%–88% mainly fructose, glucose, and sucrose), dietary fiber (~5%–8.5%) and small amount of protein, fat, ash, and high quantities of phenols [4].

The dietary fiber content of the dates has been involved in many therapeutic implication conditions such as diabetes, hyperlipidemia, and obesity and may exhibit a protective effect against hypertension, coronary heart disease (CHD), cholesterol, colorectal and prostate cancers, and intestinal disorders [2]. In addition to date's nutritional value, date fruits are a good source of phenolic compounds which is related to its high antioxidant activity. Antioxidants are known to have a preventive role against free radicals and therefore may contribute in the prevention of degenerative diseases such as cancer, inflammation, aging, and atherosclerosis [5]. Therefore, the regular consumption of these fruits and vegetables rich in phenolic compounds is usually associated with lower risk of neurodegenerative disorders and better cognitive performance in the elderly [6].

Alzheimer's disease (AD) is one of the progressive neurodegenerative disorders and a leading cause of dementia which is characterized by cognition and memory impairment [7]. Data on the exact incidence and prevalence of AD in Oman is scarce. However, as Ministry of Health estimated, the number of patients diagnosed with neurological diseases was 415 in 2008 (MOH, Oman, 2008) [8]. This number is expected to increase in the coming years as the lifestyle of people is changing and many diseases such as diabetes mellitus, hypertension, cancer, and heart disease have been on the rise and also an increase in road traffic accident, which may double the chance leading to AD.

The pathology of the AD is characterized by neurotic plaques and neuro-fibrillary tangles, combined with massive neuronal loss. The major constituents of senile plaques are 39–42-amino-acid peptides, cleaved from a larger protein called β-amyloid precursor protein (APP) [9]. Thus, the formation of the peptide, known as amyloid-beta protein (Aβ), clusters into amyloid plaques on the blood vessels and outside the neurons of the brain leading to neural death. The mechanism of the aggregation of the Aβ in patients with AD is poorly understood [10]. The amyloid hypothesis of AD envisages generation of insoluble, fibrillogenic amyloid peptides constituting the core of neuritic plaques (one of histopathological hallmarks of AD) as one of the key steps in the disease pathogenesis.

Although there are currently no proven ways to delay the onset or slow down the progression of AD, evidences suggest that diet can influence the risk of developing AD and alter Aβ levels. It follows that if the fibrillogenic nature of Aβ could be circumvented by agents or nutrients, disease prevention or amelioration could possibly be translated into reality. The key step, then, involves identifying these agents that promote solubility or arrest fibrillogenesis of toxic Aβ. Using *in vitro* thiofavin T assay, we found certain Omani fruits (pomegranate, figs, and papaya) to have such potentials (unpublished observation). In addition, antioxidant contents of these fruits may also play an important disease-modifying effect, as oxidative stress could be a mediator of cellular damage as mentioned above [11,12]. Emerging evidence suggests that the Mediterranean diet (consisting of fruits and vegetables) plays a beneficial role to reduce the development and progression of dementia [13–17]. Researchers have utilized curcumin from turmeric, pomegranate juice, blue berry juice, fruit and vegetable juices, and walnut in animal models and or human subjects with AD and have demonstrated clinical benefits with better cognitive performance [18–22].

Reports suggest that the natural products that promote solubility and inhibit *in vitro* fibrillation of Aβ may offer benefit with respect to memory functions in older adults at risk of dementia and enhance memory and motor performance in aged animals [23–26]. We propose that natural antioxidants and other protective principles (including antiamyloid contents) in the fruits might possess disease-modifying or neuro-protective effect in these types of disorders with cognitive decline. There are no previous studies exploring the benefits of Omani date palm fruits against Aβ fibrillization *in vitro*. To fill the information gap, the current study was performed. The main objective of this study was to investigate the effect of three Omani date varieties (Fardh, Kasab, and Khalas) on the *in vitro* fibrillization of the soluble Aβ with respect to their protection against AD.

MATERIALS AND METHODS

FRUITS AND EXTRACTION

Omani date palm fruits (Fardh, Kasab, and Khalas) were purchased from local farmers from Nakhal region, Oman. The edible part of dry dates (50 g) was taken and cut into small pieces with a sharp knife. Dates were crushed in 150 mL methanol:water (4:1) using a blender (Philips, HR 1737, Brazil). The extract was mixed by the stirrer at room temperature (25°C) for 15 h. The extracts were centrifuged (Sanyo, Harrier 18/80, refrigerator, MSE [UK]) at 6000 r/min for 30 min at 3°C. The extracts then were filtered and concentrated under reduced pressure at 40°C for 3 h using rotary evaporator (R-215, vacuum-controlled V850 with heating bath B-491). The crude extracts were stored at −40°C for later use.

Synthetic Aβ 1–42 was purchased from Anaspec Inc, USA and all other chemicals were from Sigma Aldrich USA. Soluble Aβ was prepared according to the method described previously [27] with modification. Synthetic Aβ was mixed with 70 μL of NH_4OH and vortexes for 1 min. Soluble Aβ (250 μM) was obtained by dissolving synthetic Aβ in 886 μL of 0.01 M phosphate-buffered saline solution (P-5368 PBS, Sigma Chemical Company) at pH 7.4.

Total Phenolic Content (Folin-Ciocalteu Assay) and Total Flavonoids Assay

Total phenolics were determined using the Folin–Ciocalteu method as described [28] with modification. The total phenol concentrations were expressed as gallic acid equivalents GAE (mg/g of dry fruit). Total flavonoids were determined as described previously and the concentrations were expressed as catechin equivalents CE (mg/g of dry fruit).

Effect of Date Palm Fruit on the Fibrillization of Soluble Aβ

In the experimental samples, 10 μL of soluble Aβ (40 μg) was incubated with different concentrations of date palm extract (DPE) (50 μg of GAE, 100 μg of GAE) in a total volume of 50 μL. The control sample had Aβ in buffer, and the blank sample had DPF in Tris-buffer (no Aβ). The degree of Aβ fibrillization in these samples was examined by Thioflavin T (ThT) fluorescence spectroscopy after 72 h of incubation. This assay is used to assess the aggregation/fibrillization of Aβ. ThT does not bind to monomers, dimers, or tetramers, but binds to higher-order aggregate [29]. The reaction mixture contained 50 μL of sample, 2 μM ThT, and 50 mM glycine-NaOH buffer, pH 9.0 in a total volume of 1 mL. Fluorescence intensity was measured at an excitation wavelength of 435 nm and emission wavelength of 485 nm. Data analysis: Statistical analysis was done by using Graph Pad Prism 5 software.

RESULTS AND DISCUSSION

Total Phenolic and Flavonoid Contents of Dates

Three Omani date varieties (Fardh, Kasab, and Khalas) were analyzed for their phenolic content and they presented a range of 194.3–234.6 mg of GAE/ 100 g of dry fruit. Fardh and Khalas varieties have more or less similar amount of polyphenols, while the Kasab variety had the lowest value as compared with Fardh and Khalas. The total flavonoids of Omani date varieties ranged from 24.28 to 33.49. Highest levels of flavonoids found in Fardh than Kasab and Khalas (Table 29.1).

Studies on other Omani date varieties had reported phenolic content values between 172 and 246 mg/100 g fresh weight [3]. Results showed that Omani dates had the highest level of phenolic content compared with dates cultivated in Iran (2.49–8.36 mg/100 g fresh weight) [5]. Date fruits are normally exposed to various environmental/nonenvironmental factors such as variety, growing conditions, maturity, season, geographic origin, fertilizers, soil type, storage conditions and amount

TABLE 29.1
Total Phenolic and Flavanoids Content of Three Omani Date Varieties

Omani Date Palm Fruits Variety Names	Total Phenols (mg of GAE/100 g)	Total Flavonoids (mg CE/100 g)
Fardh	234.6 ± 0.38	33.49 ± 0.49
Kasab	194.3 ± 0.54	27.29 ± 0.92
Khalas	231.2 ± 0.50	24.28 ± 0.09

Note: Data expressed in means ± SD ($n = 3$).

of sunlight received, temperature, humidity, moisture content, and so on. These factors have been associated with the observed differences in the content of polyphenols [3,5].

Inhibition of Aβ Fibril Formation by Omani Date Extracts

The effect of different concentrations of date extracts on the fibrillization of soluble Aβ (after 3 days of incubation) was assessed by Thioflavin T fluorescence spectroscopy. Results showed that date extracts had significantly inhibited aggregation/fibrillation of Aβ in a concentration-dependent manner (Figure 29.1).

Our results show that the maximum inhibition of Aβ fibril formation was offered by the Khalas variety extract in a dose-dependent manner (74.6% and 84.8%). Fardh (63% and 57%) and Kasab (58.6% and 41%) also offered inhibition of Aβ fibril formation, respectively, in dose-dependent manners.

Dietary approaches hold promise as effective and safe preventive interventions. Dietary factors represent the most potent determinants of metabolic health and have been shown to mitigate specific mechanisms of neurodegeneration [30]. Polyphenol consumption is important in this regard, and epidemiological studies indicate that the consumption of fruits and vegetables is associated with lower risk of neurodegenerative disorders and better cognitive performance in the elderly [31]. Recent studies suggest that blueberry supplementation beneficial with respect to memory functions in older adults at risk of dementia and enhances memory and motor performance in aged animals [23–26]. Foods containing high levels of antioxidants may also slow down the progression of AD, possibly by preventing or neutralizing the damaging effects of free radicals [32,33]. The essential fatty acids contained in fish oil (e.g., docosahexaenoic acid/DHA) and vitamin E may be neuroprotective in humans and animal models and epidemiological evidence suggests that high intake of food-based vitamin E/DHA is associated with a lower incidence of AD in humans [32–35].

Phytochemicals have beneficial effects on health and phytochemicals like polyphenols have been shown to have antioxidant properties and to suppress inflammatory and other pathways [36–37]. Phytochemicals such as curcumin and epigallocatechin-3-gallate were shown to offer neuroprotection in APPsw mice by lowering the plaque burden and oxidized protein levels [37–38]. Studies suggested that dietary supplementation with fruit extracts high in antioxidants (e.g., blueberries, strawberries, walnuts, and Concord grape juice) can decrease the enhanced vulnerability to oxidative stress that occurs in aging and these reductions are expressed as improvements in behavior and memory [25].

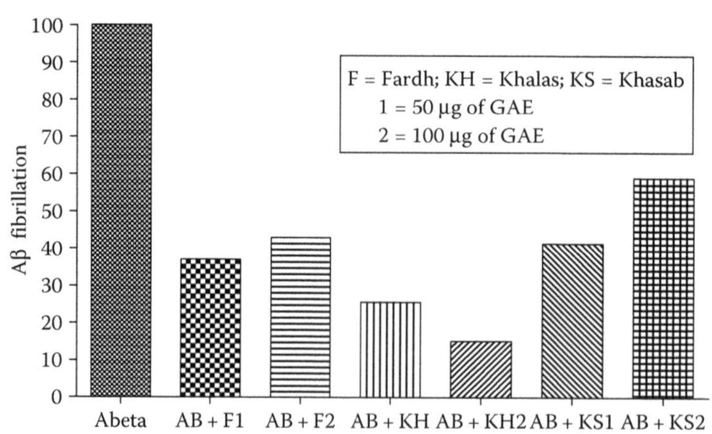

FIGURE 29.1 Percentage of Aβ fibrillation and inhibition by Omani date varieties.

Additional mechanisms involved in the beneficial effects of fruits and vegetables include enhancement of neuronal communication via increases in neuronal signaling and decreases in stress signals induced by oxidative/inflammatory stressors [25]. Blueberry or Concord grape juice supplementation in humans with mild cognitive impairment increased verbal memory performance, thus translating animal findings to humans [25]. Recent studies conducted by the authors in the United States suggest that dietary supplementation of walnut improves cognition and memory in AD transgenic mice and offers cyto-protection against Aβ-induced cytotoxicity in PC-12 cells [22,39]. Taken together, studies indicated that individuals consuming a diet containing high amounts of fruits and vegetables exhibit fewer age-related diseases such as cognitive decline, dementia, in AD, and other related conditions. Diet may be associated with risk of cognitive decline, dementia, and AD. Numerous studies suggest that a greater intake of high-antioxidant foods may increase "health span" and enhance cognitive and motor function in neurodegenerative diseases and aging. A recent study conducted in Sweden suggests that midlife fruit and vegetable consumption may reduce the risk of dementia and AD [40]. A discovery such as this would greatly benefit millions of people and help in warding off the AD and dementia epidemic that is upon us. The present proposed study is one such step toward that direction.

Sultanate of Oman is the eighth-largest producer of date palm fruit (*P. dactylifera* L.) in the world. USDA nutrient database reported that dates are good sources of minerals (Se, Cu, K, and Mg), vitamins (C, A, B6, B9, B2, B3), and dietary fiber. Furthermore, dates can be considered to be a good source of total phenolics and natural antioxidants (such as anthocyanins, ferulic acid) [41,42]. Date palm fruit can be considered as an antiatherogenic and a novel antiviral agent [43,44].

There are, in total, 12 different varieties of date palm fruits grown in Oman and we have identified that an extract of certain Omani fruits (pomegranate and figs) promotes solubility and inhibits *in vitro* fibrillation of beta amyloid (unpublished data). The conversion of fibrillar Aβ to its nonfibrillar form and inhibition of Aβ fibril formation by diet are considered to be important strategies for treating and preventing AD and other neurodegenerative diseases. Our results suggest that moderate consumption of fruits such as dates grown in Oman over the routine intake might offer protection to AD. There are no such studies conducted by using fruits grown in Oman against AD and this is the first study related with this in Oman. The outcome of the study will explore the medicinal importance of Omani fruits around the globe and also gives economic impact to Oman.

In conclusion, the active components present in Omani date palm fruits (Khalas, Fardh, and Kasab) have an effect against the fibrillation of Aβ protein and this could be able to give a novel lead in developing new therapeutic drugs that might be helpful in the treatment and or prevention against AD. But the exact mechanism is still unclear and further extensive cell culture studies, preclinical and clinical studies, are needed to investigate whether the supplementation of date palm fruits could prevent, delay the onset, or reduce the risk of AD by keeping the Aβ in soluble form.

ACKNOWLEDGMENTS

This work was partially supported by the financial assistance provided by the research Council Oman (RC/AGR/FOOD/11/01) and by the Dean's seed grant, CAMS Sultan Qaboos University, Oman. The help provided by Mrs Vandita Singh, for total polyphenols and flavonoids analysis is gratefully acknowledged.

REFERENCES

1. Allaith, A.A. 2008. Antioxidant activity of Bahraini date palm (*Phoenix dactylifera* L.) fruit of various cultivars. *Int. J. Food Sci. Technol.* 43:1033–1040.
2. AL-Farsi, M., AL-Asalvar, C., Morris, A., Baron, M., Shahidi, F. 2005. Compositional and sensory characteristics of three native sun-dried date (*Phoenix dactylifera* L.) varieties grown in Oman. *J. Agric. Food Chem.* 53:7586–7591.

3. Al-Farsi, M., Alasalvar, C., Al-Abid, M., Al-Shoaily, K., Al-Amry, M., Al-Rawahy, F. 2007. Compositional and functional characteristics of dates, syrups, and their by-products. *Food Chem.* 104:943–947.
4. Elleuch, M., Besbes, S., Roiseux, O., Blecker, C., Deroanne, C., Drira, N., Attia, H. 2008. Date flesh: Chemical composition and characteristics of the dietary fiber. *Food Chem.* 111:676–682.
5. Biglari, F., AlKarkhi, A., Easa, A.M. 2008. Antioxidant activity and phenolic content of various date palms (*Phoenix dactylifera*) fruits from Iran. *Food Chem.* 107:1636–1641.
6. Scarmeas, N., Stern, Y., Mayeux, R., Manly, J.J., Schupf, N., Luchsinger, 2009. Mediterranean diet and mild cognitive impairment. *Arch. Neurol.* 66(2):216–225.
7. Li, M., Chen, L., Hong Seng Lee, D., Yu, L., Zhang, Y. 2007. The role of intracellular amyloid b in Alzheimer's disease. *Progr. Neurobiol.* 83:131–139.
8. Ministry of Health Report, Oman 2008.
9. Zawia, N.H., Lahiri, D.K., Cardozo-Pelaez, F. 2009. Epigenetics, oxidative stress, and Alzheimer disease. *Free Radic. Biol. Med.* 46:1241–1249.
10. Chauhan, A., Pirttila, T. et al. 1998. Aggregation of amyloid b-protein as function of age and apolipoprotein E in normal and Alzheimer's serum. *J. Neurol. Sci.* 154:159–163.
11. Behl, C., Davis, J.B., Lesley, R., Schubert, D. 1994. Hydrogen peroxide mediates amyloid β protein toxicity. *Cell* 77(6):817–827.
12. Pappolla, M.A., Simovich, M.J. et al. 2002. The neuroprotective activities of melatonin against the Alzheimer b-protein are not mediated by melatonin membrane receptors. *J. Pineal Res.* 32(3):135–142.
13. Féart, C., Samieri, C. et al. 2009. Adherence to a Mediterranean diet, cognitive decline, and risk of dementia. *JAMA*. 302(6):638–648.
14. Gustaw-Rothenberg, K. 2009. Dietary patterns associated with Alzheimer's disease: Population based study. *Int. J. Environ Res. Public Health.* 6(4):1335–1340.
15. Dangour, A.D., Allen, E. et al. 2009. Fish consumption and cognitive function among older people in the UK: Baseline data from the OPAL study. *J. Nutr. Health Aging* 13(3):198–202.
16. Scarmeas, N., Stern, Y., Mayeux, R., Manly, J.J., Schupf, N., Luchsinger, J.A. 2009. Mediterranean diet and mild cognitive impairment. *Arch. Neurol.* 66(2):216–225.
17. Scarmeas, N., Stern, Y. et al. 2006. Mediterranean diet and risk for Alzheimer's disease. *Ann. Neurol.* 59(6):912–921.
18. Thomas, P., Wang, Y.J. et al. 2009. Grape seed polyphenols and curcumin reduce genomic instability events in a transgenic mouse model for Alzheimer's disease. *Mutat. Res.* 661(1–2):25–34.
19. Hartman, R.E., Shah, A. et al. 2006. Pomegranate juice decreases amyloid load and improves behavior in a mouse model of Alzheimer's disease. *Neurobiol. Dis.* 24(3):506–515.
20. Krikorian, R., Shidler, M.D. et al. 2010. Blueberry supplementation improves memory in older adults. *J. Agric. Food Chem.* 58(7):3996–4000.
21. Dai, Q., Borenstein, A.R., Wu, Y., Jackson, J.C., Larson, E.B. 2006. *Am. J. Med.* 119(9):751–759.
22. Chauhan, A., Essa, M.M., Muthaiyah, B., Chauhan, V. 2009. Dietary supplementation of Walnut improves memory deficits and learning skills in transgenic mouse model of Alzheimer's disease. *22nd Biennial Meeting of the ISN/APSN Proceedings*, Taiwan, pp. 10–12.
23. Joseph, J.A., Shukitt-Hale, B. et al. 1999. Reversals of age-related declines in neuronal signal transduction, cognitive, and motor behavioral deficits with blueberry, spinach, or strawberry dietary supplementation. *J. Neurosci.* 19(18):8114–8121.
24. Joseph, J.A., Denisova, N.A. et al. 2003. Blueberry supplementation enhances signaling and prevents behavioral deficits in an Alzheimer disease model. *Nutr. Neurosci.* 6:153–162.
25. Casadesus, G., Shukitt-Hale, B. et al. 2004. Modulation of hippocampal plasticity and cognitive behavior by short-term blueberry supplementation in aged rats. *Nutr. Neurosci.* 7:309–316.
26. Krikorian, R., Nash, T.A., Shidler, M.D., Shukitt-Hale, B., Joseph, J.A. 2009. Concord grape juice supplementation improves memory function in older adults. *Br. J. Nutr.* 23:1–5.
27. Chauhan, N., Wang, K., Wegiel, J., Malik, M. 2004. Walnut extract inhibit the fibrilization of amyloid beta-protein and also defibrillizes its preformed fibrils. *Curr. Alzheimer Res.* 1:183–188.
28. Waterman, P.G., Mole, S. 1994. *Analysis of Phenolics Plant Metabolities*. Oxford: Blackwell Scientific Publication. pp. 83–91.
29. Wisniewski, T., Castano, E M., Golabek, A., Vogel, T., Frangione, B. 1994. Acceleration of Alzheimer's fibril formation by apolipoprotein E *in vitro*. *Am. J. Pathol.* 145:1030–1035.
30. Cotman, C.W., Head, E. et al. 2002. Brain aging in the canine: A diet enriched in antioxidants reduces cognitive dysfunction. *Neurobiol. Aging* 23(5):809–818.

31. Letenneur, L., Proust-Lima, C. et al. 2007. Flavonoid intake and cognitive decline over a 10-year period. *Am. J. Epidemiol.*; 165(12):1364–1371.
32. Kostrzewa, R.M., Segura-Aguilar, J. 2003. Novel mechanisms and approaches in the study of neurodegeneration and neuroprotection. A review. *Neurotox. Res.* 5:375–383.
33. Polidori, M.C. 2003. Antioxidant micronutrients in the prevention of age-related diseases. *J. Postgrad. Med.* 49:229–235.
34. Grant, W.B. 2003. Diet and risk of dementia: Does fat matter? The Rotterdam study. *Neurology* 60:2020–2021.
35. Morris, M.C., Evans, D.A. et al. 2005. Relation of the tocopherol forms to incident Alzheimer disease and to cognitive change. *Am. J. Clin. Nutr.* 81:508–514.
36. Aggarwal, B.B., Shishodia, S. 2004. Suppression of the nuclear factor-kappaB activation pathway by spice-derived phytochemicals: Reasoning for seasoning. *Ann. N.Y. Acad. Sci.* 1030:434–441.
37. Joseph, J.A., Shukitt-Hale, B., Casadesus, G. 2005. Reversing the deleterious effects of aging on neuronal communication and behavior: Beneficial properties of fruit polyphenolic compounds. *Am. J. Clin. Nutr.* 81:313S–316S.
38. Weinreb, O., Mandel, S., Amit, T., Youdim, M.B. 2004. Neurological mechanisms of green tea polyphenols in Alzheimer's and Parkinson's diseases. *J. Nutr. Biochem.* 15:506–516.
39. Muthaiyah, B., Essa, M.M., Chauhan, A., Chauhan, V. 2011. Protective effects of walnut extract against amyloid beta peptide-induced cell death and oxidative stress in PC12 cells. *Neurochem. Res.* 36(11):2096–2103.
40. Hughes, T.F., Andel, R., Small, B.J., Borenstein, A.R., Mortimer, J.A., Wolk, A., Johansson, B., Fratiglioni, L., Pedersen, N.L., Gatz, M. 2010. Midlife fruit and vegetable consumption and risk of dementia in later life in Swedish twins. *Am. J. Geriatr. Psychiatry* 18(5):413–420.
41. USDA. *National Nutrient Database for Standard Reference*, US Dept of Agriculture; May 15, 2007.
42. Al-Farsi, M., Lee, C.Y. 2008. Nutritional and functional properties of dates: A review. *Crit. Rev. Food Sci. Nutr.* 48:877–887.
43. Sabah, A.A.J., Naji, M.A. 2010. In vitro evaluation of the antiviral activity of an extract of date palm (*Phoenix dactylifera* L.) pits on a pseudomonas phage. *eCAM* 7(1):57–62.
44. Waheed, S., Siddique, N. 2009. Evaluation of dietary status with respect to trace element intake from dry fruits consumed in Pakistan: A study using instrumental neutron activation analysis. *Int. J. Food Sci. Nutr.* 60(4):333–43.

Index

S

T

U

V

W

X

For Product Safety Concerns and Information please contact our EU
representative GPSR@taylorandfrancis.com
Taylor & Francis Verlag GmbH, Kaufingerstraße 24, 80331 München, Germany

www.ingramcontent.com/pod-product-compliance
Ingram Content Group UK Ltd.
Pitfield, Milton Keynes, MK11 3LW, UK
UKHW052218190926
475702UK00006B/1073
9781439849453